BRIEF Contents

TABLE OF

Contents

CHAPTER NINE
Index Management **321**

CHAPTER TEN
Data Integrity Constraints **357**

CHAPTER ELEVEN
Users and Resource Control **389**

Preface

This textbook is a detailed guide to basic database administration for the Oracle9*i* database. The database administrator (DBA) uses the techniques demonstrated in the book that are essential for creating the initial database and configuring the storage space, tables, users, and security for a database. The book is a basic guide because it covers some of the first concepts that a DBD needs, omitting more advanced topics, such as network configuration and performance tuning are not discussed.

Intended Audience

This book is intended to support individuals in database courses covering database administration using the Oracle9*i* database. It is also intended to support individuals who are preparing for the Oracle Fundamentals I: Oracle9*i* exam that is required for certification as an Oracle Certified Associate (OCA) or an Oracle Certified Professional (OCP.)

Prior knowledge of general relational database terminology and concepts is required. In addition, the reader should have basic knowledge of SQL (Structured Query Language.) While it is preferable that the reader have knowledge of Oracle9*i*'s SQL, the reader's experience using SQL on other databases, such as SQLServer, is acceptable. The reader should be able to write SQL commands for querying, inserting, updating, and deleting data in relational tables.

Oracle Certification Program (OCP)

This textbook covers the objectives of *Exam 1Z0-031, Oracle Fundamentals I.* The exam is the second and final exam required for individuals seeking certification as an Oracle Certified Associate (OCA). OCA Certification is a prerequisite to other exams for Oracle Certified Professional (OCP). OCA Certification is also required for taking the exam for Oracle9*i*AS Web Administrator Certified Associate. Information about registering for these exams can be found at *www.oracle.com/education/certification.*

Approach

The concepts introduced in this textbook are presented in business scenarios. Concepts are introduced and examples of real-life uses for the concept are discussed. Then students follow along with hands-on practices to drive home the concepts in every chapter. The case project at the end of each chapter works within the context of a hypothetical "real world" business: an online newspaper publishing company named Global Globe News

Company. The case studies begin with an empty database, then add tablespaces and adjusting initialization parameters. Each chapter builds on the concepts of the previous chapter, and the case study adds details to the database. By the end of the book, the student has established a completed database schema containing tables, views, indexes, and objects. The student also creates database users, roles, and profiles that provide the Global Globe's employees with varying security levels of access to the data. This allows students to not only learn the syntax of a command, but also to apply it in a real-world environment. In addition, there are several script files that generate the data in tables and the case study. These script files are available to allow students hands-on practice in re-creating the examples and practicing variations of SQL commands to enhance their understanding.

Because the first thing new database administrators often do is to create a database, the initial focus of this textbook is on the creating a running database. In Chapters 1 and 2, the database software, its components, memory structure, and background processes are introduced. Chapter 3 discusses and uses the commands for creating a new database. Each subsequent chapter zooms in on one area that the DBA must administer, such as tables, indexes, users, and roles. Each chapter covers in depth the clauses of the commands that affect the underlying structure of each component. In this way, the student becomes familiar with the many choices available when creating new database structures and learns how to select the appropriate choice in each particular situation.

To reinforce the material presented, each chapter includes a chapter summary. In addition, at the end of each chapter, groups of activities are presented that test students' knowledge and challenge them to apply that knowledge to solving business problems.

Overview of This Book

The examples, projects, and cases in this book will help students to achieve the following objectives:

- Understand key components of the Oracle9*i* database architecture and installation options.
- Use DBA tools, especially the Enterprise Management console, and understand the background processes that perform database operations.
- Create a new database instance using both manual SQL commands and database tools.
- Use Data Dictionary views to monitor database structures and activities; use SQL to manage control files.
- Understand the concepts and use SQL to create, modify, or remove redo log files, control files, and database-generated diagnostic files.
- Understand the physical and logical structures that make up the database; create, modify, and drop database tablespaces; manage undo data.
- Use SQL to create, modify, and drop many types of tables and indexes using additional clauses that control storage settings, LOB storage, and partitioning.

- Create users with SQL including clauses for storage quotas; modify and drop users.
- Understand and manage system privileges, as well as object privileges, profiles, and roles.
- Use globalization support for language-dependent behavior in the database.

The contents of each chapter build on the concepts of previous chapters. **Chapter 1** describes the key components of the Oracle9*i* database architecture and installation options. **Chapter 2** introduces DBA tools, including the Enterprise Management console, and describes the background processes that perform database operations. **Chapter 3** shows two methods for creating a new database instance: manual SQL commands and the Database Configuration Assistant tool. **Chapter 4** shows how to query the Data Dictionary views and how to use SQL to manage control files. **Chapter 5** introduces the redo log file, and describes diagnostic files available to the DBA. **Chapter 6** maps out the physical and logical structures that form the database and then moves on to help the student create tablespaces and manage undo data. **Chapter 7** describes table structures and shows how to create several types of tables with emphasis on the storage settings. **Chapter 8** shows how to create several more complex types of tables and describes how to modify or drop tables. **Chapter 9** leads the student through creation of several types of indexes, specifying storage parameters. The chapter then demonstrates monitoring, reorganizing and dropping indexes. **Chapter 10** defines data integrity and then uses SQL to create, modify, and drop constraints. **Chapter 11** shows how to create users, modify and drop users, and then how to create and manage profiles. **Chapter 12** looks at system and object privileges and shows how to use them. It then demonstrates Oracle9*i*'s auditing capabilities. **Chapter 13** describes how to create, use, and drop roles. **Chapter 14** shows how Oracle9*i* supports language-dependent behavior in the database and describes how to use NLS parameters.

Features

To enhance students' learning experience, each chapter in this book includes the following elements:

- **Chapter Objectives:** Each chapter begins with a list of the concepts to be mastered by the chapter's conclusion. This list provides a quick overview of chapter contents as well as a useful study aid.
- **Methodology:** As new commands are presented in each chapter, the syntax of the command is presented and then an example illustrates the command in the context of a business operation. This methodology shows the student not only *how* the command is used but also *when* and *why* it is used. The step-by-step instructions in each chapter enable the student to work through the examples in this textbook, engendering a hands-on environment in which the student reinforces his or her knowledge of chapter material.

Tip: This feature, designated by the *Tip* icon, provides students with practical advice. In some instances, tips explain how a concept applies in the workplace.

Note: These explanations, designated by the *Note* icon, provide further information about concepts or a syntax structures.

- **Chapter Summaries:** Each chapter's text is followed by a summary of chapter concepts, as a helpful recap of chapter contents.
- **Review Questions:** End-of-chapter assessment begins with a set of 10 to 15 review questions that reinforce the main ideas introduced in each chapter. These questions ensure that students have mastered the concepts and understand the information presented.
- **Exam Review Questions:** Seven to ten certification-type questions are included to prepare students for the type of questions that can be expected on the certification exam, as well as to measure the students' level of understanding.
- **Hands-on Assignments:** Along with conceptual explanations and examples, each chapter provides ten hands-on assignments related to the chapter's contents. The purpose of these assignments is to provide students with practical experience. In some cases, the assignments are based on tables or other structures created using a script provided with the book that the student runs just before working the assignments.
- **Case Project:** One or two cases are presented at the end of each chapter. All the case projects are based on a fictional business called Global Globe News Company and ask students to create the database structures discussed in the chapter as part of the growing database system for the fictional company. These cases are designed to help students apply what they have learned to real-world situations. The cases give students the opportunity to independently synthesize and evaluate information, examine potential solutions, and make recommendations, much as students will do in an actual business situation.

The Course Technology Kit for Oracle9*i* Software, available when purchased as a bundle with this book, provides the Oracle database software on CDs, so users can install on their own computers all the software needed to complete the in-chapter examples, Hands-on Assignments, and Case. The software included in the kit can be used with Microsoft Windows NT, 2000, or XP operating systems. The installation instructions for Oracle9*i* and the log in procedures are available at **www.course.com/cdkit** on the Web page for this book's title.

Teaching Tools

The following supplemental materials are available when this book is used in a classroom setting. All teaching tools available with this book are provided to the instructor on a single CD-ROM.

- **Electronic Instructor's Manual:** The Instructor's Manual that accompanies this textbook includes the following elements:
  - Additional instructional material to assist in class preparation, including suggestions for lecture topics.
  - Answers to end of chapter Review Questions, Exam Review Questions, Hands-on Assignments, and Case Projects (when applicable).
- **ExamView®:** This objective-based test generator lets the instructor create paper, LAN, or Web-based tests from testbanks designed specifically for this Course Technology text. Instructors can use the QuickTest Wizard to create tests in fewer than five minutes by taking advantage of Course Technology's question banks—or create customized exams.
- **PowerPoint Presentations:** Microsoft PowerPoint slides are included for each chapter. Instructors might use the slides in three ways: As teaching aids during classroom presentations, as printed handouts for classroom distribution, or as network-accessible resources for chapter review. Instructors can add their own slides for additional topics introduced to the class.
- **Data Files:** The script file necessary to insert data into the Global Globe database tables is provided through the Course Technology Web site at **www.course.com**, and is also available on the Teaching Tools CD-ROM. Additional script files needed for use in specific chapters are also available through the Web site.
- **Solution Files:** Solutions to the end of chapter material are provided on the Teaching Tools CD-ROM. Solutions may also be found on the Course Technology Web site at **www.course.com**. The solutions are password protected.

ACKNOWLEDGMENTS

To my Course Technology development team, headed up by Betsey Henkels and Tricia Boyle: please accept my heartfelt gratitude. I have learned a great deal during this process and appreciate all your patience and perseverance. My thanks and appreciation to Anne Valsangiacomo and her Production team, as well as Nicole Ashton and her Quality Assurance team for making the manuscript as technically flawless as possible. And thank you, Bill Larkin, Acquisitions Editor, for joining with me in a vision that has come to its successful conclusion with the publishing of this book. And, on a personal note, thanks to my son, Blue, and husband, Patrick, who remind me that "relationship" is not just a database term.

The following reviewers also provided helpful suggestions and insight into the development of this textbook: Michael P. Burns, DeVry University; Norma E. Hall, Manor College; James C. Homan Ed.D.; Henry Ford CC; Girish H. Subramanian, Pennsylvania State University, Harrisburg; and Arta Szathmary, Bucks County Community College.

Read This Before You Begin

To the User

Data Files

Much of the practice you do in the chapters of this book involves creating, modifying and then dropping a database structure (such as a table, index, or user). Most of the practices in the chapters and the hands-on exercises at the end of the chapters can be done without running any data files. At certain points in the book, however, you will need to load data files created for this book. Your instructor will provide you with those data files, or you can obtain them electronically from the Course Technology Web site by accessing *www.course.com* and then searching for this book's title. When you reach a point in the book where a data file is needed, the book gives you instructions on how to run each data file. The data files provide you with the same tables and data shown in the chapter examples, so you can have hands-on practice re-creating the practice commands. It is highly recommended that you work through all the examples to reinforce your learning.

The script files for Chapters 2 through 14 are found in the **Data** folder under their respective chapter folders (for example **Chapter09** and **Chapter10**) on your data disk and have the file names that correspond with the instructions in the chapter. If the computer in your school lab—or your own computer—has Oracle9*i* database software installed, you can work through the chapter examples and complete the hands-on assignments and case projects. At a minimum, you will need the Oracle9*i* Release 2, Personal Edition of the software to complete the examples and assignments in this textbook.

Using Your Own Computer

To use your own computer in working through the chapter examples and completing the hands-on assignments and case projects, you will need the following:

- **Hardware:** A computer capable of using the Microsoft Windows NT, 2000 Professional, or XP Professional operating system. You should have at least 256MB of RAM and between 2.75GB and 4.75GB of hard disk space available before installing the software.
- **Software:** Oracle9*i* Release 2 Enterprise Edition, or, at a minimum, Oracle9*i* Release 2 Personal Edition. The Course Technology Kit for Oracle9*i* Software contains the database software necessary to perform all the tasks shown in this textbook. Detailed installation, configuration, and logon information are provided at *www.course.com/cdkit* on the Web page for this title. ***Note: when prompted for the Database Identification, type ORACLASS in the Global Database Name box and in the SID box.***

- **Data files:** You will not be able to use your own computer to work through the chapter examples and complete the projects in this book unless you have the data files. You can get the data files from your instructor, or you can obtain the data files electronically by accessing the Course Technology Web site at *www.course.com* and then searching for this book's title.

When you download the data files, they should be stored in a directory separate from any other files on your hard drive or diskette. You will need to remember the path or folder containing the files, because you'll have to locate the file while in SQL*Plus Worksheet in order to execute it. (The SQL*Plus Worksheet is the interface tool you will use to interact with the database.)

When you install the Oracle9*i* software, you will be prompted to supply the database name for the default database being created. Use the name "ORACLASS" to match the name used in the book. If you prefer a different name, remember that anywhere the book instructs you to type in ORACLASS, you should type in your database name instead. You will be prompted to change the password for the SYS and SYSTEM user accounts. Make certain that you record the names and passwords of the accounts because you will need to log in to the database with one or both of these administrative accounts in some chapters. After you install Oracle9*i*, you will be required to enter a user name and password to access the software. One default user name created during the installation process is "scott". The default password for the user name is "tiger". If you have installed the Personal Edition of Oracle9*i*, you will not need to enter a connect string during the log in process. As previously mentioned, full instructions for installing and logging in to Oracle9*i*, Release 2, are provided on the Web site for this textbook at *www.course.com*.

Visit Our World Wide Web Site

Additional materials designed especially for this book might be available at **www.course.com**. Visit this site periodically for more details.

To the Instructor

To complete the chapters in this book, your students must have access to a set of data files. These files are included in the Instructor's Resource Kit. They may also be obtained electronically by accessing the Course Technology Web site at *www.course.com* and then searching for this book's title.

The set of data files consists of script files that are executed either at the beginning of the chapter or before starting the hands-on exercises. After the files are copied, you should instruct your students in how to copy the files to their own computers or workstations.

Maintain the directory structure found in the original data files: Data\Chapter01, Data\Chapter02 and so on.

You will need to provide your students with this information, which is used in several chapters:

- The passwords for the SYSTEM and SYS users on their workstation.
- The database name, which is assumed to be ORACLASS. If the database name is not ORACLASS, inform the students that they should substitute the correct name whenever the text tells them to enter ORACLASS.
- The full path for the ORACLE_BASE and ORACLE_HOME directories on their workstations. ORACLE_BASE is a variable name used in Oracle documentation and in this book to refer to the root directory of the Oracle database installation. ORACLE_HOME is the root directory of the Oracle software. In Windows, ORACLE_BASE is typically **C:\oracle** and ORACLE_HOME is typically **C:\oracle\ora92**.
- The computer name of the workstation.
- The full path names for directories where the students can create additional directories and files. The directories must be on the hard drive (not removable media, such as floppy disks). The table below describes the storage requirements for each student.

Table R-1 Storage requirements for each workstation

Chapter #	Suggested Directory Name	Storage Requirement
3	newdb	500 M
4	newctl	100 M
5	newredo1	100 M
5	newarch1	100 M
5	newarch2	100 M
6	newdata	375 M
Total		**1.3 G**

The chapters and projects in this book were tested using the Microsoft Windows 2000 Professional operating system with Oracle9*i* Release 2 Enterprise Edition and the Microsoft Windows XP Professional operating system with Oracle9*i* Personal Edition, Release 2 (9.2.0.1.0).

Course Technology Data Files

You are granted a license to copy the data files to any computer or computer network used by individuals who have purchased this book.

CHAPTER

1

ORACLE ARCHITECTURE OVERVIEW

In this chapter, you will:

- Learn about Oracle9*i* architecture and key Oracle9*i* software components
- Look at the ORACLASS database used in exercises throughout the book
- Discover differences between Oracle9*i* client and server installation options
- Learn how to use the Oracle Universal Installer
- Examine why to use OFA (Optimal Flexible Architecture)

Oracle Corporation offers many products in addition to its **RDBMS** (Relational Database Management System) software. The core product, the Oracle9*i* database, contains a suite of software that includes required components, work-saving tools, utilities, and management software that is primarily used for DBA. (**DBA** stands for both database administration and database administrator.) The Oracle9*i* database has more features for use with the Internet than prior releases. For example, a new XML DB (database) feature provides an interface within the database for easily storing, updating, and displaying XML-formatted data. The JVM (formerly called Jserver) within the database supports more internal Java programming for better performance of Java applications.

Oracle also offers related software suites, such as: **Oracle9*i* Application Server** (a Web server with special plug-ins for the Oracle database); **Oracle Financials** (a set of software designed for bookkeeping, accounting, inventory, and sales); and **Oracle JDeveloper** (an application builder that writes Java code using a Windows-like interface). This book covers only the core Oracle9*i* RDBMS software suite, which includes the key components described in this chapter.

Your journey into the world of Oracle9*i* begins as you become familiar with the basic components of the Oracle9*i* RDBMS software suite. This chapter introduces you to the components that make up that suite and to the Oracle Universal Installer, which you use when installing any of Oracle's products. The chapter closes with a discussion of the architecture strategy that Oracle uses to maximize flexibility when implementing its database and related software products.

INTRODUCTION TO ORACLE9*i* ARCHITECTURE

The Oracle9*i* RDBMS software suite (referred to as *Oracle9i* throughout the book) includes everything you need to build and maintain a relational database. The basic software runs the database engine, manages the data storage for all information in the database, and provides tools to manage users, tables, data integrity, backups, and basic data entry. The basic software includes additional tools and utilities that help you monitor the performance and security of the database.

This section analyzes the components of Oracle9*i* and describes the function of each component. Understanding how the components fit together helps you prepare for software installation. After installation, you will know which components to use for various tasks, such as creating a new user or adjusting storage space.

The Oracle9*i* RDBMS software suite can be enhanced with many extra software components that Oracle sells for additional cost. This book covers only the components included when you buy the database software. Typically, you purchase additional software to write applications that access the database. Many other vendors offer software that adds onto an Oracle9*i* database. For example, ESRI offers a complete GIS (graphic information system) software solution for computer-generated maps that stores all its data in your Oracle9*i* database.

Key Components of the Oracle9*i*

Before installing Oracle9*i*, the Installation window lists all the components it plans to install. There are over 200 components listed for the basic installation alone. Fortunately, most of these components are modules that support a smaller set of major components. Figure 1-1 illustrates a conceptual view of the Oracle9*i* database engine and the core utilities that comprise Oracle9*i*. These key components make up Oracle9*i* running on any operating system or hardware. Oracle has standardized its software so that it looks and acts identically at the user interface level on all platforms. Oracle customized its software release on each platform at the lowest levels to optimize performance based on each platform's characteristics. An example is the process that takes data from Oracle9*i* memory buffers and writes it to a hard disk.

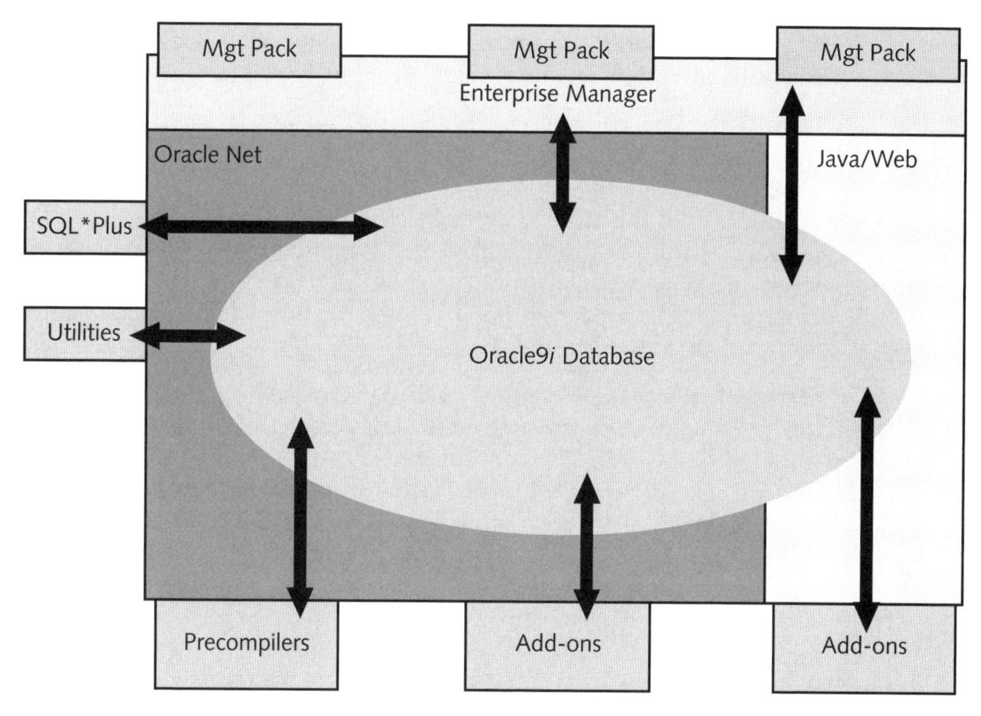

Figure 1-1 Oracle9*i* RDBMS software suite includes several core utilities and tools

Whether you download the software or order CDs, your software includes the core database suite and several "try and buy" packages. The documentation explains which portions are standard and which require the additional purchase of a license. The following list corresponds to the features shown in Figure 1-1 and briefly describes each of the core utilities and tools. All of these features are part of the Oracle9*i* RDBMS software suite.

- **Oracle9*i* database:** If you want, a preconfigured database can be created during installation. When installing an upgrade, this is not needed, but new installations need a database, and using a preconfigured database is convenient. The database itself contains a number of preset features that speed up its use. For example, there are several predefined, high-level users that aid you in creating other users and in monitoring your database statistics. The PL/SQL engine included with the Oracle software provides a specialized programming language you can use for creating triggers, packages, and procedures. The **Java Virtual Machine** (JVM) engine allows you to store, parse, and execute Java applets, servlets, and stored procedures within the database.
- **Oracle Net:** This tool provides the network link between the Oracle9*i* database and most applications that communicate with the database. Oracle Net must be configured to define ports and the protocols allowed to send and receive data for each database in your network. If you are running Enterprise

Manager, SQL*Plus, Forms, or non-Oracle products, Oracle Net is the gateway into the database.

- **Java/Web:** The other way to access the Oracle9*i* database is through Java ports. There are a number of different programs and tools installed with your Oracle9*i* database to support Web and database integration. These tools include a fully functioning **Apache Web server** containing the following: modules that hook into the database; database packages to support XML input and output; a precompiler for SQLJ (Java code with embedded SQL commands); and a gateway for PL/SQL to allow HTML documents to be delivered directly from the database to the user.
- **Oracle Enterprise Manager (OEM):** Oracle9*i* builds more enhancements into this Windows-like Database Administrator tool, which was first introduced with Oracle7. The tool integrates many utilities and monitoring tools into a single interface. Multiple databases across local and remote networks can be mapped and managed within the Enterprise Manager console. Figure 1-2 shows the console with its navigation tree displayed. Your console gives you access to wizards (for example, there are backup, recovery, and SQL analysis wizards) and management packs.

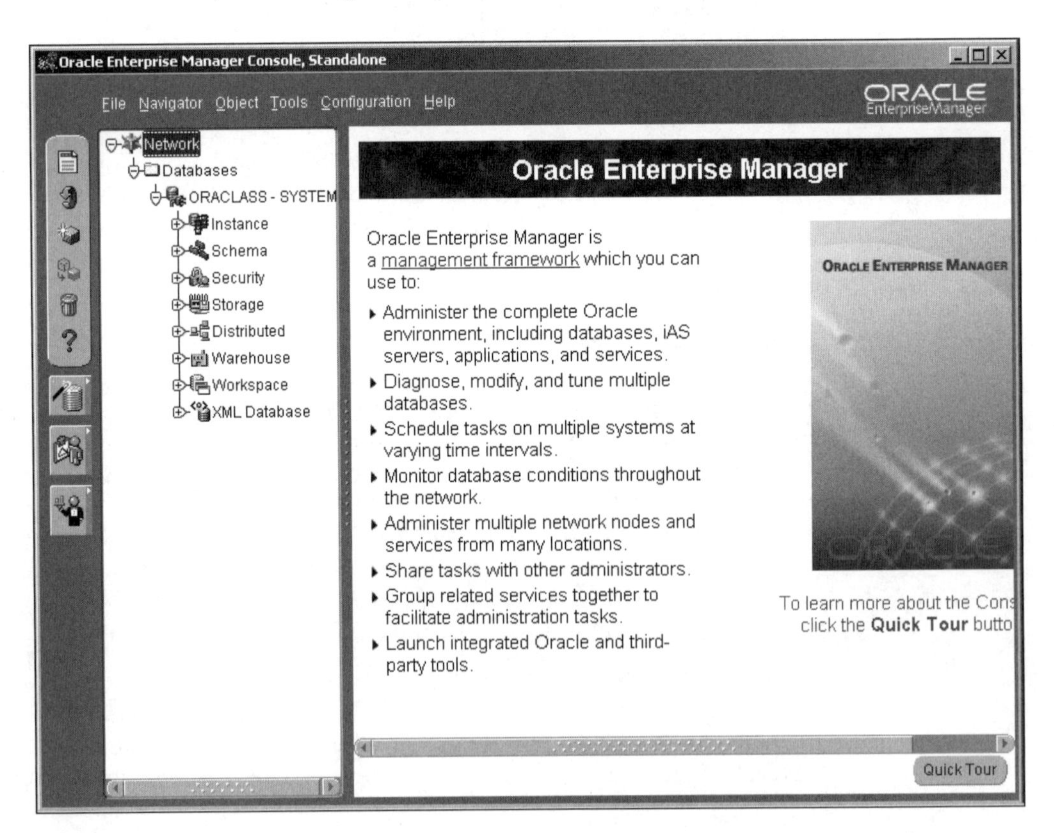

Figure 1-2 The Enterprise Manager console provides centralized access to tools

- **Management packs:** These are sets of related tools, wizards, and assistants that are added as a group to the Enterprise Manager console. The Standard Pack that comes with your Oracle9*i* installation includes management tools for starting and stopping the database, creating users, creating tables and objects, and administering storage space. Additional packs can be purchased and added on later. The Enterprise Manager also has a Web-based component that provides Internet access to the Enterprise Manager console and most of its integrated tools. This portion of Enterprise Manager uses the Java/Web support component to access the database rather than using Oracle Net.
- **SQL*Plus:** This tool allows you to create and run queries, add rows, modify data, and write reports using SQL. SQL*Plus accepts standard SQL, with all the Oracle enhancements to SQL, such as the TO_DATE function. In addition, SQL*Plus provides commands for customizing output, defining variables, and executing files. SQL*Plus in Oracle9*i* has been enhanced with commands for starting and stopping the database.
- **Utilities:** There is a group of utilities for backup, migration, recovery, and transporting data from one database to another. For example, the **Recovery Manager** automates database recovery after a failure. Export and Import (EXP and IMP) provide a command line interface to extract tables, schemas, or entire databases into files that can later be reloaded into the same database or into another database. There are additional tools for migration to a different Oracle release, for loading new data from a flat file into a table, and for transferring data from one character set to another. A **character set** provides special symbols for various languages.
- **Precompilers:** Precompilers (such as Pro*COBOL and Pro*C) support embedded SQL commands within programs in C, C++, or COBOL. The precompiler translates the SQL command into the appropriate set of commands for the program, which is then compiled and ready for executing. Precompilers save a great deal of programming time.
- **Add-ons:** Oracle offers many additional components that can enhance the database system. Some of these access the database through Oracle Net, whereas others access the database through the Java/Web connectors. A section later in this chapter describes some of the add-ons available.

As you begin working with each of these main components, the supporting modules and the connections between all the components become clearer.

Running the Database

When you install an Oracle9*i* database, you install the software components, create database files to store your data, and then start a set of background processes that allocate memory and handle database activities. Figure 1-3 shows a typical installation of the Oracle9*i* database.

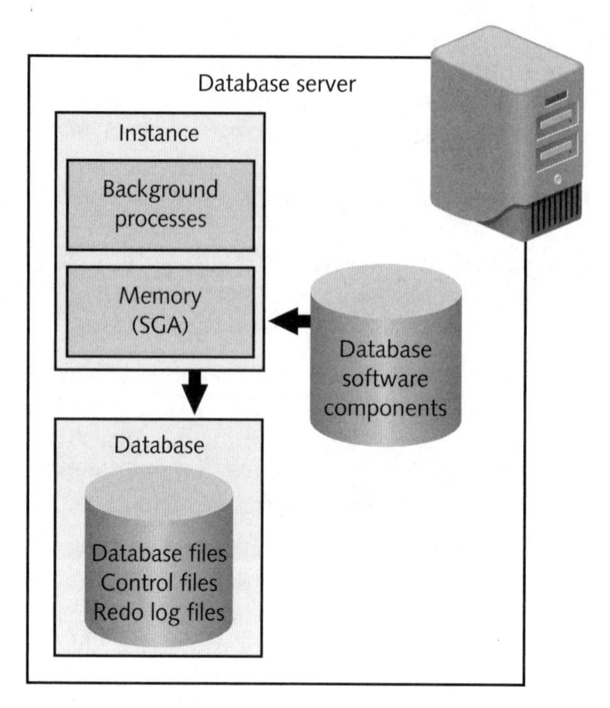

Figure 1-3 A computer containing one instance is a single-instance server

Oracle defines a **database** as the collection of operating system files that store your data. The database has three types of files: control files, database files, and redo log files. These file types are described later in this chapter.

The **database software components**, such as Oracle Net, Enterprise Manager, and the RDBMS components are installed on the computer. To use the database, you start up a **database instance**, which allocates **memory** for the database (called the **System Global Area**, or SGA) and starts up a set of background processes. **Background processes** are the programs that run on the computer. The background processes handle user interaction with the database and manage memory, integrity, and I/O for database files. Chapter 2 describes the makeup of the files, memory structures, and background processes of a database instance.

The combination of database software, a database (the files), and a database instance (the SGA and the background processes) is called a **database server**. Figure 1-3, Figure 1-4, and Figure 1-5 show three different configurations used to implement a database server.

- **Single-instance server:** This is the typical installation and contains one computer with one set of database files and one instance that accesses the files. Figure 1-3 shows this configuration.
- **Multiple-instance server:** Another installation possibility for one computer has two instances, each with its own database files. Figure 1-4 shows this configuration. The two instances run independently of one another. This is useful

when you want a dedicated instance for two development teams, for example, and you have one powerful computer to support both teams.

- **Clustered servers:** As shown in Figure 1-5, this configuration includes several computers (**database nodes**), each one containing a database instance. Another computer (**file server**) houses the database files, which are shared by all the instances. The **cluster manager** is a software application that coordinates all the instances and the tasks that each one handles. The cluster manager can reside on a separate computer or on one of the nodes. Clustered servers are complex to manage, but supply increased computing power for applications that require speed for processing massive amounts of data or complex calculations, such as astronomy or physics applications. One task can be divided into smaller pieces and divided among the instances to complete the task more quickly. Oracle Real Application Clusters is an add-on component that can manage clusters of Oracle9*i* instances.

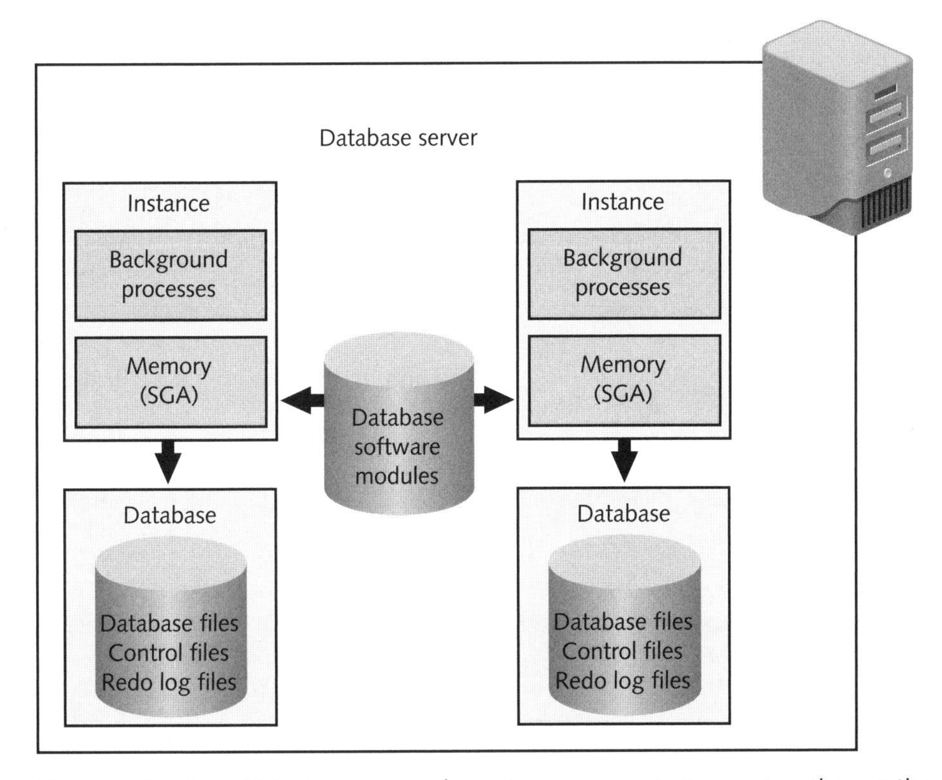

Figure 1-4 A multi-instance server hosts two or more instances together on the same computer

Clusters are beyond the scope of this book; however, it is helpful to be aware of this configuration for more advanced applications.

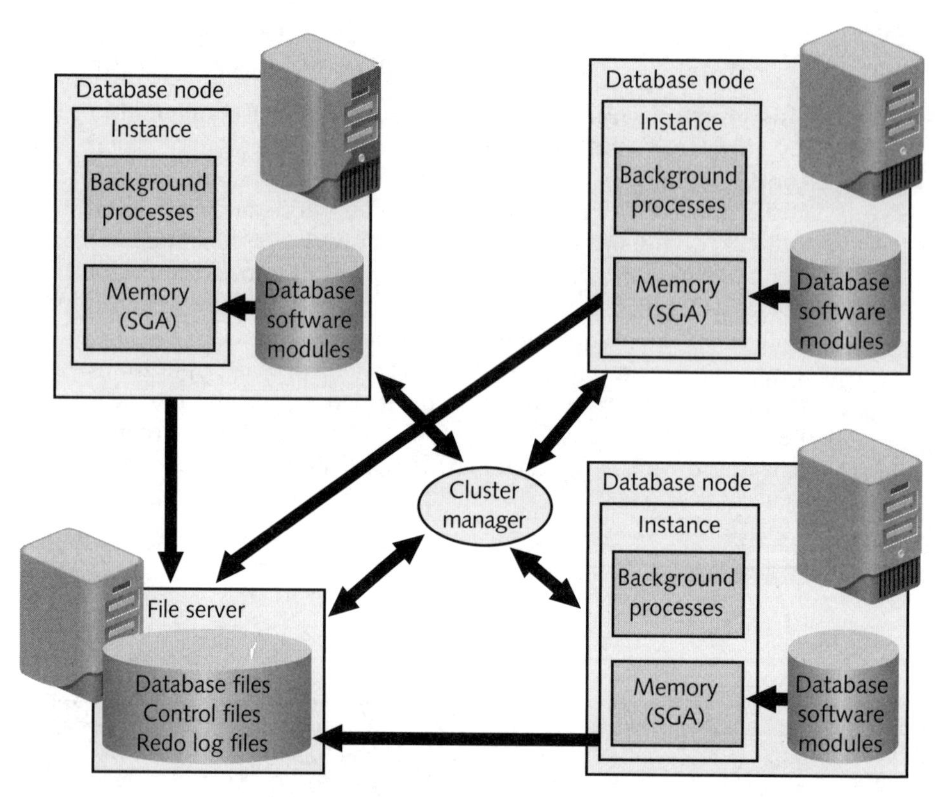

Figure 1-5 Many computers can act as one when connected in a cluster

On the single-instance server (the typical configuration for an Oracle9*i* database), how does the user interact with the database? When a user runs an application that uses the database, the application creates a **user process** that controls the connection to the database process. On the database side, the process that interacts with the user process is called a **server process**. There are two basic methods used to connect the user process and the server process:

- **Multithreaded server:** A **thread** is a link between a user process and a server process that requires processing by the database. The multithreaded server allows one server process to handle many user process threads. This method provides better efficiency for systems that have many users accessing the system at once, such as a central database serving one hundred programmers developing applications.
- **Dedicated server:** This method connects one user process with one server process. This is better for systems with fewer users or with user processes that require a great deal of processing time on the database.

If you install Oracle9*i* Enterprise Edition or Personal Edition and choose to have a predefined database created on installation, the database is configured in dedicated server mode.

Optional Additions to Oracle9*i*

Oracle9*i* can be enhanced by purchasing additional features from Oracle. Many of these features are included in the base software on a trial basis, and it is sometimes unclear which portions you have purchased and which require an additional license fee. The following list summarizes the optional features that require additional license fees.

Oracle allows the use of optional features in your test environment without restriction. However, if you begin to use these features in your production environment, you need to purchase the appropriate license.

- **Oracle Partitioning:** Builds the capability to partition tables across multiple tablespaces. High volume tables, such as historical records in data warehouses, benefit from partitioning by speeding up data retrieval for queries. **Data warehouses** are storages of data that are used to support management decisions.
- **Oracle Real Application Clusters (RACs):** Provide management tools to support database clusters. **Database clusters** combine multiple databases sharing a central memory area, so that a long-running task can be spread among several databases, speeding up processing time.
- **Oracle Spatial:** Adds programmed packages to the database to handle spatial objects. **Spatial objects** store data related to time and space in a way that allows you to calculate distance or time between objects, and to draw lines, polygons, and points on a map, and similar functions.
- **Oracle Change Management Pack:** Adds several utilities to the Enterprise Manager console for identifying and distributing database changes among multiple databases.
- **Oracle Tuning Pack:** Shows you which SQL commands use the most resources and provides automated tools to analyze and tune the commands.
- **Oracle Diagnostics Pack:** Brings expert advice into the Enterprise Manager console with graphs, database monitoring, and analysis tools. For example, you can review a graph of memory usage and then review hints on how to adjust memory-related initialization parameters to improve performance.
- **Oracle Management Pack for SAP R/3:** Enhances the Enterprise Manager console to detect and monitor SAP R/3 application servers and clients that use the Oracle 9*i* database. SAP R/3 is a popular financial planning and accounting software, similar in scope to Oracle Financials.
- **Oracle Management Pack for Oracle Applications:** Adds monitoring tools to the Enterprise Manager console to automate monitoring of Forms sessions and Concurrent Managers used when running Oracle Applications.

- **Oracle Data Mining:** Supports setting up algorithms and functions that search and retrieve data warehouse information. The Java interface for the tool is based on emerging standards for Java Data Mining (JDM).
- **Oracle Advanced Security:** Adds a security layer to Oracle Net and the database. This package encrypts outgoing data before it goes onto the network or across the Internet. It also supports special methods of user authentication such as the programs used with automated tellers, with which you must have a valid bank card in your possession and know the Personal Identification Number (PIN) to access your account data.
- **Oracle Label Security:** Is useful when high security standards must be met. By labeling individual rows with a security profile and matching that with a user's security profile, the database restricts access to rows within a table or view.
- **Oracle On-line Analytical Processing (OLAP) Services:** Allow you to combine standard file directory support with database delivery of tables or views. When you view a directory that is controlled by Oracle OLAP services, you see a list of files as you would in any directory. When you open one of these files, Oracle runs an extraction program and creates a file with the extracted data. The DBA predefines the format of the file contents.

Few database installations require all of these options. Typically, data warehouses might use partitioning and data mining to increase performance, whereas banks or accounting firms might use Oracle Advanced Security and OLAP services to control data accessibility.

INTRODUCTION TO ORACLASS DATABASE

This book has a database set up for your use throughout the chapters. There are step-by-step instructions in many of the chapters that lead you through the creation of tables, users, indexes, and more. Hands-on assignments at the end of each chapter are intended to be run on the database as well. The database is a standalone database instance on your workstation. This means that you alone have control over the database. All the workstations start out with the same basic database and build on this foundation.

The main features of the ORACLASS database are:

- The database has Enterprise Edition software, unless your instructor informs you otherwise.
- The database has a user named CLASSMATE whose password is CLASSPASS. Most of the exercises and step-by-step instructions begin with your logging in as this user. Initially, the CLASSMATE user owns no tables or other objects in the database. You add these as you go along.

- In some chapters, you are asked to run a script to create tables or other objects needed for the chapter. Each chapter contains the necessary instructions for these scripts.
- The database also has the standard users, SYSTEM and SYS, which are installed to provide you with Database Administration rights on the database. The password for SYSTEM and SYS will be provided by your instructor.

Every chapter has an ongoing case project that you develop from beginning to end. The project is the database for the Global Globe newspaper chain.

Global Globe is a national newspaper chain and you have been hired to design and build its database. You are starting from scratch, and the design elements are revealed as you work the case project at the end of each chapter. You build tables, add indexes, constraints, users, and security roles as you learn about these elements. You monitor the database activity and make adjustments to suit the ever-changing requirements of the Global Globe users and administration.

The database has already been installed and is running on your workstation now. The next section describes the software components and the database architecture that have been installed, as well as the other options available when installing a new database.

OVERVIEW OF ORACLE9*i* INSTALLATION OPTIONS

When installing Oracle9*i*, you have several high-level choices that determine the components installed on your computer. One important concept to understand before installing Oracle9*i* is the multi-tier architecture that Oracle9*i* uses. Multi-tier architecture became popular in the 1980s and has grown into an important and lasting design element of many computer systems, not just database systems. The term **multi-tier**, also referred to as **n-tier**, means that the data, processing, and user interfaces are divided into separate areas that are fairly independent of one another. Figure 1-6 shows a typical n-tier application system. The advantage of an n-tier system is that you can make changes to one tier without requiring changes to the other tiers. In a typical Oracle9*i* application system, the data tier contains the Oracle9*i* database server, the middle tier contains the application server, which may or may not be Oracle's application server (Oracle9*i*AS), and the user tier includes a browser. The user tier may have a custom application instead of a browser, or a combination of browser with Java applet, for example.

This book works almost entirely on the data tier of this architecture. You dig into the processes, memory structures, and data structures that make up the database server. This means that the software that was installed for your use is a database server configuration. The server configuration includes components that contain the database software modules needed to run the database. A second path of installation is available as well: the client configuration.

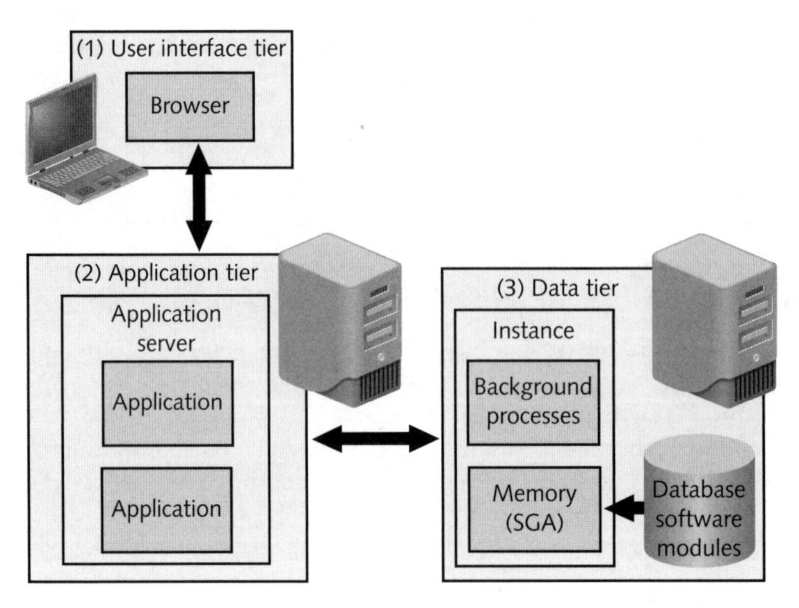

Figure 1-6 Three tiers (numbered 1, 2, and 3) compose a typical database system

To create a database server, you need the server configuration. This configuration installs all the software modules required to run the database and communicate with any application that requires data from the database. In addition, the server configuration provides you with a predefined database (if you want to use it) ready for use.

The client configuration is a combination of the user interface tier and the application server tier. It is a specialized setup for end users running database applications that interact with the database on their own workstations. The client configuration gives the user the communication modules needed to reach the database directly (skipping the application tier). It also has an option to install the Enterprise Manager components for the database administrator who wants to reach the database from a remote computer.

Figure 1-7 shows the client and server paths and the resulting components installed. Looking at Figure 1-7, you can see the decisions you must make when installing Oracle9*i* software. First, you choose between client and server installations. Within client, you select either Administrator or Runtime configuration. Within server, you choose between Personal, Standard, and Enterprise editions. In both client and server installations, you can opt for Custom installation, which allows you to select specific components individually.

Comparing Server-Side Installation Options

This section describes in more detail each of the three options for installing Oracle9*i* server components: Enterprise Edition, Standard Edition, and Personal Edition.

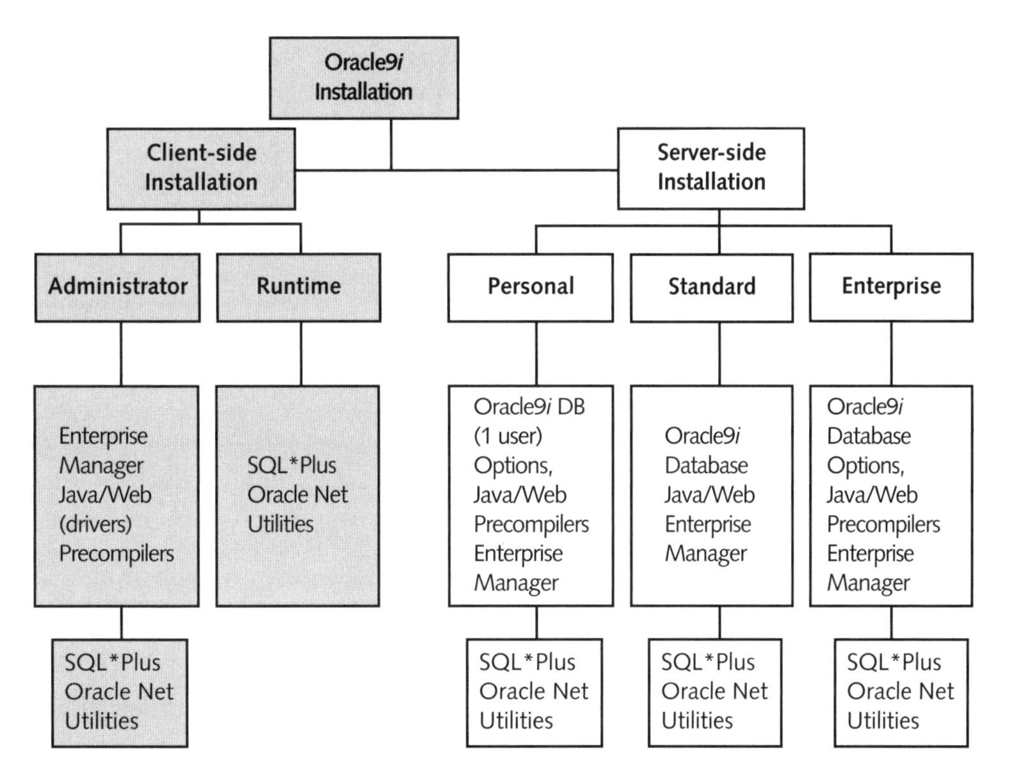

Figure 1-7 The components installed depend on which installation path you select

Enterprise Edition

When installing the server-side options, the most complete installation is the Enterprise Edition. The Enterprise Edition includes all the major components available in Oracle9*i*, as shown in Figure 1-7. This edition is intended for businesses with one or more of the following requirements:

- Multiple users connect to the database concurrently.
- Applications (Web-based or client/server) connect to the database.
- High data volume is common.
- Multiple database instances support the business and may be networked. Each instance consists of its own set of background processes and memory allocation.
- Database replication for duplicating data across multiple sites is necessary.

The Enterprise Edition runs on many different platforms, including UNIX, Windows 2000/NT, Linux, and HP.

To avoid installation errors when using Windows 2000, install Service Pack 2 before installing the Oracle software.

Your workstation should have either the Enterprise Edition or the Personal Edition installed.

Standard Edition

The Standard Edition provides basic support for multiuser database applications on a smaller scale than the Enterprise Edition. It is intended for use by a smaller number of users, such as those in a single department within a company.

Although replication, security, and Web access are available with the Standard Edition, less emphasis is placed on providing high volume, high availability capabilities. The Standard Edition cannot be upgraded with database features, such as partitioning and clustering.

Personal Edition

The Personal Edition allows access by a single user to the database instance. Two primary uses for the Personal Edition are:

- Programming, in which the programmer develops applications that run on the Personal Edition database and then are moved to the Enterprise Edition instance
- Deployment, in which a company develops and distributes an application/database package for use by single users

For programming, the Personal Edition is fully compatible with the Enterprise Edition, so the program developed using Personal Edition can be ported to the Enterprise Edition without modification. The Personal Edition is fully compatible with Enterprise Manager beginning with Oracle9*i*, Release 2. The Personal Edition can be installed on all platforms available to the Enterprise and Standard Editions.

Oracle9*i*, Release 1 (Personal Edition only) is the final release available for Windows 98.

Comparing Client-Side Installation Options

Client-side installations facilitate user accesses to a remote Oracle 9*i* database. Therefore, the software installed supports connectivity to this remote database. The two variations on a client-side installation include one for Administrators and one for Runtime use. Figure 1-7 lists the components installed for each option. For both options, the Oracle Net component on the client side handles communication with the remote database by sending requests across a network to the Oracle Net component on the server side, which passes requests on to the database.

Administrator

The Administrator option provides the user management tools, including the Enterprise Manager, to provide remote management of multiple databases. This option includes many of the same features as the Enterprise Edition option on the server side, except there is no database installed.

Runtime Option

The Runtime option is intended primarily for programmers who are developing applications on their own client machines while using a remote database as the connection to the database. This is often easier to manage than providing a database for each workstation, because the database changes need not be replicated for each programmer. The Runtime option installs basic connectivity features such as Oracle Net.

Both the Administrator and Runtime options support Windows 98 along with all the platforms in which you can install Enterprise Edition.

This book primarily covers the Enterprise Edition of the server-side installation, because it is the most complete edition of the database. However, all the exercises have been tested using both Enterprise Edition and Personal Edition, and notes have been added if differences between the two versions were found.

Introduction to Oracle Universal Installer

Regardless of which installation option you select, Oracle Universal Installer handles the job. The Oracle Universal Installer provides a common user interface on all platforms when installing any Oracle product. Figure 1-8 displays one of the main windows in the Universal Installer. This shows how you use a navigation tree to select and deselect installation options.

The Universal Installer keeps log records on your computer of previous installation activity. If you install upgrades or enhancements later, the Installer skips redundant subcomponents. This saves time during subsequent installations.

The Universal Installer also enables you to view the Oracle components and subcomponents currently installed and to uninstall any component. Subcomponents are often used in more than one major component, so the Universal Installer does not allow you to accidentally remove a subcomponent that is still required by installed components.

Oracle Home

A unique feature of the Universal Installer is its ability to install several versions of Oracle software on a single machine by setting up a separate directory structure for each version and its software components. Oracle executable files are stored in a directory tree referred to as **ORACLE_HOME**. The Oracle Universal Installer defines this directory in the UNIX environmental variables, or in Windows Registry entries. A single machine can contain more than one value for ORACLE_HOME.

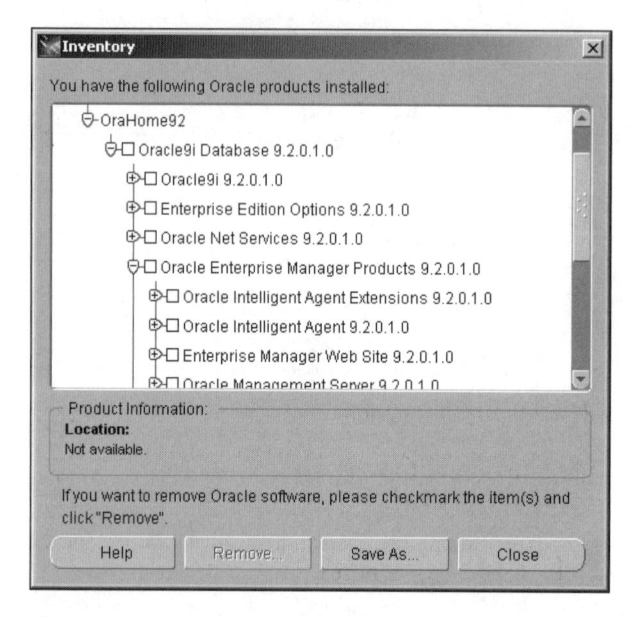

Figure 1-8 The Oracle Universal Installer shows options in a navigation tree format

On Windows platforms, you assign a name to each Oracle Home directory and use a feature called Home Selector to change the current target. On other platforms, the target is by default the last Oracle Home directory in which the Universal Installer was active.

Here is an example of using multiple Oracle Homes. Figure 1-9 shows a database server with two versions of the database software and one database instance for each version.

Looking at the figure, imagine you are the database administrator (DBA) at an e-commerce Web company. You have Oracle8*i* loaded on your Windows NT computer. You need that version to support troubleshooting and minor upgrades to your company's production system, which uses Oracle8*i* on a UNIX server. Your Oracle8*i* software resides in a directory tree starting at the **C:\ORA8I** directory, which you named **Home1**, when you defined the Oracle Home during installation. Now, you want to install an evaluation copy of Oracle9*i* without disturbing the Oracle8*i* installation. Early in the installation, you define a new Oracle Home, called **Home2**, with a root directory of **C:\ORA9I**.

You have both database instances up and running on your NT computer. You want to run SQL*Plus on the NT computer. How do you determine which instance you reach with SQL*Plus? There are two **sqlplus.exe** executable files in the NT computer system: one for each Oracle version. You must define a **target** Oracle Home, which directs any call to Oracle software to only one of the Oracle Home directories. The PATH variable contains both Oracle Home directories. Whichever comes first takes precedence. By running the Oracle Universal Installer feature called the Home Selector, you can easily change the order of the Oracle Home directories in the PATH variable. Let's say you run the Home Selector and select Home2. Now, your system uses the Oracle9*i* software. Switch to Home1, and your system now uses the Oracle8*i* software.

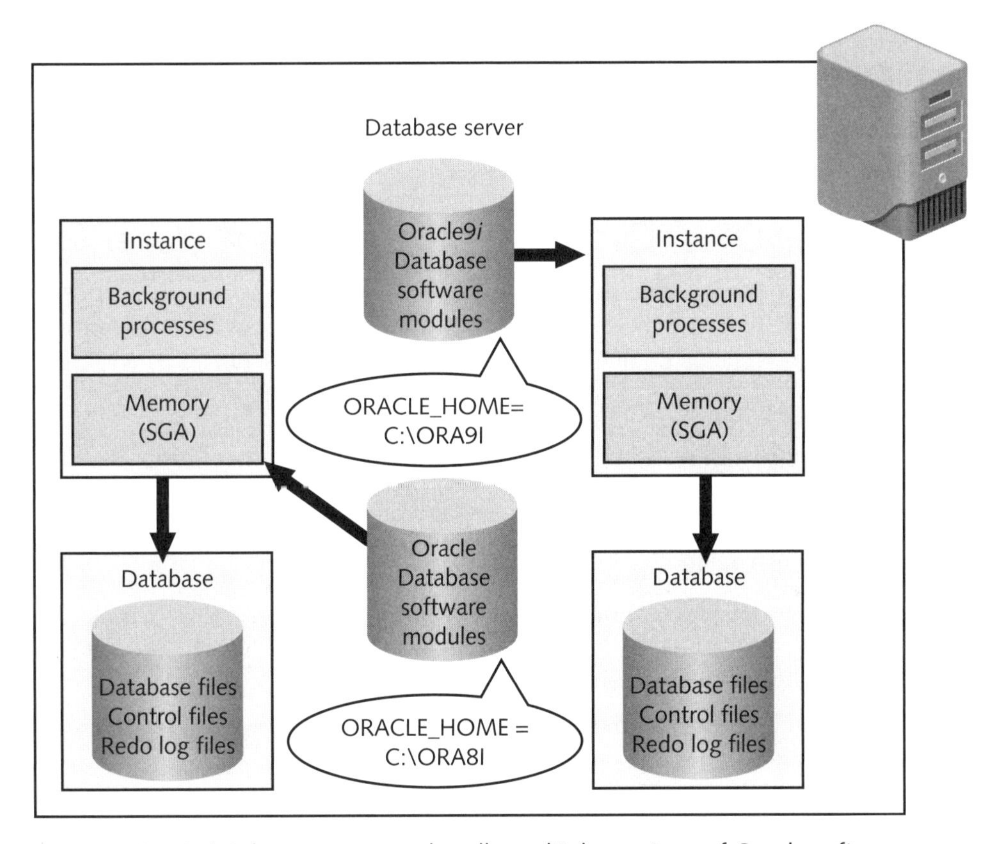

Figure 1-9 A database server can handle multiple versions of Oracle software

Silent Install

The Universal Installer provides a SILENT MODE of installation that enables you to run an installation without any human intervention. The silent install helps when you are making identical installations on multiple machines. With this process, you predefine responses to the questions asked during installation by recording the answers in a response file. Presetting the responses guarantees identical installations. In addition, predefined responses allow you to run several installations at once.

Before installation, start with the response file template provided, and modify it to fit your specifications. Then, instead of using Autostart to run the installation from the CD, use a command line, and type the command. On Windows NT, enter the following command in the directory in which Universal Installer resides:

```
setup.exe -responseFile <filename> -silent
```

Replace <filename> with the actual response filename.

On UNIX, enter this command:

```
runInstaller -responseFile <filename> -silent
```

The Universal Installer can also be used to install applications that you write yourself by preparing an installation script with the optional Oracle package called the Software Packager.

DESCRIPTION OF OFA (OPTIMAL FLEXIBLE ARCHITECTURE)

The Optimal Flexible Architecture (OFA) brings standard conventions to the directory structure and filenames for your Oracle software. OFA provides standards intended to improve performance of the database by doing the following:

- Spreading I/O functions across separate devices by separating data from software
- Improving performance by separating products into distinct directories that can be located on separate devices to reduce bottlenecks
- Speeding up administrative tasks such as backups by using naming standards for file types
- Improving detection and prevention of fragmentation in datafiles by using naming standards that quickly identify which tablespace and datafile are associated with one another

The OFA recommends a standard directory structure as well as specific naming conventions.

Directory Structure Standards

The recommended OFA standards for directory structure create order from a myriad of executable files, data storage files, and administration files. The standards allow the storing of multiple versions of the software on a single computer.

The structure begins with a root directory called **ORACLE_BASE**. All Oracle software files and database storage files reside on subdirectories under ORACLE_BASE. In Windows, the default ORACLE_BASE directory is:

```
C:\oracle
```

In UNIX, the default ORACLE_BASE directory is:

```
/pm/app/oracle
```

ORACLE_BASE houses a set of directories, in which all the Oracle software, database data, and administrative files, as well as the ORACLE_HOME directory/directories are stored. The following directories are found under ORACLE_BASE:

- **Admin/<database name>:** Stores initialization files and high level log files.
- **Oradata/<database name>:** Stores database datafiles, control files, and redo logs.
- **Product/<release number>:** On UNIX systems, stores Oracle software, such as the database engine and Oracle Net. The Product directory holds a subdirectory named after the release number of the software. For example, software for Oracle9*i*, Release 1 is in a directory named **9.1**.
- **Ora<releasenumber>:** Stores the Oracle software on Windows systems. For example, software for Oracle9*i*, Release 2 is in a directory named **Ora92**.

As you can see, the database files allow for division by function and by either database instance or software release number. Documentation is stored separately from the executable files. Executable files for Oracle9*i*, Release 2 are stored separately from executable files for Oracle8*i*, Release 1, and both sets are found below a common directory in the ORACLE_BASE directory. Datafiles for each database instance are separated, although both sets are found below the Oradata directory. All the previously listed directories contain sets of subdirectories in which actual files are stored. Subdirectories allow you to move sets of related files without losing the standardized directory structure.

The path to all the Oracle software is called **ORACLE_HOME**. ORACLE_HOME is defined as a variable within your system, so that all Oracle software subdirectories can be found using a relative address that is prefixed by the value in the ORACLE_HOME variable. For example, you might have a program that calls an executable **runnit.exe** that is found in the **bin** directory. Rather than coding the full path and filename, the program code contains the ORACLE_HOME variable like this: **$ORACLE_HOME/ bin/runnit.exe**. The variable keeps the program flexible in case the Oracle software is moved or the directory is renamed. When you have more than one software release installed on a single computer, each has its own ORACLE_HOME. The default ORACLE_HOME of Oracle9*i*, Release 1 or UNIX is:

```
/pm/app/oracle/product/9.1
```

In Windows, the default ORACLE_HOME of Oracle9*i*, Release 1 is:

```
C:\oracle\ora91
```

ORACLE_HOME contains a **bin** directory that holds most of the executables. Each subcomponent, such as SQL*Plus and Oracle Net, has its own subdirectories that contain additional files used by that subcomponent. For example, Oracle Net software has an executable in the **bin** directory and has additional files in the **network** directory. Optional add-on software from Oracle, such as interMedia, adds its own directory within the ORACLE_HOME structure. Figure 1-10 shows two directory structures, one for UNIX, and one for Windows.

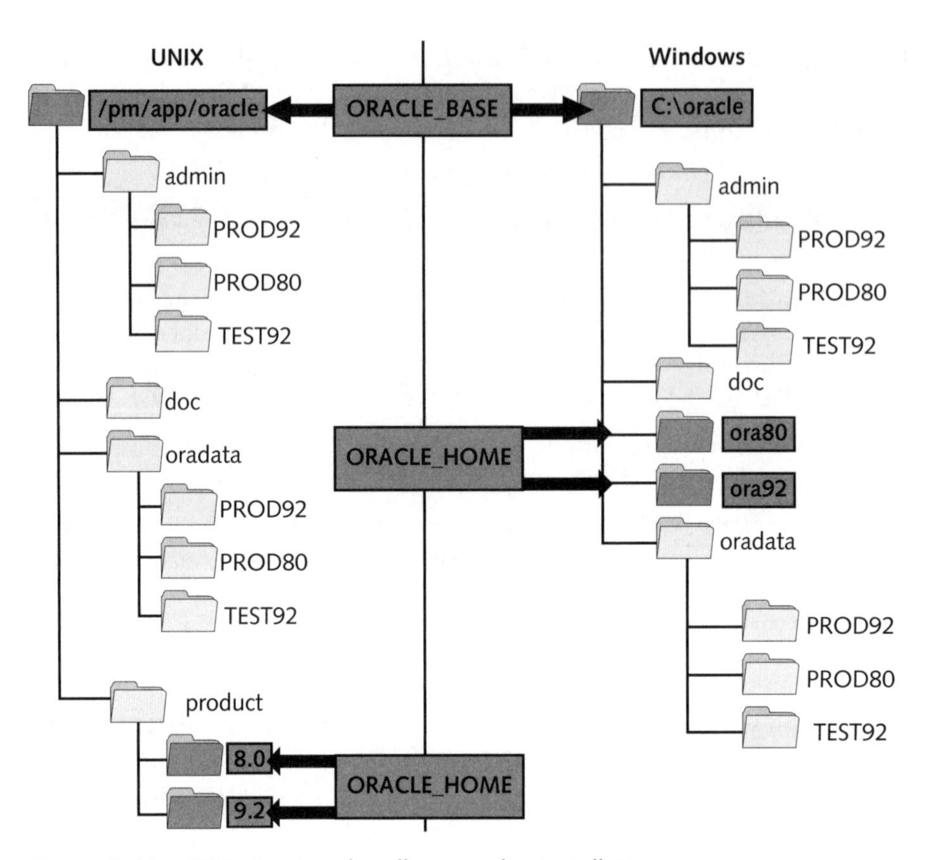

Figure 1-10 OFA structure handles complex installations

Looking at Figure 1-10, observe that both structures contain the OFA structure for a computer with two software releases (9.2 and 8.0) and three database instances (PROD92, PROD80, and TEST92). The PROD92 and TEST92 instances have the same ORACLE_HOME.

File Naming Standards

The OFA standards call for naming files according to their type and sequence of creation. Table 1-1 shows the naming standards recommended. The naming standards are only recommendations, not requirements. If you prefer to use a different naming standard, you can do so without any penalties. You can even rename the files that Oracle creates (for example, the files created when you allow the installation to create a predefined database for you). If you choose to follow the naming standards, when the database grows, you issue Oracle commands to create additional files, telling Oracle to create files that you name according to these conventions. This assures that the original files and all subsequent files follow the same naming standards.

All the files in Table 1-1 are located within the ORACLE_BASE/**Oradata** directory structure in the OFA configuration.

Table 1-1 OFA naming standards

File Type	Description	Suffix	Examples
Control file	Administrative file containing up-to-date information on the database structure, log files, and checkpoints; critical file for opening the database.	**.ctl** The filename should be controlNN, where NN is a sequence number.	**Control01.ctl** **Control02.ctl**
Redo log file	A set of files that record changes to database data. Online redo logs record changes immediately. Archive redo logs contain copies of current and past redo logs for recovery if needed.	**.log** The filename should simply be redoNN, where NN is a sequence number.	**Redo01.log** **Redo02.log**
Database files	A physical file on the computer disk containing data, such as tables and indexes. Each database file belongs to one tablespace in a database. A tablespace may use more than one database file.	**.dbf** The filename should be the tablespace name and a seq-uence number.	**System01.dbf** **Users01.dbf** **Users02.dbf** **Temp01.dbf**

The primary purpose of these naming standards is to speed up the process of identifying files. Quickly locating files helps during recovery processes and aids in routine tasks, such as backing up files, relocating files to alleviate bottlenecks, and adjusting file sizes.

CHAPTER SUMMARY

- The core Oracle9*i* RDBMS software includes a suite of products.
- The basic installation includes over 200 components and subcomponents.
- The key components of Oracle9*i* RDBMS software are: Oracle9*i* database, Oracle Net, Java/Web support and tools, Enterprise Manager, SQL*Plus, utilities, and precompilers.
- Options to Oracle9*i* increase functionality to the database or add more tools to the basic tool set provided.
- A database instance is made up of the memory area (SGA) and the background processes started on a computer.
- A database server contains a database instance and the database files that store the database data.

- A database server can contain a single instance, multiple instances, or can be a combination of multiple servers, called a cluster.
- Some of the key optional components include Oracle Partitioning, Oracle Real Application Clusters, Oracle Spatial, several Enterprise Manager packs, Oracle Data Mining, Oracle Advanced Security, Oracle Label Security, and Oracle OLAP services.
- The most complete and robust installation option is the Enterprise Edition for the server side.
- Other choices for the server side are: Standard Edition (for workgroups) and Personal Edition (for individuals).
- On the client side, you can install the Administrator option (for administration of remote databases) or the Runtime option (for running applications that access remote databases).
- The ORACLASS database is provided to you for running exercises throughout the book.
- A running case project develops a database system for a fictional newspaper chain called Global Globe.
- Multi-tier (n-tier) architecture separates different levels of software components.
- The client-side installation option runs remote applications or remote DBA tasks.
- The server-side installation option runs a database server.
- The server-side has three paths: Enterprise Edition, Standard Edition, and Personal Edition.
- The client-side installation option has two paths: Administrator and Runtime.
- Oracle Universal Installer is a user-friendly interface for installing or uninstalling all Oracle products.
- The Oracle Home Selector feature of the Universal Installer enables the installation of multiple versions of Oracle software on one computer.
- The Silent Install option prebuilds and runs an installation without any human intervention.
- Optimal Flexible Architecture (OFA) defines standard names for directory structures, database files, control files, and redo log files.
- OFA uses the ORACLE_BASE variable to define the root location for all Oracle subdirectories, including Oradata (where database files reside) and ORACLE_HOME (where Oracle executable software resides).

Review Questions

1. The Oracle9*i* Application Server is part of the Oracle9*i* database software suite. True or False?
2. List the key components that come with Oracle9*i* RDBMS software.
3. Explain the function of Oracle Net in relation to the database.
4. Explain the function of Oracle Net in relation to the utilities component.
5. Which two key components provide ways to start and stop the database?
6. Explain how a precompiler works.
7. Oracle9*i* supports multiple languages. True or False?
8. Describe the difference between the Administrator and Runtime installations.
9. The Personal Edition database is identical to the Enterprise Edition database. True or False?
10. Oracle Universal Installer CANNOT install non-Oracle applications. True or False?
11. List two situations in which you would use the Oracle Home Selector.
12. Write the command to run Universal Installer in SILENT MODE on Windows, in which your response file is named **resp0901.txt**.
13. Describe why an Oracle Flexible Architecture helps improve performance.
14. Oracle executable files are found under the ORACLE_BASE directory. True or False?
15. Oracle administrative files are found under the ORACLE_HOME directory. True or False?

Exam Review Questions—Database Fundamentals I (#1Z0-031)

1. Your database server has a multithreaded server. You have fifty user sessions. How many server sessions do you have?
   a. Fifty
   b. Less than fifty
   c. More than fifty
   d. One
2. Your ORACLE_HOME is defined as **D:\oracle\ora91** and your ORACLE_BASE is defined as **D:\orabase**. Which statement is true?
   a. This structure complies with OFA.
   b. This structure does not comply with OFA.
   c. This structure cannot be used, because it does not comply with OFA.
   d. This structure is not specified by OFA and, therefore, may or may not comply.

3. Which of these functions is handled by the key components of Oracle9*i* *and* can be enhanced with optional software that Oracle sells? Choose all that apply.
   a. Query the data
   b. Enforce user security
   c. Do backups
   d. Monitor performance
   e. All of the above
4. Which platforms can install the Personal Edition of Oracle9*i*, Release 1? Choose all that apply.
   a. Windows NT
   b. Windows 95
   c. UNIX
   d. Windows 98
   e. Linux
5. Oracle Partitioning can aid in which of these applications?
   a. An accounting application requiring complex calculations
   b. A sales projection application requiring large quantities of historical data
   c. An online store needing fast response time for many users
   d. A small company with a work group of five programmers
6. Which of these optional components enhance the database internally? Choose all that apply.
   a. Oracle Advanced Security
   b. Oracle Label Security
   c. Oracle Spatial
   d. Oracle Diagnostics Pack
   e. All of the above
7. When installing Oracle9*i* on a computer that currently has Oracle8*i*, which of the following statements are *not* true? Choose two.
   a. You must replace the Oracle8*i* software with the Oracle9*i* software.
   b. Your ORACLE_HOME is permanently set to point to the newest software.
   c. The Oracle Universal Installer handles either release of the software.
   d. A database instance can run on one or the other, but not both releases.
   e. Enterprise Manager is installed with Oracle9*i* Enterprise Edition.

8. Which components make up a database instance? Choose two.
   a. Database software modules
   b. Database files
   c. System Global Area
   d. Oracle Net
   e. Background processes
9. The Standard Edition of the database is primarily used for what purpose?
   a. Testing new versions of the database
   b. Running applications on a client workstation
   c. Developing applications on a programmer workstation
   d. Running administrative tasks
10. Which component is NOT available on a client-side installation?
    a. Enterprise Manager
    b. Precompilers
    c. SQL*Plus
    d. Utilities
    e. None of the above

HANDS-ON ASSIGNMENTS

1. Looking at only the Windows side of Figure 1-10, list the directories in the diagram that contain files related to Oracle8. Use full path names in your list. Repeat for the UNIX side. Save your work in a file named ho0101.txt in the Solutions\Chapter01 directory on your student disk.
2. Your office plans to install Oracle9*i* Enterprise Edition, version 9.0 on its server. They want to know ahead of time what the directory structure will look like. The root directory is going to be **/all_apps/oracle**. Draw a diagram, using Figure 1-10 as a guide, in which you show the location of all the key components and the database datafiles. Save your work in a file named ho0102.bmp in the Solutions\Chapter01 directory on your student disk.
3. Using your diagram from Assignment 2, what are the full path values of ORACLE_HOME and ORACLE_BASE? Save your work in a file named ho0103.txt in the Solutions\Chapter01 directory on your student disk.

4. You have just installed the Administrator client-side software on your NT computer. There is another NT computer on your network that has the Enterprise Edition server-side software installed and running. You start up Enterprise Manager and look at the list of users on the remote database. How did the information travel between the two network nodes? Draw a diagram, similar to Figure 1-1, that shows a client and a server communicating with one another. Draw all the components installed on each side, and draw a line indicating the network between the two components on each side that communicate with one another. Save your work in a file named ho0104.bmp in the Solutions\Chapter01 directory on your student disk.
5. Your e-commerce business is growing fast. You currently have the Standard Edition of Oracle9*i* on your server. Your server is having trouble with servicing more than five customers at a time. The server tends to bog down when any user searches the product table. Write a recommendation letter describing three options that help solve one or both of these problems. Save your work in a file named ho0105.txt in the Solutions\Chapter01 directory on your student disk.

CASE PROJECT

The Global Globe has an Online Applications branch and a Data Warehouse Applications branch. The Online Applications branch has just moved to a different city. This poses a problem because your central database (Enterprise Edition) was used for programmers in both divisions. Each programmer has a computer, with the Standard Edition installed (without a database), on which he or she develops applications. You see that this configuration may not be suitable for a new remote location. Come up with two different strategies for supporting programming at the remote site that *do not* require remote access to the central database. Draw diagrams (similar to Figure 1-1) of each plan showing the programmer nodes and any other database nodes at the remote site. Create a list that compares the advantages and disadvantages of each plan. Save your diagram in a file named case0101.bmp and your list in a file named case0101.txt in the Solutions\Chapter01 directory on your student disk.

CHAPTER

2

OVERVIEW OF DATABASE ADMINISTRATOR (DBA) TOOLS

In this chapter, you will:

- Identify the main DBA tools in the Oracle9*i* software suite
- Configure Oracle Net to connect to the database
- List the memory and background process components of the database instance
- Start using the Enterprise Manager

As you discovered in the previous chapter, Oracle9*i* brings with it a host of tools and utilities to aid you in working with the database. Two of the most important tools are Oracle Net and Enterprise Manager. In this chapter, you examine the many tools that are available to assist you as a database administrator (DBA). Then, you learn how to set up and configure Oracle Net on a network. After you have Oracle Net running properly, you go through the steps to configure Enterprise Manager, and then tour the main DBA tools included in Enterprise Manager.

OVERVIEW OF DBA TOOLS

In the previous chapter, you were introduced to the key components of the Oracle9*i* software suite. Within these key components, you find many tools designed to streamline the work of the DBA. Many of these tools become integrated in a central workspace, the Enterprise Manager console. You have a hands-on tour of the console later in this chapter. Some tools are stand-alone, some are reached inside the console, and others are available both independently and through the console. Table 2-1 lists the DBA tools and their uses, and also indicates whether the tool is available through the console, as a stand-alone, or both.

Table 2-1 DBA tools and their uses

Tool name	Function	Start from console?	Start from operating system?
Analyze Wizard	Collects statistics about tables or objects, validates tables, finds continuing rows	Yes	No
Backup Wizard	Creates a backup plan and schedules it with step-by-step instructions	Yes	No
Data Migration (Upgrade) Assistant	Migrates data from older versions to newer versions of the database	No	Yes
Database Configuration Assistant	Creates new database instances	No	Yes
Enterprise Manager Configuration Assistant	Configures Enterprise Manager with step-by-step instructions	No	Yes
Enterprise Manager Console	Monitors databases and provides access to database tools	No	Yes
Export Wizard	Exports data using navigator-based export tool	Yes	No
Import Wizard	Imports data exported with Export Wizard	Yes	No
Instance Manager	Starts or stops the database; views and modifies initialization parameters; monitors user activity	Yes	No
Log Miner	Queries redo log files	Yes	No
Net Configuration Assistant	Configures Oracle Net with step-by-step instructions	No	Yes
Net Manager	Configures Oracle Net	Yes	Yes
Schema Manager	Views and modifies data; views and modifies table structures, indexes, views, objects, procedures, and packages	Yes	No

Table 2-1 DBA tools and their uses (continued)

Tool name	Function	Start from console?	Start from operating system?
Security Manager	Views and modifies users, roles, and profiles; creates new users; changes passwords	Yes	No
SQL*Plus	Executes SQL commands; runs reports and queries; starts and stops the database	No	Yes
SQL*Plus Worksheet	Executes SQL commands; runs queries	Yes	Yes
Storage Manager	Views and modifies database storage	Yes	No
Summary Advisor Wizard	Generates recommendations for data warehousing, such as whether to create summary tables	Yes	No

All of the tools that can be run from the operating system are found on the Start menu in Windows in some branch of the Oracle tree of applications. In UNIX, Solaris, and Linux, you may need to start them on a command line. (In Windows, you can also start the tools from a command prompt if you want.) Table 2-2 shows the operating system commands for starting each tool in Windows and in UNIX. Solaris, Linux, and other UNIX-related operating systems also use the commands listed in the UNIX column of the table.

Table 2-2 Operating system commands for starting tools

Tool Name	UNIX command	Windows command
Data Migration (Upgrade) Assistant	Dbua	launch.exe "<oracle-home>\assistants \dbma"<oracle-home>\assistants\dbma\ dbua.cl
Database Configuration Assistant	Dbca	launch.exe "<oracle-home>\assistants\ dbca"<oracle-home>\assistants\dbca\dbca.cl
Enterprise Manager Console	oemapp console	oemapp.bat console
Enterprise Manager Configuration Assistant	Emca	emca.bat
Net Configuration Assistant	netca	launch.exe "<oracle-home>\network\tools" <oracle_home>\network\tools\netca.cl
Net Manager	netmgr	launch.exe "<oracle-home>\network\tools" <oracle_home>\network\tools\netmgr.cl
SQL*Plus	sqlplus <username>/ <password>	sqlplus <username>/<password>
SQL*Plus Worksheet	oemapp worksheet	oemapp.bat worksheet

Whether you start the tools from the command line, console, or the Windows menu, these tools give you a way to work on the database in a Windows-style environment, in which the actual Oracle commands are generated for you. However, it is important to be familiar with both automated (using a tool) and manual (using a command) methods for some tasks. Sometimes, you do not have a Windows-like interface for diagnosing and correcting a problem with the database. In these cases, you must understand how to work directly from the command line. Other times, as you become familiar with both methods, you may find that certain tasks are more quickly accomplished using manual methods. For example, changing a user's password can be done with a single command line in SQL*Plus, as shown in Figure 2-1. On the other hand, changing a user's password in Security Manager requires opening Windows, navigating through lists of users, and then typing the password twice, as shown in Figure 2-2. Of course, you must know the command by heart before the manual version is faster than the tool. Later chapters teach both manual and tool-based methods for many tasks.

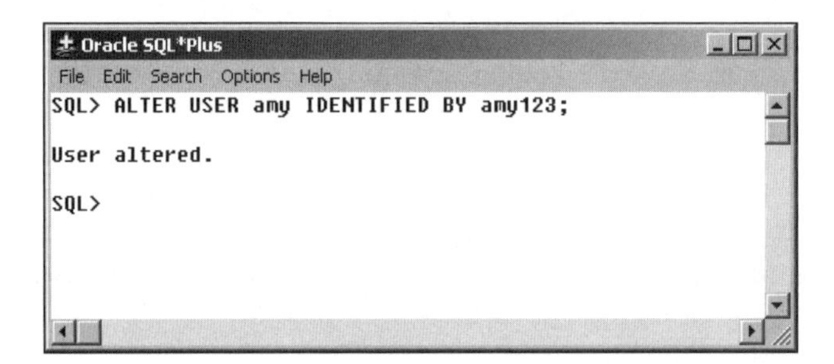

Figure 2-1 Changing a user's password manually requires one command

Some tasks can be handled by more than one tool. For example, you can export data using the Export Wizard or the Backup Manager. You can stop the database using Instance Manager or SQL*Plus. To view data in a table, you can use the Schema Manager or write a query in either SQL*Plus or SQL*Plus Worksheet.

Oracle Net is a common denominator for these tools. In every case, you must connect to the database via Oracle Net. For tools that run inside the Enterprise Manager console, a single logon gives you access for all the tools. When running a tool directly from the operating system, you must log on with a user name, a password, and a database or network name known to Oracle Net. The next section describes how to configure and test Oracle Net connections.

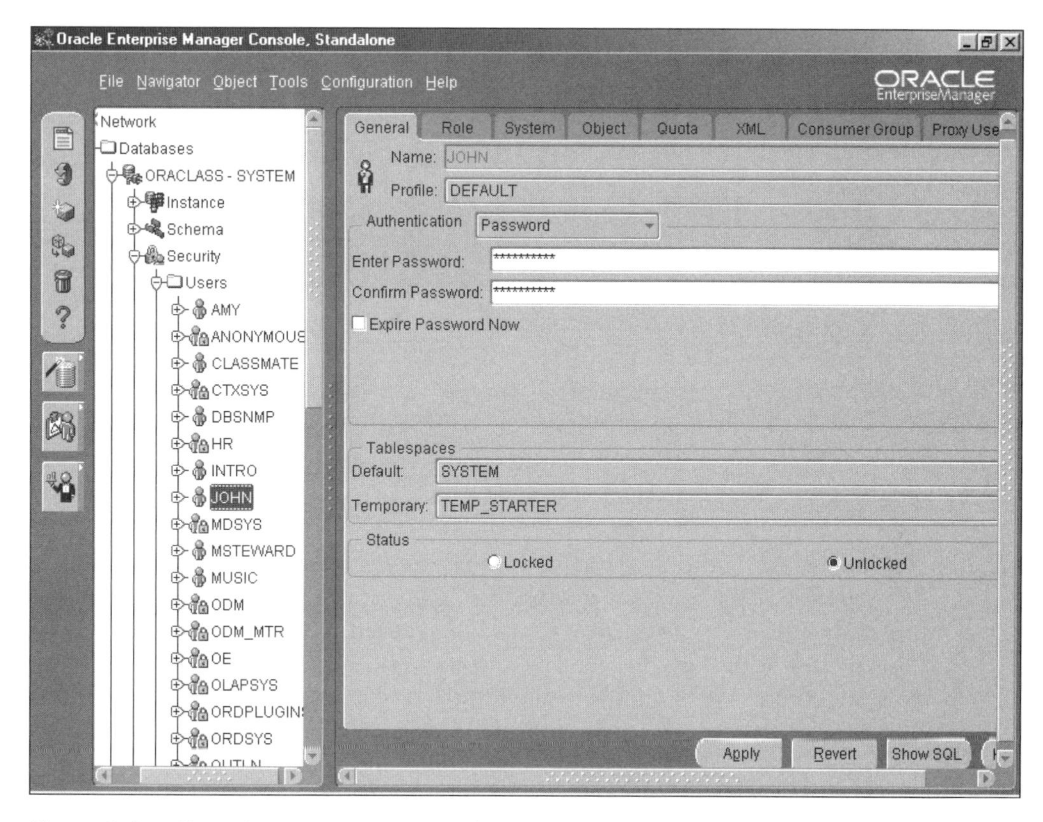

Figure 2-2 Changing a user's password in Security Manager takes more time

Configuring Oracle Net to Connect to the Database

Nearly every time you access Oracle9*i*, you go through Oracle Net. Therefore, a critical step in setting up your working environment involves properly configuring Oracle Net to access the database on which you want to work.

Overview of Oracle Net Architecture

Oracle Net is made up of several subcomponents that work together to translate your requests, such as SQL queries, into packages for the local or Internet network. Each package has a destination and an identifying code that tells the receiving end when it has collected all the packages for a single request. Figure 2-3 shows a basic diagram of a user with Enterprise Manager on his desktop computer connecting to an Oracle9*i* database on a database server across a local network.

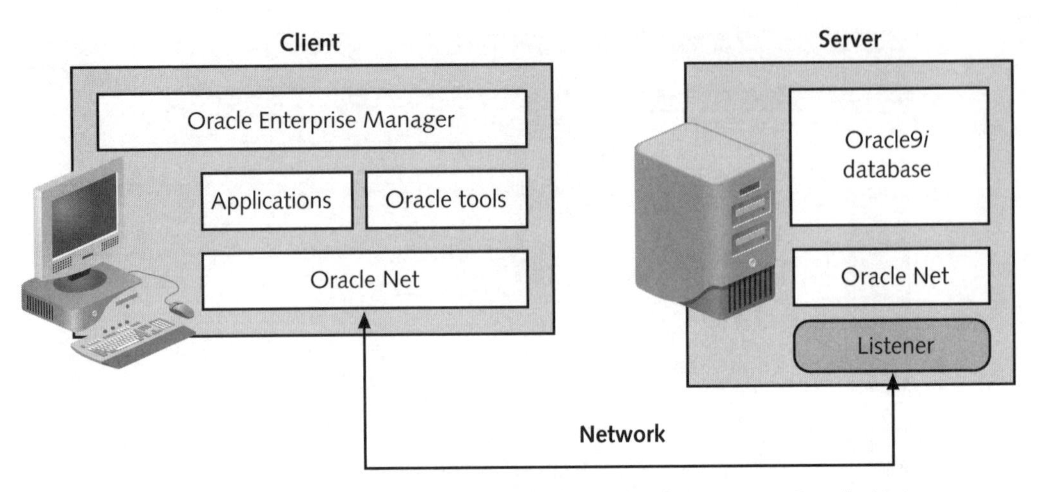

Figure 2-3 Communication between a user and the database using Oracle Net

As you see in Figure 2-3, Oracle Net resides on both the client and the server side of the network. This is necessary when using applications (such as a tax preparation application that uses Oracle to store its data) and when using Oracle tools, such as Enterprise Manager or SQL*Plus, installed on the client side. Both installations of Oracle Net must be configured so that they are synchronized to the target database. On the client side, Oracle Net accepts requests from the Oracle tool, translates them into the local network protocol (usually TCP/IP), and sends them across the network. The configuration is stored in the **tnsnames.ora** file on the client. On the server side, the computer has its own copy of the **tnsnames.ora** file that defines the database that runs on the computer and remote databases on other computers if needed. Oracle Net receives the request from the network, translates it into Oracle protocol, and sends it to the database. The database has a service, called the **Listener**, which waits for requests and responds to them. Oracle Net must know not only what computer contains the database, but also what the database's name is and to which port number its Listener is tuned. All these details about a database are stored in Oracle Net as the **service name** of the database. A **service name** is a set of information that Oracle Net uses to locate and communicate with an Oracle database. After a service name is defined on both the client and server side, any tool on the client side can reach the server side database using the service name combined with a valid user name and password. Typically, a tool prompts the user to log on and provide all three of these vital pieces of information. Figure 2-4 shows the logon screen for SQL*Plus as an example. The service name goes in the box called Host String.

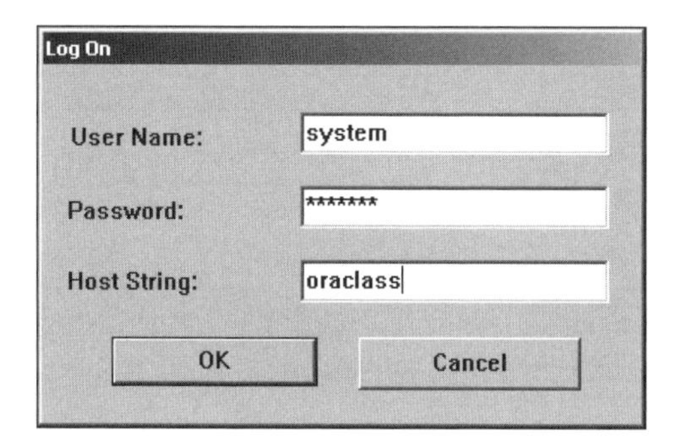

Figure 2-4 Specify the Oracle Net service name in the Service box when logging in

When you are using tools that reside on the same computer as the database, you might assume that no Oracle Net connection is needed. You have the choice to use either Oracle Net or Bequeath, a protocol allowed only when you are logged onto the database machine. Either method works equally well.

Figure 2-5 shows the two primary access points for reaching the database server from a client computer. In addition, it shows the path of communication used when you access the database while logged onto the server. The three methods are:

- **Client with Oracle Net:** When a client computer runs applications such as those written in C or COBOL, the client computer must install the client-side version of Oracle9*i*. This provides Oracle Net and its tools, so the client can configure a service name to reach the database server over the network.
- **Client with JDBC driver:** A client running a browser with a Java applet can use the JDBC thin driver to access a remote database via the Internet. The JDBC thin driver can be included in the applet, so that no additional software needs to be installed. On the server side, the database only requires a standard TCP/IP protocol program (which is generally already installed and used for most Internet communication).
- **Terminal with direct connection:** When you log directly onto the database server, you access the database directly via the Bequeath protocol. To use this, you simply leave the service name blank when logging onto any tools or utilities. You are automatically directed to the database that is running on the machine. This technique also supports a workstation computer that has its own copy of the database running. By default, you reach the database on your computer when you omit the service name during log on.

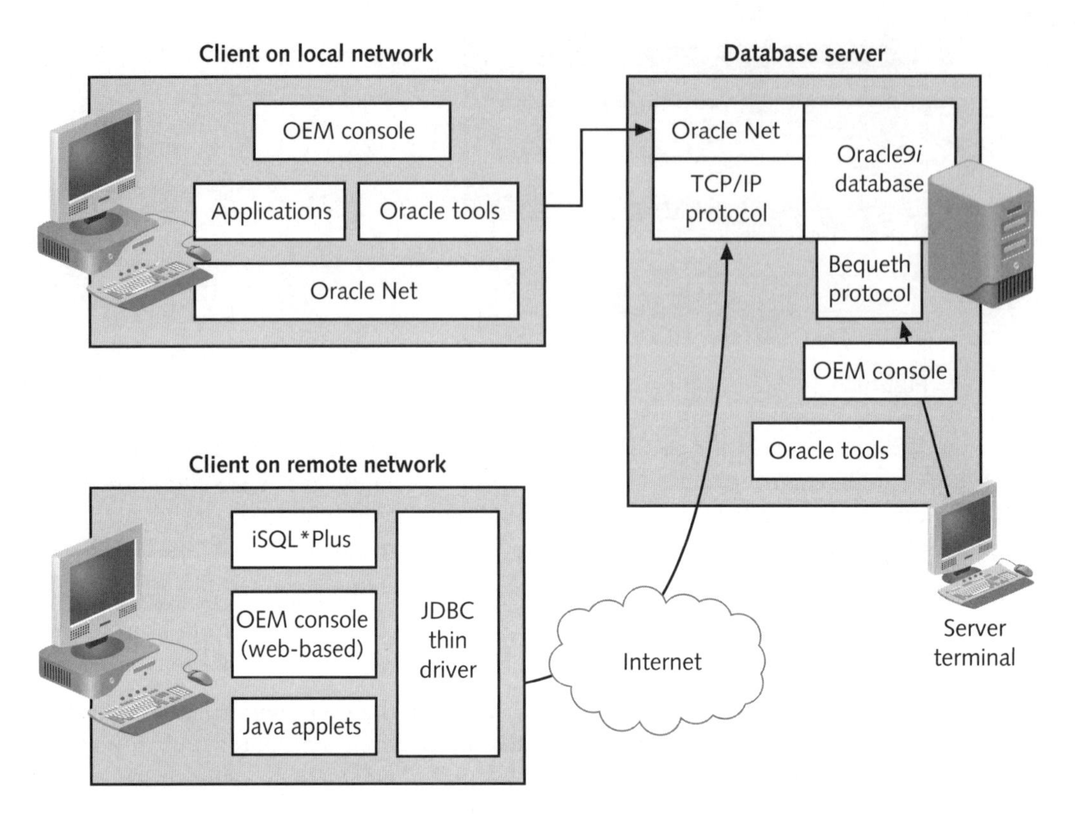

Figure 2-5 Three methods of connecting to the database

The following section leads you through the steps to set up an Oracle Net connection on the client.

Step-By-Step Configuring of Oracle Net

Your first hands-on assignment is to configure Oracle Net on your workstation so that you can use the Enterprise Manager, SQL*Plus, and other tools needed for the rest of the book. Your instructor provides you with appropriate user names, passwords, and database names.

Follow these steps to configure Oracle Net.

1. To start Net Manager in Windows, on the taskbar, click **start/Programs/Oracle OraHome92/Configuration** and **Migration Tools/Net Manager**. In UNIX, type **netmgr** on the command line. The initial screen for Net Manager appears, as shown in Figure 2-6.

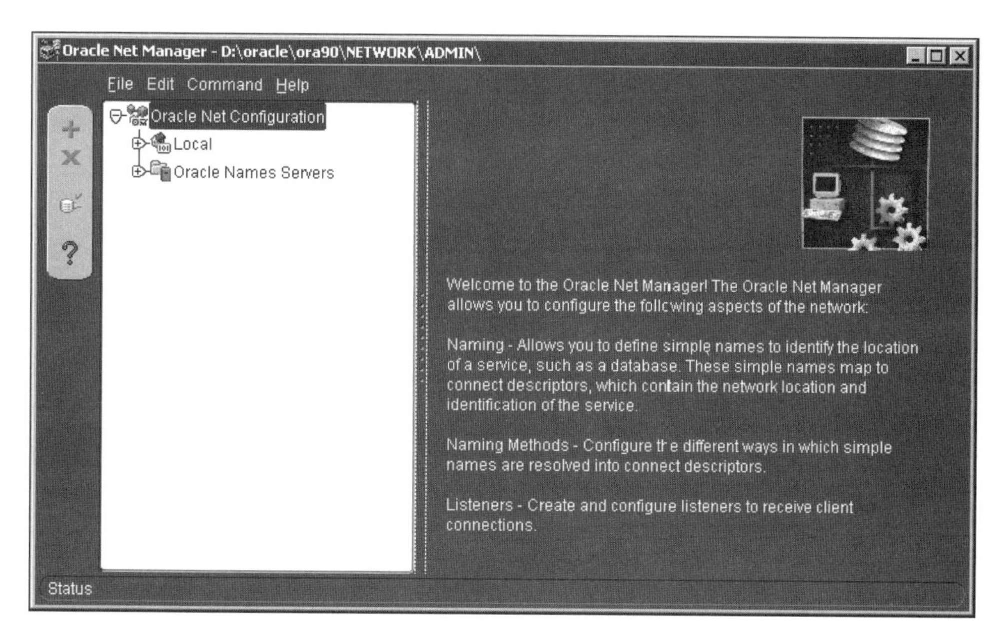

Figure 2-6 No login is required for Net Manager

2. Expand the Local node by clicking the **plus sign** to the left of the Local icon.
3. Expand the Service Naming node by double-clicking the **folder icon**. A list of database connections appears below the Services Naming node. Assume for now that these are not working for your database and that you must create a fresh service name for your database. After a service name is created and saved in Oracle Net's configuration file, any Oracle tool or utility on that computer can connect to the database using the service name.
4. Click the **green plus sign** in the left margin of the window, as shown in Figure 2-7. The Net Service Name Wizard starts up.

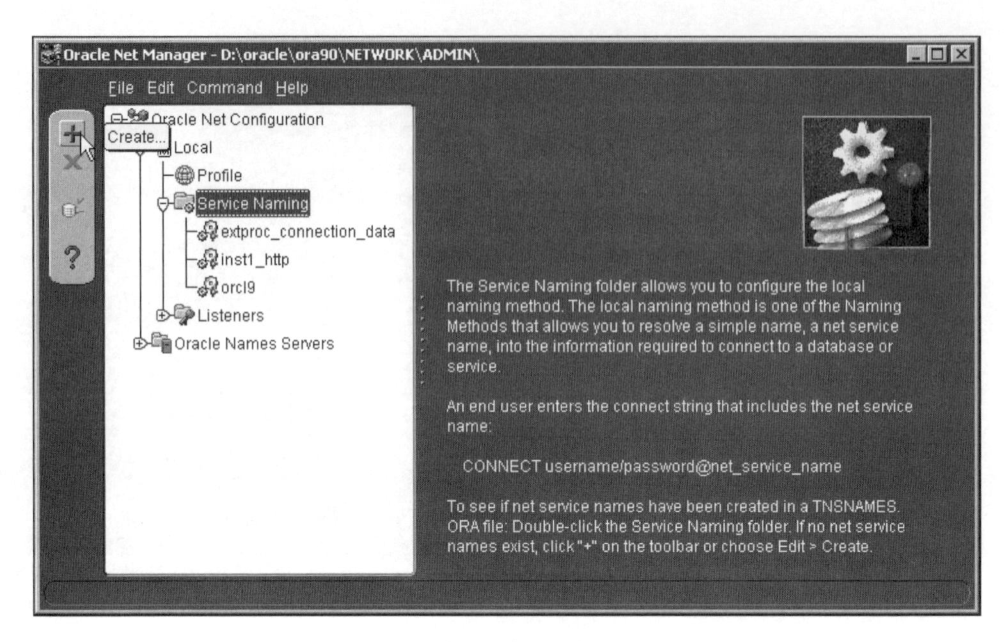

Figure 2-7 Begin creating a Service by clicking the green plus sign

5. Type the new service name, **oraclassX**, in the box as shown in Figure 2-8, and click **Next**.

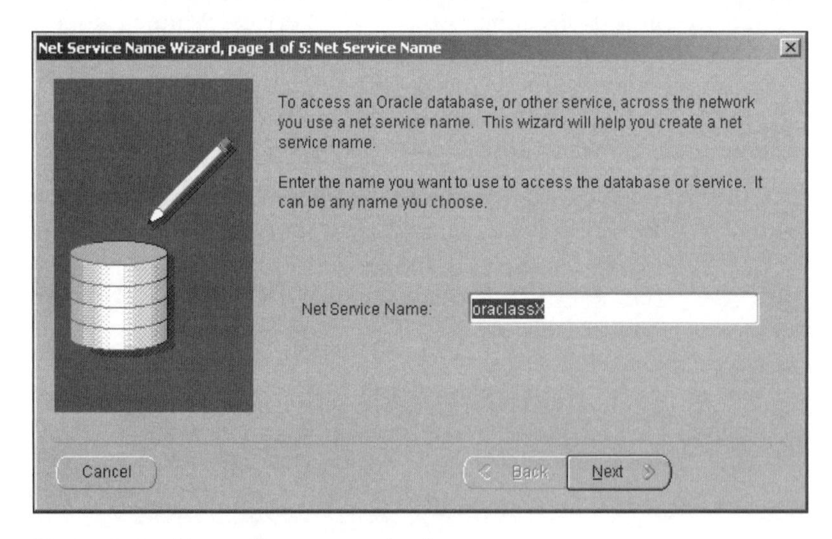

Figure 2-8 Name the new service here

6. For the connection, select **TCP/IP** (Internet Protocol) as the protocol, as shown in Figure 2-9, and click **Next**.

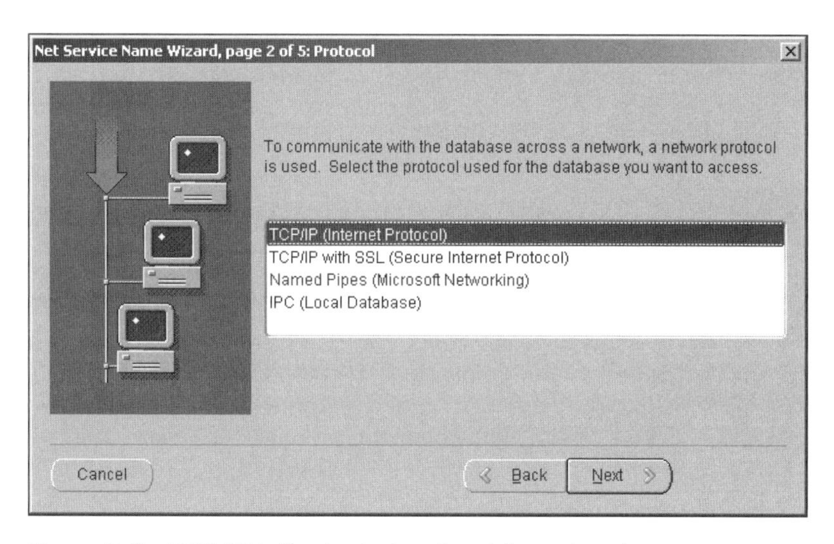

Figure 2-9 TCP/IP is the typical protocol for networks

7. Type the computer name on which the database resides in the Host Name box. *Your instructor provides this name.* Accept the default port setting of **1521**. This port setting is the default port setting for most Oracle database listeners. Figure 2-10 shows the window filled in with a computer name of student01 and port of 1521. Click **Next**.

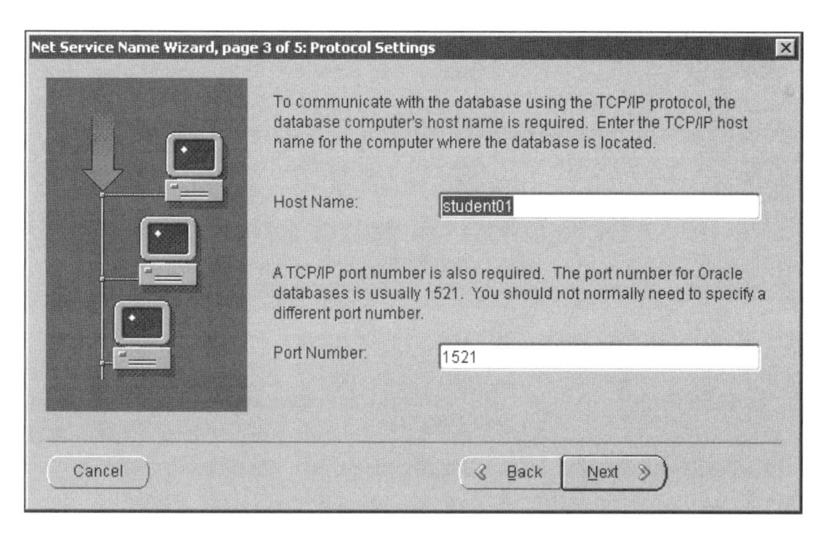

Figure 2-10 Identify the host computer where the database runs

To discover the name of a Windows computer, click **Start/Settings/Control Panel**. Then double-click **System**. Click the **Network Identification** tab. In Windows XP, click the **Computer Name** tab. The computer name appears listed in the Full computer name field.

8. Accept the default selection of Oracle8*i* or later, and fill in the Oracle9*i* database service name. *Your instructor provides this information.* Figure 2-11 shows the screen with a database service name of **class.edu**. If you are setting up a service name for Oracle8 and lower, the database **SID** (System Identifier) is used. The SID has been replaced by the database service name in Oracle8i and higher. This change gives more flexibility when implementing complex database configurations, such as clustered or parallel databases. Leave the Connection Type selection as its default "Database Default." The connect type can be "Database Default," "Dedicated Server," or "Shared Server." Choosing "Database Default" allows Oracle Net to connect according to the database setup, rather than dictating which mode to use. Click **Next** to proceed to the next window.

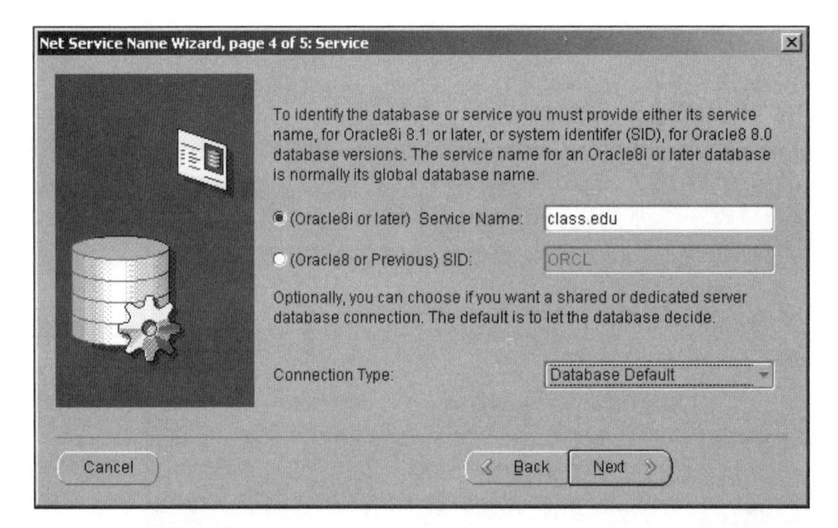

Figure 2-11 Identify the database instance by its service name

9. Click **Test**. This starts an automatic connection to the database using the new service name you created. The test logs on as user *scott* with password *tiger*. Figure 2-12 shows the test. This is the default user and password installed in Oracle databases to store demonstration tables. If you have removed *scott* or changed the password, the automated test fails. However, you can adjust the user/password combination to any valid user by clicking **Change Login**, changing the user name and password, clicking **OK**, and then clicking **Test** again.

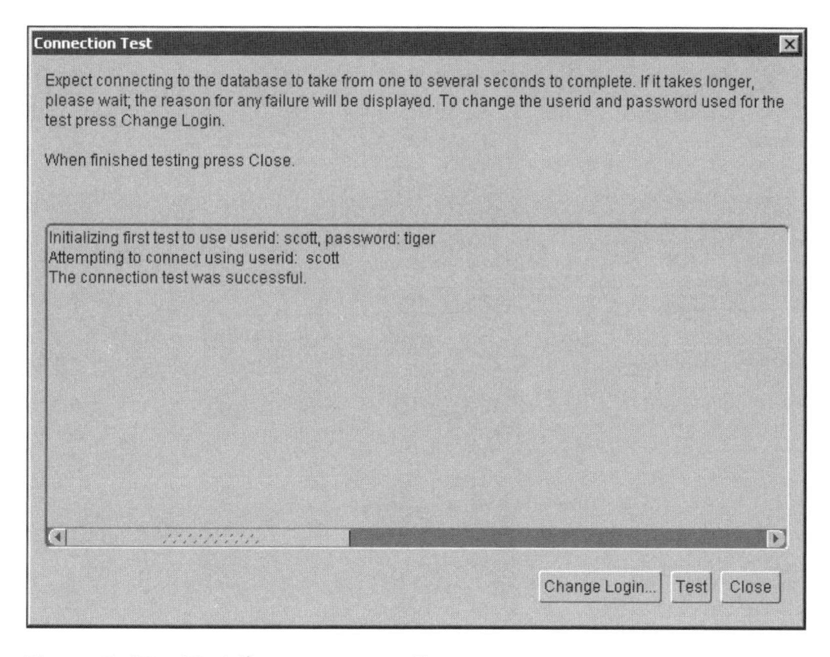

Figure 2-12 Test the new connection

10. Click **Close** to close the test window.
11. Click **Finish** to complete the definition. This returns you to the main window, in which the new service name is added to the list of names under the Service Naming folder.
12. Save the configuration by selecting **File/Save Network Configuration** from the menu. Figure 2-13 shows the new service name and its configuration.

Make changes to any service name configuration by typing over the current settings. Remember to test and save your changes. Test changes by selecting **Command/Test Service** from the menu.

The file **tnsnames.ora** stores the Oracle Net configurations that you work on when using Net Manager. This file is found in the **ORACLE_HOME/network/admin** directory.

Use caution when editing this file. You must use a plain text editor, such as Notepad, to avoid adding any characters to the file that would corrupt it.

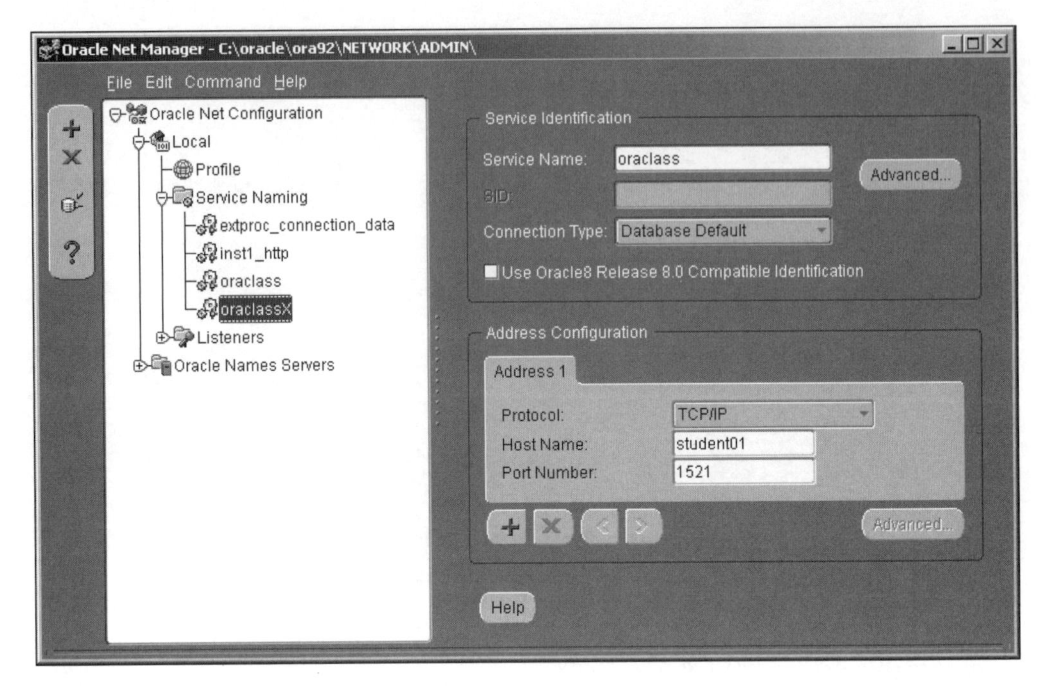

Figure 2-13 A new service must be saved to take effect

It may be helpful to think of the tools you have been working with so far as the outer shell of the database system. The next section takes you "under the hood" to look at the internal workings of the database.

ORACLE MEMORY AND BACKGROUND PROCESSES

To start this section, remember that in Chapter 1 you learned that the database instance runs on a database server and uses data inside the database. How does a user, running an application from a client machine, connect with the database? The next few sections describe this process in detail.

Connecting to the Database

The processes that support a user's activities on the database begin the moment the user logs onto the database. There are two configurations for this connection: the shared server mode and the dedicated server mode. Figure 2-14 shows an overview of the processes involved when the database is in shared server mode.

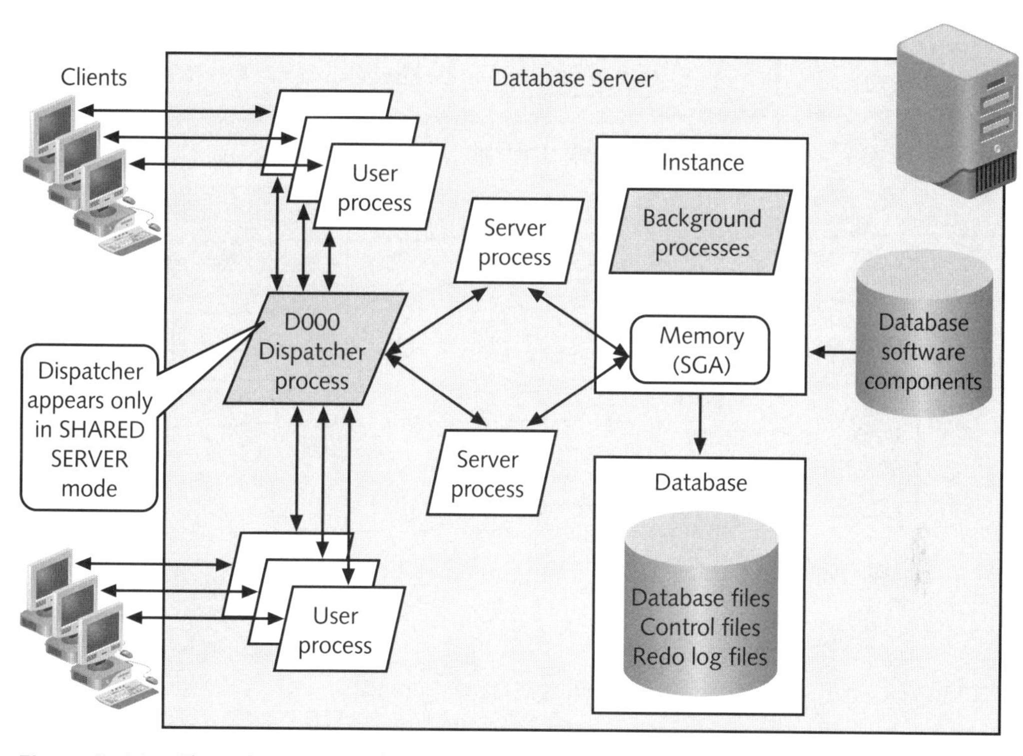

Figure 2-14 Shared server mode uses a dispatcher to distribute connections

The user's application might be a C program, a Java applet, or an Oracle tool, such as SQL*Plus. In any case, after a request for logging onto the database is received by the database server, Oracle creates a **user process**. The user process handles the communication with the user's application.

Then, Oracle creates a **server process**. In the case of a **dedicated server**, every user process has its own server process. With a shared server, several user sessions may share one server process. The dispatcher background process distributes the connections from user processes to server processes. A **shared server** uses CPU and memory more efficiently than a dedicated server by swapping out user processes during idle time. This is the best configuration for a database system with many users, especially if they are online users, because there is usually plenty of idle time between activity for sharing the server process.

Figure 2-15 shows the user and server processes for a database in dedicated server mode. Everything is the same, except there is no dispatcher to be the intermediary between the user process and the server process. Every user process has its own server process. This is the best configuration for a database system with large amounts of memory and CPU power because user processes never have to wait for a server process to be available.

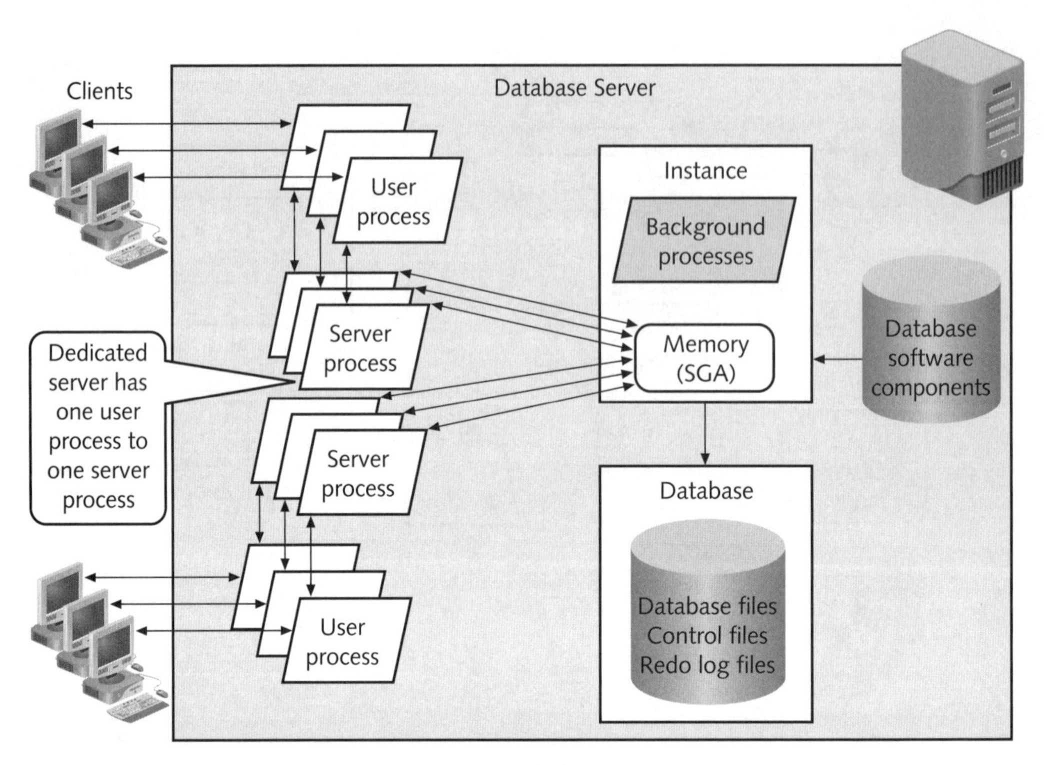

Figure 2-15 A dedicated server can have hundreds of user and server processes

The link from the user session, through the server session, and to the database instance is called a **connection**. When you log onto SQL*Plus, for example, the tool tells you that you are connected. This means that the tool has established a connection with the database instance through the user process and server process on the database server. A **session** lasts from the time you make a connection until you end the connection. When you log off, for example, or exit an application, you end your session.

Now you have established a connection. The next step occurs when you perform a task that requires the database to interact with your user process. For example, you have logged onto SQL*Plus and you now execute a query. This is where the database's memory and background processes come into play.

Memory Components and Background Processes

The two main sections of memory for the Oracle instance are the System Global Area (SGA) and the Program Global Area (PGA). Figure 2-16 shows a map of the PGA, the SGA, and the components within each.

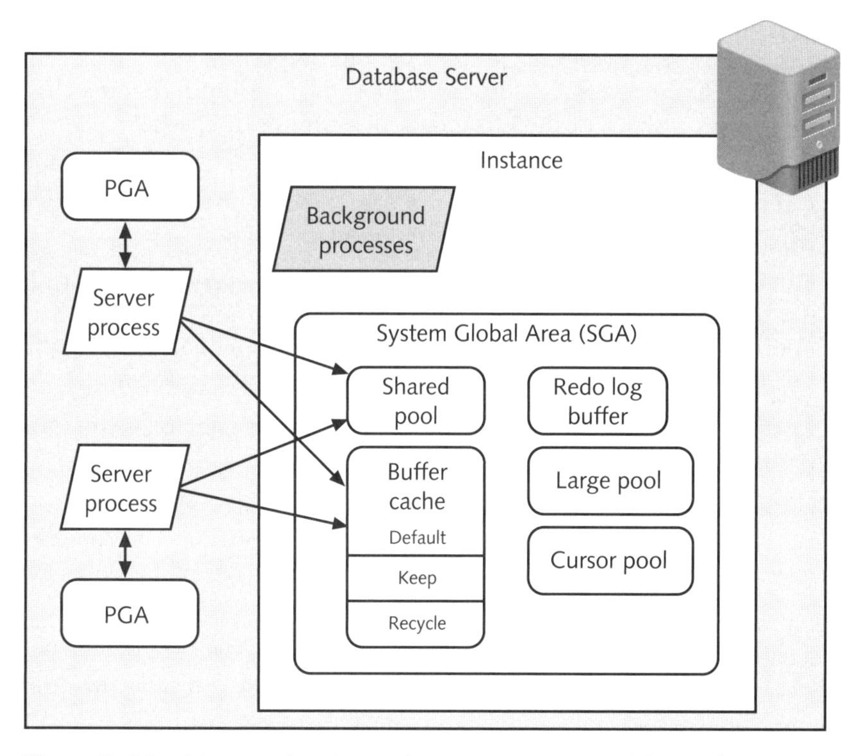

Figure 2-16 Memory has two primary components: SGA and PGA

The SGA is allocated when an instance is started and is unallocated when the instance is shut down. The memory components within the SGA are:

- **Buffer cache:** Data blocks that have been read or modified are stored here for quick retrieval. The blocks stay as long as there is room. The more frequently a block is used, the longer it stays in the buffer. The buffer blocks are reused as needed to make room for newer data blocks. In the buffer cache, the server processes obtain their data initially. The buffer cache is populated when a server process requests data (for example, when you execute a query) from the instance. If the buffer already has the block requested, the server process reads the block from the buffer. If not, the server process reads the block from disk to the buffer, and then uses the block. The block stays in the buffer until the buffer needs to reuse the space.
- **Shared pool:** This area stores parsed SQL in memory when a server process sends a request, such as a query. In the shared pool, Oracle stores results of parsing (interpreting the SQL into machine language) and the results of the Optimizer's path selection process. Oracle compares any new requests coming

in with requests already in the shared pool. If possible, Oracle reuses old requests to save time. Oracle has a list of comparison rules, which it uses to decide whether to reuse a stored request. If it cannot find a match, the new request is parsed, and the Optimizer determines the path.

- **Redo log buffer:** Whenever a change is made to data, the redo buffer stores a copy of the changed data and the original data, in case it is needed. See Chapter 5 for a discussion of the redo buffer.
- **Cursor pool:** This provides extra areas for the server processes to handle cursors (pointers that control data retrieval for groups of data). This is an optional component. It is also called the software code area. A similar pool, called the Java pool, can be added for Java application storage.
- **Large pool:** This optional memory area improves response time for background processes and for backup and recovery processes.

The PGA is broken into private chunks for each server process. This is the work area for the application code that works with the data for the application. The application's compiled code resides here along with copies of data blocks that the application is working on.

The background processes support and monitor the server processes and handle database management tasks to keep the database running efficiently and to help maintain fast performance. Figure 2-17 shows the background processes and how they interact with the SGA and the datafiles.

The following list describes how the background processes that are pictured in Figure 2-17 work within the database.

- **PMON (Process Monitor):** The process monitor process cleans up after user processes are finished. Any resources the user processes were using are restored. PMON also monitors the server and user processes and tries to restart them if they have stopped unexpectedly.
- **DBW*n* (Database Writer):** The database writer process writes buffers to datafiles. The only buffers that are written back to the datafiles are buffers that have been modified. For example, you issue an UPDATE command that modifies ten rows. The buffer containing the changed data is in the buffer cache, until the buffer needs more room. An instance has one process automatically (DBW0), and you can configure additional processes (DBW1, and so on) to speed up performance, if your system is very active. See Chapter 5 for more information.

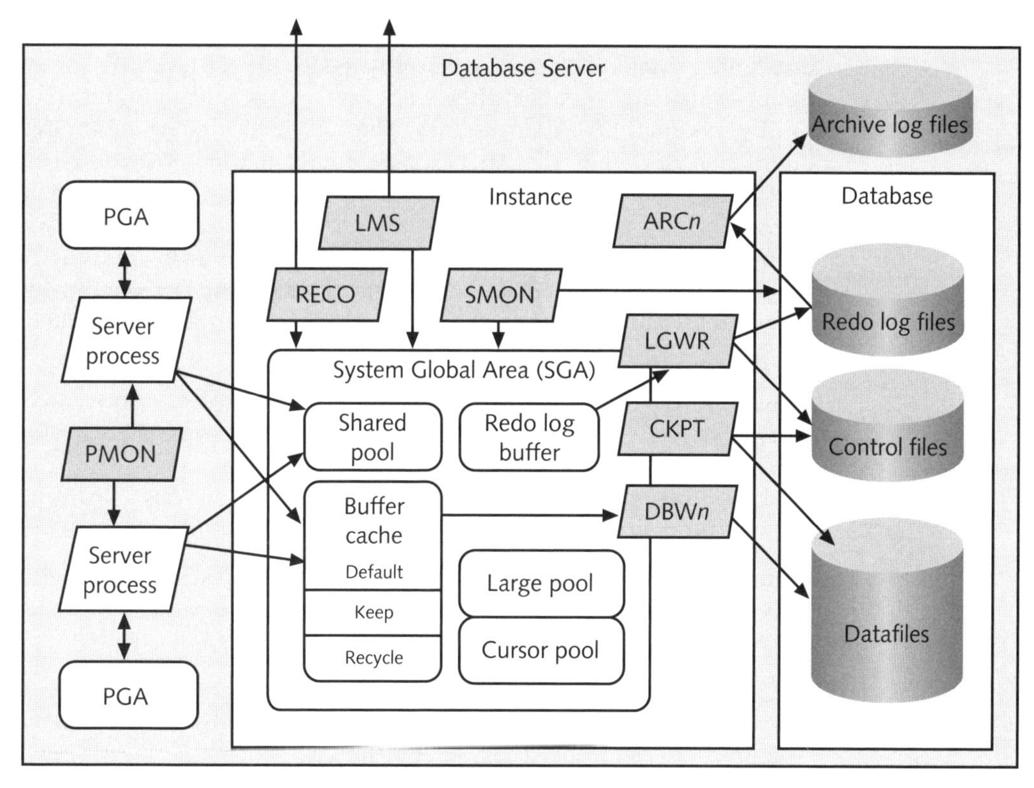

Figure 2-17 The background processes handle activity behind the scenes

- **LGWR (Log Writer):** The log writer process writes redo log buffers to the redo log files. This occurs every three seconds and also when the buffer is one third full and when the DBWn writes its buffers to the datafiles. See Chapter 5 for information about redo logs. If there are multiple files in a log group, LGWR writes to all the files at the same time, so that they are all identical. This is a safeguard for the data, because it is stored in two separate files.
- **CKPT (Checkpoint Process):** The checkpoint process does not write any data to disks. It signals the DBWn to do so by issuing a **checkpoint**. A checkpoint is assigned a **system change number (SCN)** which is written into each redo log entry, the control files, and each data block that is written back to the datafiles. A checkpoint aids in recovery by helping the recovery process determine which redo log entries must be reapplied. A checkpoint is issued every time changes are committed.
- **SMON (System MONitor):** The system monitor process handles recovery if it is needed. It also cleans up any unneeded temporary tables and restores blocks that are released (such as when you coalesce a tablespace) and makes them available for use. See Chapter 8 for information about free space and coalescing tablespaces.

- **RECO (Recoverer Process):** The recoverer process is only present in a distributed database system. A **distributed database system** has multiple instances that are used as if they were one instance. Data on either database can be modified by either instance. The recoverer process communicates with other databases when the databases are connected in a distributed system. Distributed systems allow data changes to span more than one database. The recoverer process looks for distributed processes and helps fix errors in changes that have failed because of communication problems. The recoverer process keeps trying to make the connections that are needed, making the distributed system better at recovering from communication problems.
- **ARC*n* (Archiver):** The archiver process is only present when Oracle9*i* is in ARCHIVELOG mode. See Chapter 5 for information on how archiving works. The archiver process copies the redo log files to a separate location outside the database. This preserves the redo log file data so that it can be used for recovery if needed. You can create up to ten (ARC0 through ARC9) archiver processes.
- **Dnnn (Dispatcher Process):** The dispatcher process is shown in Figure 2-14. This is the process used only in shared server mode to distribute user processes between server processes.
- **LMS (Lock Manager Server):** The lock manager server appears only when the database is part of a cluster. The LMS allows updates that affect data on more than one database at a time.
- **CJQ*n* (Job Queue Coordinator):** Not shown in Figure 2-17, the job queue processor runs jobs submitted in the background (sometimes called batch jobs) that access the database. You can have more than one, if you need to increase the number of jobs you want to run at the same time.

This section introduced you to all the internal workings of the database (the SGA and background processes) and the networking component of the database (Oracle Net). Next, you explore the front door of the database where it is easy to decipher the contents of your database: Enterprise Manager.

INTRODUCING ENTERPRISE MANAGER

Enterprise Manager works as a central clearing house of DBA tools. Use it to find your way around the database quickly without getting bogged down in SQL commands and complex data dictionary views. Much of what you need to know can easily be discovered using the Enterprise Manager. For example, perhaps you are wondering who is working on the database right now. Find out by opening the console and viewing the active sessions listed in the Instance Manager. Not only the current activity, but the database initialization parameters, the storage size, and the structural details about tables, views, and other database objects are easily visible in a navigation tree structure. The following section shows you how to start up the Enterprise Manager console.

Starting and Configuring the Enterprise Manager Console

There are two ways to run the console. The first way, called **stand-alone**, is more common in smaller organizations. The Oracle Enterprise Manager console (OEM console) in stand-alone mode connects directly to the database you want to use. Earlier in the chapter, Figure 2-5 showed three ways of accessing the database, and the console in its stand-alone mode is shown in the diagram using all three types of connections. This direct connection is convenient and requires less effort to configure. Basically, it is ready to go after installation. You can connect to multiple databases (one at a time) with this mode.

The second mode, called Enterprise Management Server mode, is shown in Figure 2-18. The Enterprise Management Server mode connects first with the Enterprise Management Server and subsequently to one or more database servers.

The **Enterprise Management Server** is a part of the Enterprise Management package that is installed along with the Enterprise edition of the Oracle9*i* RDBMS. The purpose of the Enterprise Management Server is to enhance the Enterprise Manager with additional capabilities. The Enterprise Management Server is designed to accommodate large systems with multiple database servers. In Figure 2-18, the Enterprise Management Server is shown installed on the database server, which is the way it is installed when you install the Enterprise edition of Oracle9*i*. After the database was created, a schema was added that stores all the details about the console's jobs and other settings. This schema is called the Repository. When you start the console to use the Enterprise Management Server, you must log onto the Server, not onto the database.

Another option for running the Enterprise Management Server (not shown in the figure) is to set up a separate database server dedicated to the Enterprise Management Server. This alternative creates a dedicated database for the Enterprise Management Server, making it more stable and removing the load from your primary database server.

The primary differences between the two modes are that in Enterprise Management Server mode, additional tools are added, you have access to a job scheduler, and you can set up tasks that run in the background on remote databases. For example, a job can be set up to run once a week. The job shuts down the database, executes a backup, and restarts the database. The success or failure of the job is reported in Enterprise Manager's Job History Log. The Enterprise Management Server stores information in its repository, which resides inside an Oracle9*i* database. The repository keeps track of the Enterprise Management administrators, their privileges, the jobs they create, and the results of those jobs. DBAs can even create jobs that page them when emergencies occur, such as when the database fails or runs out of space. For the purposes of this book, you only need the tools available in the stand-alone mode of Enterprise Manager.

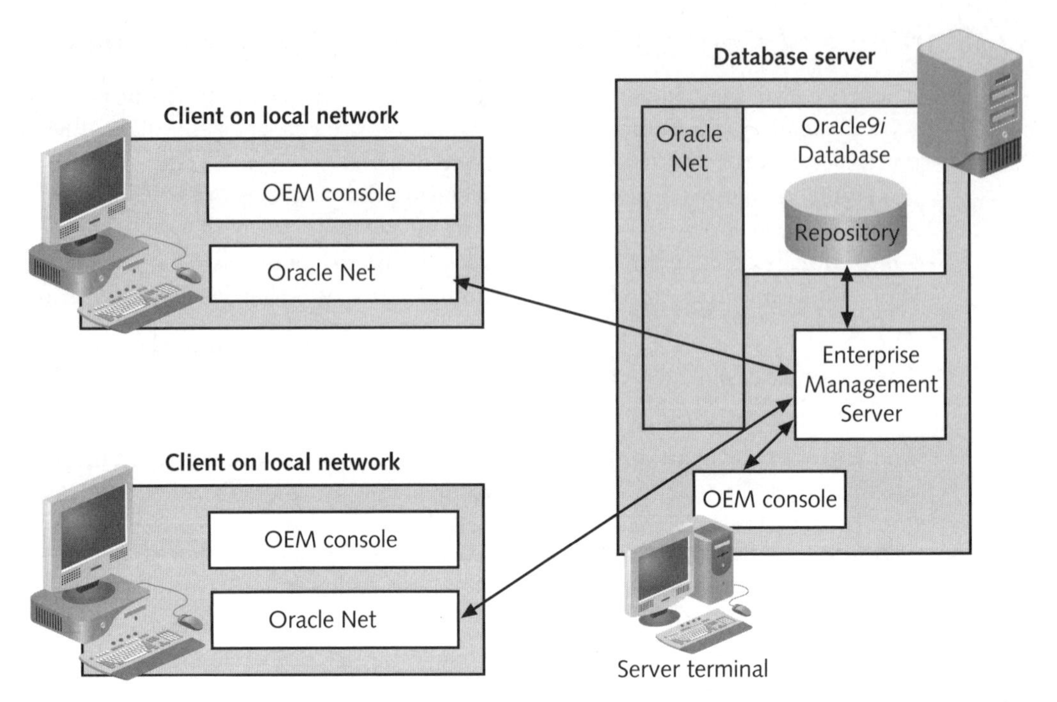

Figure 2-18 Three Enterprise Manager Consoles in Enterprise Management Server mode

To start up the Enterprise Manager console (referred to as the console throughout the book) in stand-alone mode, follow these steps:

1. Click **Start/Programs/Oracle-OraHome92/Enterprise Manager Console.** The Enterprise Manager Console Login screen appears.
2. Select the **Launch standalone** radio button, and click **OK**. Figure 2-19 shows the initial console window.
3. In the main window of Enterprise Manager console, double-click the **Databases** folder. This lists all available databases.

The ORACLASS database service name should already be listed. If it is not, perform Steps 4 and 5. Otherwise, skip Steps 4 and 5. *The next two steps are needed only when a new database service name must be added to the console.*

4. Click the **Navigator** menu, and then click **Add Database to Tree.** This starts up a wizard that adds the new database service to the console.

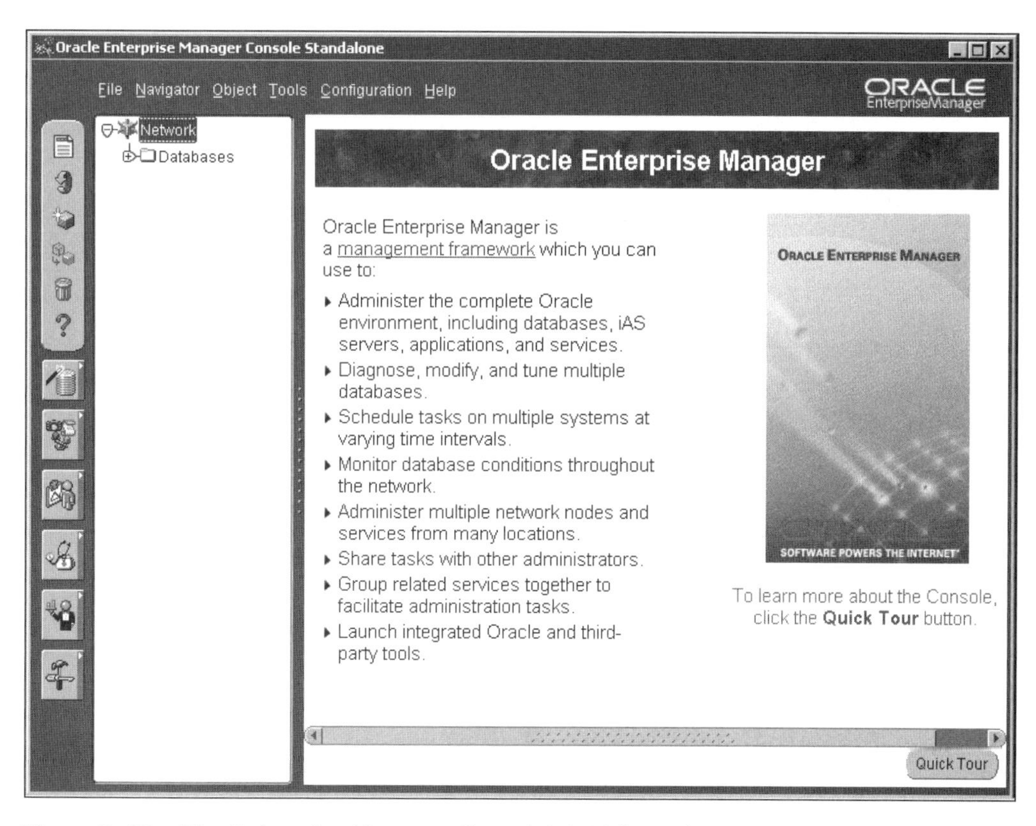

Figure 2-19 The Enterprise Manager Console's initial window

5. As shown in Figure 2-20, click the second radio button labeled "Add selected databases from your local tnsnames.ora file." Then adjust the check boxes so that only the **ORACLASS** database is checked. Click **OK** to complete the task.

   The top radio button allows you to add databases that are not defined in your local Oracle Net configuration. This is a shortcut method to add new databases, because it updates the local **tnsnames.ora** file and adds the database to the console. It is limited, however, to the TCP/IP protocol only.

6. Expand the **ORACLASS** database node by either clicking the **plus sign** to the left of the node or double-clicking the **database name**. A Database Connect Information window appears.

7. Log onto the database as **SYSTEM** as shown in Figure 2-21: SYSTEM is a predefined user in the Oracle9*i* database that has DBA privileges. When using the console, you should log on as SYSTEM, so you have appropriate privileges to perform tasks (such as creating new users and tables) within the console. *Type the password provided by the instructor.* Click the **check box** so that the information is saved in the console's preferred credentials for this database.

Clicking the check box initiates settings that make it unnecessary for you to type the user name and password; they are automatically entered when you expand the database node. Click **OK** to continue. A window opens explaining that Oracle stores encrypted passwords. Click **OK** to continue.

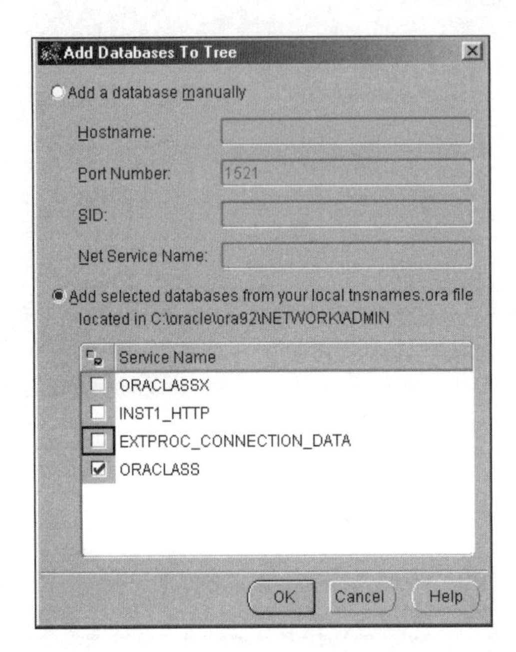

Figure 2-20 Add the new database to the console

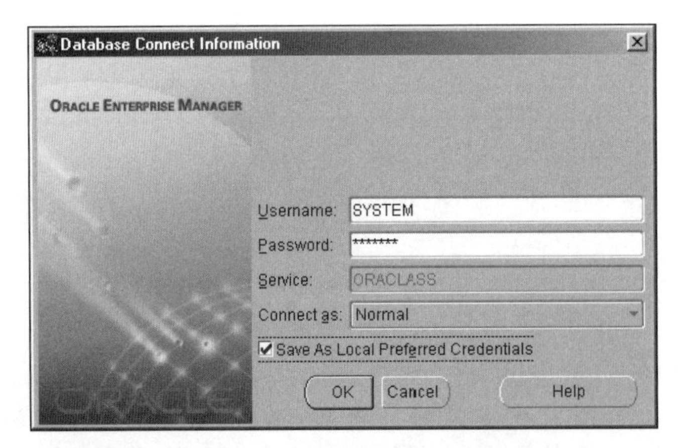

Figure 2-21 Set up the login one time and save it for the future

With the console open, begin the tutorial by following the directions in the following sections.

Viewing the Features of Enterprise Manager

Start up the console as described previously and begin a tour of the Enterprise Manager's primary DBA tools. These tools aid in the creation and maintenance of database users, files, and tables.

There are four primary tools and many additional tools found in the console. The four primary tools are:

- **Instance Manager:** Monitors activities in the database and starts and stops the database
- **Schema Manager:** Views table structures, creates new tables, indexes, views, and any other type of object stored in the database
- **Security Manager:** Creates new users, allocates storage resources to users, and changes passwords
- **Storage Manager:** Monitors storage use (tablespaces and datafiles) and adds more space as needed or adjusts settings on existing storage units

See Table 2-1 at the beginning of this chapter for a list of other tools and utilities found in the console (those marked "Yes" in the "Start from console?" column are found in the console, either in its stand-alone mode or its Enterprise Management Server mode.)

The next sections are tutorials that guide you through each of the four major console tools.

Instance Manager

You are already in the console, so follow these steps to examine the Instance Manager.

1. Double-click the **Instance icon**. The window on the right displays a description of the Instance Manager tool. All the top-level icons in the console have an introductory window similar to this one.
2. Click the **Quick Tour** button in the lower-right corner of the screen. As shown in Figure 2-22, you find a window with a presentation about the Instance Manager. Each tool within Enterprise Manager has a Quick Tour associated with it that gives you a thorough overview of the tasks that are handled by that tool.

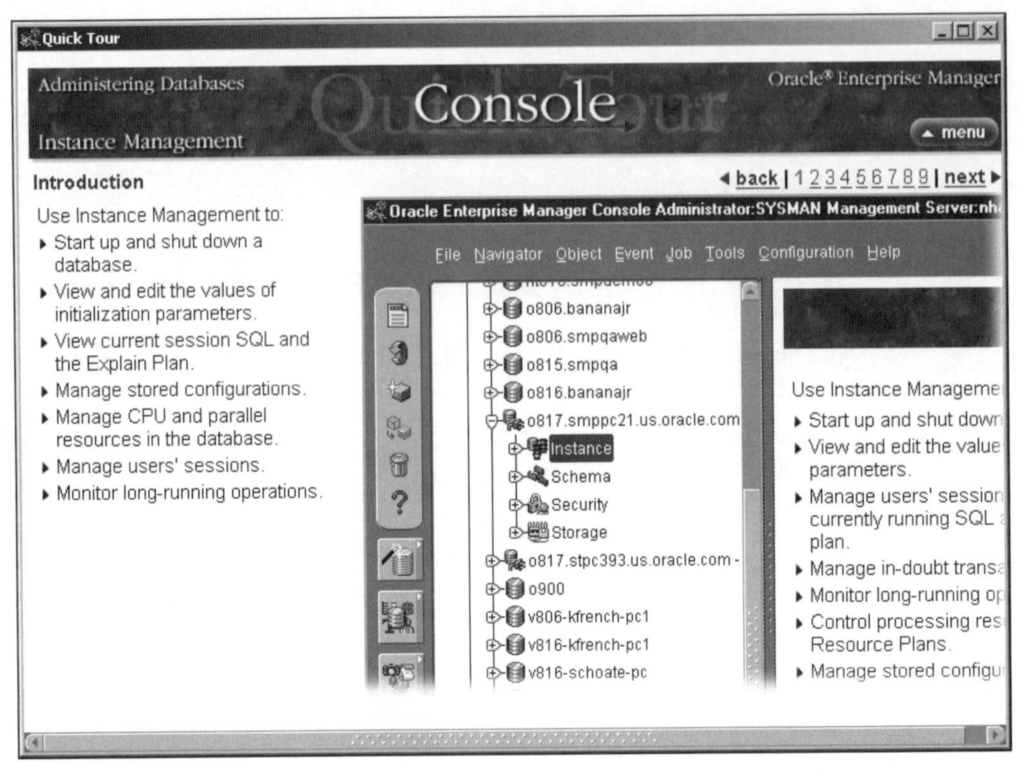

Figure 2-22 Quick Tours provide high-level descriptions of tool tasks

3. Browse the Quick Tour to become familiar with the Instance Manager. As you see, the Instance Manager's primary tasks revolve around the current state of the database. For example, you can examine a list of users and transactions currently running in the database. Exit the Quick Tour by clicking the **X** in the top-right corner.

4. Double-click the **Configuration icon**. This brings up a status window on the right side of the console. This section of the Instance Manager helps with these tasks:
   - Setting initialization parameters
   - Starting or shutting down the database
   - Monitoring and adjusting memory usage
   - Locating redo log files

5. Click the **All Initialization Parameters** button. The list displays all initialization parameters available and their current settings. Changes to parameters can be made in this list. The method of applying a change to a parameter depends on the type of parameter. Dynamic parameters (those with a check in the Dynamic column) can be reset immediately by simply clicking the Apply button that executes the change immediately. Static parameters can be changed

here, but the change must be saved in an initialization file **(init.ora)**, and then the database must be shut down and restarted to apply the change.

6. Select **audit_trail**, and then click **Description**. As you see in Figure 2-23, a short definition of the use of this parameter displays in the bottom of the window. This feature enables you to quickly determine the effect that changing a parameter might have on your database performance.

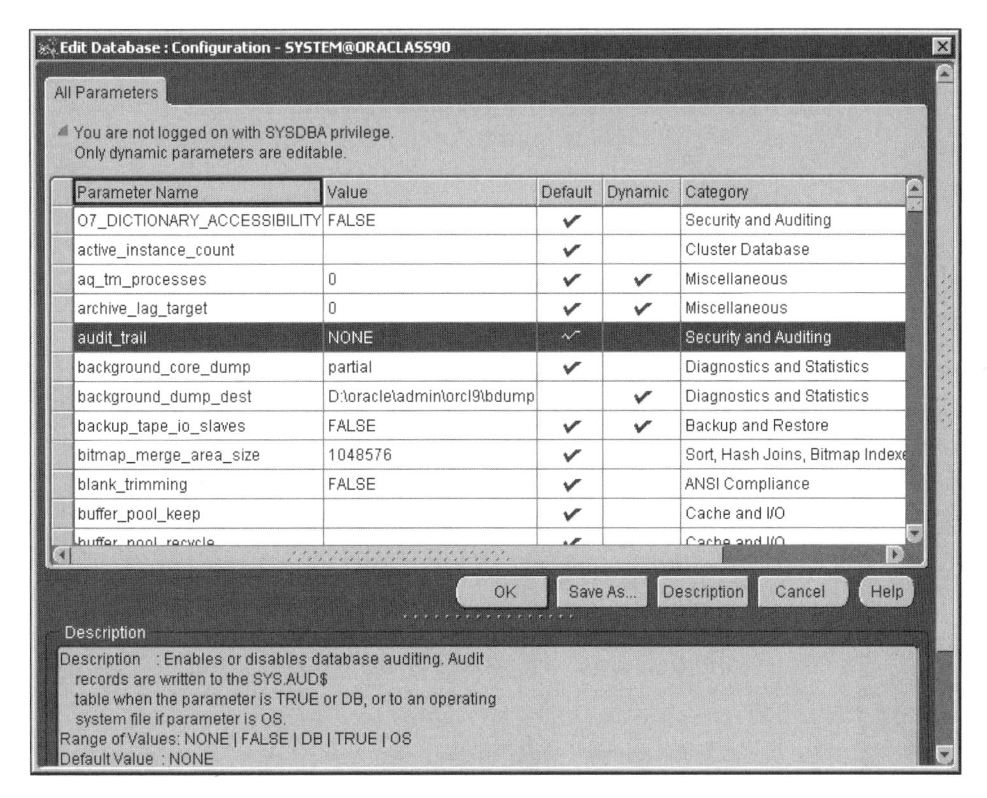

Figure 2-23 Initialization parameters have default and customized settings

7. Click the **Category** column heading, and scroll down to the **Pools** category and then to the **SGA Memory** category. As you can see, there are a large number of parameters that affect the memory usage of the database instance.

The ability to sort lists by column heading appears throughout the tools in the console.

8. Click the **Cancel** button to return to the main console window.

Changes made in the console are applied to the database through SQL commands that are built and then executed in the background. To see the commands, click **Navigator/Application SQL History** in the menu.

9. Double-click **Sessions**. This reveals a list of users with sessions started in the database. At least six background processes are also listed. The number of background processes varies depending on your database configuration.
10. Click **SYSTEM** under Sessions. The right side displays details about the session as shown in Figure 2-24. This is the session you are running while in the console. (Recall that you logged onto the database with the user name SYSTEM.) Notice the Kill Session button. This button allows you to end a user session from the console. For example, a new programmer might have created and run a query that has incorrect selection criteria. Because of the error, the query has run for twenty minutes and has slowed the database down enough to cause other users to call the DBA and complain. You, as the DBA, find the user session and stop it using the Kill Session button.

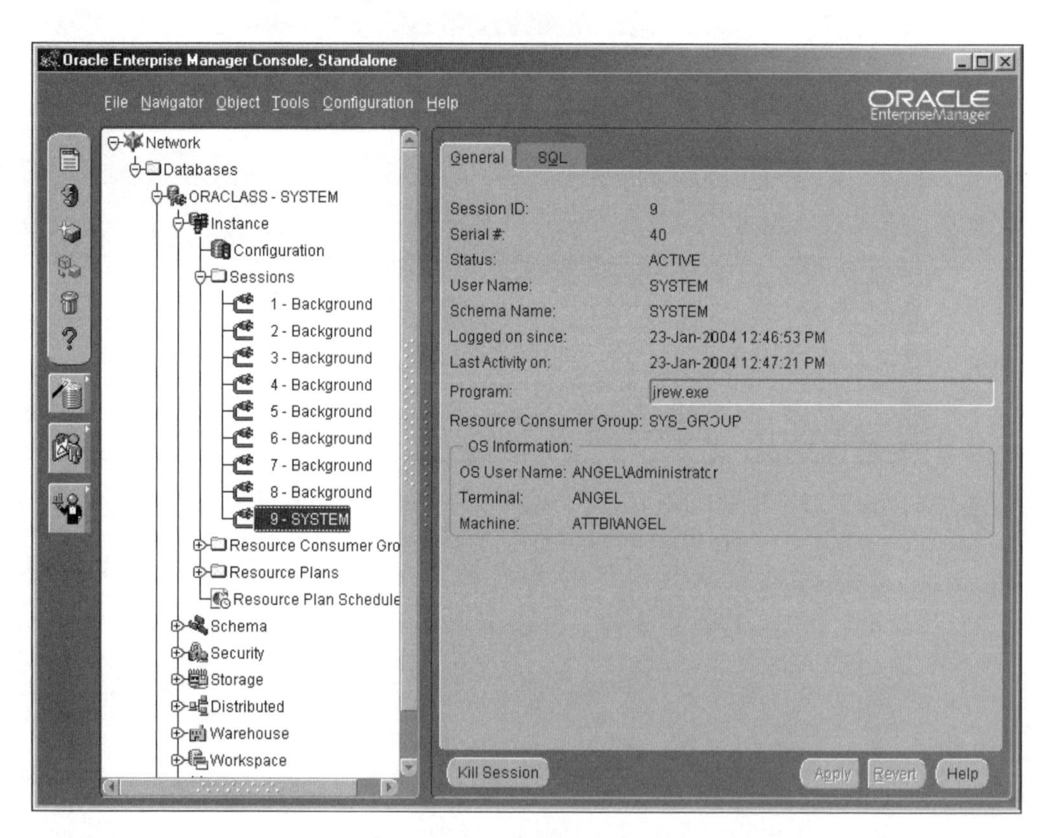

Figure 2-24 Individual users show up here when they use the database

11. Collapse the Instance Manager node by clicking the **minus sign** next to the Instance icon.

Leave the console open, and continue to the next section to examine the Schema Manager.

Schema Manager

The Schema Manager gives you a way to manipulate anything in the database that is in a schema. A **schema** is a set of database structures, such as tables, indexes, user-defined attributes, and procedures. Each schema is owned by one Oracle user and is identified by the user's name. For example, the user named Sidney creates two tables and a view. These three objects are collectively called the Sidney schema. A **database object** is not only an object table, but also any other structure held in the database, such as a relational table, an index, a PL/SQL procedure, a Java servlet, or even a customized attribute with methods and rules attached.

Begin using the Schema Manager by completing the following steps:

1. Double-click the **Schema** icon in the console. The list you see below the Schema icon, as shown in Figure 2-25, are the users that own database objects stored in the Oracle9*i* database.

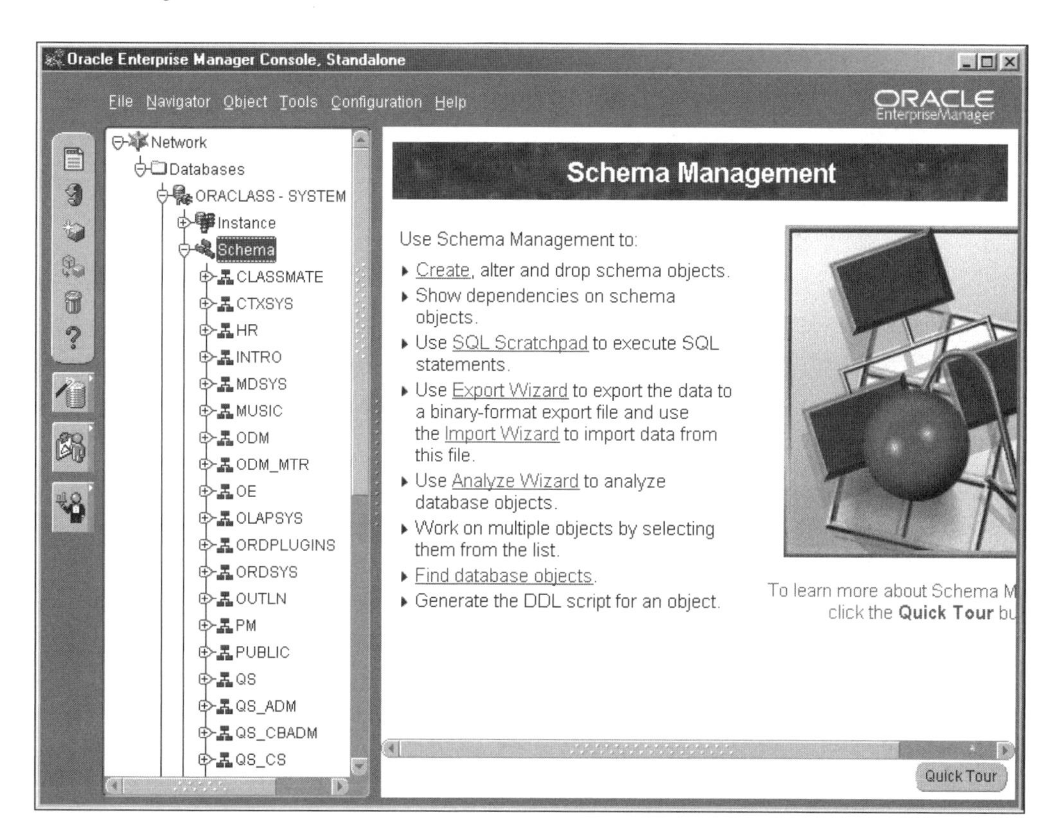

Figure 2-25 Schemas contain many types of database objects

2. Scroll down in the left window and double-click the **SYSTEM** schema. All the objects in the SYSTEM schema are listed in the right side of the console. All object types are listed in the left side.
3. Double-click the **Table** folder. A list of tables appears below the folder. In the right window, a sorted list displays all the tables owned by SYSTEM, their name, tablespace, and so on.
4. Scroll down and double-click the **HELP** table. Now, a Property window appears in the right side, as shown in Figure 2-26. The Property window shows the columns contained in the table and much more. Notice the tabs along the top named Constraints, Storage, and so on. The Property window and its related tabs allow you to perform these tasks:
   - Add a new column
   - Remove an existing column
   - Change the characteristics of a column, such as its data type or size
   - Add and remove constraints, such as primary and foreign keys
   - Adjust storage allocation
   - View current statistics such as number of rows

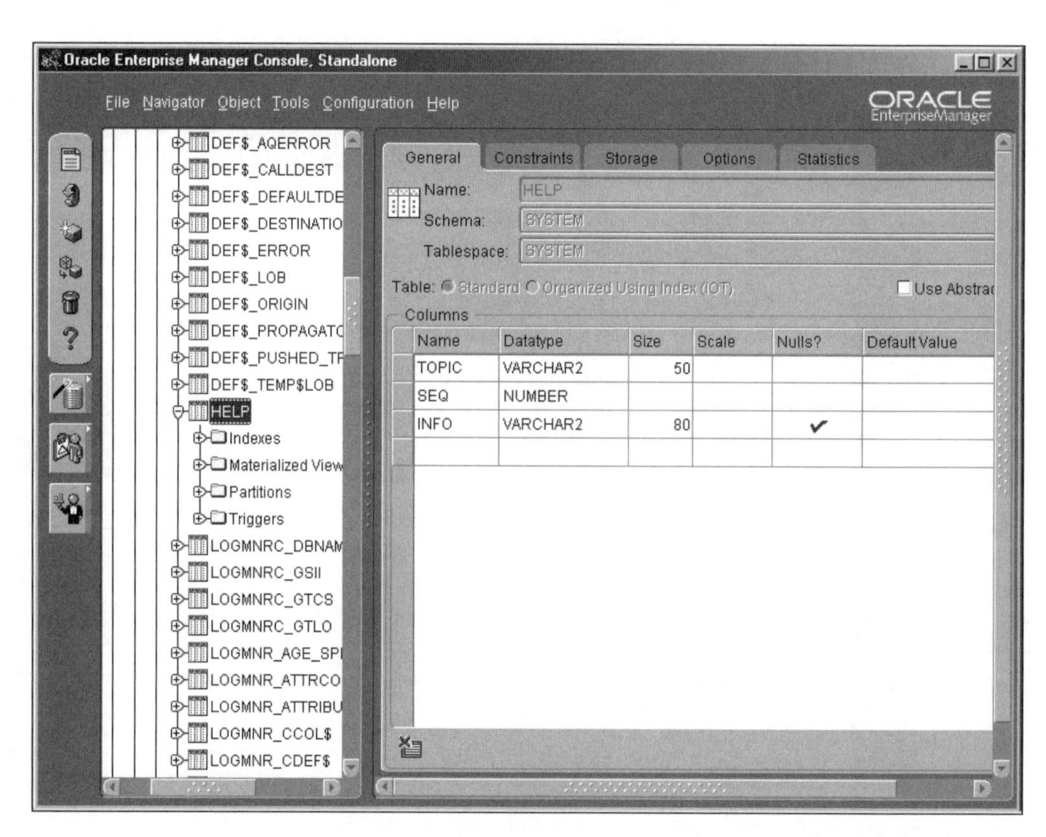

Figure 2-26 The table property sheet displays many details about a selected table

5. Click the **Indexes** folder below the HELP table in the left side of the console. This displays a list of indexes created on this table on the right side of the window. You can also view all indexes under the Indexes folder listed on the same tree level as the Tables folder.

6. Right-click the **HELP** table. A pop-up menu appears as shown in Figure 2-27. This menu displays more tasks that you can perform on a table using Schema Manager. For example, one of the menu items is Remove, which drops the table from the database. Another menu item is View/Edit Contents, which displays a spreadsheet for editing the table's data.

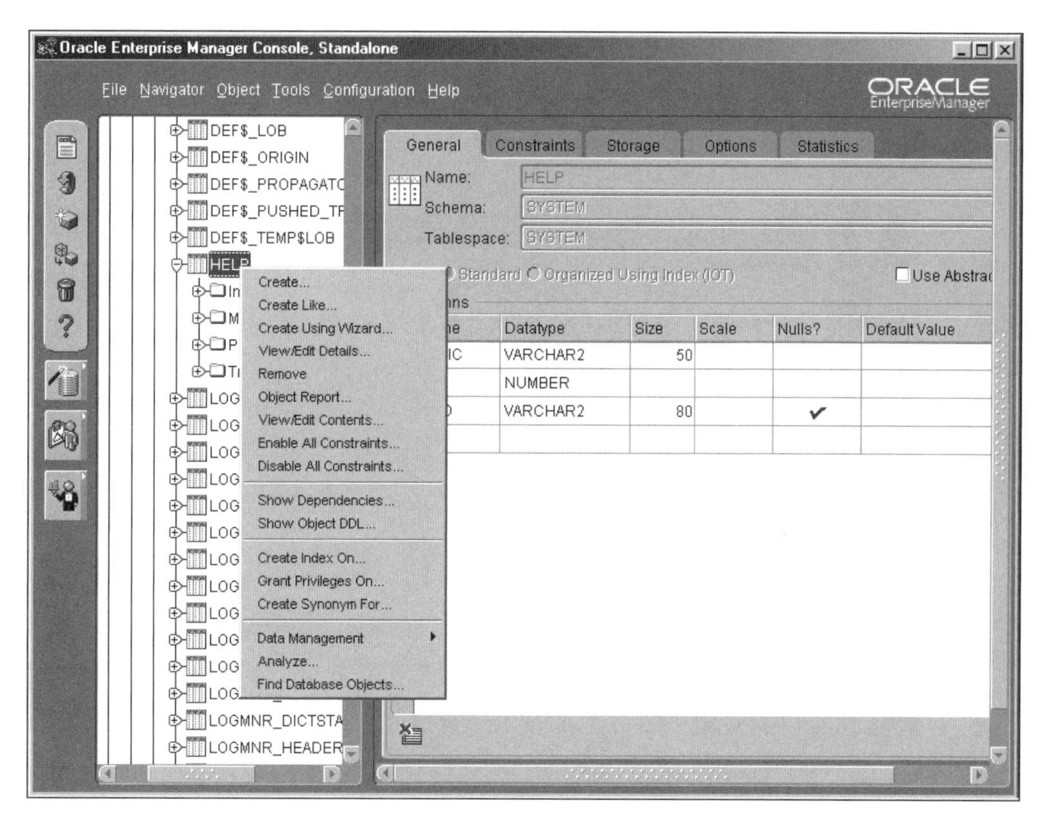

Figure 2-27 The pop-up menu lists specialized tasks for tables

7. Select **Show Object DDL** from the pop-up menu. **DDL** stands for Data Definition Language. After selecting Show Object DDL, Oracle9*i* generates the SQL command used to create the object and displays it in a window. Viewing the SQL code for tables, indexes, and other objects can be a useful learning aid. You have the option of saving the DDL to a file as well.

8. Click **Close** to return to the main console window.

9. Scroll down and right-click the **Views** folder. A new pop-up menu with different selections appears.
10. Select **Save List** in the pop-up window. A window appears giving you some choices on how your list will be generated and saved. The Save List function can be used in many areas of the console to generate a list of the items you are viewing and save them to a text file or HTML document. Figure 2-28 shows an example of the list of views saved in HTML format.

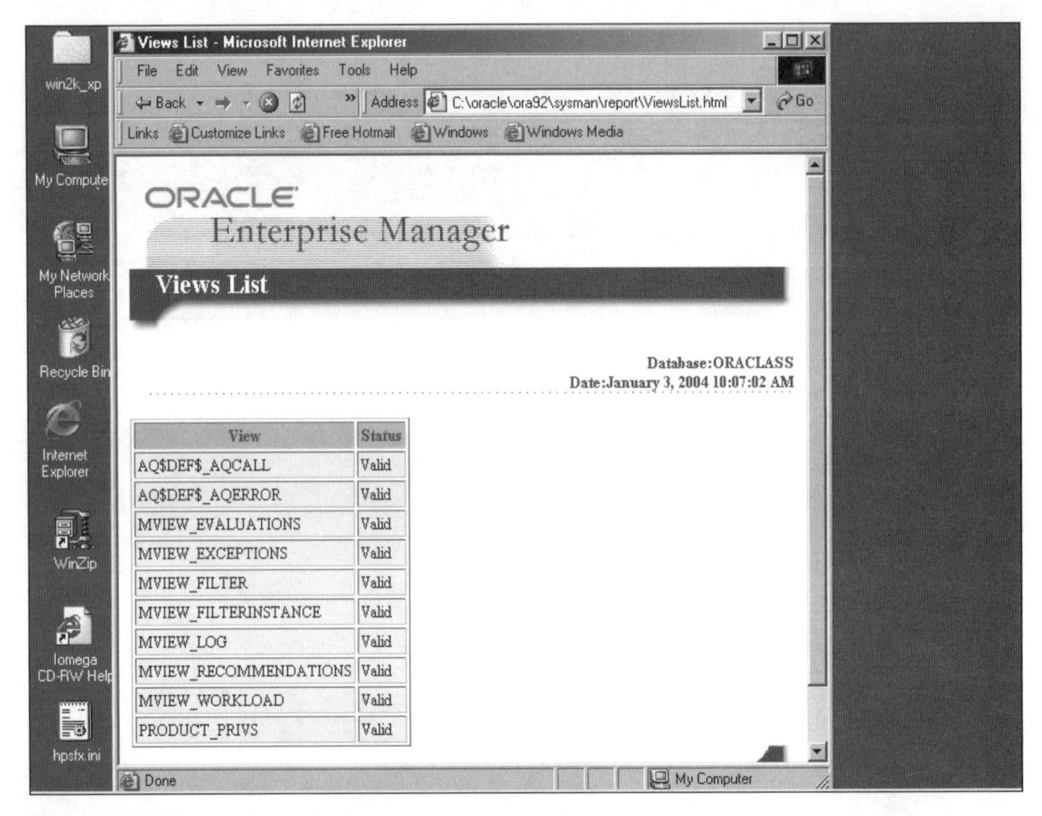

Figure 2-28 The console can automatically generate a report for you

11. Click **Cancel** to return to the console window.

Leave the console open to continue the tutorial in the next section, which covers the Security Manager.

Security Manager

The Security Manager's primary tasks involve setting up and maintaining users in Oracle9*i*. Follow these steps to examine Security Manager in the console.

1. In the console, double-click the **Security** icon. The introductory window of the Security Manager appears, displaying an overview of its purpose.
2. Double-click the **Users** folder. This displays a list of users.
3. Scroll down and select the **SYSTEM user**. A Property window appears for the SYSTEM user on the right side of the console, shown in Figure 2-29. Here are some of the tasks you can perform in this property sheet:
   - Change the password
   - Expire the password (forcing the user to change her password)
   - Modify the default tablespace where all objects created by a user are stored
   - Assign or revoke roles and privileges
   - Adjust storage limits (quotas)

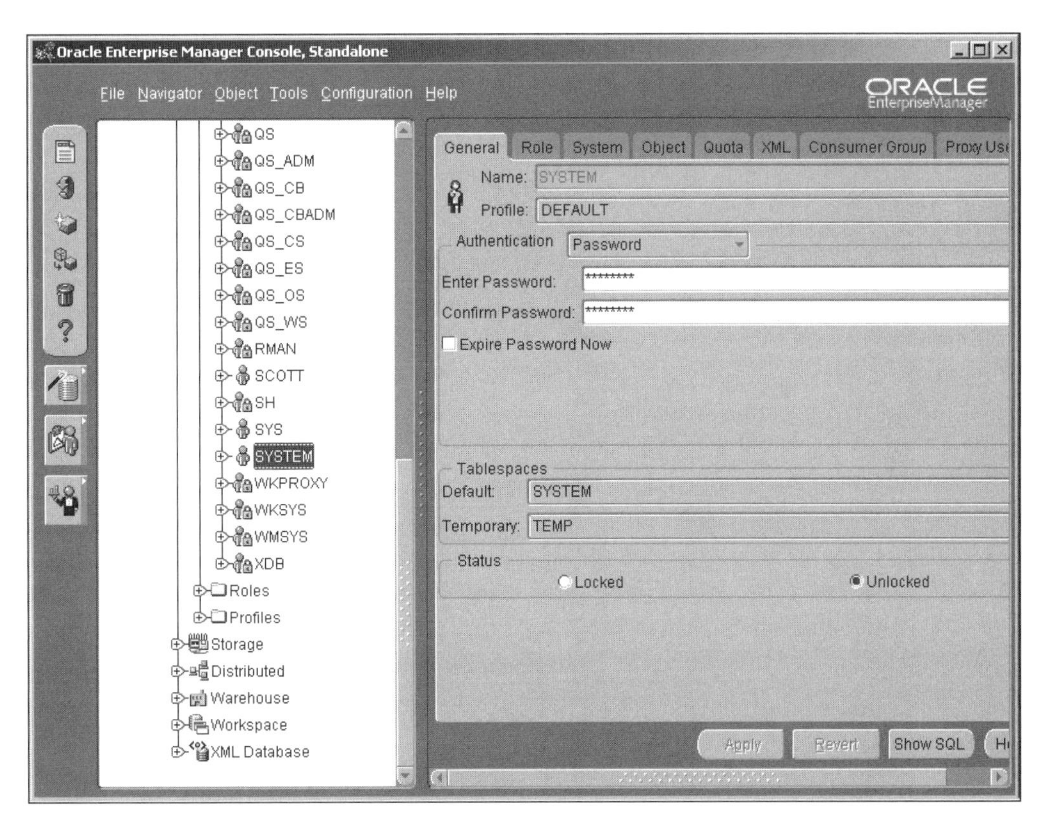

Figure 2-29 The user property sheet allows changes to passwords among other tasks

4. Double-click the **Roles** folder. This displays a long list of roles that are predefined in the Oracle9*i* database to help organize all the different permissions needed for various tasks. A **role** is a set of related privileges that can be granted as a set to another role or to a database user.

5. Scroll down in the left side of the console and select the **RESOURCE** role. The Property window for the role appears in the right side of the console. The tabs across the top represent the types of privileges that can be assigned to a role:
   - **Role**: Embed the privileges of the selected role into this role
   - **System:** Allows system-related tasks, such as creating users or tables
   - **Object:** Allows access (select, insert, update, delete, or a combination of these) to data in tables, views, or other database objects
   - **Consumer Group:** Sets the resource allocations that are inherited by any user assigned the role
6. Click the **System** tab. The privileges listed at the bottom of the window are those granted to the RESOURCE role. The RESOURCE role is assigned to any user who is allowed to create objects in the database. The privileges reflect this, as shown in Figure 2-30. For example, the CREATE TABLE privilege allows a user to create a table.

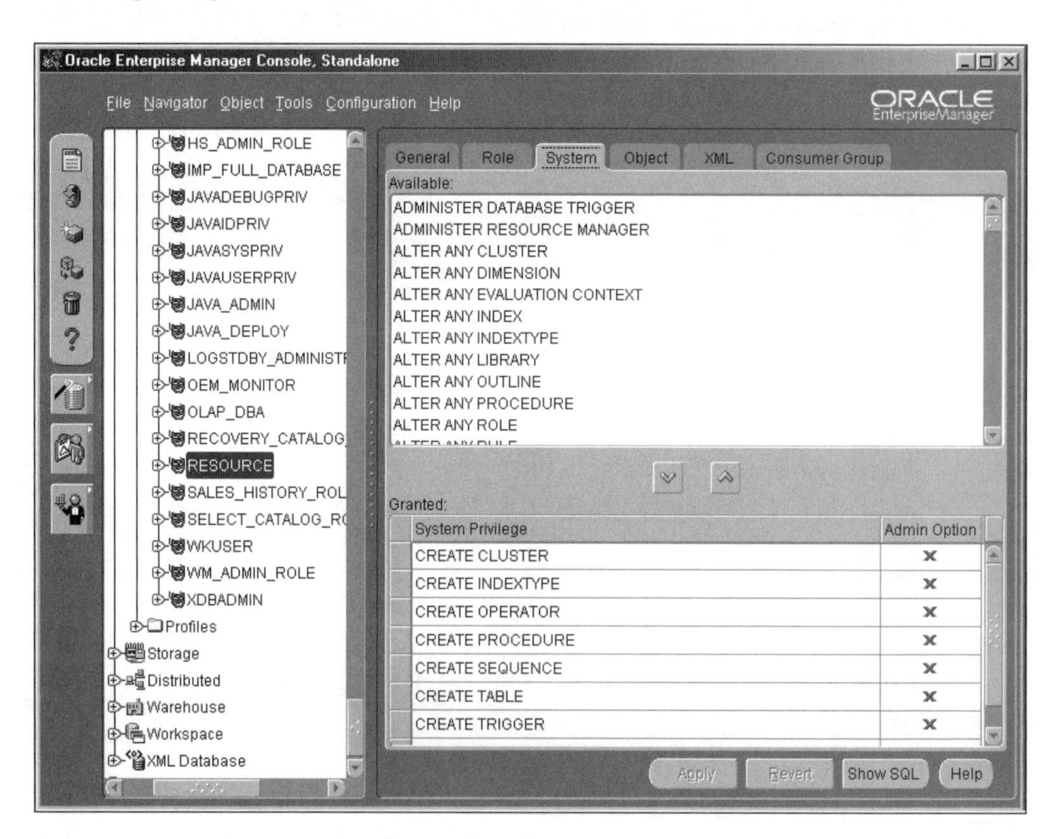

Figure 2-30 Roles are assigned sets of privileges

Keep the console as it is to continue to the next section, which describes the Storage Manager.

Storage Manager

The final primary tool you are to examine is the Storage Manager. This one, as its name implies, performs storage-related tasks in the Oracle9*i* database. Follow the steps to see the main points of interest in the Storage Manager.

1. Double-click the **Storage** icon in the left side of the console. This starts up the Storage Manager.
2. Select **Tablespaces** under the Storage icon. The right side displays data about all the tablespaces in the database. In Figure 2-31, observe the graph that illustrates the percentage of storage used in each tablespace.

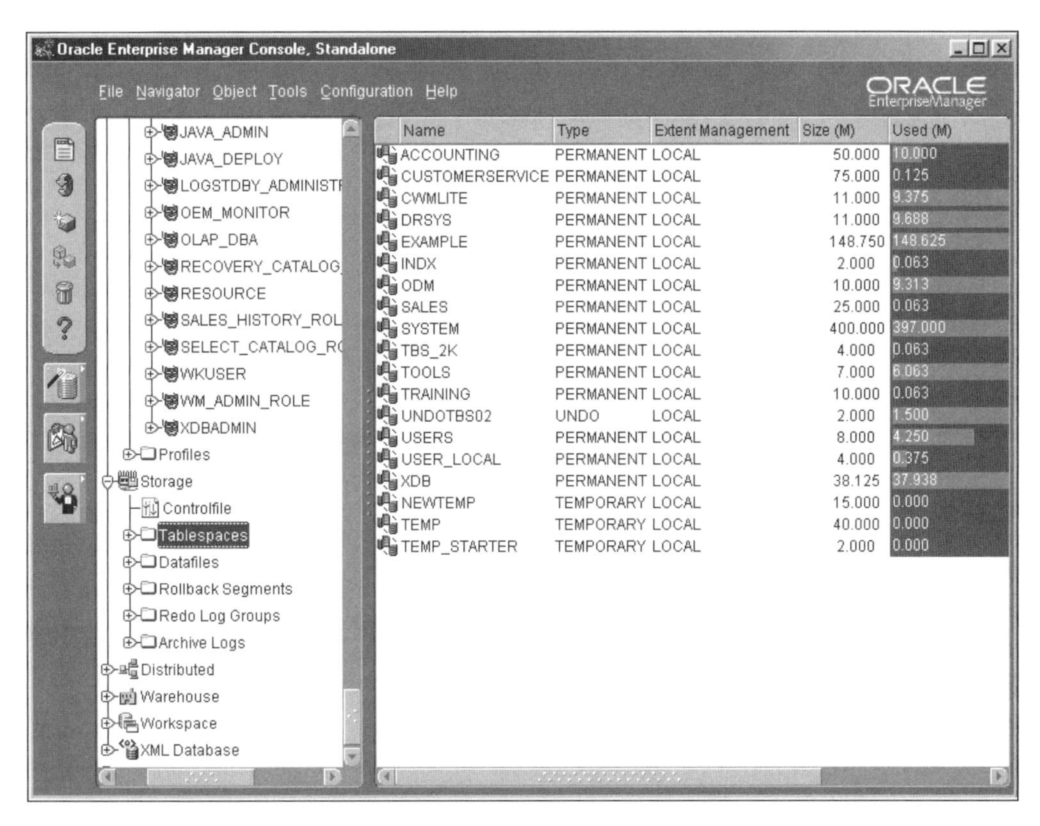

Figure 2-31 Tablespace storage information can be easily read at a glance

3. Double-click the **Datafiles** folder. A similar display of storage information appears in the right side.
4. Click the datafiles whose name ends in **TOOLS01.DBF**. A property sheet displays details about this datafiles.

5. Click the **Storage** tab. The storage screen displays as shown in Figure 2-32. Notice the check box that indicates this datafiles can be automatically extended. This feature saves a great deal of human intervention. It allows Oracle9*i* to add more space, in the predefined increments shown in this window, to the datafiles when the datafiles runs out of the allocated space. You can also set a cap on the maximum size of the datafiles to prevent it from taking up too much disk storage.

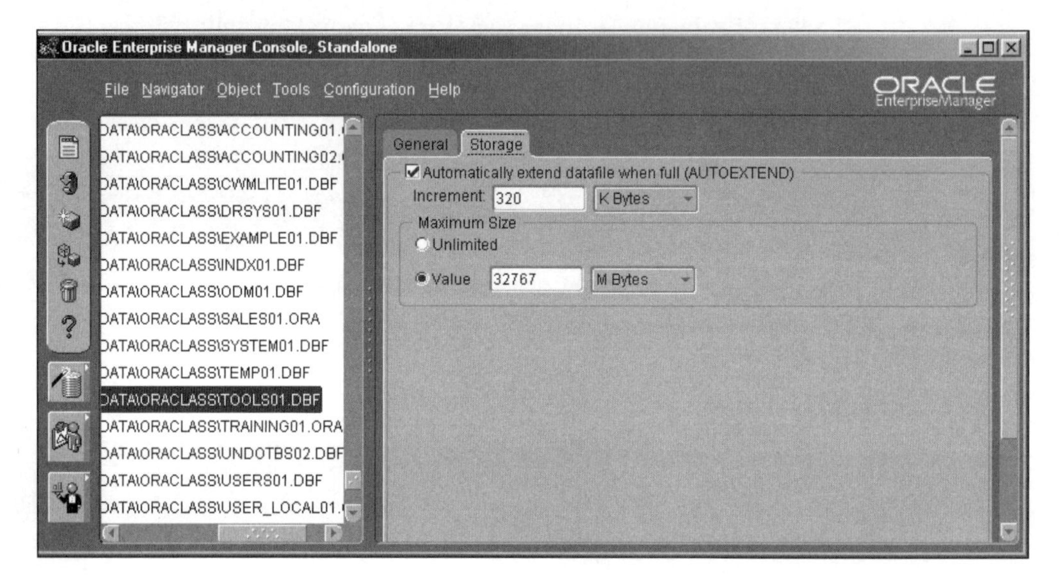

Figure 2-32 Storage definitions for data files allow the file to grow in size

6. Close the console by clicking the **X** in the top-right corner of the window.

This concludes the overview of the four primary tools in the Enterprise Manager. These tools are revisited in later chapters as related subjects are covered, such as creating a table.

Chapter Summary

- Some DBA tools are available in either stand-alone mode or inside the Enterprise Manager console, whereas others can be run both ways.
- Using command-line interfaces to manually execute commands may be necessary, even if a Windows-like tool can do the job.
- Oracle Net is a common connection point that communicates between a tool and the database.
- Oracle Net must be configured on both the client and the server when using Oracle tools on the client side.

- The Oracle service called the Listener waits to receive requests sent to the database.
- A service name defines the database's name, location, and listening port.
- Use Net Manager to configure Oracle Net.
- Net Manager contains a wizard to guide you through the steps of configuring a new service name.
- The file **tnsnames.ora** stores Oracle Net configuration settings.
- In SHARED SERVER mode, a dispatcher process determines which user process connects to each server process.
- SHARED SERVER mode has one server process for several user processes.
- DEDICATED SERVER mode has one server process for each user process.
- A server process reads data from the datafiles and places it in the buffer cache.
- The shared pool stores parsed SQL commands for possible reuse.
- The redo buffer stores all changes to data before the changes are written to a disk.
- The cursor pool provides extra memory for program cursors.
- The large pool adds more memory to improve response time.
- The PGA stores information separately for each application.
- The background processes are named PMON, DBWn, LGWR, CKPT, SMON, RECO, ARCn, Dnnn, LMS, and CJQn.
- The Enterprise Manager console has two modes: stand-alone and Enterprise Management Server.
- The Instance Manager tracks database activity, such as user sessions, initialization parameters, and memory usage.
- Use the Navigator selection from the menu to add a database to the console.
- When logging onto a database in the console, set the user and password as preferred credentials, and you need not enter them again.
- Quick Tour buttons for each tool provide overviews in HTML format.
- The Schema Manager provides details on schema objects, such as tables, views, and triggers.
- Use Schema Manager to add or drop columns, change storage, and view the data of tables.
- The Security Manager is focussed on users and privileges.
- Roles group privileges into related sets.
- Use the Security Manager to change passwords and adjust quotas of users.
- The Storage Manager displays information about datafiles and tablespaces.

Review Questions

1. The Data Migration Assistant is run from the console. True or False?
2. The Log Miner supports data warehousing. True or False?
3. The dispatcher process distributes ______________ processes among the available ______________ processes.
4. Using a tool is always preferable to using a manual method. True or False?
5. The file that stores Oracle Net configuration data is called ______________.
6. The Enterprise Manager console runs in the following two modes: ______________ and ______________.
7. The primary purpose of the Enterprise Manager console is to
   a. Gather data about remote databases
   b. Centralize access to tools and utilities
   c. Share user and server connections
   d. Automate backups
8. You have issued a query. Which memory area stores the parsed SQL command?
   a. Redo log buffer
   b. Shared pool
   c. PGA
   d. Buffer cache
9. The Storage Manager can shut down the database. True or False?
10. The ______________ Manager displays the full path name of datafiles.
11. Which of these tasks can be done by the Instance Manager?
    a. Killing a user session
    b. Dropping a table
    c. Creating a new user
    d. Running a backup
12. The ______________ process writes buffers to the datafiles.

EXAM REVIEW QUESTIONS—DATABASE FUNDAMENTALS I (#1Z0-031)

1. A procedure created by the user named HAROLDK is in which schema?
   a. SYSTEM
   b. PUBLIC
   c. The database instance
   d. HAROLDK
   e. None of the above
2. The following tasks can be handled in the Security Manager. Choose three.
   a. Assign users the ability to create tables
   b. Change a user's session
   c. Limit the resources a user can use on the database
   d. Change a table's owner
   e. Create a new user
3. A datafiles cannot be removed from the database. True or False?
4. Which of the following tools can be used to establish a connection between tools and the database? Choose two.
   a. Net Manager
   b. SQL*Plus
   c. Net Configuration Assistant
   d. Instance Manager
   e. All of the above
5. When logging onto a tool, which items are typically required? Choose all that apply.
   a. User name
   b. Port number
   c. Service name
   d. Password
   e. All of the above
6. You are configuring a new service name in Oracle Net Manager. The Test step has failed. What might be wrong? Choose three.
   a. Invalid date and time on database server
   b. Incorrect port number specified
   c. Database instance not running
   d. User SCOTT was removed or password was changed
   e. Missing or invalid NET_CONNECTION parameter

7. You log into SQL*Plus Worksheet and run a query. Which parts of the database instance are involved? Choose two.
   a. ARCn
   b. PGA
   c. Server process
   d. DBWn
   e. LGWR
8. Which of the following best describes the role of the ARCn process?
   a. Copying redo log files to a remote storage area
   b. Monitoring user processes
   c. Connecting to distributed databases to resolve connections
   d. Distributing user processes among server processes.
9. You are running the Enterprise Manager console. Your computer shuts off suddenly. Which background process cleans up the memory used by your user process?
   a. SMON
   b. PMON
   c. DBWn
   d. LMS
10. Which of the following make up the database instance? Choose three.
    a. SGA
    b. PGA
    c. Datafiles
    d. Background processes
    e. Oracle Net
    f. Database software
11. Which part of the SGA contains a record of changes made to data blocks?
    a. Shared pool
    b. Redo log buffer
    c. Large pool
    d. Program global area
12. Which of these operating system files are part of the Oracle9*i* database? Choose two.
    a. Redo log files
    b. Archived log files
    c. Control files
    d. **tnsnames.ora** files

HANDS-ON ASSIGNMENTS

Run the setup script to prepare the following assignments. To run the script, follow these steps:

a. Start up the Enterprise Manager console in standalone mode.
b Double-click the **Databases** folder.
c. Double-click the **ORACLASS** database icon.
d. Select **Tools** from the top menu, choose **Database Applications**, and then choose **SQL*Plus Worksheet**. This opens the SQL*Plus Worksheet.
e. Select **Worksheet/Run Local Script** from the top menu.
f. Navigate to the **Data\Chapter02** directory on your student disk, select **prech02.sql**, and click **Open**. This runs the script.
g. Close SQL*Plus Worksheet by clicking the **X** in the top-right corner.

Now you are now ready to work on the hands-on assignments and case projects.

1. Using the Schema Manager, find the WANT_AD table in the CLASSMATE schema and display the DDL command that created the table. Save this into a file named **ho0201.sql** in the **Solutions\Chapter02** directory on your student disk and print it. (*Hint*: Use the right mouse click.)
2. Run the Analyze Wizard on the CLIENT table in the CLASSMATE schema. Then view the results and answer these questions. Save your answers in a file named **ho0202.txt** in the **Solutions\Chapter02** directory on your student disk.
   a. How many rows are in the table?
   b. What is the average length of a row?
   c. How many empty blocks are there for the table?
3. Using the console, look at the view named WANT_AD_COMPLETE_VIEW in the CLASSMATE schema. List all the dependencies of this view. (*Hint*: Right-click the **view name** to start.) Generate a report in HTML format and save the report in a file named **ho0203.html** in the **Solutions\Chapter02** directory.
4. Look at the view named CLIENT_VIEW in the CLASSMATE schema and answer these questions:
   a. Is this view valid? If not, list the Oracle error number(s) and message(s) you find.
   b. How can this view be corrected?
   c. Fix the view.
   d. Save the DDL for the corrected view in your solutions directory in a file named **ho0204.sql** in the **Solutions\Chapter02** directory on your student disk.

(*Hint:* Compile the view to enable the Show Errors button.) Save the answers to (a) and (b) in a file named **ho0204.txt** in the **Solutions\Chapter02** directory on your student disk.

5. Practice saving a password.
   a. Change the password of the **CLASSMATE** user to **STUDENT** using the console. Test your change by connecting to **CLASSMATE** in the SQL*Plus worksheet. Explain what tool you used to change the password. Save your explanation in a file named **ho0205a.txt** in the **Solutions\Chapter02** directory on your student disk.
   b. Change the password of the **CLASSMATE** user back to **CLASSPASS** using the SQL*Plus worksheet. Save the SQL statement that you used in a file named **ho0205.sql** in the **Solutions\Chapter02** directory on your student disk.

CASE PROJECT

Using the information in the console, draw a diagram of the CLASSMATE schema. Save the diagram in a file named **case0201.bmp** in the **Solutions\Chapter02** directory on your student disk. Include the following features in your diagram:

a. All tables with the following information for each table: table name, primary key name, all columns (name, data type, size, null/not null, and default value).
b. All relationships between tables. For each relationship show whether it is one-to-many, one-to-one, or many-to-many. Label with a verb, and also list the constraint name that enforces the relationship.
c. Mark columns that are indexed with an asterisk for nonunique indexes, and with a plus sign for unique indexes.

CHAPTER

3

CREATING AN ORACLE INSTANCE

In this chapter, you will:

- Learn the steps for creating a database
- Understand the prerequisites for creating a database
- Configure initial settings for database creation
- Create, start, and stop a database instance

Database administrators often work with existing databases for some time before actually creating a new database. The DBA focuses on backup and tuning, which often require a great deal of time and effort. Still, a competent Oracle9*i* DBA must be familiar with database creation. A new database may be the best way to install a new version of the Oracle software or to set up a replicated site, for example. A few parameter settings for the database cannot be altered after database creation, making preparation and planning a critical part of database creation. This chapter describes the steps involved in planning, configuring, and creating a new Oracle9*i* database instance. Hands-on exercises give you a chance to experiment with creating new database instances on your own.

STEPS FOR CREATING A DATABASE

Creating a new database helps you understand how Oracle9*i* works and gives you experience working with parameter settings, with several of the Assistant tools provided by Oracle9*i*, and with manual commands as well. Table 3-1 lays out the steps for database creation in a checklist format.

Table 3-1 Checklist for creating a database

Main step	Detailed steps	Comment
1. Install software	__ Install Oracle9*i* RDBMS software	Choose Enterprise, Standard, or Personal Edition
	__ Install Oracle Net software	Needed for Enterprise and Standard editions (usually automatically installed with Oracle9*i* RDBMS)
	__ Install Oracle9*i* Client software	Install on client machine if you need remote access to the database on clients
2. Establish user (For UNIX only; for Windows operating systems, skip this step)	__ Create operating system user with privileges needed to create, start, and stop the database	The user may or may not be the same user who installed the software (see the specific guide to your operating system for details)
	__ Create operating system environment variables for the user	ORACLE_HOME and ORACLE_BASE are required, and more may be needed (see the specific guide to your operating system for information)
3. Confirm memory and storage availability	__ Plan the disk storage size and location	Determine how many disks are available and design the distribution of data (see Chapter 6 for details)
	__ Decide on the DBA authentication method	Choose between operating system and password file authentication
4. Determine initial settings of the database	__ Decide on the file management method	Choose between the manual and the Oracle Managed Files methods
	__ Set the initial parameters	Decide the database's name and other settings, and place them in the initialization parameters file **(init.ora)**

Table 3-1 Checklist for creating a database (continued)

Main step	Detailed steps	Comment
5. Choose the type of database installation	__ Decide on assisted or manual creation	Use the Database Configuration Assistant, or use a manual command to create the database
	__ Run the Database Configuration Assistant	If using the Configuration Assistant, follow the instructions, and set up the database
6. Create the database (this step is for manual creation only; Database Configuration Assistant handles this step for you)	__ Write and execute the CREATE DATABASE command	Using decisions made prior to this step, write an appropriate CREATE DATABASE command, and execute the command
	__ Run scripts to create data dictionary views and PL/SQL procedure	Standard scripts provided by Oracle, found in the **ORACLE_HOME\rbms\admin** directory and named **catalog.sql** and **catproc.sql**
	__ Add recommended tablespaces	Oracle recommends one tablespace for temporary tables and one for user-created tables
	__ Edit Windows Registry (if using Windows)	Add the new database name to the Registry
7. Test the database	__ Start up the database	Use console or SQL*Plus command
	__ Shut down the database	Use console or SQL*Plus command

Most of these steps are covered in this chapter, and you can try out some of the options yourself with the Hands-on Assignments at the end of this chapter.

The next section discusses the prerequisites for creating a database, which include Steps 1-5 listed in Table 3-1.

OVERVIEW OF PREREQUISITES FOR CREATING A DATABASE

Creating a database is actually a separate process that occurs after the database software has been installed. You have the option, during the software installation, to have a predefined database created immediately. You may, however, install the software only and create the database at a later time, which gives you the time to customize your database.

There are several prerequisites you must fulfill before creating a new database. First, the Oracle software must be installed on the computer. With distributed systems, the software may actually reside on a different machine than the database; however, most often, the two reside on the same machine. Second, you must be able to log on as a user with installation privileges and with the correct set of environmental variables in place. Third, the machine must have enough memory and enough disk space to install and start the database. Table 3-1 details each of these three requirements.

The details outlined in Table 3-1 are an overview of the steps for installing the Oracle9*i* software and creating an Oracle instance. Outside the classroom, you should always refer to and follow the documentation provided with your software. Oracle provides an installation guide for each operating system with the in-depth information you need to complete the tasks before, during, and after installation. The guides are found online at the Oracle Technology Network site (*otn.oracle.com*) as well as on the installation CDs.

UNIX system preinstallation tasks require more manual steps than Windows systems. For example, in UNIX, you must set the environmental variables manually and create an account for database administrators, an account for upgrades, and an account for the Universal Installer. In Windows, these tasks are either not required, or are handled automatically by the Universal Installer during installation of the software. See your operating system installation guide for detailed setup steps for UNIX.

The installation guide for your platform contains information specific to the operating system for the minimum storage and memory. For example, if you install Oracle9*i* Enterprise Edition on Windows 2000, the minimum requirements for memory and storage are:

- RAM: 128 megabytes
- Virtual memory: 200 megabytes initial and 400 megabytes maximum
- Temp space: 400 megabytes
- Storage on the ORACLE_HOME drive: 4.5 gigabytes on FAT storage system or 2.75 gigabytes on NTFS storage system
- Storage on the system drive: 140 megabytes

Configuring Initial Settings

There are three important configuration tasks that you must be familiar with before creating the database. These three tasks determine the initial settings, file management, and security for your database. In addition, permanent settings (those that cannot be altered after database creation) must be set prior to the database creation. Each of the next sections discusses one of these three tasks, which are:

1. Decide on the DBA authentication method. The most common choice is to set up a password file for authentication.

2. Decide on the file management method. Oracle recommends using Oracle Managed Files—a new feature of Oracle9*i*. Oracle Managed Files requires parameter settings that allow the database to handle creation and maintenance of control files, log files, and datafiles.
3. Set the initial parameters. These settings are stored in the **init.ora** file and include settings for the two tasks listed previously, plus many more settings.

DBA Authentication Methods

The **DBA authentication method** is the name of the method used to validate logon of users with the **SYSDBA** or **SYSOPER** role. The SYSDBA role is a predefined database role that contains all system privileges with the ADMIN option. This means that any user with SYSDBA can perform any task necessary in the database, including creation of an instance, recovery of an instance, and maintenance or creation of tables and data owned by any user in the database. The **ADMIN option** means that the user with SYSDBA role can assign any privilege to another user. The SYSOPER role has all the system privileges needed to start up, shut down, and back up the database, as well as modify database components, such as datafiles and tablespaces. The security level of users with this authority is higher than of normal users. Oracle offers two methods of security. Both methods authenticate the DBA users logging on with SYSDBA or SYSOPER roles prior to entering the database. The two methods are called operating system (OS) authentication and password file authentication.

Operating System (OS) Authentication

When using **Operating System (OS) authentication**, the user logs on without specifying an Oracle user name and password. The database retrieves the user's name from the operating system and also checks what groups the user belongs to on the operating system. The user must belong to one or both of two operating system groups specifically set up for DBAs. Use this method when you only allow local access (direct logon to the machine that has the database running) or secured remote network access to your database machine.

The two operating system groups specifically set up for DBAs are referred to as **OSDBA** (for SYSDBA privileges) and **OSOPER** (for SYSOPER privileges) in the Oracle documentation. The actual names of these groups depend on your operating system. For example, in Windows, the Universal Installer automatically creates the group for SYSDBA privileges and names it ORA_DBA. Before using the Universal Installer with UNIX, you must first manually create a group named DBA for SYSDBA privileges. On both operating systems, you must manually set up the OSOPER group if you want to use it.

To set up OS authentication, follow these steps:

1. Create an operating system user for the DBA.
2. UNIX only: Create an OSDBA group.
3. Optional: Create an OSOPER group.
4. Set the initialization parameter REMOTE_LOGIN_PASSWORDFILE to NONE.
5. Assign the operating system user to the OSDBA group or the OSOPER group.
6. Create the corresponding Oracle user in the database with the same name (see the following note).

To log onto SQL*Plus with SYSDBA privileges using the OS authentication method, type the following command on the command line:

```
SQLPLUS /nolog
```

SQL*Plus starts up without logging you onto the database. Type:

```
CONNECT /@ORACLASS AS SYSDBA
```

The slash tells the database to use OS authentication. The "@ORACLASS" identifies the database to log onto. With OS authentication, the AS SYSDBA parameter tells the database to check that your operating system name is in the OSDBA group and if so, logs you on with SYSDBA privileges. If you are not in the group, the logon fails.

The Oracle9*i* user name and the operating system user name must be identical, except when you have specified that a standard prefix be added to the OS user name. For example, by setting the OS_AUTHENT_PREFIX parameter to OPS$ (the default setting), a user who logs onto the operating system as MARTY corresponds to the database user name of OPS$MARTY. Change the parameter to a null value to eliminate a prefix.

The OS authentication method can be used for other Oracle users as well. One advantage of this method is that user names and passwords are not typed in when logging onto the database. This can be convenient for users. For example, local users at an insurance office all have their own workstations and log onto them at the beginning of the day. Using OS authentication, they are not required to log on a second time to run an Oracle-based application that calculates auto insurance estimates.

The OS authentication method can be set up for remote database access over the Internet; however, it requires more steps and should only be used when the users reach the database over a secured line, such as one with the secure sockets layer (SSL) protocol. Access via a nonsecured line opens the database to security risks, because the database and the local operating system do not have control over the user's logon procedures.

Password File Authentication

As stated previously, Oracle offers two methods of security: Operating System (OS) authentication and password file authentication. The previous section covered OS authentication, and this section covers password file authentication. Password file authentication requires you to set up a special encrypted file that contains user names and passwords of authorized DBA users. Use this method when the DBA must log on from a remote site over a nonsecure line. Typical TCP/IP networks, such as a standard local area network or the Internet, are not secure lines. Password file authentication is the proper method to use when, for example, your DBA uses the Web-based Enterprise Manager console to start or stop the database. The password file contains user names and passwords for all users that are granted the SYSDBA or SYSOPER privileges. Other users must log on directly with their own user names and passwords.

To set up password file authentication, follow these steps:

Create a new password file. The Oracle command, ORAPWD, creates the password file. The location and filename are predefined by Oracle for Windows operating systems. The file is named PWD<sid>.ORA (<sid> is the Oracle instance name) and the file is located in the ORACLE_HOME\DATABASE directory. In UNIX systems, the filename must be orapw<sid>.ora and you must specify the full path in the name parameter to define the location. Typically, the password file in UNIX systems is stored in the ORACLE_HOME/dbs directory.

1. Set the REMOTE_LOGIN_PASSWORDFILE initialization parameter to **EXCLUSIVE**.
2. Log onto the database as SYS with SYSDBA privileges (or another user with SYSDBA privileges). The SYS user is predefined in the Oracle database upon creation.
3. Create the new DBA user name if needed.
4. Grant the SYSDBA privilege or the SYSOPER privilege to the user. This action in the database automatically updates the password file to include this user name and password.

The syntax for the ORAPWD command is:

```
orapwd file=<filename> password=<pwd> entries=<maxusers>
```

Replace <filename> with the filename of the password file. Include the full path if you are using UNIX. Replace <pwd> with the actual password for the SYS user. Replace <maxusers> with the maximum number of distinct users that can be logged on as SYSDBA or SYSOPER at one time. For example, in a Windows system, you are about to create a password file named **PWDora92.ora** which will be located in the

D:\oracle\ora92\database directory. The password for SYS is GOFORIT1209, and you plan to allow no more than ten users at a time to be logged on as SYSDBA or SYSOPER. The command looks like this:

```
orapwd file=D:\oracle\ora92\database\PWDora92.ora
password=GOFORIT1209 entries=10
```

Whenever you change the password of a user with SYSDBA or SYSOPER roles, the change is automatically written to the password file.

You can log onto the database with SYSDBA or SYSOPER roles in Enterprise Manager console as well as in SQL*Plus. After logging onto the console as SYSDBA, you can start or stop the database, adjust initialization parameters, and grant SYSDBA or SYSOPER privileges to other users. You can also perform most other DBA functions, such as creating a new user, changing passwords, creating tables, and so on. To log on as SYSDBA, follow these steps:

1. Start the Enterprise Management console in stand-alone mode.
2. Double-click the **Databases folder** to view databases recognized by the console.
3. Right-click the **Database name** you want to enter, and select **Connect** from the shortcut menu. This brings up a log on window.
4. Type **SYS** in the Username box, the current password in the Password box, and select **SYSDBA** from the list in the Connect as box. Figure 3-1 shows the box filled in correctly.
5. Click **OK** to return to the Main Console window. You are now logged on as SYS with SYSDBA privileges.

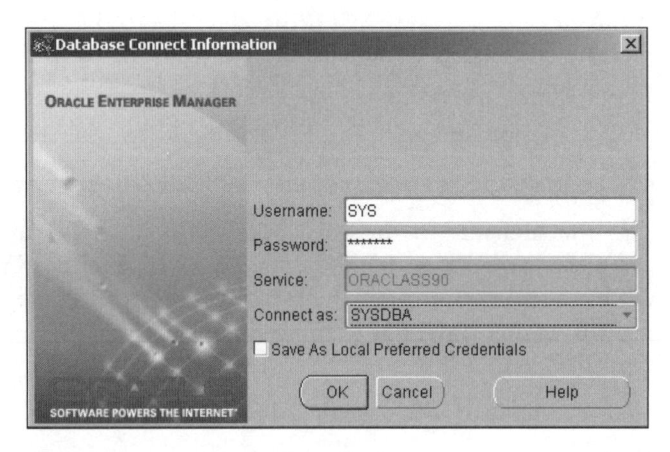

Figure 3-1 Log on as SYSDBA by reconnecting to the database in the console

After you have the password file in place and users authorized, your system has a secure method of preventing unauthorized access to the SYSDBA role. Oracle cautions that the location and name of the password file should be protected from unauthorized viewing to prevent possible attempts to hack the user names and passwords stored in the file. The file is automatically marked read-only and encrypted by Oracle.

Before creating your database, it is also important to decide how much control you need over the locations and names of datafiles, control files, and log files maintained by the database. The next section presents guidance on these decisions.

File Management Methods

An Oracle9*i* database uses several different types of files to run. As described in Chapter 1, the three types of files are control files, redo log files, and datafiles. Each type of file has unique requirements for storage, size, location, and redundancy. The two primary tasks in file management are:

- **Location of files:** A control file stores up-to-the-minute information that is critical to your database's functioning. Oracle recommends that you multiplex your control file. **Multiplexing** involves setting up at least one identical copy of the control file stored on a different physical device. Oracle handles multiplexed control files by automatically updating both copies simultaneously, so both are always kept up to date. The location of redo log files is also critical for database performance, because these files record each transaction as it is executed in the database itself. As with control files, it is recommended that you multiplex redo log files. Datafiles should be located on a different physical device than the database software to reduce bottlenecks in accessing data and database functions.
- **Addition, expansion, and deletion of files:** The data in all the tables, objects, indexes, and other database objects is stored in tablespaces. A **tablespace** is a logical data storage space. Each tablespace is mapped to one or more datafiles. The **datafiles** is a physical datafiles on the operating system. Storage requirements grow and shrink according to the activity in the database. For example, a table that stores survey results grows every day as surveys are received and entered into the database. Later, a new survey replaces the old one, and old data is purged and new data added to the table. As data storage needs change, the datafiles storing the data fill up and empty. New datafiles may be added, or old ones may be expanded. Likewise, if the volume of activity grows, the size or number of redo log files must be adjusted. Additional control files might be added to reduce the risk of losing this important file, making recovery more likely in the case of a database failure caused by physical disk damage.

There are two basic file management methods available for a new database: user-managed and Oracle Managed Files. Each has advantages and disadvantages. Oracle recommends that you implement the Oracle Managed Files method for all newly created databases, because it automates more of the DBA file-related tasks.

The next two sections describe the two choices of file management offered in Oracle9*i*.

User-Managed File Management

Earlier versions of Oracle had no options for file management. The DBA was responsible for designing the file management strategy, from file locations and names to monitoring file size and allocating more files as needed. The primary advantage of this method is the total control that you have in file management. The primary disadvantage is that many tasks involve manual intervention. Oracle8 improved the situation by adding the capability of extending the file size automatically. However, monitoring a datafile's used versus unused space requires manual tasks.

A good reason for using the user-managed method of file management is to continue with a customized file management standard that was in place for earlier versions of the database. The consistency of continuing with an established methodology may outweigh the disadvantage of having to assign the name, size, and location of your datafiles manually.

Control files are created when the database is created, based on the CONTROL_FILES parameter in the initialization parameter file. Here is an example of the parameter settings in a UNIX system that create two control files located on two different physical devices named d1 and d2:

```
CONTROL_FILES = (/d1/oracle/control01.ctl,
                 /d2/oracle/control01.ctl)
```

Redo log files are also created when the database is created. To implement user-managed redo log files, use the LOGFILES clause in the CREATE DATABASE command itself. Omitting the LOGFILES clause means that the redo log files are created as Oracle Managed Files.

The CREATE DATABASE command is described completely in a tutorial later in this chapter.

The SYSTEM tablespace and its corresponding datafiles are created when the database is created. Omitting a fully qualified DATAFILE clause in the CREATE DATABASE command causes Oracle9*i* to create an Oracle Managed Files as the datafiles for the SYSTEM tablespace. The DATAFILE clause must include the file's name, size, and location.

Clearly, Oracle9*i* makes it easier for you to choose Oracle Managed Files over the user-managed method. The next section describes this new method of file management.

Oracle Managed Files

Oracle9*i*'s new file management feature, Oracle Managed Files (OMF), automates most of the menial tasks of file management while leaving important decisions in the hands of the DBA. Oracle Managed Files handles the creation, expansion, and deletion of files as the database size changes. You, as DBA, provide the Oracle9*i* with the information it needs to know what disks to use for storage.

The advantages of using Oracle Managed Files include:

- Adherence to Optimal Flexible Architecture (OFA) naming standards
- Automatic removal of dependent datafiles when a tablespace is dropped
- Simplified syntax of the CREATE DATABASE command
- Automated expansion and addition of datafiles as storage requirements change

When you remove a tablespace, you get an error message, if there are any tables or other database objects stored in the tablespace, unless you specify INCLUDING CONTENTS in the command. See Chapter 6 for more information.

The main disadvantage of using OFA is the inability to control the exact size and name of datafiles, control files, and log files. These tasks are automated for you. You can, however, choose to create some files as user-managed files and leave others as Oracle Managed Files in the same database.

To implement Oracle Managed Files, specify values in these initialization parameters:

- DB_CREATE_FILE_DEST to define the directory where all datafiles and **tempfiles** (files used for storage of temporary data during sorting and other operations) are to be created.
- DB_CREATE_ONLINE_LOG_DEST_n to define one or more directories where control files and redo log files are to be created. The "n" on the end is replaced by a 1, 2, and so on to indicate the location of each control file and log file. By specifying more than one directory, Oracle9*i* automatically creates multiplexing control files and redo log files. This parameter is optional. If omitted, Oracle9*i* uses the same directory for control files and redo log files that you specify for datafiles.

Here is an example of settings for these parameters. Your new database is on Windows NT, and you have three physical disk drives. You have placed all the Oracle software on the C drive.

You want to use the D drive for the datafiles and one set of the control and redo log files. The E drive will contain the multiplexed copies of the control and redo log files. The two initialization parameters might look like this:

```
DB_CREATE_FILE_DEST = D:\oracle\datafiles
DB_CREATE_ONLINE_LOG_DEST_1 = D:\oracle\logfiles
DB_CREATE_ONLINE_LOG_DEST_2 = E:\oracle\logfiles
```

The directories that you specify must already exist when you create the database. When executing the CREATE DATABASE command, you must also omit the parameters that specify names for the control file, redo log file, and SYSTEM tablespace datafiles.

Oracle9*i* creates new database files as Oracle Managed Files by default, as long as the DB_CREATE_FILE_DEST parameter has been defined and you leave out the filename when creating any new files.

Oracle Managed Files have names that comply with OFA (Optimal Flexible Architecture). Standard naming patterns are used for each type of file, and unique character strings are added to guarantee unique names. For example, the third datafiles for the USERS tablespace might be named **ora_users_2ixhgl0q.dbf**.

Oracle Managed Files can be viewed in the Storage Manager like any other file. Oracle Managed Files also display in the standard Data Dictionary views that display datafiles, such as V$DATAFILE or DBA_DATA_FILES. **Data Dictionary views** provide system-related information by querying the database's internal management tables and presenting the data as views. See Chapter 4 for more information on Data Dictionary views.

The final area of preparation for creating a database involves setting the critical initialization parameters.

Set the Initialization Parameters

The previous sections described choices to make before creating a new database. Most of these choices require you to set initialization parameters. In addition to the parameters already discussed, there are several more initialization parameters that must be set before launching into database creation.

Initialization parameters are placed into a file usually called **init<sid>.ora** and located in the ORACLE_BASE/admin directory structure. Replace <sid> with the Oracle instance name. The **init<sid>.ora** file contains a list of all the initialization parameter values that the database uses for creation and also for subsequent start ups. There are over 200 initialization parameters. Any parameters not listed in the initialization file are automatically set to their default values. Oracle9*i* software comes with a sample initialization file to serve as a template for your file. Figure 3-2 shows part of the file. The format of the file is simple: Parameter names are followed by an equal sign, and then the value of the parameter follows; comment lines are marked with a leading # sign; blank lines are ignored.

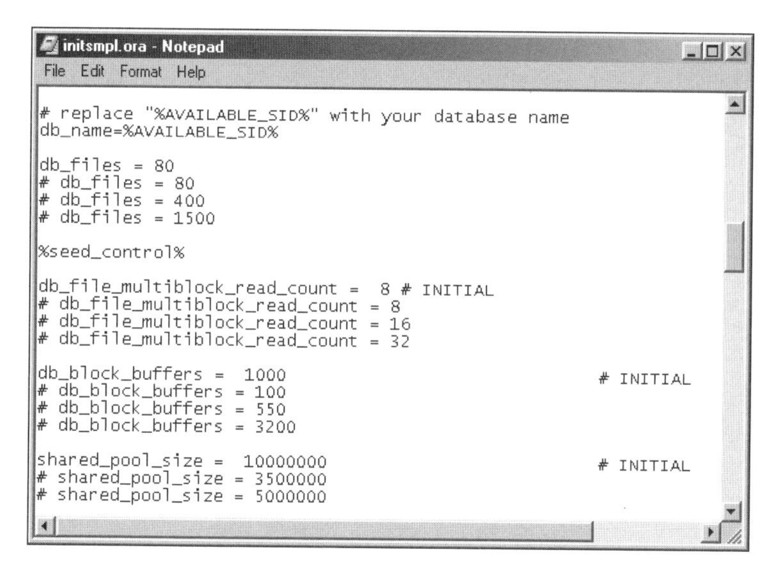

Figure 3-2 The initialization file has a plain text format

The most important parameters to set are those that cannot be modified after the database has been created. All parameters fall into one of five classes:

- **Static (Permanent):** Parameters that must be set prior to database creation and cannot be changed after creation.
- **Static (Adjustable):** Parameters that stay in force as long as the database is running. You can change these parameters by modifying the initialization parameter file and then stopping and restarting the database.
- **Dynamic (System only):** Parameters that can be changed without stopping the database. You can change the parameter by using the ALTER SYSTEM command. Changes remain in force as long as the database is running.
- **Dynamic (System and Session):** Parameters that can be changed for a session or for the entire database. Change the parameter for your session by using the ALTER SESSION command. Change the parameter system-wide by using the ALTER SYSTEM command.
- **Derived:** Parameters that Oracle9*i* sets based on the values of other parameters. Ordinarily these are not modified, although some can be set so that the derived value is overridden.

The parameters that are static and permanent must be set in the initialization parameter file before the database is created and never changed after that point. These are:

- CPU_COUNT: The number of CPUs in the database machine. This parameter is automatically set by Oracle on database creation and should not be changed. If a new CPU is added to your computer, Oracle automatically adjusts this parameter to the correct number.

- DB_BLOCK_SIZE: The block size used by Oracle when allocating storage space. Typically, the block size is 2048 or 4096 bytes, although it can be set as high as 32,768 in some operating systems.

The following parameters are typically left unchanged after database creation, even though they can be changed.

- DB_DOMAIN is the logical location of the database in the network. This is not required for a database; however, it is recommended if your company has more than one database and uses a network.
- DB_NAME is the identifying name of the database. It is up to eight characters and is also named in the CREATE DATABASE command.
- COMPATIBLE is the Oracle release number you want to emulate in the database. Ordinarily, you set this to the current release (9.2.0.1.0 for example). One reason to set the parameter to a lower release number (8.1.2 for example) is to prevent the use of new features, so that you can return to an earlier release more easily. This may be prudent when upgrading a production system with limited down time to compensate for unexpected problems.

To summarize this information, the following Table 3-2 lists all the initialization parameters discussed in this chapter.

Table 3-2 Important initialization parameters

Parameter	Description	Example
REMOTE_LOGIN_ PASSWORDFILE	Determines the authentication method for logon with SYSDBA or SYSOPER privileges	NONE: use OS authentication EXCLUSIVE: use Password File SHARE: use shared Password File (for multiple instances)
OS_AUTHENT_ PREFIX	Defines the prefix added to operating system name; for OS authentication only	OPS$: default value Null: for no prefix
CONTROL_FILES	Set the location and name of one or more control files	**(/d1/oracle/control01.ctl, /d2/oracle/control01.ctl)**: defines two files Null: use OMF (Oracle Managed Files)
DB_CREATE_ FILE_DEST	Defines the location of OMF data files and tempfiles	**D:\oracle\ora92\datafiles**
DB_CREATE_ONLINE _LOG_DEST_n	Defines the location of OMF control files and redo log files	**E:\oracle\ora92\logs** If null, use value in DB_CREATE_FILE_DEST
CPU_COUNT	Defines the number of CPUs on database machine	Automatically set by Oracle9*i*; do not change

Table 3-2 Important initialization parameters (continued)

Parameter	Description	Example
DB_BLOCK_SIZE	Defines the block size used by Oracle9*i* to allocate space	2048: 2K block size Do not change after database creation
DB_DOMAIN	Defines the domain where the database is located	OHIO.TOYSTORE.NET
DB_NAME	Defines the database name	ORA9*i* Global database name combines the DB_DOMAIN and the DB_NAME, such as ORA9*i*.OHO.TOYSTORE.NET
COMPATIBLE	Defines the Oracle release number	9.2.0.1.0

You have now reviewed all the prerequisites for preparing to create a database. Continue to the next section for some hands-on practice in creating an actual database.

CREATING A DATABASE

After you have determined all the settings and reviewed the prerequisites for your operating system, it is time to create the new database. You have two techniques from which to choose: the Database Configuration Assistant and the CREATE DATABASE command.

The Database Configuration Assistant guides you through database setup and creation using windows with instructions for each step of the way. This is a good choice for newer DBAs and tends to create a more standardized database setup.

The Database Configuration Assistant is automatically started during the installation of Oracle9*i* software, if you opted to have a seed database created as part of your installation process. A **seed database** is a database with predefined size, initialization parameters, and datafiles. Using this option saves time and allows for a fast start after software installation.

The CREATE DATABASE command gives you greater flexibility in settings; however, you must be familiar with the syntax of the command and carefully write and execute the command. This is a good choice for seasoned DBAs. It can also help if you have multiple databases to create at different sites, because you can save your settings.

The next sections guide you through the steps for each of these methods as well as steps to start and stop the database with automated and manual techniques.

Create a New Database Using the Database Configuration Assistant

Make sure you have completed all the prerequisite steps prior to starting the Database Configuration Assistant.

For this exercise, make sure you have downloaded and unzipped the files for Chapter 3 and placed them on your student disk in the Data/Chapter03 directory.

Follow these steps to create a new database.

1. Start up the Database Configuration Assistant. In Windows, click **Start/Programs/Oracle -Oraclehome92/Configuration and Migration Tools/Database Configuration Assistant**. In UNIX, type **dbca** at a command prompt and press **Enter**. The Welcome window displays (not shown).
2. Click **Next** to continue. The Step 1 window displays as shown in Figure 3-3.

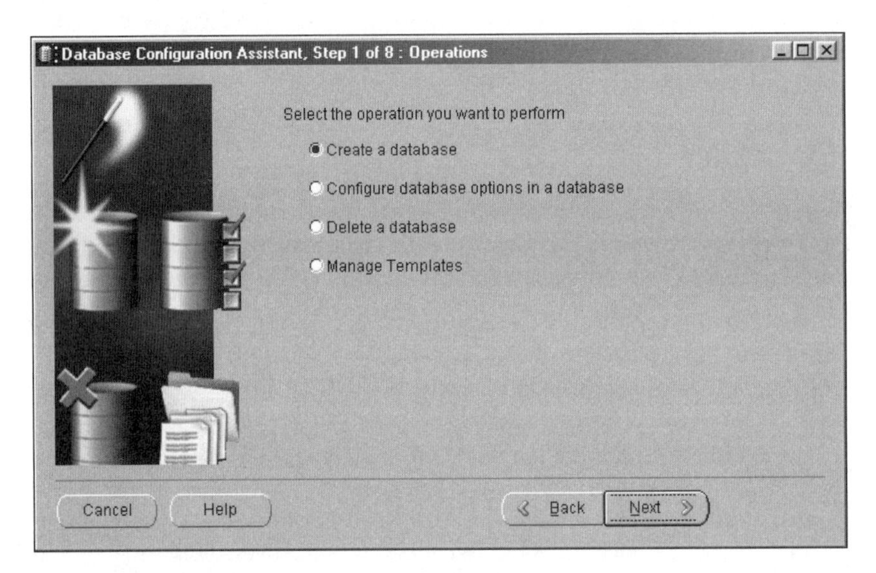

Figure 3-3 Choose to create a new database

3. Select **Create a database** radio button, and click **Next** to continue.
4. Figure 3-4 shows the options you can choose for creating database types. The four options are:
   - **Data Warehouse:** The database is optimized for queries and storing static data. It has a larger block size, larger sort area size, and a hash area defined for complex queries.
   - **General Purpose:** The database is tuned for a mixture of static data and changing data.

- **New Database:** The initial settings of the database are the same as those of the General Purpose database; however, you can customize the tablespaces and datafiles before creating the database.
- **Transaction Processing:** The database is most efficient for transactions in which users update, insert, or delete data. Initial settings are the same as those of the General Purpose database.

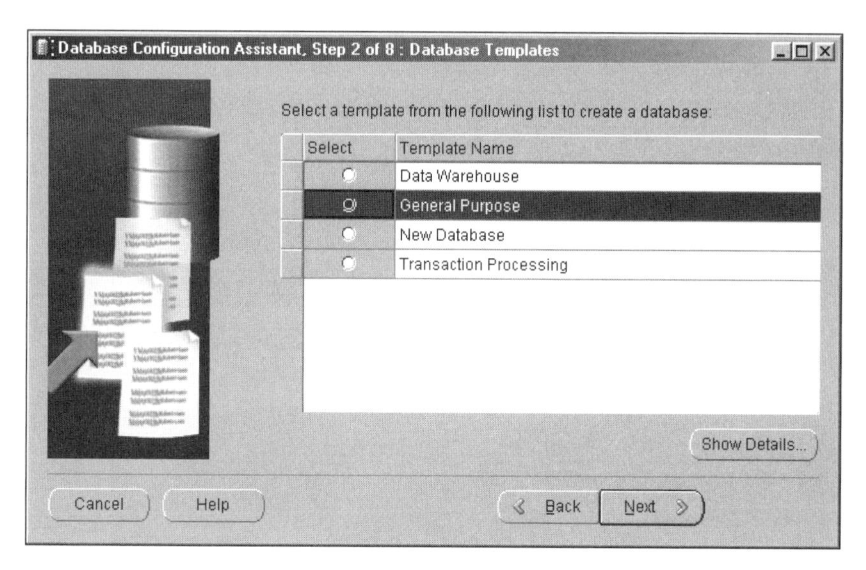

Figure 3-4 Choose variations on basic database settings

5. Select **General Purpose**, and click **Next** to continue.
6. Type **trial01.classroom** in the Domain name box. (The SID (System Identifier) is automatically filled in as **trial01**, as you see in Figure 3-5.) Click **Next** to continue.
7. Select **Dedicated Server Mode**, and click **Next** to continue, as shown in Figure 3-6. The Dedicated Server Mode works well for small workgroups, giving each user a distinct set of resources while working in the database. The other option, Shared Server Mode, works well when there are many concurrent users. Shared Server Mode shares the resources to make the best use of idle time and to reuse memory buffers whenever possible.

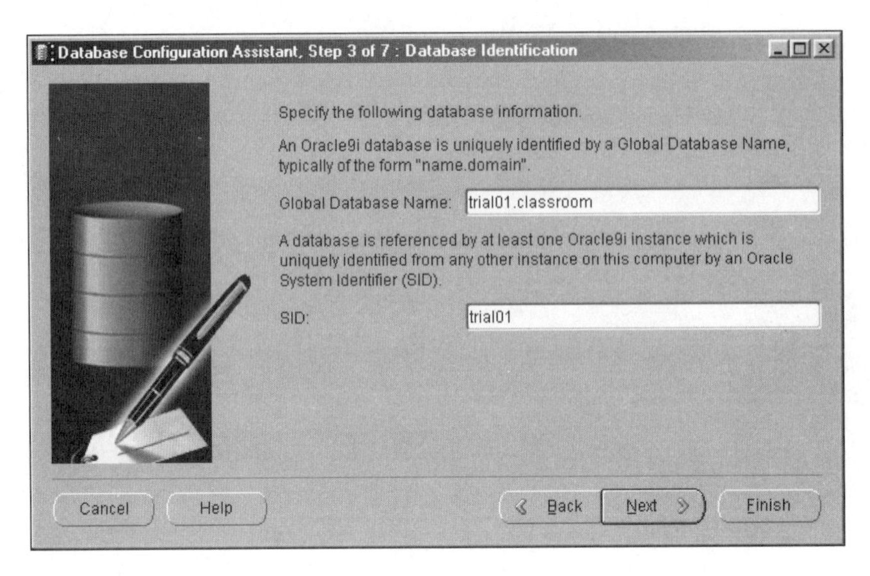

Figure 3-5 Give your database a unique name

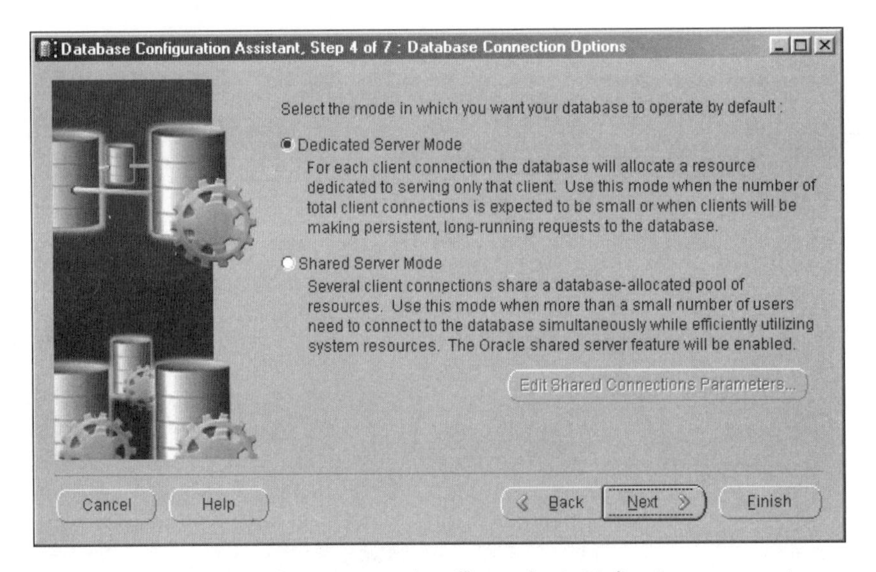

Figure 3-6 Dedicated Server Mode affects the initialization parameters

8. Review the available settings on this page. Here you have the ability to modify the memory size, archive mode, and many initialization parameters.
9. Select the **Typical** radio button in the Memory tab as shown in Figure 3-7.

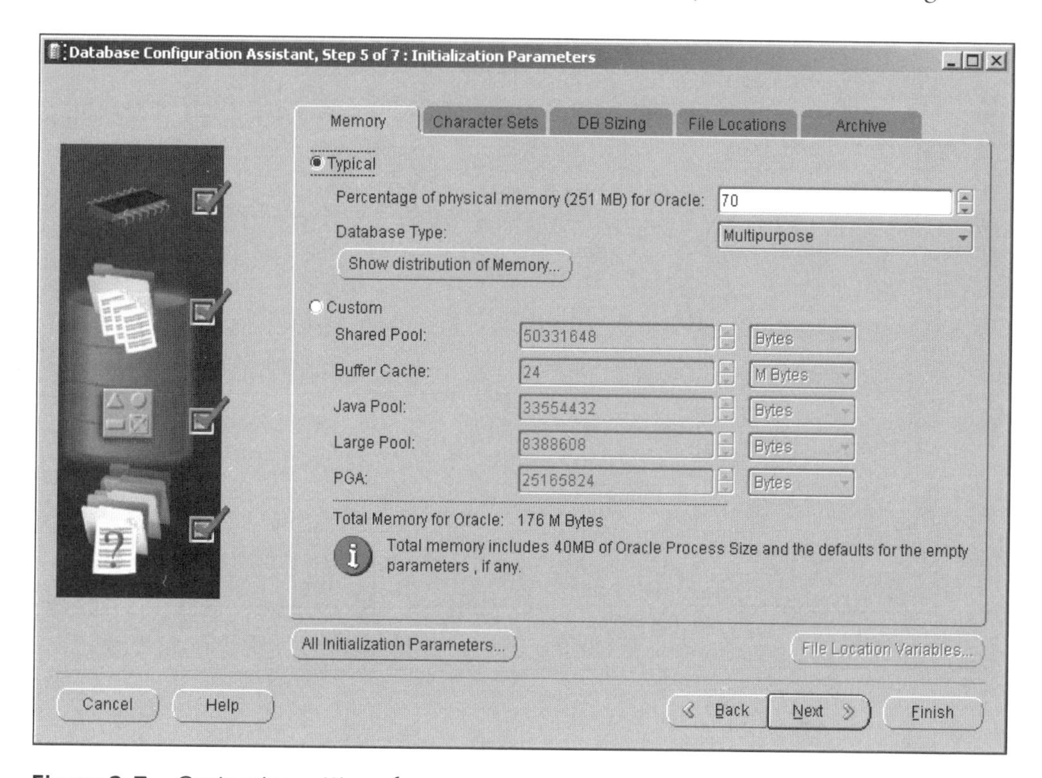

Figure 3-7 Customize settings for memory

10. Click the **All Initialization Parameters** button. This brings up the window shown in Figure 3-8 with a list of parameters. You can sort the parameters as you could within Instance Manager, by the column headings. For example, to view all the parameters that will be included in the **inittrial01.ora** file being built for the database, click the **Included (Y/N)** column heading.

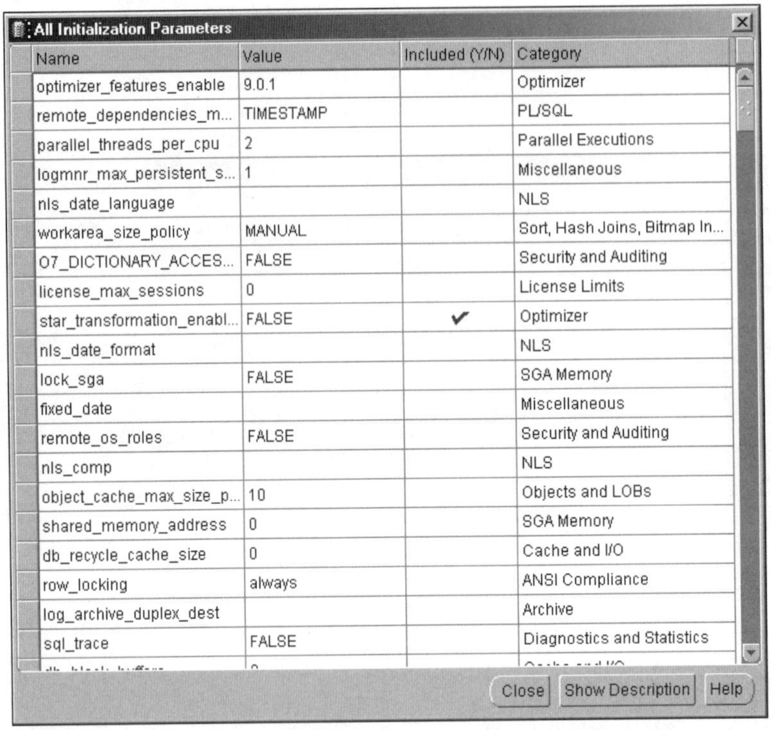

All Initialization Parameters

Name	Value	Included (Y/N)	Category
optimizer_features_enable	9.0.1		Optimizer
remote_dependencies_m...	TIMESTAMP		PL/SQL
parallel_threads_per_cpu	2		Parallel Executions
logmnr_max_persistent_s...	1		Miscellaneous
nls_date_language			NLS
workarea_size_policy	MANUAL		Sort, Hash Joins, Bitmap In...
O7_DICTIONARY_ACCES...	FALSE		Security and Auditing
license_max_sessions	0		License Limits
star_transformation_enabl...	FALSE	✔	Optimizer
nls_date_format			NLS
lock_sga	FALSE		SGA Memory
fixed_date			Miscellaneous
remote_os_roles	FALSE		Security and Auditing
nls_comp			NLS
object_cache_max_size_p...	10		Objects and LOBs
shared_memory_address	0		SGA Memory
db_recycle_cache_size	0		Cache and I/O
row_locking	always		ANSI Compliance
log_archive_duplex_dest			Archive
sql_trace	FALSE		Diagnostics and Statistics

Close | Show Description | Help

Figure 3-8 Initialization parameters are set or revised by typing in the boxes

11. Click the **Name** column to resort the parameters in alphabetical order. Then scroll down to the **os_authent_prefix** parameter. Change the parameter to null value by highlighting the box that contains **ops$** and erasing the value. Select the parameter for inclusion in the initialization file by clicking the **Included (Y/N)** box until it contains a check mark. Figure 3-9 shows the window after you have made these changes.

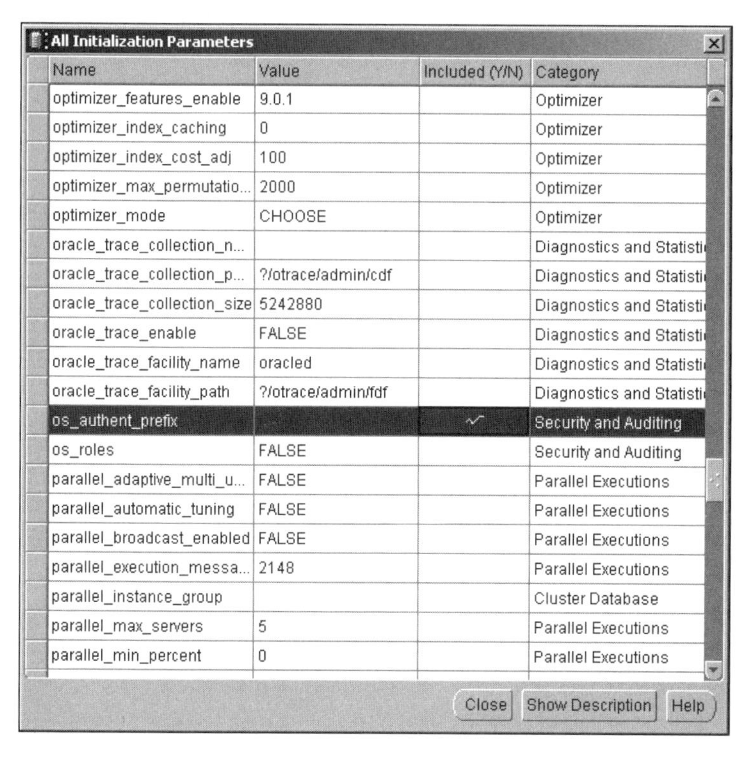

Figure 3-9 A check mark indicates the parameter will be included in the **inittrial01.ora** file

Parameters that are not checked and still show values in the window are displayed with their default values.

12. Click **Close** to return to the main window.
13. Click **Next** to continue. This brings up the Database Storage window.
14. Click the **Datafiles** folder. On the right side you find a list of datafiles and their directory locations, as shown in Figure 3-10.

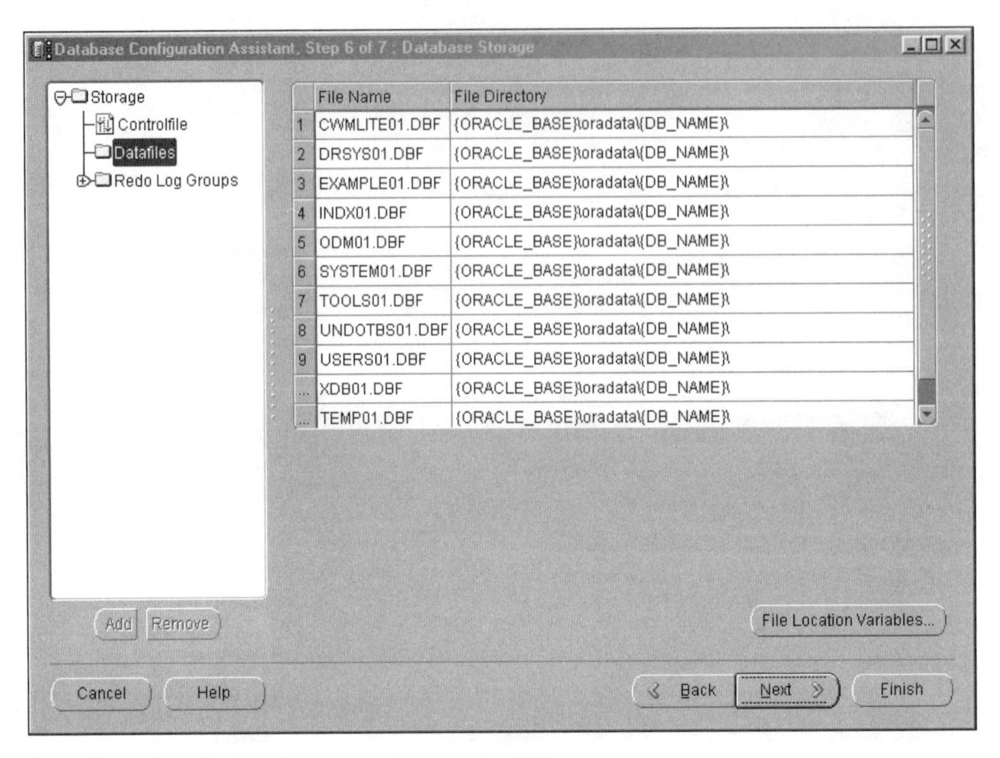

Figure 3-10 Data file locations contain variables

15. The directory names use variables to specify the exact directory to be used. Recall that ORACLE_BASE and ORACLE_HOME are standard variables set up when the database software was installed. It is possible to change the names and locations of the datafiles by typing over the existing values. For this example, accept the default values as they display.

16. Click **Next** to continue. You come to the final screen before the database is created. Here you can create the database, create a template that can be used as the beginning point for another database to be created later, or do both. If you choose to save a Database Template, a DBCA template is created that contains all the changes you have made to the standard selections. This template is added to the list of templates available the next time you use the Database Configuration Assistant. Accept the default, in which only the Create Database check box is marked as in Figure 3-11.

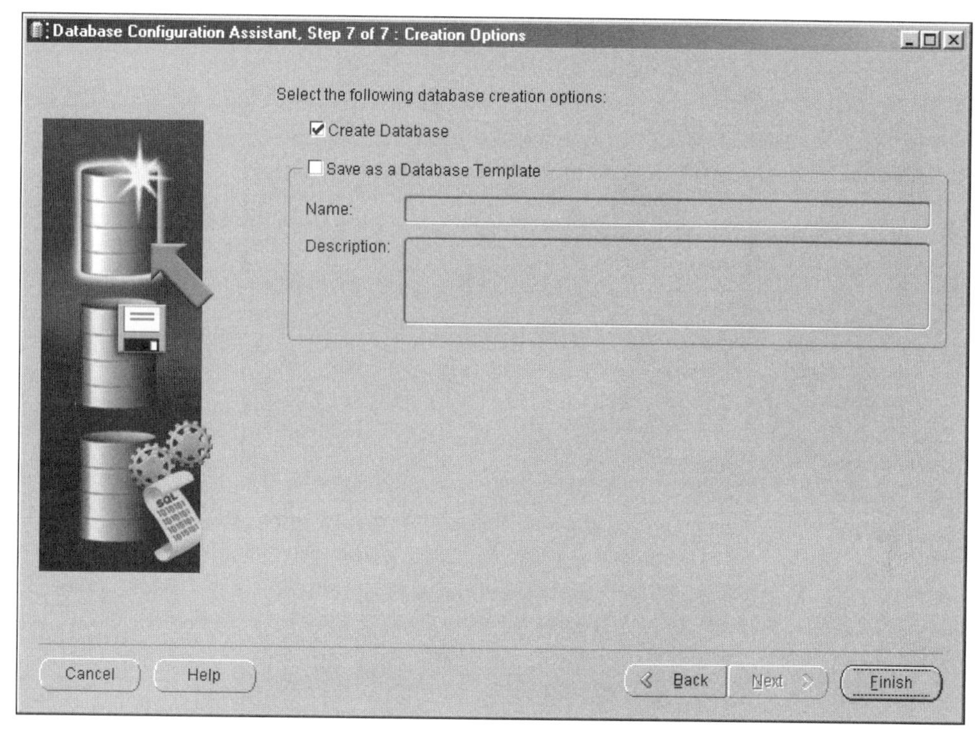

Figure 3-11 The Assistant is ready to create the database

17. Click **Finish** to begin the creation process. The Summary window displays, displaying all the settings to be used for the new database.
18. Click **OK** to continue. The Assistant starts up a background SQL process and executes the commands it needs. The process takes about 10 or 15 minutes to complete.
19. A progress window appears to display the steps, as shown in Figure 3-12. Each step gets checked off as it is completed.

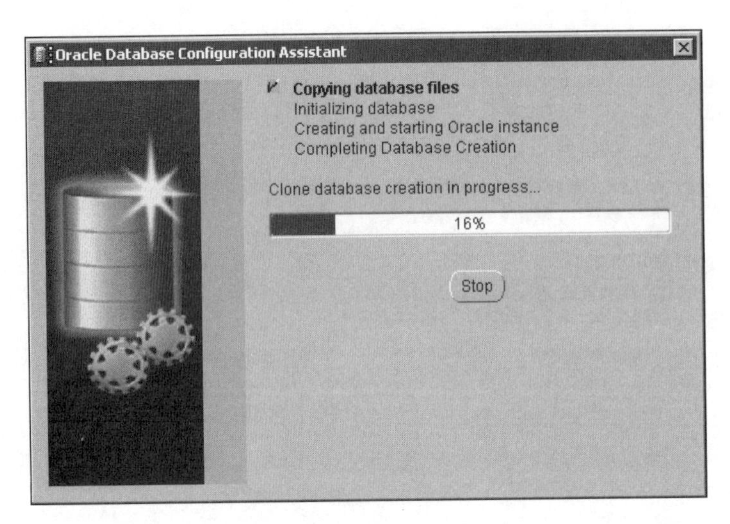

Figure 3-12 Your progress is charted as the Assistant goes to work

20. A window appears asking you to change passwords for the SYS and SYSTEM users. Type in the passwords you want for each user, as shown in Figure 3-13.

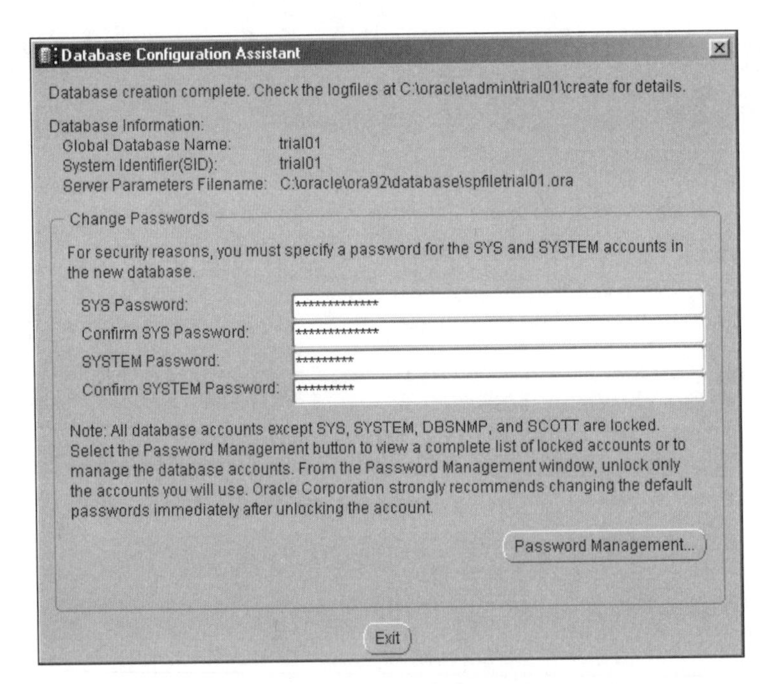

Figure 3-13 Changing the passwords improves security on a new database

21. Click **Exit** to end the Database Configuration Assistant session.

You have completed the creation of a new database. This database takes up approximately 1.2 gigabytes of storage space. Oracle9*i* is not streamlined; however, it is the most powerful database in its class, handling massive data at high speeds. For example, the Amazon.com Web site, which has thousands of concurrent users searching its book database, is run by an Oracle database.

Because you have used a Windows-like tool to create your database, the next section shows you how to start and stop your new database with another Windows-like tool: the Instance Manager in the Enterprise Manager console.

Starting and Stopping the Database with Instance Manager

One of the last steps that the Database Configuration Assistant performs when creating a new database is starting up the database. Now, it is ready to use. Or is it? Recall that in order to access any database except the Personal Edition database, the Oracle Net connection must be made between your software (such as SQL*Plus or the console) and the database. This means that you must handle the Oracle Net configuration before you can access or shut down the database using the console.

The database service runs as a set of background programs in the computer. These automatically stop without error when the computer is shut down. Usually, the database service also automatically starts up when the computer is started up.

For this section, you use the Net Manager to configure the new database's Named Service, then you add the database to the console, and finally, you use the console's Instance Manager to shut down and restart the database.

The steps for configuring a database in Net Manager were covered in Chapter 2. Review the material again if needed and configure the new database. Here are some additional notes:

- Name the Net Service Name **trial01**.
- The Database Service Name is **trial01.classroom**.
- Use the default port and protocol settings of **1521** and **TCP/IP**.
- Ask your instructor for the correct settings for Host Name.

Figure 3-14 shows the Net Manager with the new database configured as a Named Service. Your settings will be similar.

Now that Oracle Net is properly configured, the database must be added to the console. You did this in Chapter 2 as well, so refer to the steps in the section titled "Starting the Enterprise Manager Console," and repeat the process for the **trial01** database. Figure 3-15 shows the console with the trial01 database in place.

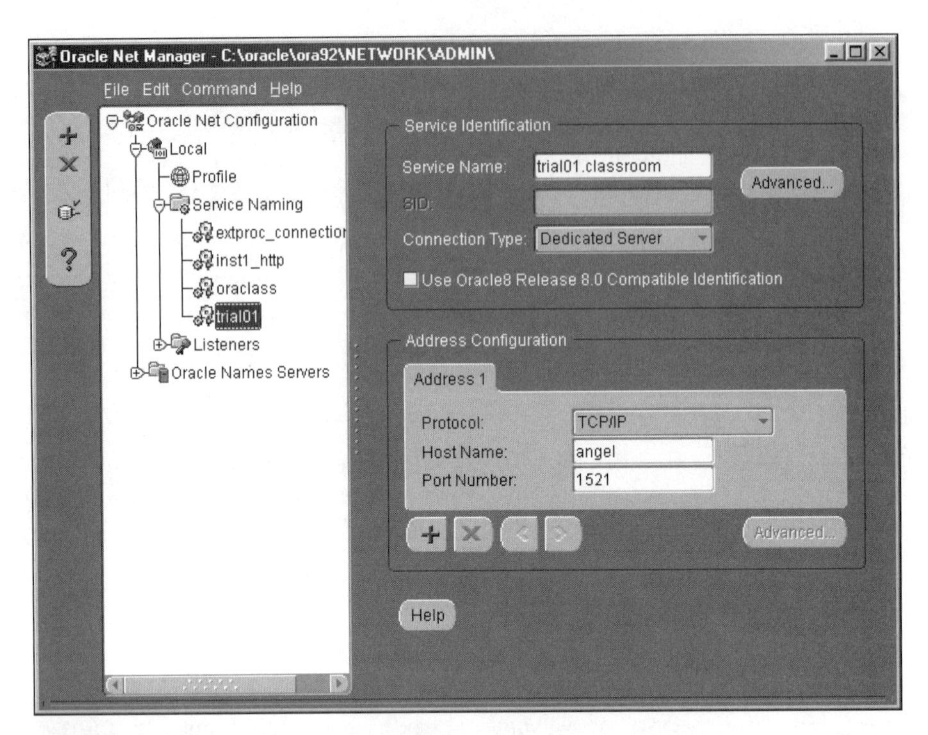

Figure 3-14 Net Manager after configuring the **trial01** database

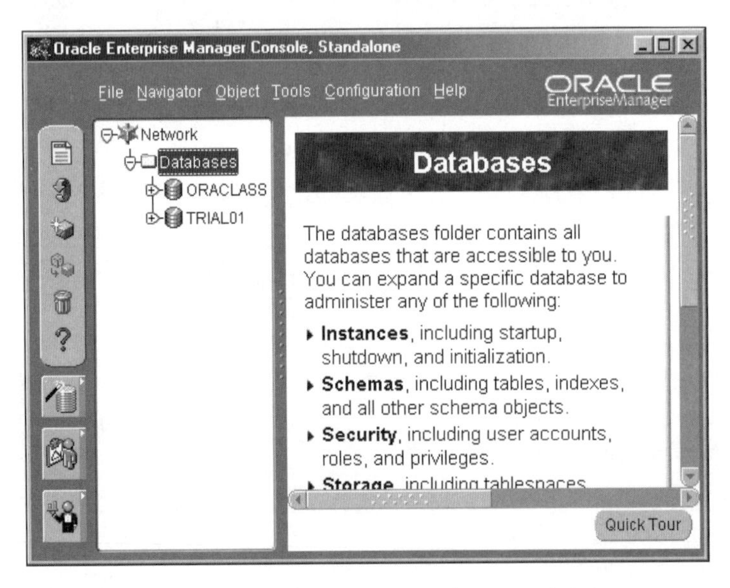

Figure 3-15 The console recognizes the new database

Follow these steps to view the status of the database, and shut it down.

1. Right-click the **trial01 database icon**, and select **Connect** from the shortcut menu. The Database Connection Information window displays.
2. Type **SYS** in the Username box, the current password for SYS in the Password box, and then select **SYSDBA** in the Connect as box. Figure 3-16 shows the window at this point. Click **OK** to continue. This logs you onto the database as the SYS user with SYSDBA privileges.

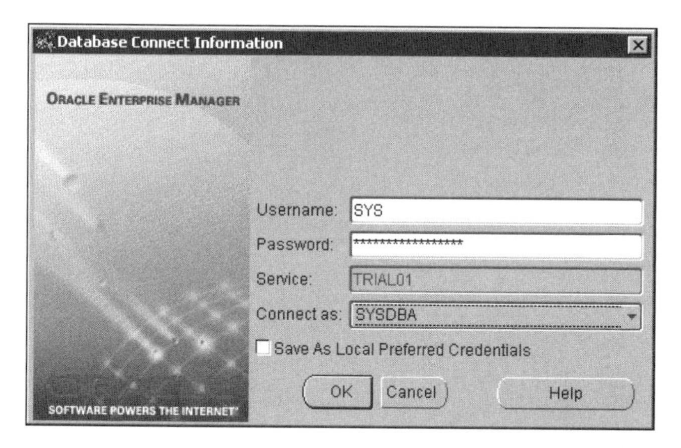

Figure 3-16 Do not save these settings as default, because they are only for special tasks

3. Double-click the **Instance icon**, and then click the **Configuration icon** to view the status of the database, as you see in Figure 3-17.
4. To begin the shut down process, click the **Shutdown** radio button, scroll down, and then click the **Apply** button in the lower-right corner of the screen. The Shutdown Options window displays as shown in Figure 3-18.
5. Accept the default setting of **Immediate** shutdown mode, and click **OK** to continue. A status window displays, showing the progress of the shut down procedure.

6. After the status window displays the message, "Processing Completed," as shown in Figure 3-19, click **Close** to return to the Console window. The stoplight symbol now displays a red light, indicating the database is shut down.

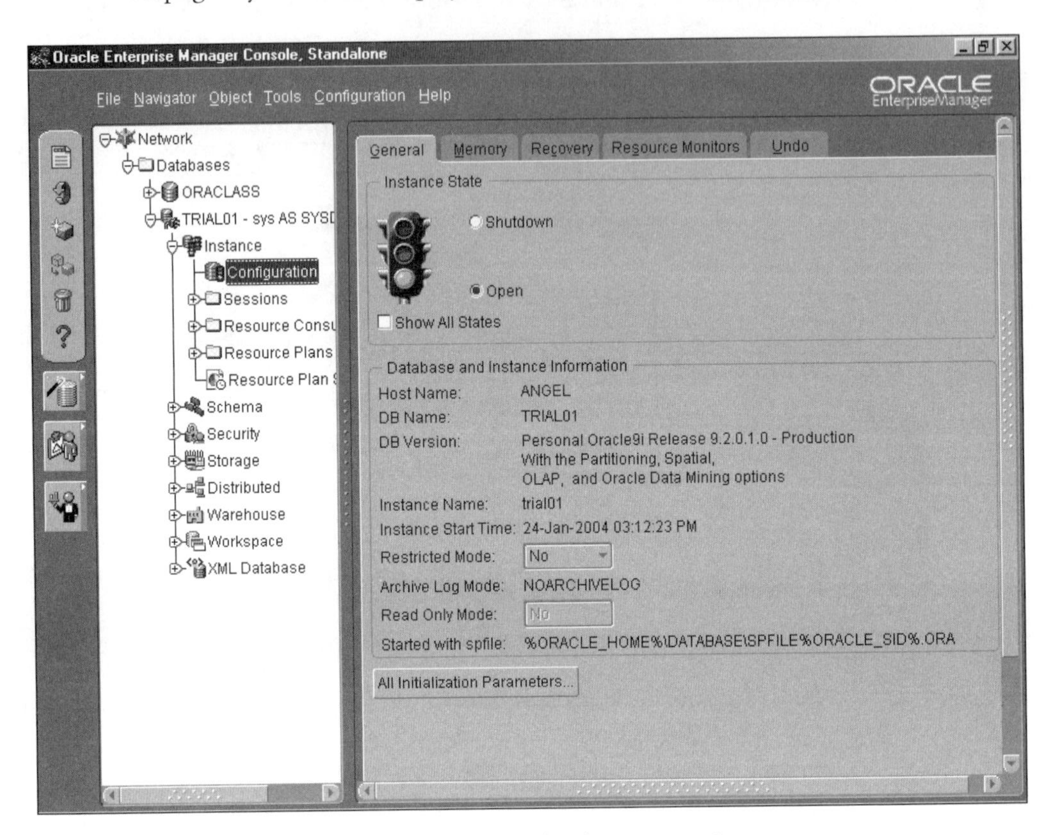

Figure 3-17 A green light signifies that the database is running

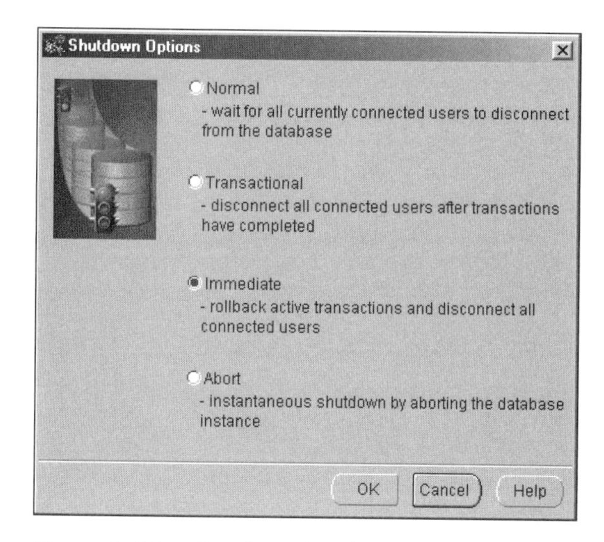

Figure 3-18 The four shutdown options are explained in this window

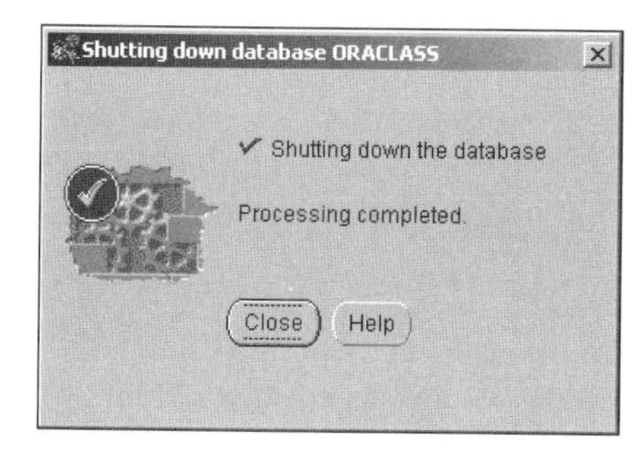

Figure 3-19 This status window displays progress during shutdown

7. To open the database again, select the **Open** radio button next to the stop light, then scroll down, and click **Apply**. The Startup Options window displays.
8. Accept the defaults to use the original initialization parameters set when you created the database. Figure 3-20 shows the window. Notice that the selection states "Use spfile for startup." An spfile (or server parameter file) is a copy of the **init<sid>.ora** and is saved in a special format in a file named **spfile<sid>.ora**, or spfile.ora. The spfile stores a persistent set of initialization parameters that can be used by remote DBAs who want to start up the database. Only a user with SYSOPER or SYSDBA privileges can modify the spfile, which is done using the ALTER SYSTEM command. Click **OK** to begin the process of starting up the database.
9. A status window with animated gears displays showing the status of the start up process. When the "Process Completed" message displays, click **Close** to return to the console.

10. Exit the console by clicking the **X** in the top-right corner of the window.

The database is running again, and your job is done.

The automated method of creating a database is convenient but highly interactive and somewhat restrictive. Using a SQL command gives you some additional flexibility in settings, as well as the option to reduce the amount of hands-on interaction needed.

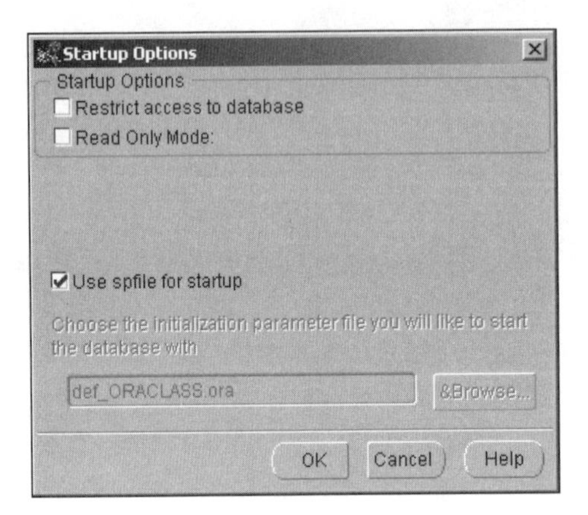

Figure 3-20 The Startup Options allow you to select a customized initialization parameter file

Creating a Database Manually

You have worked through the automated method for creating a database. The Database Configuration Manager is convenient; however, it does not provide enough flexibility for some database needs. For example, a company that stores satellite images that add 500 Megabytes a day onto the database size requires much larger tablespace and datafiles extension sizes than are provided by the Database Configuration Assistant.

Creating a database manually involves several steps. Some steps are the same as those you used to create a database with the Assistant, whereas others take the place of the steps that are automated by the Assistant. Review Table 3-1 at the beginning of this chapter to see what steps come before creating the database manually.

The new Oracle9*i* feature of changing the SYS and SYSTEM passwords has not been implemented into the CREATE DATABASE syntax yet. So, the databases created manually still have the default passwords (CHANGE_ON_INSTALL for SYS and MANAGER for SYSTEM) in force when they are created.

The database you are about to create requires about 500 megabytes of space. This is quite small as Oracle databases go, but it is sufficient to gain experience with the CREATE DATABASE command, and takes slightly less time to create than a larger database. After you are ready to create the database manually, follow these steps.

1. Create a directory structure for your database files. You need three directories: one for control files and redo log files; one for datafiles; and one for administration files. For your class, the three directories are named: **newlog**, **newdata**, and **newadmin**. Ask your instructor for the appropriate location of these three directories. They must reside on the hard drive of your computer.
2. Copy the **inittrial02.ora** file in the Chapter03 data directory into the **newadmin** directory you created.
3. Edit the **inittrial02.ora** file in the **newadmin** directory, making changes to fit your environment:
   - Change the path of background_dump_dest, core_dump_dest, and user_dump_dest parameters to the full path of your **newadmin** directory.
   - Change the path of the control files to match the full path of your **newdata** directory.
   - Change the path in the DB_CREATE_FILE_DEST parameter to match the full path of your **newdata** directory.
4. Save the changes and close the file.
5. Start a Command prompt window.
6. Create a password file in the **database** directory by typing the command shown below, modifying <Z:\zzz> to reflect the full path name of ORACLE_HOME on your computer. For example, the location of ORACLE_HOME on Windows 2000 might be **D:\oracle\ora92**. Note that the command should be typed as a single line with a space before the word password.

   ```
   orapwd file="<Z:\zzz>\database\pwdtrial02.ora"
   password=change_on_install entries=5
   ```

   If you are using UNIX, the command is:

   ```
   orapwd file="<Z:/zzz>/dbs/pwtrial02.ora"
   password=change_on_install entries=5
   ```

   Again, replace <Z:\zzz> with the full path of ORACLE_HOME.

Creating the password file is not required if your initialization parameters specify that OS authentication is to be used rather than password file authentication.

7. Adjust the environment variable for ORACLE_SID by typing this command:

   ```
   SET ORACLE_SID=trial02
   ```

8. Create a database service for the new database. A **service** in Windows is a background process. The database service is a set of background programs that run the database engine. The following command uses an Oracle program for

creating a Windows service for the Oracle database. Type the command in the Command prompt window, and then press **Enter**. Replace **<X:\xxx>** with the actual path of the **newadmin** directory. Type the command on a single line.

```
oradim -new  -sid trial02 -startmode a -pfile
<X:\xxx>\newadmin\inittrial02.ora
```

In UNIX, no service is required. Instead, you normally add to the existing script executed when the computer reboots. The addition checks the status of the database and then starts the database if needed by issuing the DBSTART command. To find an example of the script, go to the Oracle Technology Network online documentation page, select Oracle9*i* Documentation, and open the Oracle9*i* Installation Guide Release 9.2 for UNIX and search for DBSTART.

9. Connect to SQL*Plus without logging onto a database by typing this command and pressing **Enter**:

```
sqlplus /nolog
```

10. Copy the file named **createdb.sql** from the **Data\Chapter03** directory into the **newadmin** directory.
11. Open the **createdb.sql** file you just added to your **newadmin** directory in Notepad or your own editor. Figure 3-21 shows the command inside the file. The steps continue after a brief discussion of the CREATE DATABASE command.

```
createdb.sql - Notepad
File Edit Format Help
REM *****************************
REM       Revise all occurrances of
REM            <X:\xxx>    to a real path name.
REM *****************************
REM       Run script when logged in as:
REM            SYS AS SYSDBA
REM *****************************
set echo on
spool <X:\xxx>\newlog\CreateDB.log
startup nomount pfile="<X:\xxx>\newadmin\inittrial02.ora";
CREATE DATABASE trial02
MAXDATAFILES 100
MAXINSTANCES 1
MAXLOGFILES 5
MAXLOGMEMBERS 5
DATAFILE '<X:\xxx>\newdata\system01.dbf' SIZE 325M AUTOEXTEND ON
      NEXT 10240K MAXSIZE UNLIMITED
UNDO TABLESPACE UNDOTBS DATAFILE '<X:\xxx>\newdata\undotbs01.dbf'
      SIZE 25M AUTOEXTEND ON NEXT  5120K MAXSIZE UNLIMITED
CHARACTER SET WE8MSWIN1252
NATIONAL CHARACTER SET AL16UTF16
LOGFILE GROUP 1 ('<X:\xxx>\newdata\redo01.log') SIZE 100M,
GROUP 2 ('<X:\xxx>\newdata\redo02.log') SIZE 100M;
```

Figure 3-21 The CREATE DATABASE command has many parameters

The syntax of the CREATE DATABASE file contains numerous, optional parameters. The following syntax contains the primary parameters needed for creating the database.

To see the complete set of parameters, go to the Oracle9*i* online documentation, select **SQL and PL/SQL syntax and examples**, and search for CREATE DATABASE.


```
CREATE DATABASE <databasename>
MAXDATAFILES <n>
MAXINSTANCES <n>
MAXLOGFILES <n>
MAXLOGMEMBERS <n>
DATAFILE '<path>/<datafilename>' <storage_settings>
UNDO TABLESPACE <undo> DATAFILE '<path>/<file>'
       <storage>
CHARACTER SET <charsetname>
NATIONAL CHARACTER SET <ncharsetname>
LOGFILE GROUP <n> ('<path>/<file>','<path>/<file>', ...)
GROUP <n> ('<path>/<file>','<path>/<file>', ...);
```

The lines in the syntax are described here:

- **CREATE DATABASE <databasename>:** The name of the database; must be identical to the value of DB_NAME in the initialization parameters.
- **MAXDATAFILES <n>:** An optional setting that determines the maximum number of datafiles that can be open when the database is open. This determines the size of the control file (where space for the names of all open files is reserved).
- **MAXINSTANCES <n>:** The maximum number of instances of this database that can be open at the same time; minimum is 1 and maximum depends on your operating system.
- **MAXLOGFILES <n>:** The maximum number of log groups. See Chapter 5 for more details.
- **MAXLOGMEMBERS <n>:** The maximum number of members in each group. See Chapter 5 for more information.
- **DATAFILE '<path>/<datafilename>' <storage_settings>:** The path and filename of the datafiles for the SYSTEM tablespace. You can specify more than one datafiles for the SYSTEM tablespace, or you can omit this parameter to allow Oracle9*i* to create a 100 megabyte Oracle-managed file in the directory you name in the DB_CREATE_FILE_DEST initialization parameter. The <storage_settings> contains the datafiles's storage parameters and generally includes SIZE and EXTENT values for the datafiles; see Chapter 6 for more details on storage settings.

- **UNDO TABLESPACE <undo> DATAFILE '<path>/<file>' <storage>:** The name of the tablespace Oracle9*i* uses to store undo transaction data; must be identical to the value of the UNDO_TABLESPACE initialization parameter. If the UNDO_TABLESPACE parameter is omitted in both the CREATE DATABASE and the **init<sid>.ora** file, undo data is stored in redo log files. See Chapter 5 for details on redo logs.
- **CHARACTER SET <charsetname>:** The character set the database uses when storing data in the database. English/American is WE8MSWIN1252.
- **NATIONAL CHARACTER SET <ncharsetname>:** The national character set the database uses when storing data in columns that are of data type NCHAR, NCLOB, or NVARCHAR2. See Chapter 14 for more information on character sets; default is AL16UTF16.
- **LOGFILE GROUP <n> ('<path>/<file>','<path>/<file>', ...):** The group number for log files; usually sequential starting with 1. Each group has at least one file associated with it; repeat the LOGFILE GROUP phrase once for each log file group. You must have at least two LOGFILE groups. If you omit the <path><file> portion of this clause, Oracle9*i* creates an Oracle Managed file in the directory named in the DB_CREATE_ONLINE_LOG_DEST_n initialization parameter or the DB_CREATE_FILE_DEST initialization parameter (when the former is not specified). See Chapter 6 for more details on log file groups.

The CREATE DATABASE command becomes elegantly simple when you use OMF settings. When the DB_FILE_DEST initialization parameter is set, Oracle9*i* knows to locate all the datafiles. Adding the DB_ONLINE_LOGFILE_DEST_n parameter tells Oracle9*i* where to place the control files and redo log files. In fact, if you omit both of these parameters from the **init<sid>.ora** file, Oracle9*i* can still use OMF and places files in a default directory. To invoke OMF, simply omit all of the filenames and locations in the CREATE DATABASE command. For example, the following command is valid, and results in the creation of the SYSTEM tablespace, online redo logs, and control files (the number of redo logs and control files depends on the initialization parameters). Each tablespace and redo log file is 100 megabytes.

```
CREATE DATABASE testOFM
MAXINSTANCES 1
CHARACTER SET WE8MSWIN1252
NATIONAL CHARACTER SET AL16UTF16;
```

At its minimum, the CREATE DATABASE command requires only one parameter: the database name.

Having reviewed the CREATE DATABASE command in detail, you are ready to continue the process of editing the **createdb.sql** file you have opened in your editor. Follow the steps to make changes to the file, and then run the commands using SQL*Plus.

1. With the **createdb.sql** file open in your editor, change every occurrence of "X:\xxx" to the name of the directory in which you created the **newadmin**, **newlog**, and **newdata** directories. For example, if the three directories are under **C:\mytest**, then change <X:\xxx> to **C:\mytest**.
2. Save the file, and close the editor you used to work on the file.
3. Return to the SQL*Plus session that you opened earlier. If you closed the window, open a new Command prompt window, execute the following commands to set the environment variable, and start SQL*Plus again:

   ```
   SET ORACLE_SID=trial02
   sqlplus /nolog
   ```

4. Connect to the SYS user with the SYSDBA role by typing this command and pressing **Enter**. Use the current password of the SYS user.

   ```
   CONNECT SYS/CHANGE_ON_INSTALL AS SYSDBA;
   ```

5. Run the **createdb.sql** file to start the database service, and create the database instance. This command takes a few minutes to run. Type the following command, changing the <X:\xxx> to the full path of the newadmin directory:

   ```
   START <X:\xxx>\newadmin\createdb.sql
   ```

   SQL*Plus replies: "Database created."

6. Stay logged onto SQL*Plus for the next set of steps.

There are a few more details to complete before your database is ready for use: adding two tablespaces, adjusting the default use of the tablespaces, and running two scripts for implementing standard Data Dictionary views and PL/SQL processes. **PL/SQL** is the Oracle-specific procedural programming language embedded in the database. It combines procedural programming structures, such as IF-THEN-ELSE logic, and loops with SQL.

Oracle recommends adding at least two more tablespaces to your database right away: one for temporary space (space used by the database when it needs room for processing data, such as performing sorts or formatting query results); and one for tables that are created under other schemas (schemas other than SYS and SYSTEM). The names for these two tablespaces when Oracle9*i* creates them in its seed database are TEMP and USERS.

The default settings for creating new tables and for using temporary space should be adjusted so that the SYSTEM tablespace is reserved for tables owned by SYS or SYSTEM and other internally-generated objects. This helps separate your database into functional areas by tablespace, making it easier to plan backups, export schemas, and move data to a different tablespace.

You must run two Oracle-provided scripts on any new database that you created manually. The first script, **catalog.sql**, creates all the Data Dictionary views that are used by the DBA, as well as by the Enterprise Manager and other utilities, to discover information about schemas, objects, and activity in the database. The second script, **catproc.sql**, generates the standard internal packages required to run PL/SQL in the database. To perform all three of these previously listed tasks, follow these steps:

1. Using your operating system commands, or Windows Explorer, copy the **tablespace.sql** file from the **Data\Chapter03** directory into the **newadmin** directory you created earlier.
2. Open the new **tablespace.sql** file with your editor (such as Notepad), and change every occurrence of <X:\xxx> to the directory that contains the **newadmin** directory.
3. Change every occurrence of <Z:\zzz> to the path of ORACLE_HOME on your computer. Remember, ORACLE_HOME always symbolizes the directory where the Oracle software resides. For example, your ORACLE_HOME might be **C:\oracle\ora92**, and so, you would replace <Z:\zzz> with **C:\oracle\ora92**.
4. Save the file, and close your editor program.
5. Return to the SQL*Plus session that you opened earlier. If you closed the SQL*Plus session, perform the next step; otherwise, skip to Step 7.
6. If you need to create a new SQL*Plus session, open a Command prompt window and execute the following commands to set the environment variable and start SQL*Plus again:

```
SET ORACLE_SID=trial02
SQLPLUS /nolog
```

7. Connect as the SYSTEM user. The SYSTEM user has the DBA role and can create tablespaces. Type this command, replacing <service> with the actual database service name on your computer. Run the command by pressing **Enter**.

```
CONNECT <service>;
```

8. Run the **tablespace.sql** file. The process takes several minutes to complete. Type the following command, changing the <X:\xxx> to the full path of the **newadmin** directory.

```
START <X:\xxx>\newadmin\tablespace.sql
```

SQL*Plus replies, "Table space created."

9. Exit SQL*Plus by typing **exit** and pressing **Enter**.

can you do without the comforts of a navigation tree and the console's handy code generators? It is important to be prepared for unusual situations by becoming familiar with the manual methods of starting and stopping the database.

Even if the database was created with the Database Configuration Assistant, it can be started or stopped manually, and vice versa. Therefore, you may use whichever method is convenient at the time.

Follow these steps to shut down a running database using SQL*Plus:

1. To start a Command Prompt window, click **Start**, point to **Programs**, point to **Accessories**, and then click **Command Prompt** in Windows. In UNIX, go to a $ prompt, and make sure your environment variables include the ORACLE_HOME path.
2. Start up SQL*Plus without logging on by typing this command and pressing Enter:

   ```
   sqlplus /nolog
   ```

   You should see the SQL> prompt awaiting your next command.
3. Connect as SYS with SYSDBA privileges by typing the following command, using the current password for SYS and pressing **Enter**.

   ```
   CONNECT SYS/<password> AS SYSDBA
   ```

4. Type SHUTDOWN IMMEDIATE and press **Enter**. SQL*Plus displays the following lines:

   ```
   Database closed.
   Database dismounted.
   ORACLE instance shut down.
   ```

There are three options for shutdown:

- **ABORT:** Shuts down without saving any pending transactions or rolling back any unsaved changes. All users are logged off. This method is the fastest; however, you should reserve its use for situations in which the database is not functioning properly and you are unable to use SHUTDOWN IMMEDIATE successfully.
- **IMMEDIATE:** Waits for pending transactions to complete or users to log off. This is faster and convenient for environments in which you are confident that you won't disturb users. It can also help you if you have tried SHUTDOWN NORMAL and are waiting too long because of some outstanding transactions.

- **NORMAL:** Waits for all users to log off before shutting down. This may take longer, but it ensures a smooth and invisible shutdown. If you start this type of shutdown and decide you don't want to wait, cancel the command by holding the **Control** key and typing **C**. This cancels the shutdown and returns you to the SQL> prompt. Then you can type SHUTDOWN IMMEDIATE to shut down without waiting.

To start up the database, follow the previous Steps 1 through 3, type this command, and then press **Enter**. Remember to replace <X:\xxx> with the path to your **newadmin** directory:

```
STARTUP PFILE=<X:\xxx>\newadmin\inittrial02.ora
```

The PFILE parameter is needed, because you have located the **inittrial02.ora** file in a nondefault directory. By default, Oracle9*i* looks for a file named **init<SID>.ora** (**<SID>** is replaced by the database name) in the **ORACLE_HOME\database** directory.

SQL*Plus responds with status information about the database and ends with the statement: Database started. Figure 3-24 shows an example of the status messages you see when issuing the STARTUP command.

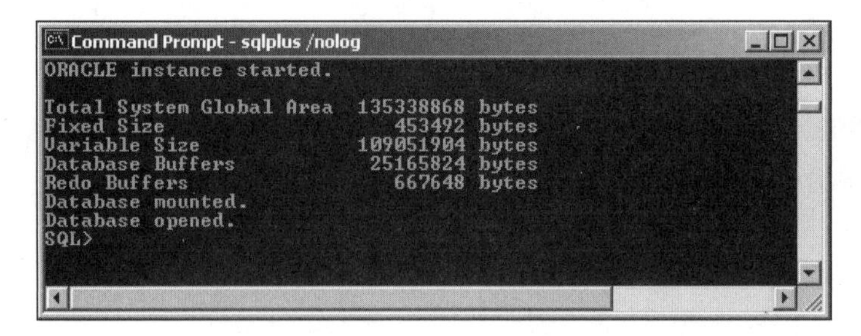

Figure 3-24 The STARTUP command displays memory storage size

After the database has been opened, it is ready for use.

To use your manually-created database with Enterprise Manager, you must also set up the appropriate Net Service Name, as described in this chapter's earlier section named, "Starting and Stopping the Database with Instance Manager."

The next chapter discusses more details you need in preparation for using your new database.

CHAPTER SUMMARY

- Installing the Oracle9*i* database software is actually separate from creating a database, although often both occur during one installation session.
- Databases can be created manually or by using the Database Configuration Assistant.
- The operating system-specific installation guide describes minimum requirements for installing a new database.

- The DBA authentication method determines how Oracle9*i* validates users logging on with SYSDBA or SYSOPER privileges.
- Operating system (OS) authentication relies on the operating system's security to validate the user's name and password and authorization group.
- The REMOTE_LOGIN_PASSWORD_FILE parameter is set to NONE for OS authentication.
- Password file authentication stores user names and passwords and group membership in an encrypted file on the operating system.
- Set REMOTE_LOGIN_PASSWORDFILE to EXCLUSIVE for password file authentication.
- The ORAPWD utility generates the password file for SYSDBA, and then the database maintains it with changes to passwords.
- Control files are multiplexed, meaning that each subsequent control file is an exact copy of the first control file.
- Because control files are important, they should be located on different physical devices to guard against damage.
- You can prevent bottlenecks in data access by placing data on several physical devices, which spreads the demand across more resources.
- Oracle Managed Files eases the DBA's ongoing problem of monitoring and controlling the growth of datafiles.
- User-managed file management offers more detailed control over datafiles than Oracle Managed File, but also requires more manual maintenance tasks.
- The Oracle Managed Files (OMF) method automates removal of dependent datafiles when a tablespace is dropped.
- OMF handles datafiles names, size, and creation, whereas you, as the DBA, handle only the location of the files.
- The parameter of DB_CREATE_FILE_DEST initialization sets the location of datafiles when using OMF.
- The DB_CREATE_ONLINE_LOG_DEST_n initialization parameters set the location of control files and redo log files when using OMF.
- OMF uses OFA (Optimal Flexible Architecture) as its file naming standard.
- Initialization parameters are divided into three major types: static, dynamic, and derived.
- Static parameters are either permanent (never changing) or adjustable by changing the value in the **init.ora** file and then restarting the database.
- Dynamic parameters can be modified for the entire system without shutting down the database and sometimes can be modified for the duration of your session without affecting the entire system.

- Derived parameters are set by Oracle9*i* based on the values of other parameters.
- The Database Configuration Assistant leads you through several steps to create a new database.
- You can choose from four types of database configurations in the Assistant, including Data Warehouse, Transaction Processing, New Database, and General Purpose.
- Dedicated Server mode works well for small workgroups, whereas Shared Server mode works better for larger groups that have many users accessing the database simultaneously.
- The Assistant provides an opportunity to customize memory size and initialization parameters, if you choose to do so.
- Whether you can or cannot adjust tablespace size and datafiles locations depends on the database type you choose.
- After creating the new database, use Net Manager to set up the Net Service name for the database.
- After the Net Manager and the console are able to reach the new database, you can use the Instance Manager to start up and shut down the database.
- To create a database manually, your first step is to set up a directory structure for the files that are to be created.
- Create a password file to implement password file authentication when the new database is created.
- A database service must be started if you are using Windows, but is not required if you are using UNIX or Solaris.
- The CREATE DATABASE command actually generates the datafiles, control files, and so on.
- After the database is created, it has the SYSTEM tablespace and (in the case of the sample used in this book) a tablespace for the undo logs.
- All manually created databases should also have a tablespace for temporary storage (usually called TEMP) and a tablespace for other schemas (usually called USERS).
- All manually created databases must set up the standard data dictionary views and PL/SQL processes by running the **catalog.sql** and the **catproc.sql** scripts provided by Oracle.
- Register the new manually created database in the Windows Registry (if you are running Windows).
- The SHUTDOWN command in SQL*Plus requires you to log on as SYSDBA.
- SHUTDOWN IMMEDIATE is faster than SHUTDOWN NORMAL.
- SHUTDOWN ABORT is used only when the database has errors and does not shut down with NORMAL or IMMEDIATE.

REVIEW QUESTIONS

1. All initialization parameters can be modified after database creation. True or False?
2. List the tools you can use to create a database.
3. ORACLE_HOME and ORACLE_BASE are used only in Windows operating systems. True or False?
4. The ________________ describes system-specific details such as memory requirements.
5. List these steps in chronological order from start to finish.
   a. Write and execute the CREATE DATABASE command.
   b. Start up the database.
   c. Set the initial parameters.
   d. Install Oracle9*i* RDBMS software.
   e. Shut down the database.
6. Which of the following steps occurs while using the Database Configuration Assistant?
   a. Setting MAXINSTANCES
   b. Selecting database type
   c. Logging onto SQL*Plus
   d. Saving old data
7. When using OS authentication, the OS user must be the same as the Oracle user name when the ________________ initialization parameter is null.
8. When in the console, you must log on with SYSDBA privileges to start the database. True or False?
9. Which of these are primary tasks in file management? Choose all that apply.
   a. Choosing the location of files
   b. Selecting the tablespace names
   c. Deleting old files and adding new ones
   d. Reviewing log files
   e. All of the above
10. Explain how user managed file management gives more control to the DBA than the Oracle Managed Files method.
11. The DB_CREATE_FILE_DEST defines the OFA standard for the database. True or False?
12. To add a remark to the **init.ora** file, begin the line with this symbol: ________________.

13. List the five types of initialization parameters.
14. You must include the CONTROL_FILE parameter in the **init.ora** file when creating a database manually. True or False?
15. The following utility creates a database service in Windows.
    a. ORAPWD
    b. DBSTART
    c. ORADIM
    d. SVRMGR
16. The NATIONAL CHARACTER SET parameter of the CREATE DATABASE command sets the character set used in the following data types. Choose all that apply.
    a. CHAR
    b. VARCHAR2
    c. NUMBER
    d. NCLOB
    e. All of the above
17. Data Dictionary views are automatically created by running the ________________ script, and PL/SQL procedures are created by running the ________________ script.
18. In Windows, you set the value of ORACLE_SID in the Windows Registry. True or False?

EXAM REVIEW QUESTIONS—DATABASE FUNDAMENTALS I (#1Z0-031)

1. Choose the phrase that best describes a DBA authentication method.
    a. Sets the maximum number of instances to run simultaneously
    b. Determines which users are allowed to log on
    c. Specifies how to validate users who log on with special privileges
    d. Enables remote access to the database
2. Which of the following initialization parameters affect DBA authentication? Choose all that apply.
    a. REMOTE_LOGIN_PASSWORDFILE
    b. COMPATIBILITY
    c. OS_AUTHENT_PREFIX
    d. ORAPWD
    e. All of the above

3. Which of the following database components are defined in the CREATE DATABASE statement? Choose three.
   a. **init.ora** file
   b. Undo tablespace
   c. Temporary tablespace
   d. SYSTEM tablespace
   e. Log groups
   f. SYSTEM password
4. When creating a database manually, which of the following statements are true regarding the control file? Choose three.
   a. The control file is multiplexed by default.
   b. The directory in which the control file is created is automatically created.
   c. The CONTROL FILE clause must be included in the CREATE DATABASE statement.
   d. The size of the control file is affected by MAXDATAFILES setting.
   e. There can be up to ten control files named.
5. Your database is primarily in English; however, you want to be able to store Chinese character data in some fields. What should you do in the CREATE DATABASE statement?
   a. Nothing: adjust the NLS CHARACTER SET in the init.ora file
   b. Nothing: the default setting for NLS CHARACTER SET handles this
   c. Set the NLS CHARACTER SET to Chinese
   d. Set the CHARACTER SET to Chinese
6. Examine the following statement.

```
1 CREATE DATABASE PROD02
2 MAXLOGFILES 5
3 MAXDATAFILES 100
4 SYSTEM TABLESPACE DATAFILE 'D:\ora92\sys01.dbf' SIZE 325M
5 UNDO TABLESPACE UNDOTBS
6 DATAFILE 'D:\ora92\undo01.dbf' SIZE 25M
7 CHARACTER SET WE8MSWIN1252
8 LOGFILE GROUP 1 ('D:\ora92\redo01.log') SIZE 50M,
9 GROUP 2 ('D:\ora92\redo02.log') SIZE 50M;
```

   Which line has an error?
   a. Line 7
   b. Line 8
   c. Line 4
   d. Line 5
   e. Line 6

7. Examine the following statement.

```
CREATE DATABASE PROD02
MAXLOGFILES 5;
```

   Which of the following statements are true? Choose two.

   a. The statement will fail.

   b. The SYSTEM tablespace will be 500 M in size.

   c. The statement will create Oracle-managed datafiles.

   d. The control file may or may not be multiplexed.

   e. The control file will be 100 M in size.

8. You are administering an Oracle9*i* database. Which of these tools should you use if you need to shut down the database using a command line?

   a. Oracle Database Configuration Assistant

   b. SQL*Plus

   c. Telnet

   d. Enterprise Manager

9. You are configuring the password file for the PROD01 database. Which of these commands creates a password file named orapwPROD01.ora located in the /usr/pwd directory and allows no more than fifteen additional DBAs to be assigned the SYSDBA role?

   a. orapwd file=/usr/pwd/orapwPROD01.ora password=trueblue2002 entries=16

   b. orapwd file=/usr/pwd/orapwPROD01.ora password=trueblue2002 entries=15

   c. orapwd pwdfile=/usr/pwd/orapwPROD01.ora entries=16

   d. orapwd file=/usr/pwd/orapwPROD01.ora maxpassword=16

10. You are preparing to create a second database instance. You copy the init.ora file and prepare to use it for the second database instance. Which initialization parameter must be changed so that it does not conflict with the first database instance?

    a. REMOTE_LOGIN_PASSWORDFILE

    b. DB_CREATE_FILE_DEST

    c. COMPATIBLE

    d. DB_NAME

HANDS-ON ASSIGNMENTS

1. You are about to create a new database for the Statistics Department of a large insurance firm. The database functions primarily as a storage area for data on past insurance clients and their claim history. Each month, any closed accounts are loaded into the database. There are 25 offices around the country that plan to load data and use

the database for gathering statistics. You want to use the Database Configuration Assistant. Using the figures and steps in the chapter to help remind you of the choices available, what selections would you make while running the Assistant? Explain your reasoning briefly. Save your work in a file named ho0301.txt in the Solutions\Chapter03 directory on your student disk.

2. You are the DBA at a factory. A new Oracle9*i* database will soon replace the current database. The old database had serious performance problems that seemed to be caused by an overload on the disk drive's I/O channel. Your analysis has also shown that two important and large tables are frequently accessed: the CAR_PART table and the ENGINE_PART table. You have a new computer with three large-capacity disk drives: F, G, and H. Describe how you plan to alleviate the I/O problem in the new Oracle9*i* database using the user-managed file management method. Save your work in a file named ho0302.txt in the Solutions\Chapter03 directory on your student disk.

3. Using the situation described in Hands-on Assignment 2, describe how you would solve the same problem using the Oracle Managed Files method. Save your work in a file named ho0303.txt in the Solutions\Chapter03 directory on your student disk.

4. Look at the following lines from an **init.ora** file, and list the errors you find; then rewrite the code with corrections, and place it in the **Solutions/Chapter03** directory on your student disk with the name **ho0304init.ora**.

```
#
 Cache and I/O
#
db_block_size=4096, db_domain="detroit.usa"
remote_login_passwordfile=EXCLUSIVE
control_files=("D:\newlogs\control99.ctl")
maxinstances=2
compatible=901
#
 Database Name
#
db_name=prod901.detroit.usa
instance_name=trial02
```

5. Using either Oracle documentation (the Oracle9i Database Reference book) or the Initialization Parameter window of the Instance Manager, complete Table 3-3. The first row is filled in as an example.

Table 3-3 Initialization Parameters

Name	Class	ALTER SYSTEM?	ALTER SESSION?	Default value
BLANK_TRIMMING	Static	No	No	False
DB_BLOCK_BUFFERS				
GLOBAL_NAMES				
OPEN_CURSORS				
FIXED_DATE				
JAVA_POOL_SIZE				

6. You created a database named trial02 in this chapter. Change the initialization parameter PGA_AGGREGATE_TARGET to **25M** using the Enterprise Manager console. (Recall from Chapter 2 that you can modify initialization parameters in the Instance Manager under the Configuration icon.) To test your results, log onto SQL*Plus and type the SQL*Plus command SHOW PARAMETERS before and after your change.
7. Again using the trial02 database, change the JOB_QUEUE_PROCESSES TO **12** using the ALTER SYSTEM command. Use SHOW PARAMETERS to test your results before and after the change. Save your SQL script in a file named **ho0307.sql** in your **Solutions/Chapter03** directory.
8. Add a new tablespace named INDX to the trial02 database. Make it an Oracle Managed file. Save the command you used to create the tablespace in the Solutions/Chapter03 directory in a file named **ho0308.sql**. What is the name of the datafiles that was created? Where is the datafiles located?
9. Look at the following CREATE DATABASE command, and find the errors. Then write a corrected version, and save it in the Chapter03 solutions directory in a file named **ho0309.sql**.

```
CREATE DATABASE ultradb FOR UPDATE OF paralleldb
MAXINSTANCES 25
DATAFILE TABLESPACE 'ORACLE_BASE\oradata\system01.dbf'
CHARACTER_SET US7ASCII
LOGFILE GROUP A ('D:\oracle\logs\redoA.log') SIZE 50M,
        GROUP B ('D:\oracle\logs\redoA.log') SIZE 50M;
```

10. Using the Database Configuration Assistant, compare the settings for the Data Warehouse model and the Transaction model. What is different and why? Save your work in a file named ho0310.txt in the Solutions\Chapter03 directory on your student disk.

CASE PROJECT

When you arrive at the Global Globe office, you find that your boss, the Senior DBA has left for Hawaii on vacation. You have one day to create a new database on the 50 programmer workstations in your office. You need help. You decide that the only way to get it all done is to create a script that each programmer can run on his or her own workstation. You know, however, that if the instructions for running the script get longer than a few short sentences, no one will do it correctly! Your job is to set up a foolproof database creation script. Earlier in this chapter, you created a **newadmin** directory, and you have the finished versions of an initialization parameter file (**inittrial02.ora**) and two scripts (**createdb.sql** and **tablespace.sql**) that you used to create the trial02 database. Adjust the parameters and the scripts, so that all the tablespaces, control files, and redo log files are Oracle Managed Files. Then create a batch file that runs both scripts. (*Hint*: to start SQL*Plus and run a file inside SQL*Plus, use this command format: SQLPLUS <user>/<password>@<dbname> @filename.sql.) Adjust the database name to trial03. The script must run without any user intervention. (You may, for the sake of this exercise, assume that the directories you need exist on every workstation, the programmer has administrator authority on his or her own workstation, and that the command to adjust the Windows Registry is not required, because you have not seen how to revise the Windows registry from the command line.) *Hint*: You will have to add this line to the **createdb.sql** script:

```
CONNECT SYS/CHANGE_ON_INSTALL AS SYSDBA
```

Save the scripts and batch file in the **Solutions/Chapter03** directory with these names: **case03createdb.sql**, **case03inittrial03.ora**, **case03tablespace.sql**, and **case03.bat**.

Clean up your workstation by using Oracle Database Configuration Assistant to delete the trial01, trial02, and trial03 databases. The remaining chapters work with the ORACLASS database instance that was installed before you started the course.

CHAPTER

4

DATA DICTIONARY VIEWS AND CONTROL FILES

In this chapter, you will:

- Use the data dictionary components and views
- List useful dynamic performance views
- Manage and multiplex control files
- Use OMF to manage control files
- Create new control files
- View control file data

Upon first inspection, it may seem overwhelming to absorb all the components found in the Oracle9*i* database. With each chapter, however, you learn and work with new components, building your understanding bit by bit. This chapter digs into several important areas of the database that you need to understand to manage the Oracle9*i* database. First, you examine the data dictionary views that display everything from top-level information (users and tablespaces, for example) to minute details (such as execution plans for queries and average column data size in a table). Next, you look closely into managing the control files, making adjustments, and using the latest Oracle9*i* features with your control files. Future chapters delve into the details of other important management areas, such as redo logs and diagnostic files.

LOOKING AT IMPORTANT DATA DICTIONARY COMPONENTS

Data dictionary views reside in the database like any other views. The views are based on tables owned by the SYS user that are updated automatically by the Oracle9*i* database. The SYS schema owns both the underlying tables and the data dictionary views. Some of the views are available for anyone to query, whereas others are reserved for DBAs only. A user needs the SELECT CATALOG privilege to be able to query the data dictionary views, and this role is part of the CONNECT role that is usually given to all users. See Chapter 12 for more information on roles. To take a quick look at a list of data dictionary views, follow these steps:

1. Start up the Enterprise Manager Console. In Windows, click **Start**, and then click **Programs/Oracle - OraHome92/Enterprise Manager Console**. In UNIX, type **oemapp console** on the command line. The Enterprise Manager Console logon screen appears.
2. Select the **Launch standalone** radio button and click **OK**. The console appears.
3. Start up the SQL*Plus Worksheet by clicking **Tools/Database Applications/SQL*Plus Worksheet** from the top menu in the console. A background SQL*Plus process starts, and then the SQL*Plus Worksheet window appears. If the background process appears in front, simply minimize it (click the minus sign in the top-right corner) so that you can see the worksheet.
4. Connect as the SYSTEM user. Click **File/Change Database Connection** on the menu. A logon window appears. Type **SYSTEM** in the Username box, the current password in the Password box, and **ORACLASS** in the Service box. Leave the connection type as "Normal." Click **OK** to continue.
5. Type the following commands in the top pane and click the **Execute** icon to run the query. The Execute icon looks like a lightning bolt and is on the left side of the window:

```
SET LINESIZE 100
SET PAGESIZE 60
COLUMN COMMENTS FORMAT A40 WORD_WRAP
SELECT * FROM DICTIONARY
ORDER BY TABLE_NAME;
```

Figure 4-1 shows part of the results of the query.

6. Remain logged on for the next exercise.

As you can see, there are over one thousand views to learn about. Who can possibly remember all those names? No one! That's why data dictionary view posters are so popular as a quick visual reference. Go to any Oracle convention and you undoubtedly see people handing out posters with lists of hundreds of data dictionary view names, categorized by various headings like "DBA," "Schema," "Performance," "Dynamic," and so on. The data dictionary views are so important to an Oracle DBA that you find these posters plastered in office after office.

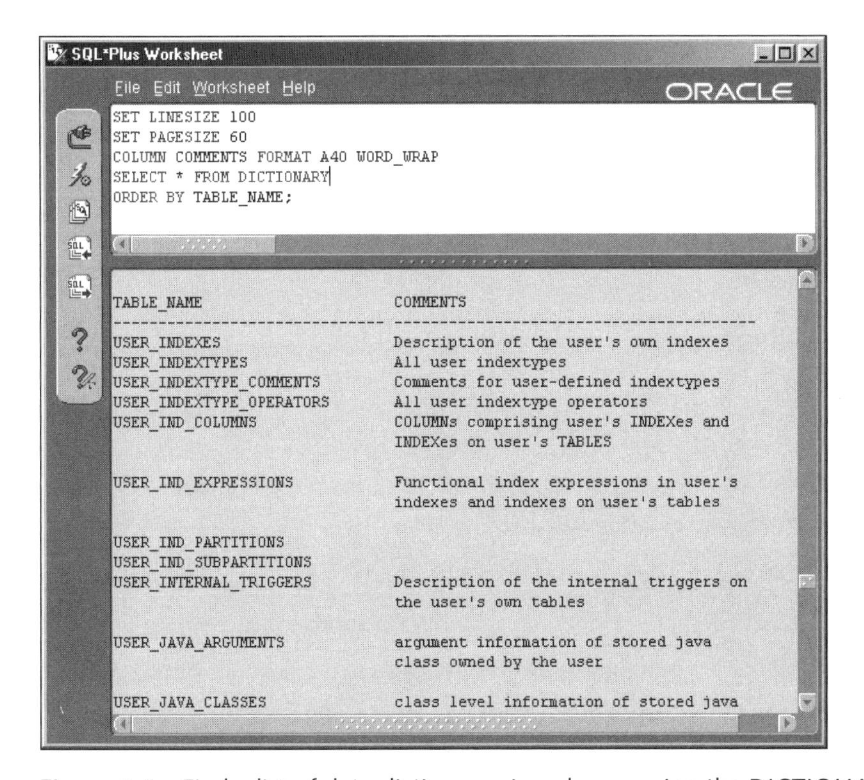

Figure 4-1 Find a list of data dictionary views by querying the DICTIONARY view

In the previous chapter, you saw that **data dictionary views** provide system-related information by querying the database's internal management tables and presenting the data as views. Like any ordinary view, data dictionary views can be queried. Figure 4-2 shows the results of querying the ALL_TAB_COLUMNS view, which contains the names of all columns in all the tables that the current user owns or is able to view.

Data dictionary views cannot be updated. The internal database system updates the underlying tables. The only exception to this is the **AUD$** file, which can be directly updated by the DBA. See Chapter 12 for information on auditing and the **AUD$** file.

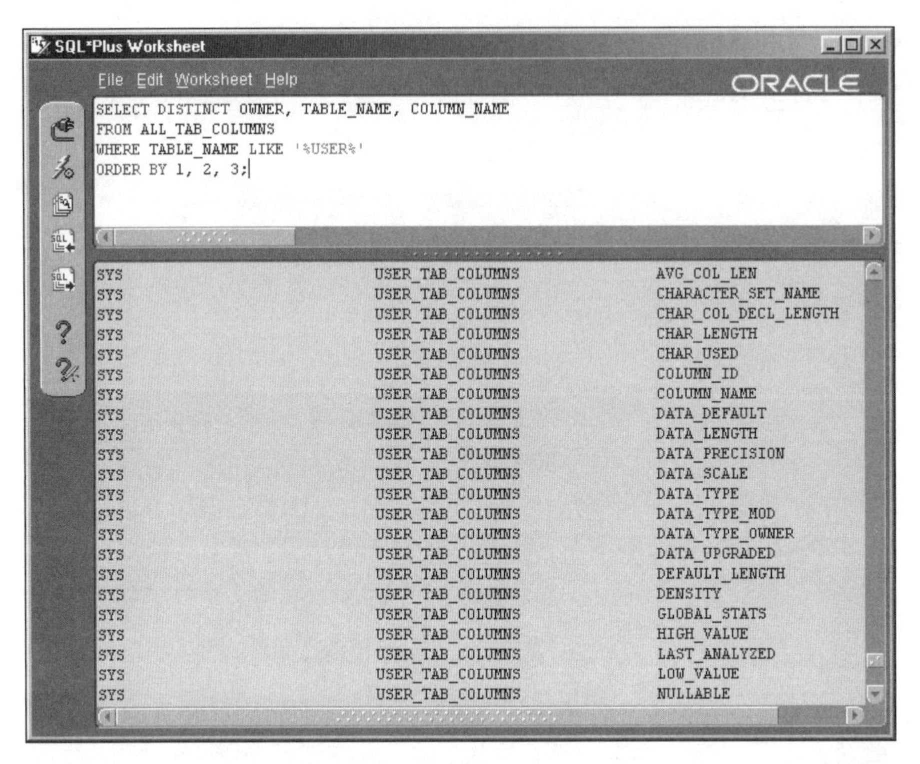

Figure 4-2 Data dictionary views contain details about a wide variety of objects

Oracle's online documentation has a list of data dictionary views. To access the list, go to Oracle Technology Network at *otn.oracle.com*. Access the **Documentation** section, choose **Oracle9*i* Database**, and then select **Oracle9*i* Database Online Documentation Release 2**. Then choose **Catalog views and Data Dictionary views** in the list of options. The alphabetical list shows each view, with a link to its definition and a link to examples found in the documentation.

Different types of data dictionary views are defined by their prefixes. Table 4-1 shows the prefixes, descriptions, and examples.

Table 4-1 Data dictionary views

Type	Prefix	Description	Examples
Static data dictionary views	USER	Views focused on a user's own objects	USER_TABLES USER_VIEWS
	ALL	Views about objects a user either owns or can query	ALL_TABLES ALL_TAB_COLUMNS
	DBA	Views available to DBAs only, showing information about all objects in the database (more detailed than the USER_ and ALL_ versions of the views)	DBA_TABLES DBA_USERS DBA_SYNONYMS

Table 4-1 Data dictionary views (continued)

Type	Prefix	Description	Examples
Dynamic performance views	V$	Views displaying current activities in the database	V$LOCK V$ACCESS
	GV$	Views that combine activities across multiple instances (for clustered databases using Oracle Real Application Clusters)	GV$LOCK GV$ARCHIVE

Generally, views with prefixes of USER, ALL, and DBA are in sets, such as USER_TABLES, ALL_TABLES, and DBA_TABLES. Each view has nearly identical columns. The only difference is that the USER version omits the OWNER column, because by definition the OWNER is the current user. In addition, the USER version sometimes omits columns to simplify the view. Similarly, V$ and GV$ views are in sets so that every V$ view has its corresponding GV$ view with an additional column (INST_ID) for the instance number. There are also a few stray data dictionary views that do not begin with these prefixes. Most of them are carried over from prior versions of the database and should not be used.

For easier access, all the views except those starting with the DBA prefix have public synonyms and public permission to query defined. A **public synonym** is a unique name for an object that allows any user to use the object without prefixing it with the owner name. The views starting with DBA require the owner, SYS, as a prefix. For example, to query the DBA_TAB_COLUMNS table, log on as SYSTEM (a DBA user) and type:

```
SELECT TABLE_NAME, COLUMN_NAME
FROM SYS.DBA_TAB_COLUMNS
WHERE ROWNUM < 200;
```

A second restriction on the views beginning with DBA is that you must have DBA privileges in the database to query the views. The users SYS and SYSTEM are the default DBA users; however, you can create more users as needed by logging on as SYS or SYSTEM and assigning the DBA role to any user.

Oracle9*i* is the final release in which the SYS and SYSTEM users exist as default DBA users. Subsequent database releases will prompt you so that you can assign the DBA user names and passwords during creation of the database.

USING DATA DICTIONARY VIEWS

Data dictionary views give you a quick look into the database and all that it contains. In fact, the Enterprise Manager Console itself uses these views to display information on its screen. Follow along with these steps to discover how the console uses data dictionary views.

1. Navigate to the **Enterprise Manager Console** window you started up earlier in the chapter.
2. Double-click the **Databases** folder. A list of databases appears below the folder.
3. Double-click **ORACLASS**. The tools appear below the ORACLASS icon.
4. Double-click the **Schema** icon. A list of schemas appears.
5. Double-click the **SYSTEM** schema name. A list of schema object types appears.
6. Double-click the **Tables** folder. A list of tables appears. Figure 4-3 shows the console at this point.

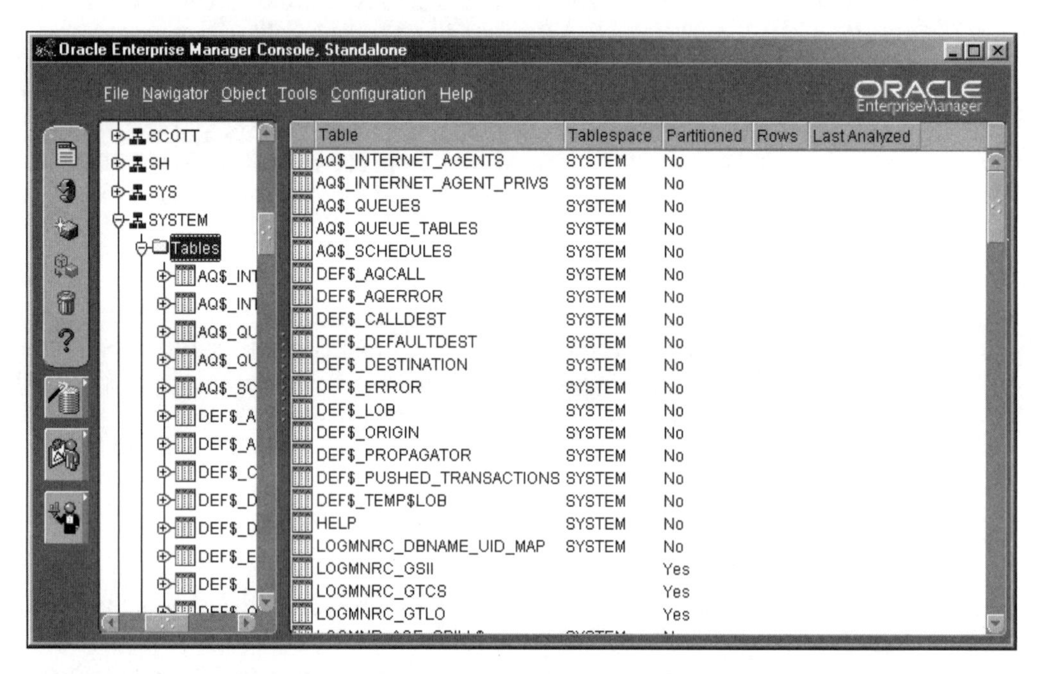

Figure 4-3 Navigating in the console is similar to navigating in Windows Explorer

7. Double-click the **HELP** table in the right pane. The property sheet for the table appears. Figure 4-4 shows the Properties window.

   Several data dictionary views were queried to build the Properties window, including DBA_TAB_COLUMNS, DBA_COL_COMMENTS, and DBA_REFS. It is sometimes useful to see the actual queries and other commands executed by the console, especially if you plan to write your own SQL commands to manually perform steps you have used with the console.

8. Close the property sheet by clicking the **Cancel** button in the bottom-center of the box.

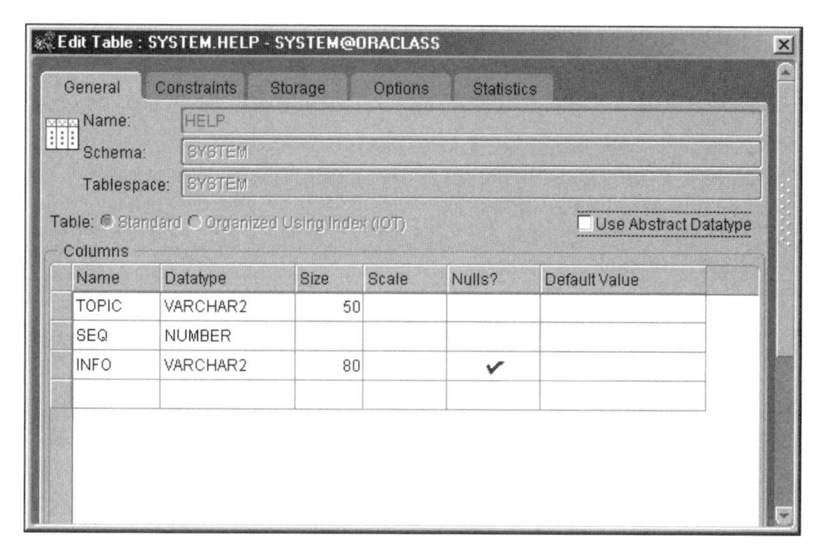

Figure 4-4 Property windows like this one are based on queries of the data dictionary views

9. Click **Navigator/Application SQL History** in the Console menu to view recent SQL commands executed in the console.
10. Scroll to the bottom of the window, and you see the queries that built the Properties window for the HELP table. Figure 4-5 shows the window that appears with SQL commands listed.
11. Close the Application SQL History window by clicking the **Close** button.
12. Remain logged on for the next practice.

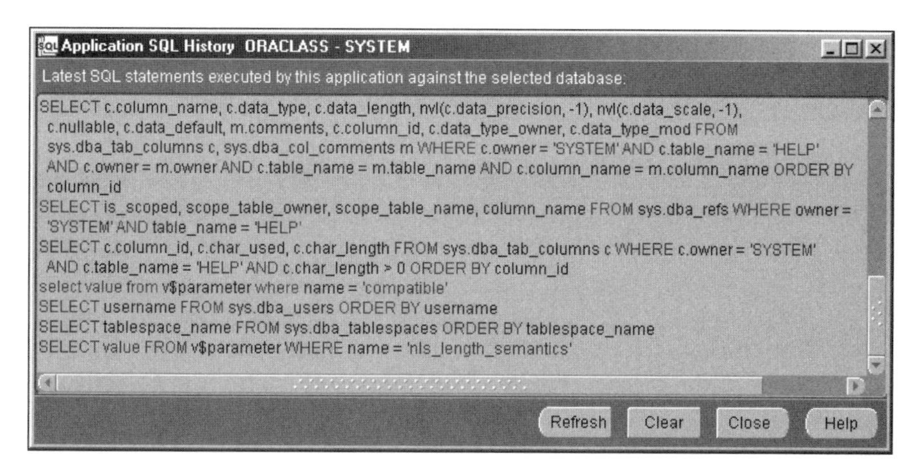

Figure 4-5 Scroll to the bottom of the window for the most recent commands

Use data dictionary views to help you build queries, debug SQL problems, and review naming conventions. In the following steps, you will see how to use data dictionary views. The data dictionary views can be queried like any other view in Oracle9*i*.

Use the SQL*Plus DESCRIBE command (abbreviated to DESC) to retrieve a list of column names and data types for any table or view, including the data dictionary views. For example, use the following steps to try out the DESCRIBE command.

1. Return to your SQL*Plus Worksheet window.
2. Run the DESCRIBE command for the DBA_COL_COMMENTS data dictionary view by typing this command and clicking the **Execute** icon:

```
SET LINESIZE 100
DESC DBA_COL_COMMENTS
```

In the preceding code, the first line adjusts the width of the display to 100 characters. The second line abbreviates the DESCRIBE command to DESC and then names the view to be described. Figure 4-6 shows the results.

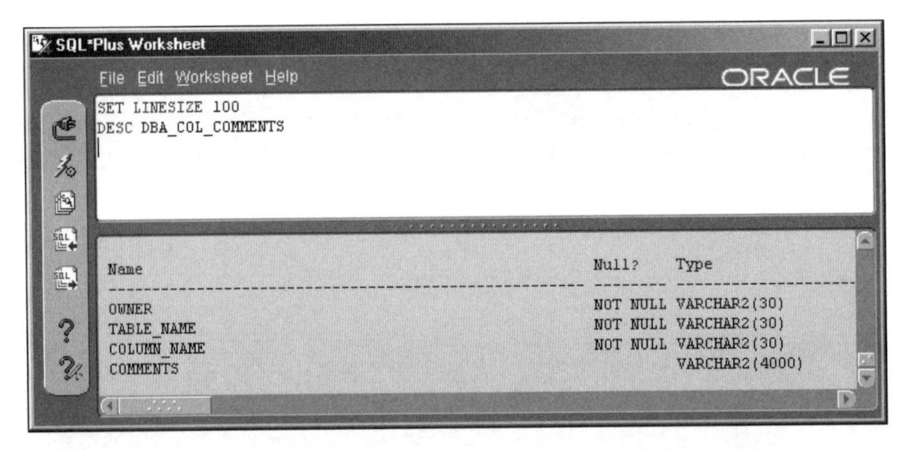

Figure 4-6 Refresh your memory of column names with DESCRIBE

3. The DESCRIBE command works on many other objects as well, such as tables, object tables, and object types. Imagine that you are writing a query that joins four tables and you have misspelled one of the table names. You are logged onto SQL*Plus Worksheet as the SYSTEM user. You know that the table's schema is QS. Type the following query to retrieve a list of all the tables in the schema. Execute the query by clicking the **Execute** icon.

```
SELECT TABLE_NAME FROM DBA_TABLES
WHERE OWNER = 'QS'
ORDER BY TABLE_NAME;
```

Figure 4-7 shows the results. Now you can return to the query you were writing and use the correct table names.

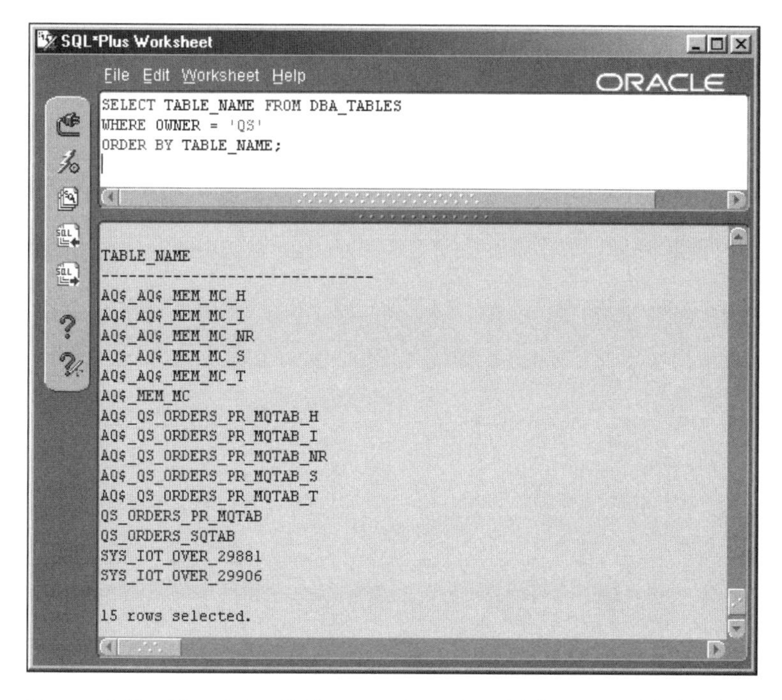

Figure 4-7 Use data dictionary views to obtain current table information

4

4. As a second example, suppose you just made changes to the production database that involved adding and removing columns in existing tables. You know that some views were affected. Type the following query to find a list of all views that are now invalid in the production database. Execute the query by clicking the **Execute** icon. If there are any invalid views, they are displayed.

```
SELECT OWNER, OBJECT_NAME, CREATED
FROM SYS.DBA_OBJECTS
WHERE OBJECT_TYPE = 'VIEW'
AND STATUS = 'INVALID'
ORDER BY OWNER, OBJECT_NAME;
```

5. In a final example, assume that your manager needs a report on the naming standards for database columns, and he wants you to include statistics on how well all the departments that develop database systems are complying with the naming standards. Your naming standards require that columns contain certain suffixes, such as _DATE or _NUM according to their data type. To develop some statistics, you write a query on the data dictionary views that lists all the suffixes found for each data type across all schemas in the database and how many times each suffix is used in a column name. The query is prepared

and you are now ready to run it. Click **File/Open** on the SQL*Plus Worksheet menu. A Navigator window appears where you can select a local file to load into the worksheet.

6. Navigate to the **Data\Chapter04** directory and select the **datatypes.sql** file. Click **Open** to load it into the worksheet.
7. Run the query by clicking the **Execute** icon. The query uses the COLUMN_NAME column of the USER_TAB_COLUMNS and finds suffixes that start with an underscore. Figure 4-8 shows the results.
8. Remain logged on for the next exercise.

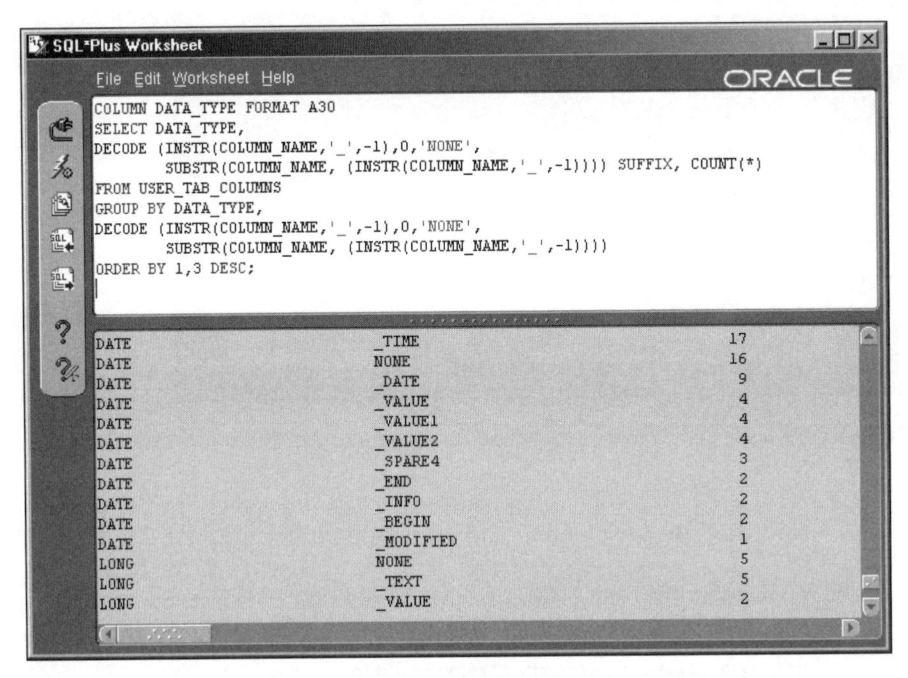

Figure 4-8 Columns of the DATE data type have eleven different suffixes

There are data dictionary views for information about users, tables, objects, indexes, roles, and so on. To help you get started in finding what information is available through the data dictionary views, here is a list of some frequently used views:

- **USER_TABLES**: Tables you own
- **USER_VIEWS**: View name and the query that created the view
- **ALL_DEPENDENCIES**: Dependencies between objects, such as indexes and foreign key constraints
- **USER_ERRORS**: Errors found in views, procedures, and other objects you own

- **USER_INDEXES**: Indexes you own
- **USER_IND_COLUMNS**: Columns in indexes you own
- **DBA_SOURCE**: Source code for all stored objects, such as functions, Java source, and triggers
- **USER_TAB_PRIVS**: Table privileges that were granted to you; for example, you were granted SELECT on the CUSTOMER table by the CUSTACCT schema
- **ALL_TAB_PRIVS_MADE**: All grants given out by you for any table, plus grants by another user for tables you own
- **USER_TAB_PRIVS_MADE**: All grants you issued for tables you own
- **DBA_USERS**: All users in the database and information about them
- **PRODUCT_COMPONENT_VERSION**: Current version numbers and names of all installed components

All these data dictionary views are static in nature, because they show information that has been registered in the database and does not change without specific action by a user. For example, a user might execute an ALTER TABLE command to add a new column to a table. The related data dictionary views reflect that change. The next section discusses dynamic performance views, which change each time you query them, because they are based on in-the-moment activities.

Useful Dynamic Performance Views

The dynamic performance views all begin with the prefix V$ and have a counterpart view of the same definition beginning with the prefix GV$. As you remember from Table 4-1, V$ views are of current activities in the database, and GV$ views combine activities across multiple instances. V$ and GV$ views query activity-oriented system tables and display information about the state of your database. These views answer questions like these: What session is using the most CPU time now? How many users are currently logged on? How much memory is being used, and by which transactions? In addition, some of the views contain historical data that is collected periodically, so you can assess trends in activities.

Documentation of the dynamic performance views is found online in the same location as the data dictionary views. Figure 4-9 shows part of the contents page listing the V$ views.

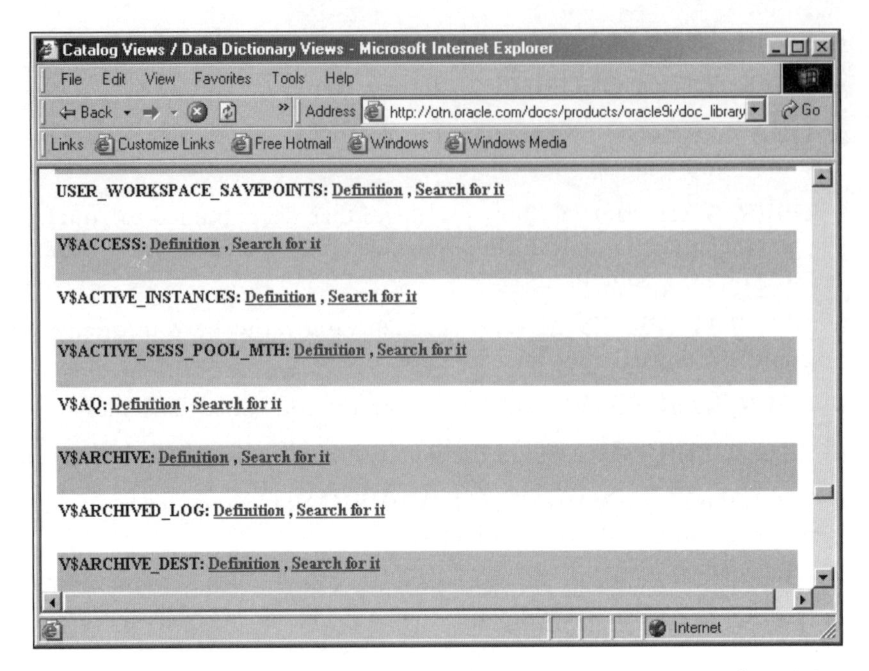

Figure 4-9 The documentation for V$ views applies to both V$ and the GV$ counterpart views

You will find that the primary use for the dynamic performance views is tuning the database system. Here are some of the key views available for gathering information to help tune the database system:

- **V$SYSSTAT**: Displays the volume of data that is changed and the number of transactions executed, among other important statistics
- **V$SQL**: Shows you the actual SQL commands that are executed, how many times the commands are reused or shared, and how much CPU each command consumed
- **V$SESSTAT**: Shows the resource consumption by session to help identify long-running or memory-consuming sessions
- **V$SESSION_WAIT**: Records information on sessions that were required to wait for resources (a common cause of slow response time)
- **V$FILESTAT**: Shows the number of reads, writes, and the time used to perform I/O activities on files
- **V$DATAFILE**: Lists the FILE# and FILE_NAME of all control, log, and datafiles

Oracle provides three options for viewing and gathering statistics that you can use while tuning the database. Although this book does not cover the details of tuning a database, the options are listed here for your information:

1. **BSTAT/ESTAT scripts**: You can find these two scripts in the **ORACLE_HOME\rdbms\admin** directory in files **UTLBSTAT.SQL** and **UTLESTAT.SQL**. They produce a simple report of database activity between two points in time. Oracle states that these scripts produce the basic, minimum statistics needed for monitoring a database.
2. **STATSPACK**: STATSPACK takes the basic statistics from the BSTAT/ESTAT scripts and adds to them the capability to save the statistics over time. The STATSPACK report provides much more information than BSTAT/ESTAT and is ideal for detecting performance bottlenecks.
3. **The Diagnostic Pack**: This optional addition to the Enterprise Manager uses the dynamic performance views to generate sets of statistics, histograms, and tuning advice. The Diagnostic Pack's set of tools, like STATSPACK, saves data in a repository set of tables, so that trends can be analyzed over time. The Performance Monitor (part of the Diagnostics Pack) displays a set of graphs and statistics in an easy-to-read panel and provides detailed drill-down capabilities.

To read more about tuning the database, go to Oracle Technology Network online documentation site for Oracle9*i* database documentation, select List of books, and then select the *Oracle9i Database Performance Tuning Guide and Reference*.

The data dictionary views and dynamic performance views provide a wealth of information about the contents and use of the database. The next section examines one of the critical management components of the database: the control file.

INTRODUCTION TO THE CONTROL FILE

Recall in Chapter 1 that the control file was defined as an administrative file containing up-to-date information on the database structure, log files, and checkpoints. Furthermore, Chapter 1 asserted that the control file is critical for opening the database. Chapter 3 showed you how to use the CONTROL_FILE initialization parameter to create one or more control files when you created a new database. This chapter shows you how to manage, multiplex, add, rename, relocate, and replace control files.

The control file contains this information about the database:

- **The database name**: Defined either in the CREATE DATABASE command or the DB_NAME initialization parameter
- **Names and locations of associated datafiles and online redo log files**: Defined by the CREATEDATABASE command and then updated whenever a new tablespace or redo log file is created

4

- **The timestamp of the database creation**: Logged when the database is created
- **The current log sequence number**: Updated whenever the current log group changes (see Chapter 5 for details)
- **Checkpoint information**: Time-stamped records identifying the most recent changes made to the database, which are matched inside the database structure itself (these are critical for recovery; see Chapter 5 for more information)

The control file is like your credit card: If your credit card gets damaged, you still have your account information, but you must replace the credit card before you can access that information. Likewise, if the control file is damaged, the database may be intact, but you cannot open the database until the control file is restored, recovered, or re-created.

Because the control file is so critical, Oracle recommends keeping redundant (duplicate) copies online at all times, in other words multiplexing the control file. **Multiplexing** means making redundant copies of files to guarantee that if the original file is damaged, another can take its place immediately. The next section shows how to manage control files, including maintaining multiple copies.

Managing and Multiplexing the Control Files

Now that you have created the database and it is up and running, what else needs to be done with the control files? The first thing you should do is multiplex the control file to ensure against total loss of the file. You can create up to eight copies of the control file. After that, the control file is self-managing unless you make any of these types of changes to the database:

- Add a new control file
- Rename or relocate one or more control files
- Replace a damaged control file

All three of these tasks have very similar procedures. Steps for each task are described in the following sections.

It is always a good idea to make a backup of your control files before you begin changing them.

Adding a New Control File

A new control file can be added at any time. For example, you may have created the database with two control files, and then later that day, after reading up on the importance of the control file, decide you really need a third copy on a third device that is more protected from power surges. Or, you may have just arrived on a new job and discovered that the database only has one control file. Your first task for the new assignment is to multiplex the control file. Follow these steps to add a new control file:

1. Shut down the ORACLASS database. You may use either the Console or SQL*Plus, as described in Chapter 3.
2. Open the Windows Explorer and locate the control file named **CONTROL1.CTL**. It should be in the **ORACLE_BASE\oradata\ORACLASS** directory.
3. Copy the control file and place the copy in a directory on your hard drive. Ask your instructor which directory to use. Use the Edit menu in the Windows Explorer. For UNIX, use the standard operating system copy command (**cp**) to make a copy of the file in a new location.
4. Rename the copy **CONTROL04.CTL**. Use the File menu in the Windows Explorer to complete this task. In UNIX, use the standard operating system command (**mv**).
5. Open the **init.ora** file for the ORACLASS database with your Notepad editor. The file is located in **ORACLE_BASE\admin\ORACLASS\pfile** directory.
6. Change the CONTROL_FILES parameter in the database's initialization parameter file by adding the new control file's name to the list of control files. There should be commas separating each file name. The lines should look similar to the following (your actual path will differ):

```
CONTROL_FILES=("D:\oracle\oradata\ORACLASS\CONTROL01.CTL",
"D:\oracle\oradata\ORACLASS\CONTROL02.CTL",
 "D:\oracle\oradata\ORACLASS\CONTROL03.CTL",
"C:\Studentfiles\henry\CONTROL04.CTL")
```

7. Save the file and close Notepad.
8. Restart the database. Use the Console or SQL*Plus as described in Chapter 3.

The database automatically begins multiplexing all the control files listed in the CONTROL_FILES parameter. It verifies that all the files are identical upon startup, which is why you made a copy rather than starting with a blank file.

If any of the control files are mismatched, the database does not open.

Renaming or Relocating an Existing Control File

There are a number of practical reasons for renaming or relocating a control file. For example, you may want to change the names of control files to comply with new naming standards your department has implemented. Or you may want to relocate a control file, because the database system has grown in size and now has a larger hard drive that is a safer location for one of the control files.

To relocate an existing control file, follow these steps:

1. Shut down the ORACLASS database. You may use either the Console or SQL*Plus, as described in Chapter 3.
2. Open the Windows Explorer and locate the control file you just created, which is named **CONTROL04.CTL**.
3. Use the **File** menu in the Windows Explorer to rename the file **CONTROLX.CTL**. For UNIX, use the standard operating system move command (**rm**) to rename the file.
4. Open the **init.ora** file for the ORACLASS database with your Notepad editor. The file is located in the **ORACLE_BASE\admin\oraclass\pfile** directory.
5. Change the CONTROL_FILES parameter in the database's initialization parameter file by modifying the name of the control file to match its current name.
6. Save the file and close Notepad.
7. Restart the database. Use the Console or SQL*Plus as described in Chapter 3.

To relocate a control file, follow all the previous steps with these exceptions: in Step 3, move the file instead of renaming the file; and in Step 5, change the file location instead of the file name in the CONTROL_FILES parameter.

Replacing a Damaged Control File

The loss of one control file can be quickly fixed without any additional recovery steps as long as you have a second copy. The control file may have been corrupted by a power interruption that unexpectedly shut down the computer. The file also may have been accidentally deleted or damaged when a hard drive failed. Whatever the reason, the database does not start until the control file is restored.

If a control file is missing or damaged, Oracle9*i* issues this error message when starting up the database:

```
ORA-00205: error in identifying control file, check alert
log for more details
```

In the case of a control file error, you must first shut down the database, fix the control file, and then restart the database.

You can fix a control file by replacing the faulty or missing control file with a copy of a valid control file. In this next set of steps, you simulate a missing control file by deliberately deleting one of the control files. Then, you fix the problem by replacing the deleted file with a copy of one of the other control files.

1. Shut down the ORACLASS database. You may use either the Console or SQL*Plus, as described in Chapter 3.
2. Open the Windows Explorer and locate the control file named **CONTROLX.CTL**. This is the control file you renamed in the previous section. In UNIX, use the File Manager or the **cd** command to navigate to the directory.
3. Delete the control file. ***This step is done only to simulate a missing control file.*** Use the File menu in the Windows Explorer to remove the file rather than renaming the file (for UNIX, use the system remove (**rm**) command to delete the file).
4. Restart the database. Use the Console or SQL*Plus as described in Chapter 3. The database startup fails with the ORA-00205 error message.
5. Shut down the ORACLASS database again.
6. Open the Windows Explorer and locate the control file named **control01.ctl**. It should be in the **ORACLE_BASE\oradata\ORACLASS** directory. In UNIX, use the File Manager or the **cd** command to navigate to the directory.
7. Copy the **control01.ctl** file and paste it into the directory where you originally placed the **CONTROLX.CTL** file, then rename the copy to **CONTROLX.CTL** to restore the file. In UNIX, use the **cp** command and name the new file **CONTROLX.CTL**. ***These instructions show you how to fix a missing or damaged control file. You replace it with a copy of a valid control file.***
8. Restart the database. Use the Console or SQL*Plus as described in Chapter 3. The database starts up normally.

To remove a control file, follow these steps. When you complete these steps, your database is back to its original configuration of control files.

1. Shut down the ORACLASS database. You may use either the Console or SQL*Plus, as described in Chapter 3.
2. Open the Windows Explorer and locate the control file named **CONTROLX.CTL**. This is the control file you restored in the previous section. In UNIX, use the File Manager or the **cd** command to navigate to the directory.
3. Delete the control file. Use the File menu in the Windows Explorer to remove the file rather than renaming the file (for UNIX use (**mv**) command to relocate rather than rename the file.
4. Open the **init.ora** file for the ORACLASS database with your Notepad editor. The file is located in the **ORACLE_BASE\admin\ORACLASS\pfile** directory.

5. Change the CONTROL_FILES parameter in the database's initialization parameter file by removing the CONTROLX.CTL file from the list of control files.
6. Save the file and close Notepad.
7. Restart the database. Use the Console or SQL*Plus as described in Chapter 3. The database starts up normally.

So far, the examples you have been working with involve control files that are managed the traditional way. With Oracle9*i*'s new Oracle Managed File (OMF) option, how do you accomplish the same tasks? The next section describes the main differences between the traditional and the OMF methods.

Using OMF to Manage Control Files

Recall from Chapter 3 that OMF handles the names and the locations of files created in association with the database. Here are the criteria for OMF control files:

- The initialization parameter DB_CREATE_FILE_DEST must be specified.
- The initialization parameter DB_CREATE_ONLINE_LOG_DEST_n can be specified (optional).
- The initialization parameter CONTROL_FILES must be null.

You must set these parameters prior to issuing the CREATE DATABASE command, or during the process of creating the database with the Database Configuration Assistant. If you specify a value for DB_CREATE_ONLINE_LOG_DEST_*n* (where the *n* is actually a "1" for the first location, a "2" for the second location, and so on), a control file is created in each directory. Otherwise, the control file is created in the **DB_CREATE_FILE_DEST** directory. Table 4-2 illustrates some examples of parameters and how the control files are created when the CREATE DATABASE command is executed.

Table 4-2 Parameters for creating OMF control files

Initialization parameters	Number of control files	Control file locations
DB_CREATE_FILE_DEST="**D:\oracle\oradata**" CONTROL_FILES not specified (null)	1	**D:\oracle\oradata**
DB_CREATE_FILE_DEST="**D:\oracle\oradata**" DB_CREATE_ONLINE_LOG_DEST_1="**C:\stable\ora**" DB_CREATE_ONLINE_LOG_DEST_2="**D:\secondary\ora**" CONTROL_FILES not specified (null)	2	**C:\stable\ora** **D:\secondary\ora**

Whichever way you choose to create the control files, there are times when you must re-create the control file from scratch.

Creating a New Control File

As you have seen, the control file is critical to the health of the database. Occasionally, the remotely possible happens: Despite your best efforts to safeguard the files, all the control files are damaged or lost. If you have multiplexed the control files across several devices, this rarely happens. Nevertheless, what do you do in this case? Your database cannot function until the control file or files are restored.

Two other occurrences warrant creating new control files:

- **Changing the value of MAXDATAFILES, MAXLOGFILES**, or **MAXLOGMEMBERS**: Recall from Chapter 3 that MAXDATAFILES, MAXLOGFILES, and MAXLOGMEMBERS were set when the database was created. These clauses affect the size of the control file and can be changed only by re-creating the control file and specifying the changed values of the clauses to be modified. If you omitted one or more of these clauses when creating the database, the values were set to a default that depends on the operating system. If your system reaches the current maximum value, the only way to change the value is through re-creating the control files with new settings.
- **Change the name of the database**: Make sure that you modify the DB_NAME initialization parameter to match the name you specify in the CREATE_CONTROLFILE statement while the database is shut down.

If your database was damaged, you may have to go through other steps to recover the database. Database recovery is not covered in this book. Oracle's online documentation contains details on how to recover a damaged database.

Assuming that the database is undamaged, but the control files are all either missing or damaged, these are the general steps for creating a new set of multiplexed control files. Do not actually perform these steps now. They are listed for information only.

1. Gather a list of all datafiles, including their full paths. There is at least one datafile for every tablespace. You can find this by looking in the directory where you store your datafiles, such as **ORACLE_HOME/oradata**. Include the names of Oracle Managed Files as well as traditionally managed files.
2. Gather a list of all redo log files, including their full paths and group number. These files may be in the same directory as the datafiles, or in a separate directory if you specified that during database creation.
3. Build the CREATE CONTROLFILE command and save it in a plain text file. The syntax of the command looks like this:

```
CREATE CONTROLFILE REUSE SET DATABASE dbname
LOGFILE
```

```
        GROUP n 'X/xxx/logfilename' SIZE mmm,
        GROUP 2 'X/xxx/logfilename' SIZE <mmm>
NORESETLOGS
DATAFILE
    '<X/xxx>/<filename>', ....
MAXLOGFILES <nn>
MAXLOGMEMBERS <nn>
MAXLOGHISTORY <nn>
MAXDATAFILES <nn>
MAXINSTANCES <nn>
ARCHIVELOG
CHARSET <charsetvalue>;
```

4. Start the database in NOMOUNT mode.
5. Run your CREATE CONTROLFILE command.
6. Start up the database again.

Here are some points to help you understand the syntax:

- The REUSE clause is optional. It tells Oracle9*i* to overwrite any existing control files it finds. If Oracle9*i* finds existing control files and there is no REUSE clause, Oracle9*i* returns an error.
- All the MAX phrases are optional and should be used to change the original or default setting of any MAX phrase. These phrases specify the maximum number of log files, log members, and so on. The MAXLOGHISTORY phrase applies only to Oracle9*i* Real Application Clusters.
- You can specify either <dbname> or SET DATABASE <dbname>. Using DATABASE simply identifies the database to connect with the control files. Using SET DATABASE renames the database.
- List all the redo log groups and files that are members of the groups in the LOGFILES phrase.
- The NORESETLOG tells Oracle9*i* to read the log files and save their settings from the last time they were used. An alternative is to specify RESETLOG; however, this is usually used only when the log files must be recovered.
- List all the datafiles in the DATAFILE phrase.
- ARCHIVELOG specifies that the redo log files should be archived before reusing, making recovery easier. The alternative, NOARCHIVELOG, is the default setting.
- Do not specify CHARSET unless your database was originally created in a character set other than the default, US7ASCII (American English).

As an example, here is a CREATE CONTROLFILE command that renames the database to TEST2004 and changes the maximum number of datafiles to 500:

```
CREATE CONTROLFILE REUSE SET DATABASE TEST2004
LOGFILE
     GROUP 1 'D:\oracle\newlogfiles\redo01.log',
     GROUP 2 'G:\shared\oracle\newlogfiles\redo02.log'
NORESETLOGS
DATAFILE
    'D:\oracle\oradata\system01.dbf',
    'D:\oracle\oradata\users01.dbf',
    'D:\oracle\oradata\users02.dbf',
    'G:\shared\oracle\oradata\accounting01.dbf'
MAXDATAFILES 500;
```


The following command can save you much time in writing the CREATE CONTROLFILE command for your database. The trick is that you must use it while the database is open. In other words, if you wait until you have lost your control files, this command cannot help you. Follow these steps to try this command now:

1. Start up SQL*Plus by opening a command line and typing this command, replacing <password> with the current password for SYSTEM *provided by your instructor.*

   ```
   sqlplus system/<password>@<ORACLASS>
   ```

2. Type the following command and press **Enter** to execute it.

   ```
   ALTER DATABASE BACKUP CONTROLFILE TO TRACE;
   ```

SQL*Plus replies, "Database altered." The command writes the CREATE CONTROLFILE command for you in a trace file. Trace files are located in the directory named in the USER_DUMP_DEST initialization parameter. Figure 4-10 shows part of the contents of the trace file. You can use this file, modify any parts that need changing, and save it as your script for restoring the control files.

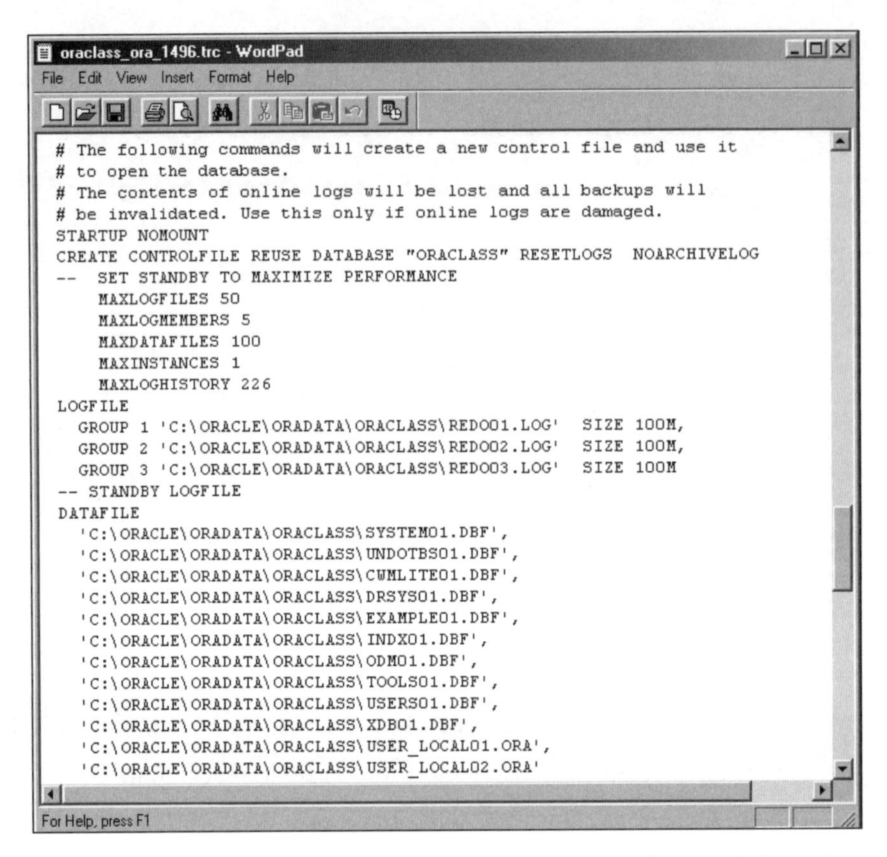
oraclass_ora_1496.trc - WordPad

```
# The following commands will create a new control file and use it
# to open the database.
# The contents of online logs will be lost and all backups will
# be invalidated. Use this only if online logs are damaged.
STARTUP NOMOUNT
CREATE CONTROLFILE REUSE DATABASE "ORACLASS" RESETLOGS  NOARCHIVELOG
--  SET STANDBY TO MAXIMIZE PERFORMANCE
    MAXLOGFILES 50
    MAXLOGMEMBERS 5
    MAXDATAFILES 100
    MAXINSTANCES 1
    MAXLOGHISTORY 226
LOGFILE
  GROUP 1 'C:\ORACLE\ORADATA\ORACLASS\REDO01.LOG'  SIZE 100M,
  GROUP 2 'C:\ORACLE\ORADATA\ORACLASS\REDO02.LOG'  SIZE 100M,
  GROUP 3 'C:\ORACLE\ORADATA\ORACLASS\REDO03.LOG'  SIZE 100M
-- STANDBY LOGFILE
DATAFILE
  'C:\ORACLE\ORADATA\ORACLASS\SYSTEM01.DBF',
  'C:\ORACLE\ORADATA\ORACLASS\UNDOTBS01.DBF',
  'C:\ORACLE\ORADATA\ORACLASS\CWMLITE01.DBF',
  'C:\ORACLE\ORADATA\ORACLASS\DRSYS01.DBF',
  'C:\ORACLE\ORADATA\ORACLASS\EXAMPLE01.DBF',
  'C:\ORACLE\ORADATA\ORACLASS\INDX01.DBF',
  'C:\ORACLE\ORADATA\ORACLASS\ODM01.DBF',
  'C:\ORACLE\ORADATA\ORACLASS\TOOLS01.DBF',
  'C:\ORACLE\ORADATA\ORACLASS\USERS01.DBF',
  'C:\ORACLE\ORADATA\ORACLASS\XDB01.DBF',
  'C:\ORACLE\ORADATA\ORACLASS\USER_LOCAL01.ORA',
  'C:\ORACLE\ORADATA\ORACLASS\USER_LOCAL02.ORA'
```

For Help, press F1

Figure 4-10 The trace file contains a complete CREATE CONTROLFILE command

3. Oracle recommends that you shut down the database at this point and run a backup of all the files, including the control files you just created.

4. Log off SQL*Plus by typing **EXIT** and pressing **Enter**.

You can use the trace file generated by Oracle9*i* to easily write a valid CREATE CONTROLFILE command.

You find out what the control files track in the next section.

Viewing Control File Data

The control file is made up of record sections. **Record sections** are lists of information by categories within the control file. Some record sections aid in file identification, including records for names and locations of datafiles, tablespaces, temp files, redo log files, and the name of the database. Other record sections aid in recovery activities and contain information on database activity, such as the status of a datafile, details on the most recent backup performed by Recovery Manager, and recovery checkpoints.

Oracle9*i*'s background processes update the information in the control file whenever certain activities take place. For example, the log writer process (LGWR) updates the control file records whenever a new log sequence number is started. See Chapter 5 for more information on the redo log processes.

There are four dynamic performance views that display the high-level contents of the control file.

- V$CONTROLFILE lists the names of the control files in use by the database.
- V$CONTROLFILE_RECORD_SECTION shows data held in the record section.
- V$PARAMETER displays initialization parameter values, including the current value of the CONTROL_FILES parameter.
- V$DATABASE lists current checkpoint numbers and control file sequence numbers.

Details contained in the record sections are spread out in many V$ dynamic performance views. Table 4-3 shows a list of some of these views.

Table 4-3 Control file views

View	Description
V$ARCHIVED_LOG	Archive of log information if the database is in ARCHIVELOG mode
V$DATAFILE	Details about datafiles
V$TABLESPACE	Names and number of tablespaces
V$LOG	Online redo log group information
V$LOGFILE	Redo log file information

The Recovery Manager's LIST command also displays many details contained in the control file that are used during database recovery. The Recovery Manager is not covered in this book.

You can query any of the views listed previously using SQL. For example, the query in Figure 4-11 looks at the datafile information found in the V$DATAFILE view.

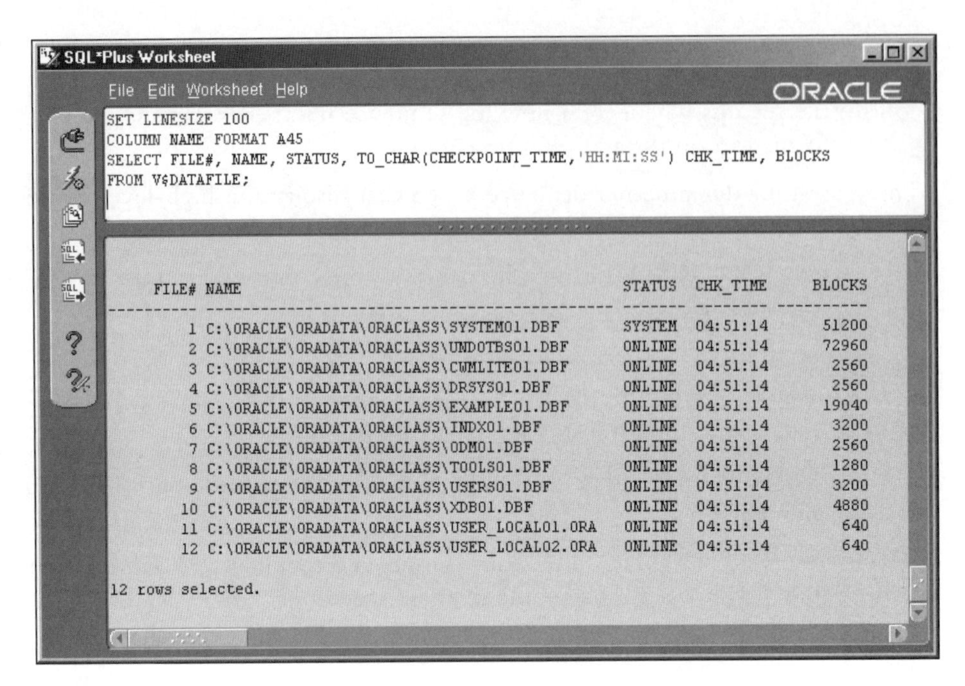

Figure 4-11 Use the TO_CHAR function to display times stored in DATE data types

In the next chapter, you examine how to maintain and view the redo log files.

CHAPTER SUMMARY

- Data dictionary views are owned by the SYS schema and are based on tables owned by SYS.
- Data dictionary views can be queried but never updated.
- Data dictionary views have prefixes of USER, ALL, DBA, V$, and GV$.
- Views prefixed by DBA are for users with DBA privileges in the database.
- Data dictionary views supply information about the structure of the database.
- Dynamic performance views begin with the V$ and GV$ prefixes.
- Dynamic performance views store current activity-oriented data.
- Dynamic performance views are used by STATSPACK and the Enterprise Manager Diagnostic Pack to track performance trends.
- Control files track the current datafiles, online redo log files, checkpoints, and log group number.

- By creating more than one control file, you multiplex the control files.
- Add extra control files by copying an existing control file and updating the CONTROL_FILES parameter.
- Rename or relocate an existing control file by moving or renaming the file and updating the CONTROL_FILES parameter.
- Replace a damaged or lost control file by replacing it with a copy of an undamaged control file.
- You can designate control files to be Oracle Managed by leaving the CONTROL_FILES parameter null and using the DB_CREATE_FILE_DEST parameter instead.
- Add DB_CREATE_ONLINE_LOG_DEST_*n* parameter values to create multiplexed Oracle Managed control files.
- You must create entirely new control files, when all control files are damaged or lost, or when certain database parameters must change.
- Use the CREATE CONTROLFILE command to create all new control files for an existing database.
- Specifying the SET DATABASE clause in the CREATE CONTROLFILE command renames the database.
- Some control file record sections contain information used for database recovery.
- Other control file record sections contain locations and names of files.
- Several V$ views query the control file, such as V$ARCHIVED_LOG.

Review Questions

1. The ______________ view displays a list of data dictionary views.
2. The SYSTEM schema owns the data dictionary views. True or False?
3. Which of the following views can a user with DBA privileges query?
   a. USER_COLUMNS
   b. DBA_GRANTS_MADE
   c. V$LOGS
   d. ALL_USERS
   e. All of the above
4. The views prefixed with USER and prefixed with ALL are the same structure except that the ______________ column is missing in the USER views.
5. The views with the ______________ prefix span multiple instances.

6. Explain the difference between the DBA_TABLES view and the USER_TABLES view.
7. Define public synonym and explain what it is used for.
8. Explain how Enterprise Manager's Schema Manager uses data dictionary views.
9. Give two examples of when you might query data dictionary views.
10. Describe how to multiplex control files.
11. When you add a datafile, you must update the control file also. True or False?
12. List two reasons for multiplexing control files.
13. List the initialization parameters used for Oracle Managed control files and describe how they are used.
14. The STARTUP command fails if there are no control files. True or False?

EXAM REVIEW QUESTIONS—DATABASE FUNDAMENTALS I (#1Z0-031)

1. The following traits describe V$ views. Choose three.
   a. Cannot be updated
   b. Records activities
   c. May contain data from multiple instances
   d. Available only to the DBA
   e. Available to any user
2. Control files contain these pieces of information about the database. Choose three.
   a. Time stamp of database creation
   b. Time stamp of last datafile update
   c. Names of datafiles
   d. Checkpoint name
   e. Checkpoint time stamp
3. A database has these initialization parameters:

   ```
   CONTROL_FILES=("D:\oracle\oradata\ORACLASS\CONTROL01.CTL")
   DB_CREATE_FILE_DEST='D:\oracle\oradata\datafiles'
   ```

   Which statement is true of this database? Choose two.
   a. The control file is multiplexed.
   b. The control file is an Oracle Managed File.
   c. The control file is not multiplexed.
   d The control file is not an Oracle Managed File.
   e. The control file is in the same directory as the datafiles.

4. Which of the following features are handled by Oracle9*i* *only* when you use Oracle Managed Files? Choose two.
   a. Location of the control file
   b. Size of the control file
   c. Multiplexing of the control file
   d. Name of the control file
   e. The control file is in the same directory as the datafiles

5. Which of these initialization parameters is used for Oracle Managed Files? Choose two.
   a. DB_CREATE_ONLINE_FILE_DEST
   b. DB_CREATE_ONLINE_LOG_DEST
   c. DB_CREATE_FILE_DEST
   d. DB_CREATE_ONLINE_LOG_DEST_n
   e. CONTROL_FILES
6. Which of these dynamic performance views would you use to view the names of the control files? Choose two.
   a. V$CONTROLFILE_RECORD_SECTION
   b. V$CONTROLFILE
   c. V$PARAMETER
   d. V$DATABASE
   e. V$CONTROLFILE_NAME
7. You want to list all of the indexed columns for objects that you own. Which data dictionary view would you query?
   a. ALL_COL_INDEXES
   b. USER_INDEXES
   c. USER_IND_COLUMNS
   d. USER_TAB_COLUMNS
8. The following lines show the commands and Oracle's responses. What should you enter next?

```
SQL>CONNECT SYS/mypwd@PROD AS SYSDBA;
Connected.
SQL>STARTUP;
Total System Global Area  135338868 bytes
Fixed Size                   453492 bytes
Variable Size             109051904 bytes
```

```
Database Buffers             25165824 bytes
Redo Buffers                   667648 bytes
ORA-00205: error in identifying controlfile, check alert log
for more info
SQL>
```

a. STARTUP NOMOUNT;

b. ALTER DATABASE BACKUP CONTROLFILES TO TRACE;

c. SHUTDOWN IMMEDIATE;

d. SHUTDOWN ABORT;

9. What is the best method for moving a control file?

   a. Copy the control file to a new location; shut down the database; modify the CONTROL_FILES parameter; and start up the database.

   b. Issue the ALTER DATABASE RENAME FILE command.

   c. Shut down the database; move the control file; modify the CONTROL_FILES parameter; and start up the database.

   d. Shut down the database; delete the control file; start up the database in NOMOUNT mode; issue the CREATE CONTROLFILE command; and start up the database.

10. What is the greatest advantage of multiplexing the control file?

   a. Less maintenance involved

   b. Less possibility of failure

   c. Faster I/O time

   d. Better archiving

Hands-on Assignments

Before starting the hands-on assignments, prepare the database by running the **setup.sql** script in the **Data\Chapter04** directory on your student disk. You must log on as SYSTEM to run the script.

1. You want to make a report of table attributes. This report consists of a series of queries on data dictionary views in which you specify the table name, and the queries return details about that table. Your goal is to have information on the report that is similar (in content, not format) to the information you see when you look at the Schema Manager's table property sheet in the console. Include as much information as you can to match the information you see for the CLASSMATE.WANT_AD table when looking at the General tab of the property sheet (table name, column names, data types, and so on). Refer to Figure 4-4 to see an example of the property sheet. Review the data dictionary views and select the views you think would produce the best results. Then write a series of queries that display the columns you want displayed on the report.

Use COLUMN commands to adjust the headings and column width so the report is more readable. Make sure the columns are listed in the same order as they appear on the property sheet. Write the queries to report the details of the WANT_AD table of the CLASSMATE schema. You should log on as CLASSMATE to run the queries.

Save the script in a file named **ho0401.sql** in the **Solutions\Chapter04** directory on your student disk.

2. Write a query, based on data dictionary views, reporting a table's name, number of rows, average row length, column names (in the order they appear in the table), the average column length, and the high and low values found in each column (the values are in hexadecimal format). When you are logged on as CLASSMATE, run the query. The query should only report on the CLASSMATE schema. Save the query in a file named **ho0402.sql** in the **Solutions\Chapter04** directory on your student disk.
3. Using V$SQL, write a set of queries that display the top ten SQL commands run in the system. Write one query for each of these criteria:
   - Most memory used (add all memory components)
   - Most CPU time
   - Most elapsed time
   - Most rows processed

   Save all the queries in a file named **ho0403.sql** in the **Solutions\Chapter04** directory on your student disk.
4. Your office has new equipment that allows you to multiplex your control files (currently, you only have one control file named **control01.ctl** on a single disk). The database name is TEST01. The SYSTEM user's password is "MYPASS". The current control file is located in the **D:\oracle\control1** directory. The new disk drives are labeled E and F. The directory structure is the same as the D drive. Write the steps and the commands needed to multiplex the control file and move it from its current location to the new drives. Use SQL comment delimiters (such as "/* ... */" or REM) for the text portions of the file and SQL commands for the command portions. Save your work in a file named **ho0404.sql** on the **Solutions\Chapter04** directory on your student disk.
5. To prepare for a disaster, you decide that you should have a CREATE CONTROLFILE command ready to go with all the right information to match your current database settings. Use the ALTER DATABASE BACKUP CONTROLFILE command to generate a SQL script for the CREATE CONTROLFILE command. Edit the file so that it can be used if all the control files exist but are damaged. Save the command in a file named **ho0405.sql** in the **Solutions\Chapter04** directory on your student disk.

6. Copy the **ho0405.sql** file into a new file named **ho0406.sql**. Now open the new file and modify it so that all files are Oracle Managed Files. In a comment at the top of the script, list the initialization parameters that must be set to support OMF, and place the files in the same directories in which they are currently found. In this case, assume that all the control files are lost (so they cannot be reused). Save the file in the **Solutions\Chapter04** directory on your student disk.
7. List the data dictionary views that you think would be most helpful to a programmer who is writing applications that select, update, and delete rows from tables in the database. Explain briefly why you chose each view. Save your answer in a text file named **ho0407.txt** in the **Solutions\Chapter04** directory on your student disk.

CASE PROJECT

The Global Globe database has multiplexed control files. You are gathering statistics about the database's current state. Write one or more queries to answer these questions about the ORACLASS database.

- Which file has had the most physical read activity?
- List the minimum time for a single I/O process, and the maximum time for a single write process for all datafiles.
- What is the most recent checkpoint number?
- When was the database created?

CHAPTER

5

The Redo Log Files and Diagnostic Files

In this chapter, you will:

- Learn to describe redo log files, groups, and members
- Manage redo log groups and members
- Learn to describe and configure diagnostic files

The control files that you studied in the previous chapter and the redo log files you study in this chapter are closely connected. Both have a role in database recovery, can be multiplexed, and can be created and maintained with Oracle Managed Files. This chapter describes how redo log files are loaded with information, what information they hold, and how to maintain them.

You must also monitor and manage the diagnostic files. In this chapter, you find out what these files contain and how to use them.

INTRODUCTION TO ONLINE REDO LOG FILES

As you saw in Chapter 1, the **redo log files** record changes to database data. These files are also called online redo log files because they are open and available ("online") whenever the database is up. This section adds to the definition.

Redo log files capture details of database transactions and information about structural or other changes to the database including:

- **Checkpoints:** Checkpoints occur during database shutdown, during a redo log switch (discussed later in this chapter), and when a checkpoint is forced by a CHECKPOINT command.
- **Changes:** When a session issues a command (such as INSERT, UPDATE, and DELETE) that changes the database, or a COMMIT command, the redo log file captures information *before the changes are actually committed*. The redo log file captures the **data manipulation language (DML)** and **data definition language (DDL)** commands and information on how to undo the commands (see Chapter 6).
- **Datafile changes:** When a new datafile is added or an old one is removed, it is recorded in the redo log file. In addition, when data has changed in the datafile, an entry is made in the redo log file recording that the datafile has changed.

Figure 5-1 shows the components of the database involved with the online redo logs as described in the list that follows:

- **Redo log group:** A set of one or more online redo log files
- **Online redo log member:** A file containing data blocks written by the LGWR process for use in database recovery
- **Archived redo log file:** An offline copy of an online redo log file written by the ARCn process
- **Redo log buffer:** The Memory storage of changed data blocks
- **LGWR background process:** A process that writes data blocks from the redo log buffer to the online redo log file
- **ARCn background process:** A process that copies a completed online redo log file to an archived redo log file
- **CKPT background process:** A process that initiates flushing of buffers, causing the LGWR to write data blocks from the redo log buffer to the online redo log file

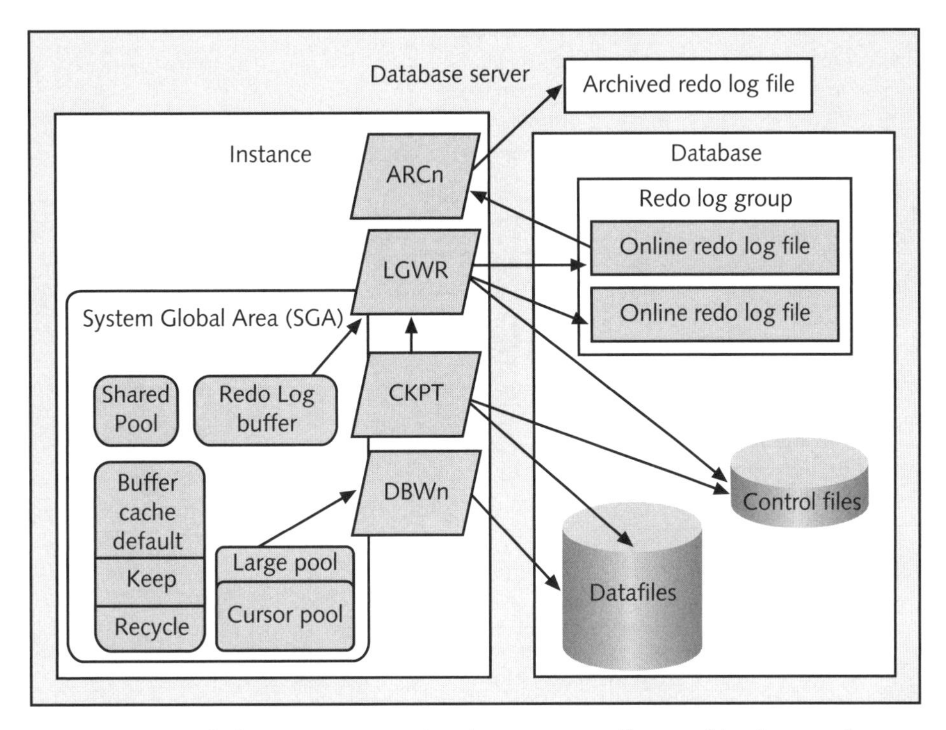

Figure 5-1 Redo log components involve memory, files, and background processes

A database must have at least two **redo log groups** containing at least one file each. Each file in a redo log group is considered a **member** of that group. Often, each group contains more than one file (discussed later in this chapter). Files in the same group are identical, just as multiple control files are identical. So, when a redo log group is recording data, the data is written to every file in the group. If one of the files in the group becomes damaged or unavailable, the other files continue to receive data. As long as one file remains in the current redo log group, the database continues to function. If the LGWR process cannot find any valid redo log files in the current group, the LGWR process instructs the database to shut down immediately. The database also shuts down if the LGWR process cannot find any valid files in the group that will become current after the switch.

The groups are used in a round-robin fashion, as shown in Figure 5-2. The numbers in Figure 5-2 show these steps:

1. The first redo log group receives data until it is full. All the files in a redo log group are identical in size. They are all written simultaneously, so they contain identical data as well. A redo log group is full when the member files are full.

2. When the first group is full, a redo log switch occurs, and the second group gets data.

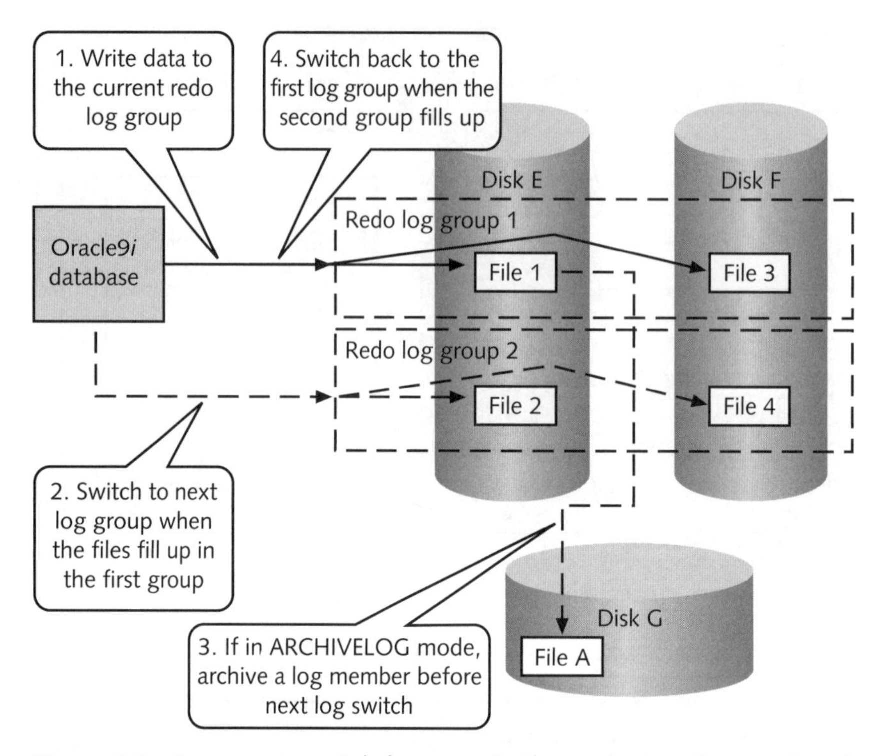

Figure 5-2 Log groups switch from one to the next when the members become full

3. If the database is in ARCHIVELOG mode (set by an initialization parameter), a file in the first redo log group is copied to an archive file while the second redo log group is being used.
4. When the second group fills up, another log switch occurs, and the first log group is reused, writing over any old data. If the database is in NOARCHIVELOG mode (the default), all the data of the first log group is lost.

The Purpose of Redo Log Files

The sole purpose of redo log files is to aid in database recovery. The redo log files keep a shorthand list of database changes as they are committed. If the database loses some changes due to a power outage, for example, the recovery process restores the changes by using the redo log files. This is why the redo log files receive the change information before the actual datafiles are updated. When an unexpected database failure occurs, the process of saving database changes may be interrupted, so the database itself may contain some datafiles that are updated and some that are not. The redo log files contain enough information to restore the lost changes.

In minor failures, such as a power outage that lasts a few minutes, the redo log files are automatically checked during database startup, and data is restored. However, in major failures, such as the loss of an entire disk due to hardware failure, data would not be saved from the online redo logs alone. You would also need a full backup of the database and archived redo log files that begin after the date of the backup.

If you want to fully safeguard your database so that you can recover from any failure, you should put your database in ARCHIVELOG mode.

5

The following steps outline the recovery procedure for problems such as a disk drive crash. Detailed steps that you can use in a real situation are covered in Oracle's online documentation on backup and recovery. The steps shown here provide a frame of reference on which you can build.

1. Before database failure, and on a regular schedule, perform full database backup procedures.
2. Adjust initialization parameters so that the database starts up in ARCHIVELOG mode as part of routine startup.
3. Upon media failure (such as a disk drive failure), restore the database from the database backup.
4. Apply archived redo log files from the oldest to the newest.
5. Apply online redo log files until all changes have been restored.
6. Apply **undo data** (data stored in the undo tablespace or undo segments that identifies uncommitted changes) to remove any uncommitted changes applied by the online redo log files. See Chapter 6 for details on undo data.

The end result is a database containing every change restored up to the minute the disk failed.

The Structure of Redo Log Files

The redo log files store a variety of information from the database activity. All this information is recorded in the redo log buffer in the System Global Area (SGA). The entire contents of the redo log buffer is written by the Oracle LogWriter (LGWR) background process to the online redo log file when any of these events occur:

- A transaction issues a COMMIT command.
- The redo log buffer reaches the point of being one-third full.
- There is more than one megabyte of *updated* records stored in the redo log buffer (not including *deleted* or *inserted* records).
- A checkpoint occurs.

The redo log file contains sets of **redo records**. A redo record, also called a **redo entry**, is made up of a related group of **change vectors** that record a description of the changes to a single block in the database. A single transaction may generate many redo entries (one for each block affected), and each of those redo entries may in turn contain many change vectors (one or more for each record in the block, depending on the change). For example, you write and execute an UPDATE command that looks for all the ten-digit zip codes stored in your CUSTOMER_ADDRESS table and then takes out the hyphen between the first five and the last four digits. This change affects data in many blocks within the table, so there are many redo records written to the redo log buffer. Here is another example: When you type the COMMIT command, the redo log buffer's redo records are moved to the redo log files. If the buffer becomes more than one-third full before you commit your transaction, your redo records are written to the log files. In recovery, these changes are applied to the database as if they had been committed and then rolled back using the undo data.

MANAGING REDO LOG FILES

Like control file maintenance, there are several tasks necessary for keeping the redo log files configured and functioning well. For example, you may notice that the log files fill up very quickly and, therefore, you may decide to change the size of the files.

Before digging into the SQL commands and initialization parameter settings, look into the background processes involved in the redo log.

Figure 5-3 shows a diagram of the redo log processes. In the figure, a user has issued an UPDATE command followed by a COMMIT command to save the updated rows. The COMMIT triggers the LGWR background process to write all the data from the log buffer into the current log group.

The numbers on the figure correspond to these technical notes:

1. When a user submits changes to the database, the changes are stored in the redo log buffer within the SGA memory area of the database.
2. When the user executes a COMMIT command, the user's server process initiates the commit process. The first step in the commit processes saves the redo log buffer to the redo log files.
3. The LGWR background process retrieves the buffer and writes it to the current redo log group. It also writes a commit record, identifying the transaction that was committed. In the diagram, Redo log group 1 is the current group. A copy of the buffer and the commit record is written to each file in Redo log group 1. In the diagram, the two members of Redo log group 1 are called **redo01.log** and **redo03.log**, and they reside on separate disk drives.

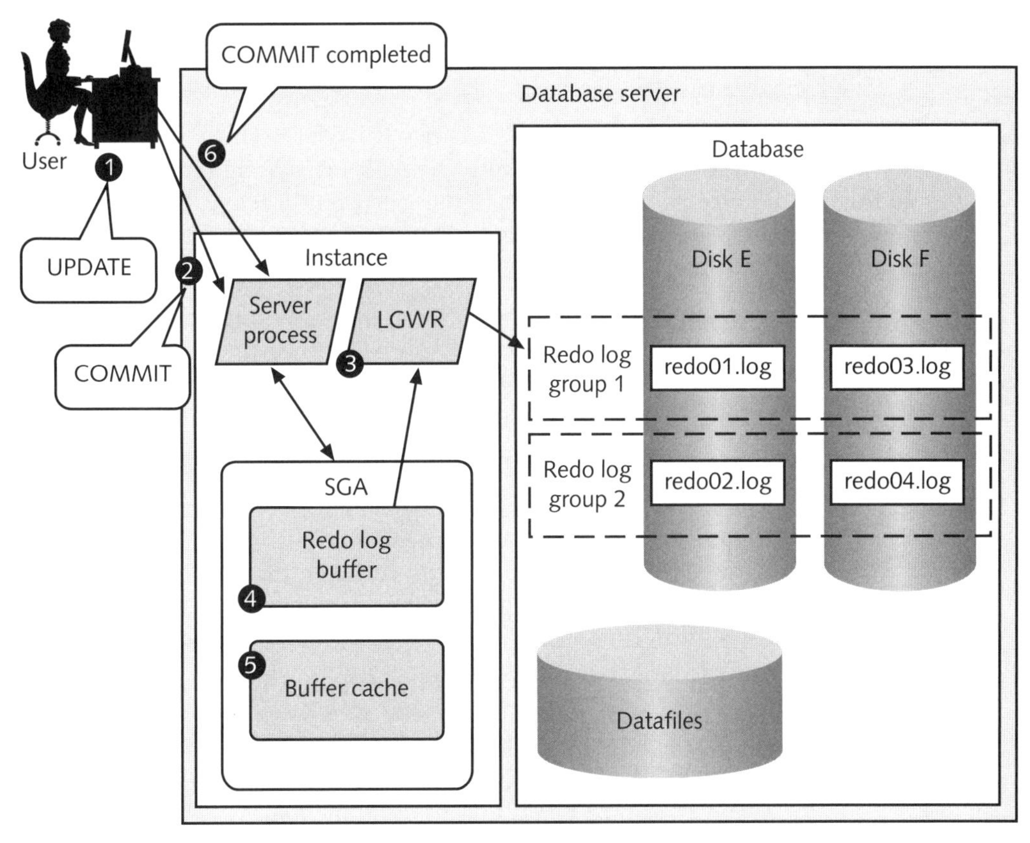

Figure 5-3 Committing a change requires writing to the redo log file

4. After the buffer and the commit record have been written to the redo log group, the buffer contents are deleted (the process of copying the buffer and then deleting its contents is called **flushing** the buffer).
5. The changed data blocks are kept in the buffer cache. They are not written to the datafiles at this point. This is called a **fast commit**. The data is written from the buffer cache to the datafile later (see the next section on checkpoints for the details).
6. After the redo buffer data has been written successfully to the redo log group, Oracle tells the server process that the commit succeeded.

The next section describes events that trigger activities involving the redo log files.

Log Switches and Checkpoints

As you saw in the previous section, a **log switch** is an event in which the LGWR process stops writing to one log group and begins writing to another log group. A log switch is triggered when a log group fills up with records. Figure 5-2 outlined the steps in a log switch.

Adding more files to a log group does not change the volume capacity of the log group. Rather, it multiplexes the files, allowing you to store the same redo log data in several locations.

Sometimes, you may need to manually trigger a log switch. Do this when you want to perform maintenance on the log group that is currently active. Maintenance can only be done while a log group is idle. Follow these steps to cause a log switch before the current log group is full:

1. Start up the Enterprise Manager console. In Windows, on the Taskbar, click **Start** then click **Programs/Oracle - OraHome92/Enterprise Manager Console**. In UNIX, type **oemapp console** on the command line. The Enterprise Manager Console login screen appears.
2. Select the Launch standalone radio button, and click **OK**. The console appears.
3. Start up the SQL*Plus Worksheet by clicking **Tools/Database Applications/SQL*Plus Worksheet** from the top menu in the console. A background SQL*Plus process starts, and then the SQL*Plus Worksheet window appears. If the background process appears in front, simply minimize it (click the **minus sign** in the top-right corner) so that you can see the worksheet.
4. Connect as the SYS user. Click **File/Change Database Connection** on the menu. A logon window appears. Type **SYS** in the Username box, the **current password** for SYS in the Password box *as provided by your instructor*, and **ORACLASS** in the Service box. Select the connection type **SYSDBA** in the Connect as box. Click **OK** to continue.
5. Type and execute the following command to switch log groups:

   ```
   ALTER SYSTEM SWITCH LOGFILE;
   ```

6. The system displays "System altered" in the bottom pane of the worksheet. Remain logged on for the next exercise.

When you execute a COMMIT command, you have seen that it causes the LGWR to write the log buffer to the current log group. One of the details contained in the redo entry is the System Change Number or SCN. The **SCN** is a sequential number that is incremented for each change that modifies the physical database files. The current SCN is stored in the control file. The SCN may not actually be incremented when you issue a COMMIT command, because the change may stay in a memory buffer for efficiency. It is more efficient to write all the changes at once from the memory buffer. The event in which buffers are written to the datafiles is called a checkpoint.

A **checkpoint** is a moment in time when the CKPT background process signals that all used memory buffers are to be written to disk. Used memory buffers are called **dirty buffers**. In addition, a log switch triggers a checkpoint. The checkpoint process involves incrementing the SCN value and writing the checkpoint information to the control file and the datafile headers. Following this process, all dirty buffers are written to their respective disks. Memory buffers are written to datafiles by the Oracle DataBase Writer (DBWR) background process, and log buffers are written to log files by the Oracle Log Writer (LGWR) background process.

When a log switch triggers the checkpoint, the CKPT process must complete its work before LGWR switches to the next redo log group. This keeps the redo buffer synchronized with the memory buffer. If the LGWR is waiting for the CKPT process, the log switch may be slowing down response time. If this occurs, add another redo log group, so that there is time for the first group to be archived before it is needed again.

Checkpoints aid in database recovery. If the database shuts down unexpectedly, the automatic recovery process looks at the control file SCN (written by the checkpoint process) and begins rolling forward through changes found in the redo log file that are associated with a higher SCN number. This speeds up the recovery process by skipping data that has already been written to the physical disk.

The redo log files are so critical to recovery that Oracle recommends always multiplexing them. The next section describes how.

Multiplexing and Other Maintenance

Recall from Chapter 4 that multiplexing is the practice of maintaining multiple identical copies of a file to reduce the risk of loss. When multiplexing the redo log files, you simply add new file members to each group. As you saw in Figure 5-3, a group can contain many files, and the files can be located on several disk drives. Here are some additional points about redo log groups and multiplexing redo log files:

- All the files in a redo log group must be the same size.
- The LGWR process writes concurrently to all the files in one redo log group.
- The LGWR process never writes to two different redo log groups at the same time.
- If one or more files are damaged within a redo log group, LGWR writes to the remaining file or files and does not stop any database operations.
- If all the files in a redo log group are damaged, LGWR stops all database operations until a successful log switch occurs (DBA intervention is required to issue a manual log switch).
- If a log switch is to a group that is pending archive, LGWR waits until the group is archived (an automatic process) and then continues the switch.
- If a log switch fails because all the redo log files in the group to which the database is switching are damaged, the database shuts down and must be recovered after restoring the redo log files (DBA intervention is required).

The next sections present several important tasks associated with redo log groups and group members.

Adding a Member to a Group

One reason to add member files to an existing redo log group is that the CREATE DATABASE command did not multiplex the redo log groups. Another reason is that one of the member files was damaged or lost because of accidental deletion or because a disk failed. The file should be replaced.

Assume that you have created a new database that has two redo log groups with one file each. To multiplex the groups by adding one more file to each group, follow these steps:

1. You are logged onto the Enterprise Manager Console and logged onto the SQL*Plus Worksheet as SYS from a previous section in this chapter.
2. Switch to the console, and double-click the **ORACLASS** database. A list of available tools appears below the database.
3. Double-click the **Storage** icon. A list of storage components appears.
4. Double-click the **Redo Log Groups** folder. Figure 5-4 shows what the screen looks like. Your database may have more than two redo log groups.
5. Click the first redo log group listed below the Redo Log Groups folder in the left pane. The property sheet appears for this redo log group. Figure 5-5 shows the property sheet. Yours has a different file name. This illustrates how you can use the console to discover the number of redo log groups and the number of files contained in each group. You can also query dynamic performance views to find this information, as shown later in this chapter.

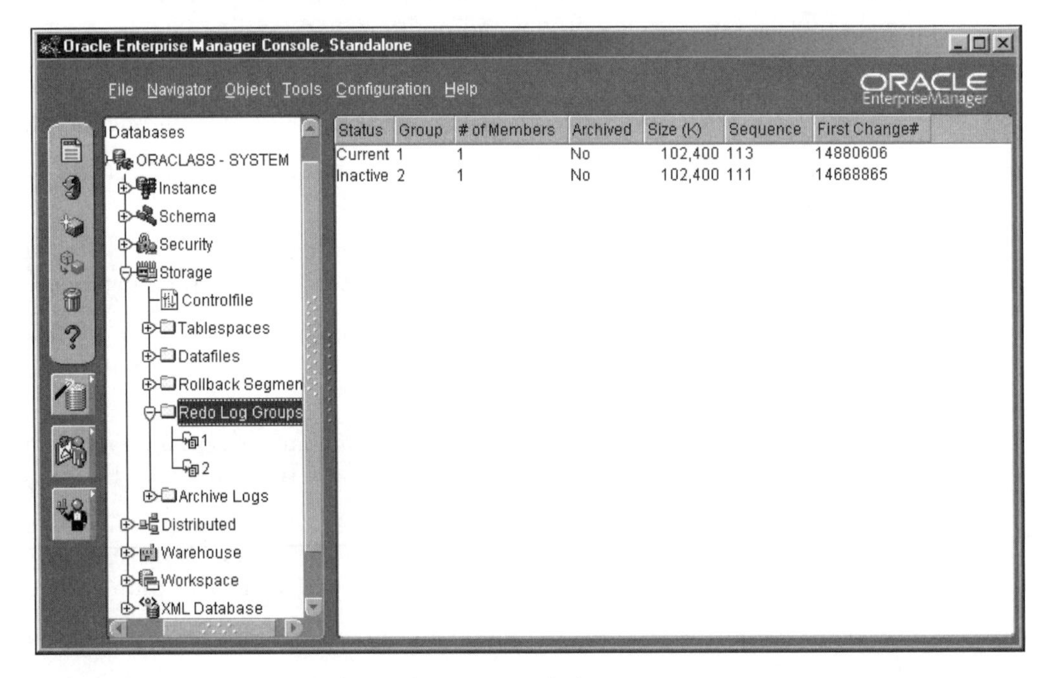

Figure 5-4 Your database has at least two redo log groups

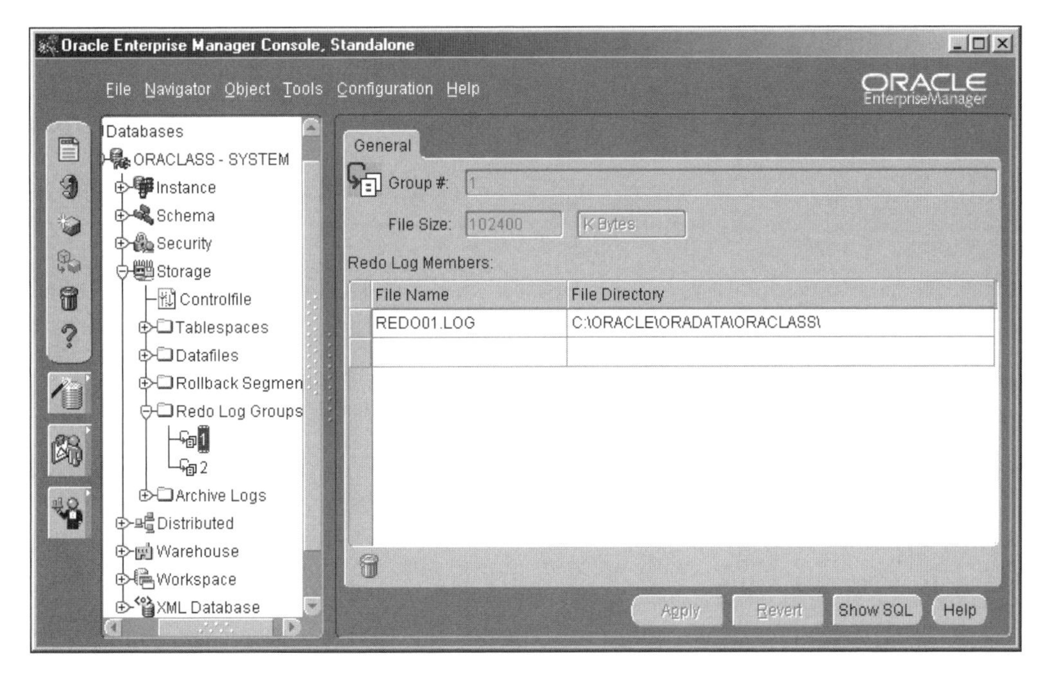

Figure 5-5 The redo log group has one file

6. Switch to the SQL*Plus Worksheet session in preparation for adding a new member to each group.

7. In this example, add a new file named **redo0102.log** to Group 1 and a file named **redo0202.log** to Group 2. You will place these new log files in a directory on the hard disk. Ask your instructor for the path of the directory you should use. Type and execute the following two commands to add a new member to each group. (Replace <X:\xxx> with the actual directory path.)

```
ALTER DATABASE ADD LOGFILE MEMBER '<X:\xxx>\redo0102.log'
TO GROUP 1;
ALTER DATABASE ADD LOGFILE MEMBER '<X:\xxx>\redo0202.log'
TO GROUP 2;
```

   The worksheet replies, "Database altered." By adding a second file to each group, you have multiplexed the redo logs.

8. Remain logged onto the console and the SQL*Plus Worksheet for the next exercise.

No size is specified for the new member, because all members must be identical in size. Therefore, Oracle9*i* uses the size of the existing members to size the new member.

Adding a New Group

Continuing with the example database, you now have two groups with two files each. Imagine that the alert log contains warning messages stating that the LGWR process has to wait for the ARCn process. **ARCn** represents one or more archiver background processes. The first archiver process is named ARC0; the second (if there is one) is ARC1. There can be up to ten archiver processes. For more information, see the section on archiving a redo log group later in this chapter. You detect the warning message occurring several times a day. To correct the problem, you must create a new redo log group so that the archive process has some lead time to archive the inactive group. To accomplish this, follow these steps:

1. Type and execute the following command in SQL*Plus Worksheet to add a new group. It is not a requirement that all groups contain the same size files or the same number of files. You choose to create a group that matches the other two groups (two files of 20 MB each). You add the new group in the same directory used to add the members in the previous command, so replace <X:\xxx> with the appropriate path:

```
ALTER DATABASE
   ADD LOGFILE GROUP 3('<X:\xxx>\redo0301.log',
   '<X:\xxx>\redo0302.log') SIZE 20M;
```

2. Remain logged on the next exercise.

The GROUP *n* parameter of the ADD LOGFILE phrase is not required; however, it makes maintenance of groups easier. Always use sequential numbers when numbering the groups to save space in the control file.

Renaming or Moving a Redo Log File

If your system acquires an additional disk drive, you may find that moving one member of each redo log group to the extra drive reduces the risk of losing redo logs. This is especially true if two or more members of each group reside on the same physical device. Renaming a redo log file may be done to keep up with new naming standards or to simply make the file names more meaningful to you.

Renaming or moving a redo log file is similar to working with the control file: You must work with the operating system and with Oracle9*i* to accomplish the task. For this exercise, imagine that you choose to rename all the members within the third redo log group that you created in the previous section. The new group has two members; therefore, you are renaming two files. Follow these steps to rename them.

Oracle recommends that you back up your database before working on the redo log files in the event you need to recover from an error. Oracle also recommends that, after you have successfully completed the change, you back up the control file.

1. Shut down the ORACLASS database by typing the following command in your SQL*Plus Worksheet session and clicking the **Execute** icon. Remember, the SHUTDOWN command can be used only when you are logged on as the SYS user in SYSDBA mode.

   ```
   SHUTDOWN IMMEDIATE
   ```

2. Locate the two members of the new redo log group. In Windows, open Windows Explorer and locate the redo log files you created. They are named **REDO0301.LOG** and **REDO0302.LOG**. In UNIX, open a command prompt, and use the **cd** command to reach the correct directory.

3. Rename each file to match a new naming pattern. The pattern is:

   REDO_GRn_Mn.log

   So, for example, rename **REDO0301.LOG** to **REDO_GR3_M1.LOG**. Use the File menu in Windows Explorer. If you are in UNIX, use the standard operating system copy command (**mv**) to move the file to its new file name. For example, the following command renames one file:

   ```
   mv REDO0301.LOG REDO_GR3_M1.LOG
   ```

4. Switch back to the SQL*Plus Worksheet. Then start up the database in MOUNT mode by typing and executing this:

   ```
   STARTUP MOUNT
   ```

5. Alert Oracle9*i* of the renamed files by typing and executing this command (replacing <X:\xxx> with the appropriate directory path):

   ```
   ALTER DATABASE
     RENAME FILE '<X:\xxx>\REDO0301.LOG',
                 '<X:\xxx>\REDO0302.LOG'
              TO '<X:\xxx>\REDO_GR3_M1.LOG',
                 '<X:\xxx>\REDO_GR3_M2.LOG';
   ```

6. Open the database for normal use by typing and executing this:

   ```
   ALTER DATABASE OPEN;
   ```

7. Remain logged on for the next exercise.

Dropping Redo Log Members or Groups

There are many reasons to drop redo log members or groups. To name just a few:

- A disk containing redo log members has failed, and you do not want to replace the files.
- Tuning recommendations suggest reducing the number of redo log groups.
- Individual log members have become corrupted, and you want to replace them later.

When dropping redo log members, remember that there must be at least one member in every group. In addition, a database must have at least two groups, so you cannot drop a group if it is one of the last two groups.

The Oracle9*i* rules state that an online redo log group (or member of a group) can only be dropped when the group is inactive. That means that LGWR process is not using the current group. If the database is in ARCHIVELOG mode, the group must also have been archived. The V$LOG data dictionary view shows group status. The group must have a status of INACTIVE and an archived flag of YES to be dropped.

Follow these steps to check the status of one member, and then drop the file:

1. Type and execute the following query in SQL*Plus Worksheet to determine the status of the third redo log group. Figure 5-6 shows the results. You may have more than two members in each group. As you can see, Group 3 has a status of UNUSED, meaning it has never been used. The archived status is YES. Even though this group is not archived, Oracle9*i* initializes a new group by setting the archived status to YES.

```
SELECT GROUP#, STATUS, ARCHIVED, MEMBERS FROM V$LOG;
```

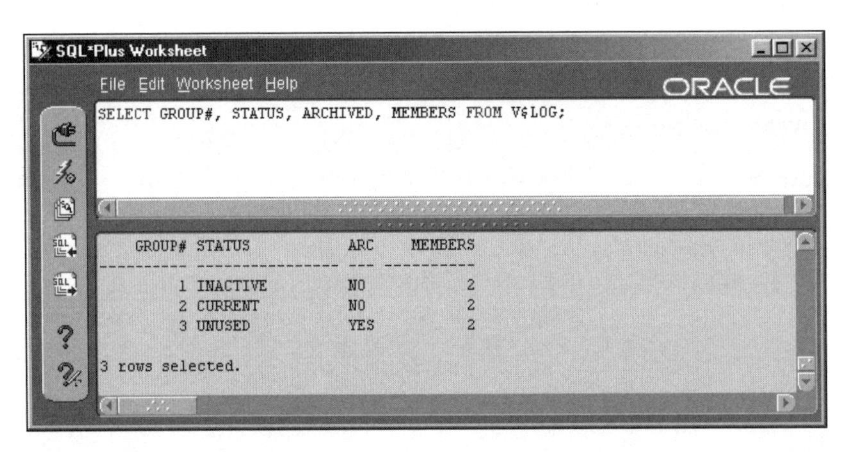

Figure 5-6 You can drop a member of a group only with either INACTIVE or UNUSED status

2. Type and execute the following command to drop the file named **REDO_GR3_M1.LOG** from Group 3. (Replace <X:\xxx> with the actual file path.)

```
ALTER DATABASE DROP LOGFILE MEMBER
 '<X:\xxx>\REDO_GR3_M1.LOG';
```

3. Using your operating system, delete the file that you dropped. The database does not drop the file unless the file is an Oracle managed file. In Windows, use Windows Explorer to locate and select the file, and then click **File/Delete** from the menu. In UNIX, open a command prompt, and use the **cd** command to navigate to the appropriate directory. Then type **rm REDO_GR3_M1.LOG**, and press **Enter**.

4. Now, drop the entire third group by typing the following command in SQL*Plus Worksheet.

```
ALTER DATABASE DROP LOGFILE GROUP 3;
```

5. Using your operating system, delete the remaining file that was in Group 3. In Windows, use Windows Explorer to locate and select the file, and then click **File/Delete** from the menu. In UNIX, open a command prompt, and use the **cd** command to navigate to the appropriate directory. Then type **rm REDO_GR3_M2.LOG**, and press **Enter**.

If the redo log file that you want to drop is in the active group, force a log switch, and then force a checkpoint before attempting to drop the file. You can also force archiving if the database is in ARCHIVELOG mode.

6. Type and execute the following command in SQL*Plus Worksheet to force a log switch:

```
ALTER SYSTEM SWITCH LOGFILE;
```

7. Repeat the query on V$LOG. You see that Group 2 is now current, but Group 1 has a status of ACTIVE. This means that files in the group cannot be dropped, because some transactions that have dirty buffers (changed data not written to the datafile) have redo records in the group, even though a log switch has occurred. To force these outstanding records to be flushed from the buffer, thus changing the status of the group from ACTIVE to INACTIVE, type the following command and execute:

```
ALTER SYSTEM CHECKPOINT;
```

8. Close your SQL*Plus worksheet session by clicking **X** in the upper-right corner.

9. Close your console session by clicking **X** in the upper-right corner.

Redo log files occasionally become corrupted and interfere with database functioning. If this happens, one option is to drop the group and re-create it. However, in some cases (such as when there are only two groups remaining), a simpler solution is available: clearing the group. Clearing the group removes all the corrupted data. The syntax of the command is:

```
ALTER DATABASE CLEAR UNARCHIVED LOGFILE GROUP <n>;
```

UNARCHIVED is an optional clause used when the group is not yet archived, and you want to clear it without archiving it. Replace <n> with the actual group number you want to clear.

Archiving a Redo Log Group

As you know by now, archiving a redo log group ensures that you can perform a complete, up-to-the-minute recovery of your database in the event of catastrophic failure, such as the loss of a disk drive that held your SYSTEM tablespace. Other advantages of archiving include:

- Point-in-time recovery to restore data lost due to user error (such as deleting the contents of an entire table).
- The ability to query archived log files with LogMiner for auditing to find out who performed that DELETE command. With **LogMiner**, you can view and parse the archived redo log files. LogMiner is part of the Oracle Server.
- The ability to update a standby database (replace the failed database immediately with a ready copy).

A **standby database** is a database clone of the current database. It is kept current with the existing database by applying changes stored in the archived redo logs. If the current database is unrecoverable, the standby database replaces it, speeding up recovery time tremendously.

A database in **ARCHIVELOG mode** automatically archives redo log files. The opposite (no archiving of redo log files) is a database in **NOARCHIVELOG mode**. There are several ways to get your database running in ARCHIVELOG mode:

- Specify the ARCHIVELOG parameter in the CREATE DATABASE command.
- Set the initialization parameter LOG_ARCHIVE_START=TRUE and restart the database in MOUNT mode. Then issue the command ALTER DATABASE ARCHIVELOG, and open the database.
- Restart the database in MOUNT mode, issue the command ALTER DATABASE ARCHIVELOG, and then open the database. Finally, set the LOG_ARCHIVE_START parameter to TRUE while the database is running, by issuing the command ARCHIVE LOG START while logged on as a SYSDBA.

This command will not work in ORACLASS, because you are not in ARCHIVELOG mode. However, if you were in ARCHIVELOG mode, and you wanted to force an archive, you would use the following command:

```
ALTER SYSTEM ARCHIVELOG CURRENT;
```

In an ARCHIVELOG mode database, you can archive all noncurrent logs by stating:

```
ALTER SYSTEM ARCHIVELOG ALL;
```

There are several initialization parameters that control the archiving processes. Table 5-1 briefly describes the name and function of each one.

The only required initialization parameter to achieve successful ARCHIVELOG mode is the LOG_ARCHIVE_DEST_n parameter. (If you are running Personal Oracle, this parameter is called LOG_ARCHIVE_DEST.) The directory must exist.

Table 5-1 Initialization parameters for ARCHIVELOG mode

Parameter	Description	Example
`LOG_ARCHIVE_DEST_n 'LOCATION = '<directory>' MANDATORY or OPTIONAL REOPEN = <sec>`	Destination directory where archived log files are written. The directory must exist. n is the sequence number of the archive process (ARCn). Specify MANDATORY (must succeed) or OPTIONAL (okay to fail). If open failed, specify REOPEN with the number of seconds to wait before trying again (recommended if you specify MANDATORY). You can have up to ten processes (1–10) to speed up the archiving activity. Valid only for Enterprise Edition.	`'LOCATION= D:\oracle\oradata' MANDATORY REOPEN 50`
`LOG_ARCHIVE_DEST` and `LOG_ARCHIVE_DUPLEX_DEST`	Primary and secondary destinations for archiving. Valid if you are using Personal Oracle9*i*. Deprecated for Enterprise Edition (use LOG_ARCHIVE_DEST_n).	`F:\remote\oracle\log`
`LOG_ARCHIVE_DEST_STATE_n`	The state of each destination in LOG_ARCHIVE_DEST_STATE_n. Values can be: enabled (ready for use); defer (do not use); and alternate (use if other destination fails).	`Enabled`
`LOG_ARCHIVE_FORMAT`	A standard naming pattern (format) given to the name of archive log files with variables. Valid variables are: %s (log sequence number); %S (log sequence number, fixed length, zero padded); %t (thread number); %T (thread number, fixed length, zero padded).	`ar%S.log` (becomes, for example, **`a00125.log`**) `arch_%t_%s.log` (becomes, for example, **`arch_1_125.log`**)
`LOG_ARCHIVE_MAX_PROCESSES`	Maximum archive processes to start on database startup. Value can be 1–10. The processes are named ARC0–ARC9. The actual number in use is automatically changed depending on needs. Default is 1.	5

Table 5-1 Initialization parameters for ARCHIVELOG mode (continued)

Parameter	Description	Example
LOG_ARCHIVE_MIN_SUCCEED_DEST	Set minimum destination success. 1–10 if using LOG_ARCHIVE_DEST_n. 1 or 2 if using LOG_ARCHIVE_DEST. Limited by how many destinations you have named, whether they are optional or mandatory, enabled or deferred. Default is enabled. Ignored if lower than the number of enabled mandatory destinations.	5
LOG_ARCHIVE_START	Start up archiving. TRUE = set database in automatic ARCHIVELOG mode on startup FALSE = set database in manual ARCHIVELOG mode. Ignored if no LOG_ARCHIVE_DEST_n and no LOG_ARCHIVE_DEST values are specified (this means the database is in NOARCHIVELOG mode).	TRUE

To see the values currently set for these parameters, follow these steps:

To see the values currently set for these parameters, you could type **SHOW PARAMETERS LOG** in a SQL*Plus or SQL*Plus Worksheet.

You can type and execute this command in SQL*Plus Worksheet to display all parameters with LOG in their names. Figure 5-7 shows part of the results.

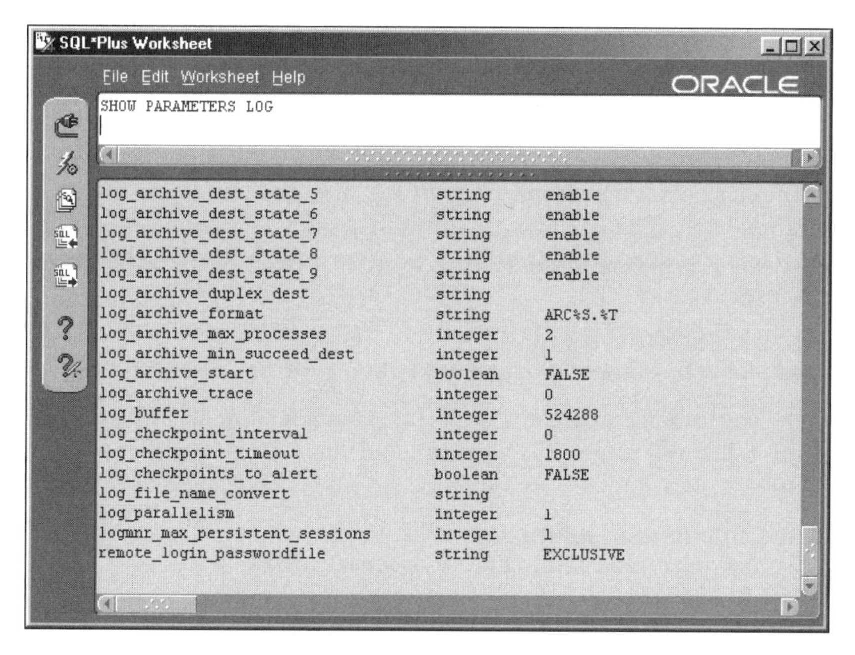

Figure 5-7 Display all initialization parameters containing "LOG"

As with control files and datafiles, it is possible to use the Oracle Managed Files method with redo log files. The next section discusses your options.

Using OMF to Manage Online Redo Log Files

When using OMF to manage online redo log files, all you need to do is ensure that the appropriate initialization parameters are set up, and then simply omit the file names and paths in all commands associated with managing the redo log files.

The same initialization parameters you used with OMF for control files work for the log files. You can store all the files in a single directory (datafiles and all) by specifying DB_CREATE_FILE_DEST. Or, you can spread the control files and redo log files into their own directory or directories by specifying one or more DB_CREATE_ONLINE_LOG_DEST_n. If you have more than one destination, each new destination tells Oracle9*i* to create a new multiplexed set of redo log files.

Here are some examples of initialization parameters, SQL commands, and OMF operations with the redo log groups and members.

Do not attempt to run the commands shown in this section. They are only examples and will cause errors if run in the ORACLASS database.

Begin by specifying the initialization parameters. For example:

```
DB_CREATE_ONLINE_LOG_DEST_1 = 'D:\oracle\logs'
DB_CREATE_ONLINE_LOG_DEST_2 = 'E:\oracle\logs'
```

With the initialization parameters done, the CREATE DATABASE command can easily be used to set up Oracle-managed redo log files. You can leave out the LOGFILE parameter completely. The following command creates a user managed datafile for the SYSTEM tablespace (due to the DATAFILE clause) and creates Oracle managed log files and control files:

```
CREATE DATABASE NEWDB92
DATAFILE 'D:\oracle\data\system01.dbf' SIZE 250M;
```

Alternatively, you can include the LOGFILE parameter but specify only a file size, as shown in the following example. Do not include any details on groups or datafiles for the log members:

```
CREATE DATABASE NEWDB92
DATAFILE 'D:\oracle\data\system01.dbf'
LOGFILE 10M;
```

The previous statement creates Oracle managed redo logs, control files, and a user managed SYSTEM datafile. The default size of OMF log files is 100 MB. Including "LOGFILE 10M" overrides the default size. Figure 5-8 illustrates the redo log groups and members created in this example.

Oracle Managed Files always begin with a template. The exact template depends on your operating system, but it usually looks like this: "o1_mf_%g_%u_.log", where %g is the group number, and %u is a unique eight-character string generated by Oracle9*i* to ensure that the file name is unique.

You can add an Oracle managed log group to a database whose log groups are already OMF by using the following ALTER DATABASE command. (You must be logged on with SYSDBA privileges to execute the command.)

```
ALTER DATABASE ADD LOGFILE;
```

The database creates a new group and adds one member of the new group into each directory specified in DB_CREATE_ONLINE_LOG_DEST_n. In the example you have been using, in which two parameters were defined, the new group would have two members.

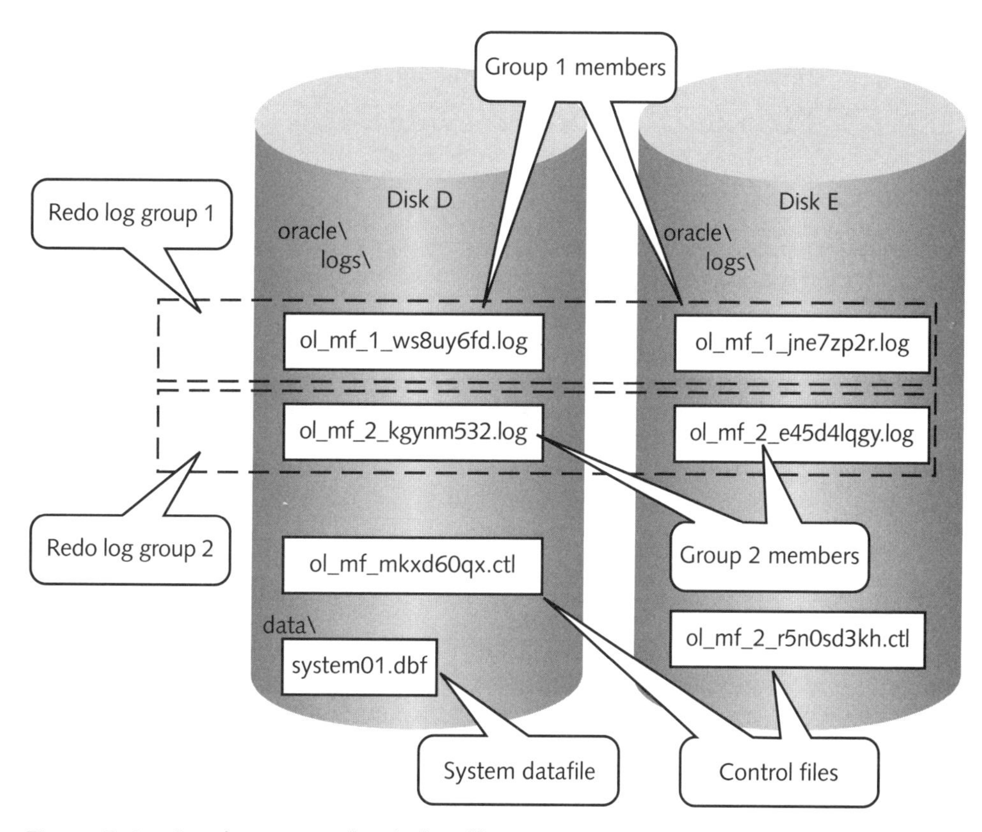

Figure 5-8 Oracle-managed redo log files automatically form two groups

If you define only the DB_CREATE_FILE_DEST and no values for DB_CREATE_ONLINE_LOG_DEST_n in the initialization parameters, all the OMF files are created, including log files, in that one directory. If you omit both the DB_CREATE_FILE_DEST and the DB_CREATE_ONLINE_LOG_DEST_n parameters, Oracle9*i* does not create any OMF files.

Finding Redo Log Information in Dynamic Performance Views

The dynamic performance views display a variety of data about the online redo log files. Figure 5-6 shows a query of V$LOG dynamic performance view.

The status of a redo log group can be:

- **UNUSED:** Never used, such as just after being added, or after being cleared
- **CURRENT:** Currently in use by LGWR (the CURRENT status implies the group is also ACTIVE)
- **ACTIVE:** Needed for crash recovery, but not currently in use by LGWR

- **CLEARING:** In the process of being re-created after the ALTER DATABASE CLEAR LOGFILE command has been used
- **CLEARING_CURRENT:** Currently being cleared of a closed thread (not a normal state, occurs if there is an error during log switching)
- **INACTIVE:** Not needed for crash recovery

Table 5-2 shows the dynamic performance views that contain details on the redo log groups or their member files.

Table 5-2 Dynamic performance views about redo logs

View name	Partial list of columns	Description
V$LOGFILE	GROUP# STATUS MEMBER	Log group members, including file names. Status can be: INVALID: bad file; STALE: incomplete contents; DELETED: no longer available for use; BLANK: file in use
V$THREAD	THREAD# CURRENT_GROUP# CHECKPOINT_CHANGE#	In a database with one instance, THREAD# is always 1; shows current System Change Number (SCN), current redo log group number, and other up-to-date log details
V$LOG	GROUP# STATUS	Group details, including status of each group
V$LOG_HISTORY	SEQUENCE# FIRST_CHANGE# RECID	Information about archived logs from the control file; each archived log file has a control file record ID (RECID)
V$ARCHIVED_LOG	RECID NAME BLOCKS	One record for each archived log file, including the file name and time stamps of the archive
V$ARCHIVE_DEST	DESTINATION BINDING STATUS	Location of archived log files; binding can be MANDATORY (successful archive to this destination is required) or OPTIONAL (successful archive is not required)
V$ARCHIVE_PROCESSES	PROCESS STATE	The status of the ARCn processes, including their status and state; STATE can be IDLE or BUSY

As always, query the dynamic performance views like any other view.

The next section looks at another area in which information about the functioning and health of the database is stored: the diagnostic files.

OVERVIEW OF DIAGNOSTIC FILES

When a database is running, it may encounter errors caused by problems in a user's session or problems in the background processes. The database has a set of files that logs errors, warnings, or major events that occur in the database.

The three types of files that log database information to diagnose problems are:

- **Alert log file:** Contains notices of major events, such as the database startup and shutdown
- **Background trace file:** Logs error messages from any of the background processes
- **User trace file:** Logs error messages from user transactions, such as SQL errors if requested by the user

The files listed previously are automatically created by the database. The alert log file is created (if it does not already exist) when the database starts up. The other two files are created only when an error is encountered.

The default location of these files depends on the operating system. You can specify the location, size, and tracing level by setting these initialization parameters:

- **BACKGROUND_DUMP_DEST:** Location of the alert log file and the background trace files. This can be changed while the database is running with the ALTER SYSTEM command.
- **USER_DUMP_DEST:** Location of the user trace files: Logs error messages from user transactions, such as SQL errors. Change is allowed with the ALTER SYSTEM command.
- **MAX_DUMP_FILE_SIZE:** Maximum size (in kilobytes or megabytes or operating system blocks) of trace files. Specify UNLIMITED (the default) to place no limits on trace file size. Change the size for your session with ALTER SESSION or for the running database with ALTER SYSTEM.
- **SQL_TRACE:** Setting of TRUE means that all user sessions have trace files generated with information on their SQL activities. FALSE means no user sessions have trace files generated. The default is FALSE. Be cautious when setting this parameter to TRUE, because it generates numerous files quickly. It is preferable to use the ALTER SESSION command to turn on tracing for specific sessions one at a time.
- **TIMED_STATISTICS:** Setting of TRUE means that SQL trace functions can track the timing of SQL, such as CPU time and elapsed time. FALSE means that timed statistics are not gathered.

- **STATISTICS_LEVEL:** Level of statistics. This is a new parameter in Oracle9*i* Release 2. The parameter determines the level of statistics gathered for all the trace activities in the database. The BASIC level suppresses all statistics. The default level is TYPICAL, which includes timed statistics, and advisories from many processes, such as buffers and share pool sizing. ALL includes everything in TYPICAL, plus timed operating system statistics and row source execution statistics.

The alert log file is named **alert.log** and resides in the directory that is defined by the **BACKGROUND_DUMP_DEST** parameter. If you do not define BACKGROUND_DUMP_DEST, the **alert.log** file is placed in a default directory that is operating system dependent. The purpose of the alert log is to record messages that you ordinarily see if you are the systems operator watching the server's monitor. The information is recorded in the alert file so it can be reviewed at any time. The information is written in chronological order and consists of informative messages, as well as certain types of serious errors in the alert log:

- Internal errors, deadlock errors, and block corruption errors
- DBA commands, such as ALTER DATABASE, STARTUP, SHUTDOWN, CREATE TABLESPACE, and so on
- Certain error messages about dispatcher processes
- Errors found during automatic refresh of materialized views

Figure 5-9 shows an example of the contents of an alert log file.

The background trace files are created by each of the background processes, if the process encounters an error. The file names always contain the name of the background process. For example, a trace file generated by the LGWR process might be named **LGWR0205.log**. Trace files are located in the directory you specify in the BACKGROUND_DUMP_DEST initialization parameter. Monitor the trace files to determine if the background processes contain problems. For example, if the log writer background process (LGWR) found a corrupted member of a redo log group, it would write that error to the trace file.

The utility TKPROF reads trace files and loads the data into a readable file or even loads it into a database table, so you can query the results. Read about TKPROF in the online documentation book titled *Oracle9i Database Performance Tuning Guide and Reference*.

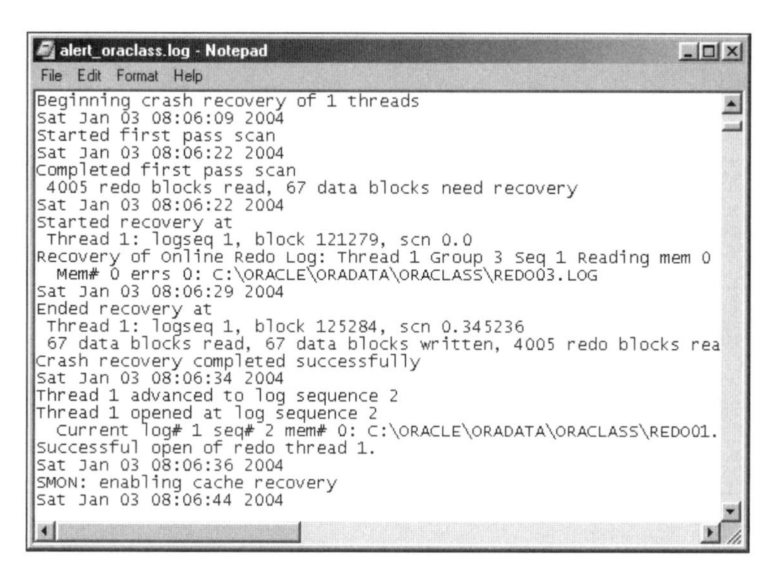

Figure 5-9 Routine notices as well as serious errors appear in the alert log file

Perform the following steps to view the current initialization parameters that are related to the trace files.

1. Start up the SQL*Plus Worksheet. In Windows, click **Start** on the Taskbar, and then click **Programs/Oracle - OraHome92/Application Development/SQLPlus Worksheet**. In UNIX, type **oemapp worksheet** on the command line. The standard Enterprise Manager login screen appears.
2. Type **SYSTEM** in the Username box, the current password for SYSTEM in the Password box, **ORACLASS** in the Service box, and leave the Connect as box displaying "Normal". Click **OK** to continue. The SQL*Plus Worksheet window displays. If the background SQL window overlays the worksheet, simply click the **minus sign** in the upper-right corner of the SQL window so that you can see the Worksheet window.
3. Type and execute these commands to check the current setting of statistics and the location of trace files.

```
SHOW PARAMETERS STATISTICS
SHOW PARAMETERS DUMP
SHOW PARAMETERS TRACE
```

Figure 5-10 shows the results. Your results will be similar, but the directory names may be different.

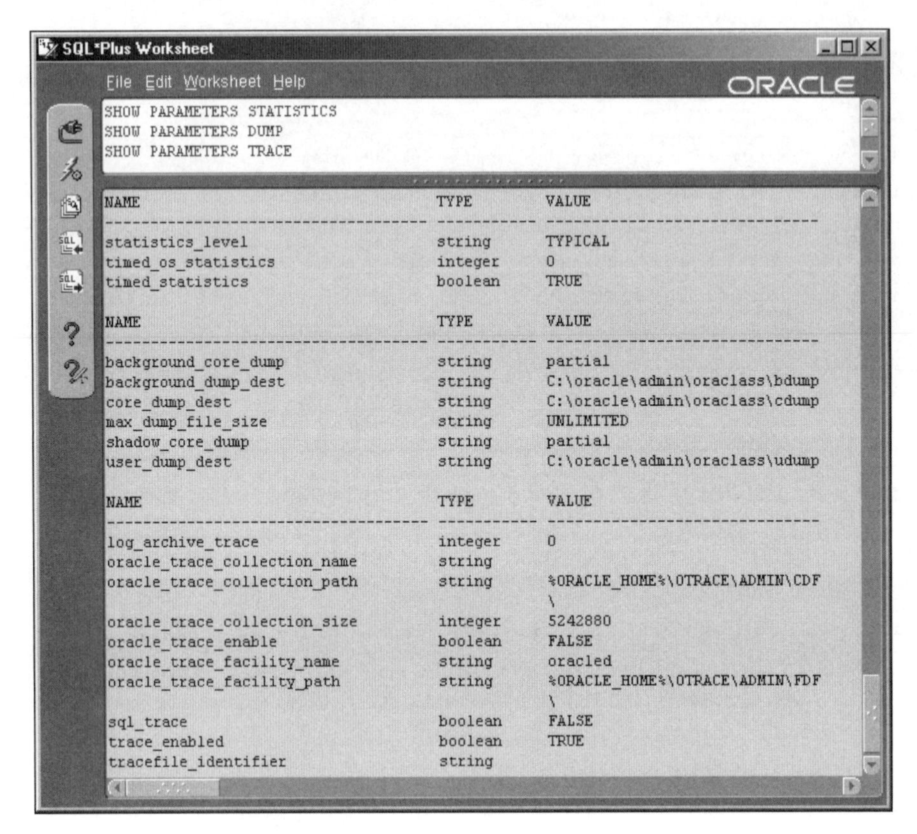

Figure 5-10 Statistics and tracing must be adjusted to gather information on SQL tasks

4. Log off SQL*Plus Worksheet by typing and executing **EXIT**. Then click **OK** in the dialog box.
5. Locate the directory in which background trace files are located. Examine the files in the directory. Note the file names, which tell you which process had errors or warnings.

The add-on for Enterprise Manager called Diagnostics Pack contains some excellent tools for gathering and reporting statistics. Third-party software is also available for diagnostics on Oracle databases.

Keep an eye on the **alert.log** file for early warnings of problems.

Chapter Summary

- Redo log files are also called online redo log files.
- Redo log files primarily contain information on database changes.
- Redo log groups always contain at least one file each.
- At least two redo log groups must exist for a database to function.
- One group at a time is active, and changing to another log group is called a log switch.
- Groups are reused sequentially, and log file data is lost unless the log group is archived.
- Change information is recorded in the redo log before updating the datafile.
- Redo logs can be used to recover from minor failures such as power outages.
- Serious database damage requires archived redo logs, a valid control file, and a database backup for recovery.
- The redo log buffer is part of the SGA.
- Redo log files contain redo records or redo entries made up of change vectors.
- A single transaction may generate many redo entries.
- When a redo log group contains more than one file, it is multiplexed.
- The SCN is incremented every time the database changes.
- A checkpoint flushes dirty buffers to be written to disk.
- When redo log files are multiplexed, one damaged file does not cause a system error.
- The command ALTER DATABASE ADD LOGFILE MEMBER ... TO GROUP n creates a new redo log file.
- ALTER DATABASE ADD LOGFILE GROUP creates a new redo log group.
- Renaming redo log files must occur with the database in NOMOUNT mode and is done with the ALTER DATABASE RENAME FILE command.
- ALTER DATABASE DROP LOGFILE MEMBER removes a redo log file from the database, but does not remove the physical file (except if it is an Oracle Managed File).
- ALTER DATABASE DROP LOGFILE GROUP removes an entire redo log group.
- ALTER DATABASE CLEAR LOGFILE GROUP removes corrupted data from a group.
- Place a database in ARCHIVELOG mode with the ARCHIVELOG parameter in the CREATE DATABASE command.

- After the database is created, change to ARCHIVELOG mode with the ALTER DATABASE ARCHIVELOG command or by setting LOG_ARCHIVE_START to <TRUE>.
- The only required parameter for ARCHIVELOG mode is LOG_ARCHIVE_DEST_n or the deprecated parameter (now only for Personal Oracle) LOG_ARCHIVE_DEST.
- LOG_ARCHIVE_FORMAT initialization parameter provides a naming pattern for archived redo log files.
- View current parameter settings with the SHOW PARAMETERS command.
- Use DB_CREATE_FILE_DEST or DB_CREATE_ONLINE_LOG_DEST_n to implement OMF with redo log files.
- Specifying more than one DB_CREATE_ONLINE_LOG_DEST_n parameter causes OMF to create multiplexed redo log files.
- The V$LOG dynamic performance view displays redo log group status, with CURRENT meaning the group is in use.
- The V$LOGFILE shows redo log member status, with blank status indicating the file is in use.
- The alert log file should be monitored regularly to detect errors.
- The background trace files record errors that occur within the background processes.
- The user trace files log errors from user transactions.
- BACKGROUND_DUMP_DEST and USER_DUMP_DEST initialization parameters set the location of the background trace files and the user trace files.

Review Questions

1. Describe the difference between ARCHIVELOG mode and NOARCHIVELOG mode.
2. When is the redo log buffer written to the redo log file? (Choose all that apply.)
   a. A transaction changes data.
   b. The redo log buffer becomes one-third full.
   c. A redo log group is archived.
   d. The database starts up.
3. There is one change vector per data block in the redo log entry. True or False?

4. Look at Figure 5-3. Which of the following statements are true? (Choose 2.)
   a. LGWR fills **Redolog3.log** before writing to **Redolog2.log**.
   b. A log switch changes from Drive E to Drive F.
   c. LGWR fills **Redolog1.log** before writing to **Redolog3.log**.
   d. A log switch changes from Group 1 to Group 2.
5. You have determined that the ARCn process is causing the LGWR process to wait while it archives a log group. What should you do?
6. You have two redo log groups. Each group has three members that are 150 megabytes in size. The members reside on the G, E, and F drives. You issue this command:

```
ALTER DATABASE ADD
LOGFILE MEMBER 'G:\ora\logs\redo0102.log'
      SIZE 200M TO GROUP 1;
```

   The statement will fail because:
   a. All redo log groups must have the same number of members.
   b. You cannot add a member to a redo log group.
   c. The size is incorrect.
   d. The size cannot be specified in this command.
7. You have two multiplexed redo log groups, and all the members reside on a single disk. Explain why this defeats the purpose of multiplexing.
8. When a checkpoint occurs, the ________________ background process writes ________________ buffers to disk.
9. Define SCN.
10. If all the members of the current redo log group become damaged, what happens?
11. When is it appropriate to use LOG_ARCHIVE_DUPLEX_DEST?
12. To find the number of the current redo log group, query the ________________ view.
13. Set the location of the alert log file with the ________________ parameter.

Exam Review Questions—Database Fundamentals I (#1Z0-031)

1. Redo log files store information about these types of events. Choose three.
   a. Checkpoints
   b. Archived redo logs
   c. UPDATE commands
   d. Database shutdowns
   e. Log switches

2. A database is in ARCHIVELOG mode, and there are two redo log groups. Group 1 is current and filled. Group 2 was just archived. What happens next? Place these events in order of occurrence.

   1. Archive Group 2
   2. Log switch to Group 2
   3. Log switch to Group 1
   4. Archive Group 1
   5. Fill Group 2

   a. 1, 2, 5, 4, 3

   b. 2, 4, 1, 5, 3

   c. 2, 4, 5, 3, 1

   d. 1, 2, 4, 5, 3

3. A database is in ARCHIVELOG mode, and there are two redo log groups. Group 1 is current and Group 2 has been archived. Then Group 1 becomes full. In what order do the following five events occur?

   1. Archive Group 2
   2. Log switch to Group 2
   3. Log switch to Group 1
   4. Archive Group 1
   5. Fill Group 2

   a. 1, 2, 5, 4, 3

   b. 2, 4, 1, 5, 3

   c. 2, 4, 5, 3, 1

   d. 1, 2, 4, 5, 3

4. Your database is in NOARCHIVELOG mode. You have updated some data in the database. Suddenly, the power fails. You did not have a chance to commit your updates. Which of these statements is true regarding the redo log file?

   a. Your changes were recorded in the redo log buffer but not in the redo log file.

   b. Your changes will be restored only if they were recorded in the redo log file.

   c. The redo log file will be reset, and your changes will be lost.

   d. Your changes will not be restored even if they were recorded in the redo log file.

5. Your database has two redo log groups on one disk. You want to multiplex the groups. Which of the following actions is the best strategy?
   a. Create a new redo log group on a separate disk.
   b. Move one group's members to a separate disk.
   c. Move one group to a separate disk.
   d. Create a new member in each group on a separate disk.
6. You issue the following command, which completes successfully:

```
ALTER DATABASE
   ADD LOGFILE GROUP 3 SIZE 500K;
```

   What do you know about the database? Choose three.
   a. The database is in ARCHIVELOG mode.
   b. The database already has two redo log groups.
   c. The redo log files are managed by OMF.
   d. The database is started and in NOMOUNT mode.
   e. The database is started and in MOUNT mode.
7. The following initialization parameters are set:

```
DB_CREATE_FILE_DEST='D:\oracle\data'
DB_CREATE_ONLINE_LOG_DEST_1 = 'D:\oracle\logs'
DB_CREATE_ONLINE_LOG_DEST_2 = 'E:\oracle\logs'
```

   You execute the following command:

```
CREATE DATABASE TESTXYZ;
```

   Which of the following statements is true?
   a. The statement fails because it is missing the LOGFILES clause.
   b. The statement succeeds, and the redo log files are multiplexed.
   c. The statement fails because it is missing the DATAFILE clause.
   d. The statement succeeds, and two redo log files are created.
8. It is 11:00 AM on June 24, 2004. Your database has a single instance and is about to archive the redo log file assigned log sequence 21. The LOG_ARCHIVE_FORMAT is **my%S%t.log**. What is the new archive log file named?
   a. my210001.log
   b. my000240001.log
   c. my211100.log
   d. my000211.log

5

9. If there is no value set for the directory that is defined by the BACKGROUND_DUMP_DEST parameter, where do the background trace files reside?
   a. USER_DUMP_DEST
   b. ORACLE_HOME
   c. The location defaults to a directory whose name is operating system dependent.
   d. If the parameter is not set, no trace files are created. Thus, there is no location.
10. Which of the following statements correctly describes renaming a log file from '**D:\oralog\redo01.log**' to '**E:\oralog\redo01.log**'?
    a. SHUTDOWN IMMEDIATE; move file to new destination; STARTUP MOUNT; RENAME FILE...; STARTUP OPEN;
    b. SHUTDOWN IMMEDIATE; move file to new destination; STARTUP NOMOUNT; RENAME FILE...; STARTUP OPEN;
    c. SHUTDOWN IMMEDIATE; move file to new destination; STARTUP MOUNT; RENAME FILE...; ALTER DATABASE OPEN;
    d. Move file to new destination; SHUTDOWN IMMEDIATE; STARTUP MOUNT; RENAME FILE...; ALTER DATABASE OPEN;

HANDS-ON ASSIGNMENTS

1. Your office database was created by the previous DBA who left for parts unknown. You are reviewing the database setup. You want to know the answers to these questions:
   1. What are the redo log group numbers, and how many members does each group have?
   2. What directory or directories holds the redo log group members?
   3. Are there any archived redo log files? If so, where are they located? How many files are there?

   Write SQL commands or queries to discover all these answers. Save your SQL script in a file named **ho0501.sql** in the **Solutions\Chapter05** directory on your student disk.
2. Continuing with the database left to you by a wayward DBA in Assignment 1, you discover the following facts:
   - All the members of redo log Group 1 reside on the same drive in this directory: **F:\group1**.
   - Group 2 members are on the same drive as well, in: **F:\group2**.

- Group 1 has two members, and Group 2 has three members.
- Each member is named MEM<gg><mm>.log (where gg is the group number and mm is the member number). For example, the first member of Group 2 is: **MEM0201.log**.

Assuming that the files are not OMF, write all the SQL commands required to spread the members across the F, G, and H drives and add another member to Group 1. Keep the member names the same, but use a directory named **oracle\redolog** in each drive. Save your script in a file named **ho0502.sql** in the **Solutions\Chapter05** directory on your student disk.

3. Use the online documentation to determine which of the initialization parameters in this chapter are static, which are dynamic, which can be changed using ALTER SESSION, and which can be changed using ALTER SYSTEM. Make four lists (one for each category). Some parameters appear on more than one list. Save your findings in a file named **ho0503.txt** in the **Solutions\Chapter05** directory on your student disk.
4. Study Figure 5-11, and list all the initialization parameters with the values you can determine from the figure. Write the CREATE DATABASE command for this database. Assume the name of the database is CHAP5. Save the initialization parameters and the CREATE DATABASE command in a file named **ho0504.sql** in the **Solutions\Chapter05** directory on your student disk.

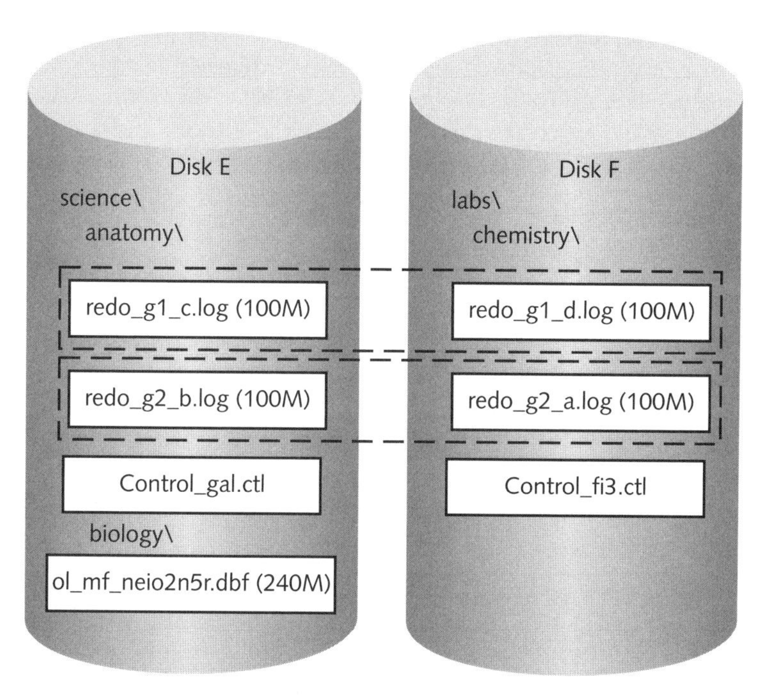

Figure 5-11 System diagram of a theoretical database

5. Looking at Figure 5-11, write the SQL command to add a new redo log group with two members that are 250 kilobytes in size and match the distribution of the other two groups.

   Then write a query to confirm that the changes have taken effect by displaying the status of every redo log member and its group association and location. Save your work in a file named **ho0505.sql** in the **Solutions\Chapter05** directory on your student disk.

6. Find the trace files for the background processes. Look in the files, and find three error, warning, or informational messages. They begin with "ORA-" in most cases. For each message, list:

   - The file name in which the message was found
   - The message text
   - The date and time of the message
   - The background process that generated the message

   Save your list in a file named **ho0506.txt** in the **Solutions\Chapter05** directory on your student disk.

7. You plan to create a new database named PRODOMF that uses OMF for the control files, redo log files, and datafiles. You have three drives: E, F, and G. You want to multiplex the redo log files and control files across F and G and store datafiles in E. The log files should be 15 megabytes in size, and the SYSTEM datafile should be 500 megabytes. Create a partial **init.ora** file to set all the parameters you need when preparing to create the database. Write the CREATE DATABASE command to create the database. Save the script and a file named **ho0507.sql** in the **Solutions\Chapter05** directory on your student disk.

8. Add archiving to the scripts you created for the previous assignment by adjusting the CREATE DATABASE command. Then write steps to turn on ARCHIVELOG mode three different ways. Be sure to include shutting down and restarting the database if needed. Save the script and a file named **ho0508.sql** in the Solutions\Chapter05 directory on your student disk.

9. Experiment with missing redo log files. Your task is to test the documentation on how the database behaves when one or more online redo log files are missing. A log switch can happen as long as there is one good file remaining in the next redo log group. The database shuts down if a log switch begins, and there are no good files in the next redo log group. To test this, perform the following steps:

   1. Determine which redo log group is current.
   2. Using your operating system (not Oracle) to rename all the redo log files of the non current group by adding a letter "x" to the end of the file name. For example, if the file name is **redo0101.log**, the file should be renamed **redo0101x.log**.
   3. Force a log switch. What happens?

4. Rename the redo log files back to their original position. Restart the database. Which redo log group is now current?

Save all your commands and queries in a file named **ho0509.sql** in the **Solutions\Chapter05** directory on your student disk.

10. After completing the Assignment 5-9, the database was deliberately broken and restored. Review the log files and find the errors that were generated. Make a list of all the errors. Look up the errors in Oracle documentation and determine if the solution information is relevant to what you know caused the error. Write your opinion. Save the list of errors, a brief list of Oracle's suggested solutions, and your opinion of Oracle's suggested solutions in a file named **ho0510.txt** in the **Solutions\Chapter05** directory on your student disk.

CASE PROJECT

At the Global Globe Newspaper Company, your Enterprise Edition database currently is in NOARCHIVELOG mode. After reading this chapter, you convince your MIS Manager that the database must be made more fail-safe by changing it to ARCHIVELOG mode. Your manager orders the systems operator to allocate two new disk drives for your use. You plan to use them for the archived redo log files. Now, you must plan all the steps needed to complete the task.

Ask your instructor for two directories in which you should place the archive logs. If needed, create the directories. These two directories simulate two disks in the Global Globe Newspaper Company's database server. You want the database to automatically archive the log files on both the new drives (directories) in subdirectories called **oracle\logs**. Both archive locations must successfully be written for the archive to succeed. None of the directories exist yet. You think that four archive processes should be used when the database starts up.

Copy the current **init.ora** file into a file named **case0501init.ora** in the **Solutions\Chapter05** directory on your student disk. Modify it to include the parameters needed to put all these requirements in place. Then create a checklist of steps you should perform in the operating system and in the database to put all the changes into effect and begin archiving the redo log files. Remember to specify the new **init.ora** file you created during startup. (*Hint*: One of the steps requires you to shut down the database.) Perform the steps, and verify with queries that the database is in ARCHIVELOG mode, the archive directories, and the number of processes. Force an archive of the redo log group. Check the archived log files with a query. Finally, restart the database so it returns to its NOARCHIVELOG mode. Confirm with the SHOW PARAMETERS ARCH command that archiving is stopped. Save the script in a file named **case0501.sql** in the **Solutions\Chapter05** directory on your student disk.

CHAPTER

6

BASIC STORAGE CONCEPTS AND SETTINGS

In this chapter, you will:

- Differentiate between logical and physical structures
- Create many types of tablespaces
- Configure and view storage for tablespaces and datafiles
- Use undo data

Previous chapters have focused on creating the database and the processes that keep it running smoothly. Beginning with this chapter, the focus shifts to the internal structure of the database. Topics include the components that you set up to hold user data, as well as the monitoring of user resource consumption and security access.

This chapter begins with the foundation structures that the database uses to store data. You learn about the logical structures and how they map to the physical structures. You see how to configure these structures and what they are used for. You learn the internal structures needed to speed up performance. The chapter ends with a case project in which you set up structures for use in later chapters, as you add actual data, tables, and users to the database.

INTRODUCTION TO STORAGE STRUCTURES

The Oracle9*i* database has an internal set of structures that are used to store all the data, users, constraints, data dictionary views, and any other objects you want to create in the database. These structures also contain metadata maintained internally by the database. **Metadata** is data that tells Oracle9*i* about all the structures that store data in the database. For example, the table containing the list of each table's columns, their data types, lengths, and so on, is a table of metadata. The data dictionary views primarily display metadata.

Recall from Chapter 1 that the database server has several components. The database software is installed on the server. The **instance** is made up of the memory (SGA) and the background processes, and the **database** is made up of database files, control files, and redo log files. Figure 6-1 shows the components of the database server.

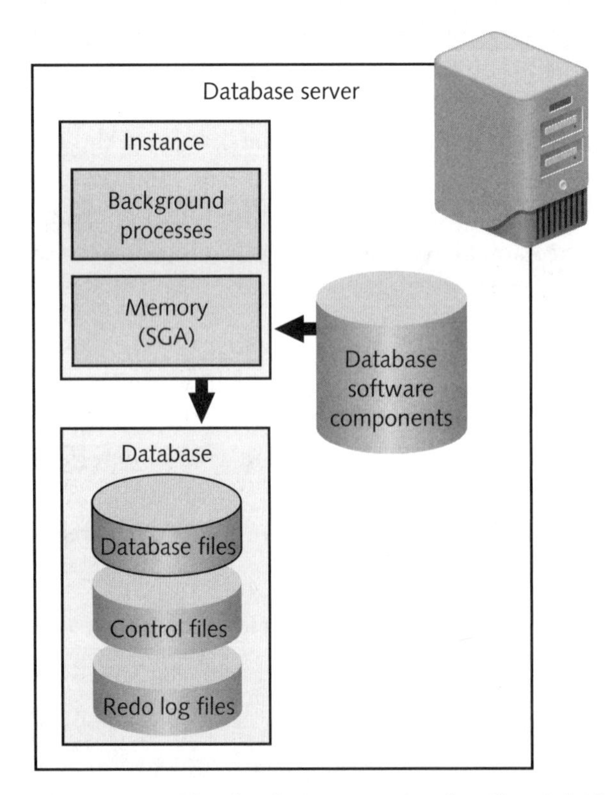

Figure 6-1 The database server is a familiar sight by now

Previous chapters examined all the components of the database except database files. This chapter takes a close look at this final component. The database files are not just ordinary files. Like other files, database files have a physical structure. But they also have a logical structure that forms the building blocks of the database system.

Logical Structure Versus Physical Structure

Physical structures are composed of operating system components and have a physical name and location. Physical structures can be seen and manipulated in the computer's operating system. **Logical structures** are composed of orderly groupings of information that allow you to manipulate and access related data. Logical structures cannot be viewed or modified outside the database. Logical structures are always associated with one or more physical structures.

Physical structures include:

- **Datafiles:** A datafile is a physical file on the computer. It is sometimes called a database file. It is the primary physical structure that contains all the data you store in the database. When you create a new database, one datafile is required, the datafile that will store the SYSTEM tablespace (a logical structure). Figure 6-2 shows several datafiles in the database. Like other files, datafiles have a size and location you can see in the operating system. A datafile can be made up of several fragmented sections in different locations on the operating system. This allows Oracle9*i* to change the size of the datafile. In Figure 6-2, the datafile named **system01.dbf** has two sections. The size of a datafile is defined in bytes, but the smallest unit with which the operating system works is called an operating system block. An **operating system block** is made up of a group of **contiguous** bytes (bytes right next to one another) within the file. The operating system block size and the logical database block size (see the bullet on data blocks that follows) are closely related and should be considered when setting the size of a datafile. The operating system block defines the units used by the operating system to handle I/O. For example, when you save a document on a disk, the data is placed onto the physical disk one block at a time, rather than one byte at a time. Handling data in blocks makes I/O efficient. The operating system block size varies from machine to machine. For example, Advanced Interactive eXecutive (AIX) systems have a block size of 4 K.
- **Redo log files:** As you saw in Chapter 5, redo log files contain redo entries that record changes to the database to allow recovery of changed data. Redo log files are handy when the database loses the data because of a failure of some kind.
- **Control files:** Chapter 4 covers the structure and contents of control files. The control files are essential for database startup and recovery. The control files contain the names and locations of all the datafiles and redo log files. Changes to the physical structure of the database, such as adding or dropping a datafile, are automatically updated in the control file.

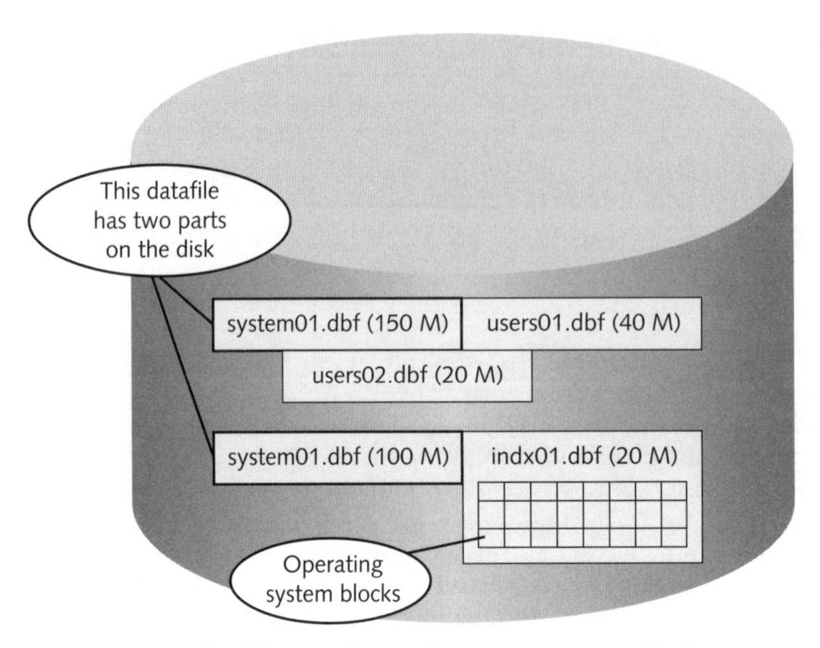

Figure 6-2 A datafile is made up of operating system blocks

Logical structures are shown in Figure 6-3 with example names of each component.

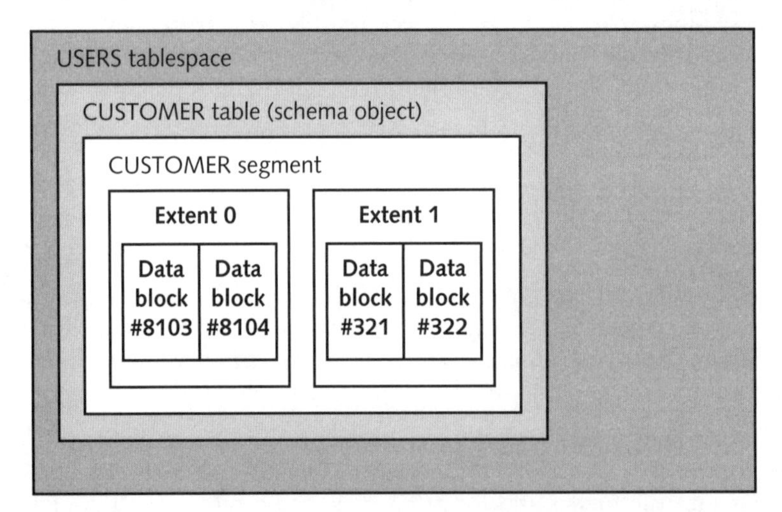

Figure 6-3 A tablespace is the largest logical structure and a data block is the smallest

Logical structures include these components, from smallest to largest:

- **Data block:** A data block is the smallest of the logical units. A data block is made up of a set number of physical bytes in a physical file. Like operating system blocks, Oracle reads and writes the size of its data block in chunks.

You set the exact number of bytes per data block when you create the database. A data block is typically 4 K or 8 K. It can't be smaller than 2 K or larger than 32 K. Data blocks are made up of one or a multiple of operating system blocks, because this creates more efficient use of the operating system for I/O operations. Set the block size to optimize the features of your database server. For example, if your server has plenty of memory and fast disk drives, you could safely set the data block size at four times the operating system size to reduce overall I/O operations and improve response time. On the other hand, if your system has less memory or slower disk drives, set Oracle's data block size exactly equal to the operating system block size to best use the available resources. The initialization parameter DB_BLOCK_SIZE contains the block size. After a database is created, you cannot change the block size without re-creating the database; however, you can create tablespaces (see the "Tablespaces" item later in this list) that have a nonstandard block size. Each data block is either assigned to one extent or is available **free space** in the datafile. Free space is defined as the group of data blocks in the datafile that are available for future allocation to an extent. Each extent contains many data blocks.

- **Extent:** An extent is a contiguous group of data blocks that are assigned to a segment (see the "Segment" item later in this list). When more space is needed for an object, such as a table, the space is allocated in the form of one extent. Extents can be different sizes, depending on the storage parameters set for the object or the tablespace (when not specified by the object). For example, you might create a CUSTOMER table and set the storage parameters, so that the first extent (the initial extent that was allocated upon creation of the table) is 20 M and the next extent (the extent allocated when there is no more room in the first extent) is 5 M. Extents always belong to one segment. Each segment can contain more than one extent.
- **Segment:** A **segment** is the set of extents that make up one schema object within a tablespace. Segments have several different uses, including being the storage holders for a schema object within a tablespace. Segments are discussed in more detail later in this chapter. Each segment belongs to one schema object. Each schema object has only one segment, except for partitioned tables and partitioned indexes, which have one segment per partition.
- **Schema object:** A **schema object** includes the wide variety of objects that can be created by users in the Oracle9*i* database. Tables and indexes are probably the most common and well-known types of schema objects. Other types of objects include views, procedures, Java-stored procedures, object tables, and sequences. Each schema object must be contained within one tablespace, with the exception of partitioned tables and partitioned indexes. In the case of partitioned tables and indexes, each partition must be contained within one tablespace, but the partitions for one partitioned table or index may reside on different tablespaces.

- **Tablespace:** A tablespace is the largest logical object. As you saw in Chapter 3, creating a database requires creating at least one tablespace: the SYSTEM tablespace. A database may contain a large number of tablespaces. Each tablespace contains one or more datafiles. Tables, indexes, and other objects are created within a tablespace. The storage capacity of a tablespace is the sum of the size of all the datafiles assigned to that tablespace. A datafile can be associated with only one tablespace within one database. The contents of the datafiles are stored in encrypted and specialized record structures readable only by the Oracle9*i* database. You cannot open that file and view the data as you would a regular text file; however, you can view all the data stored in the datafile by using SQL to query the logical components (the tables, for example) that are stored in the datafile.

Look again at Figure 6-3. The examples listed in the figure illustrate typical examples of names of each logical component. The tablespace example is the USERS tablespace. The schema object example is the CUSTOMER table. Within the CUSTOMER table is the CUSTOMER segment. Segments have the same name as their parent schema object (or the name of their parent partition in the case of partitioned tables and indexes). The CUSTOMER segment has two extents identified as Extent 0 and Extent 1. Extents do not have specific names and are always numbered sequentially starting at zero for each segment. Looking at Extent 0, there are two data blocks: Block 8103 and Block 8104. Each data block is numbered sequentially starting with zero, but the numbering is relative to its position in the datafile, not the extent. This is where the logical and the physical components intersect.

Figure 6-4 shows the connection between the logical components and the physical components of the database. As shown in this figure, a datafile is always associated with one tablespace. In addition, an extent is always associated with a contiguous group of data blocks within one datafile. There is no direct correlation between a segment and a datafile. A segment can be spread among several datafiles as long as the datafiles belong to the tablespace that contains the segment. Likewise, there is no direct correlation between a schema object and a datafile. Data blocks are made up of one or more operating system blocks, depending on the size of each type of block.

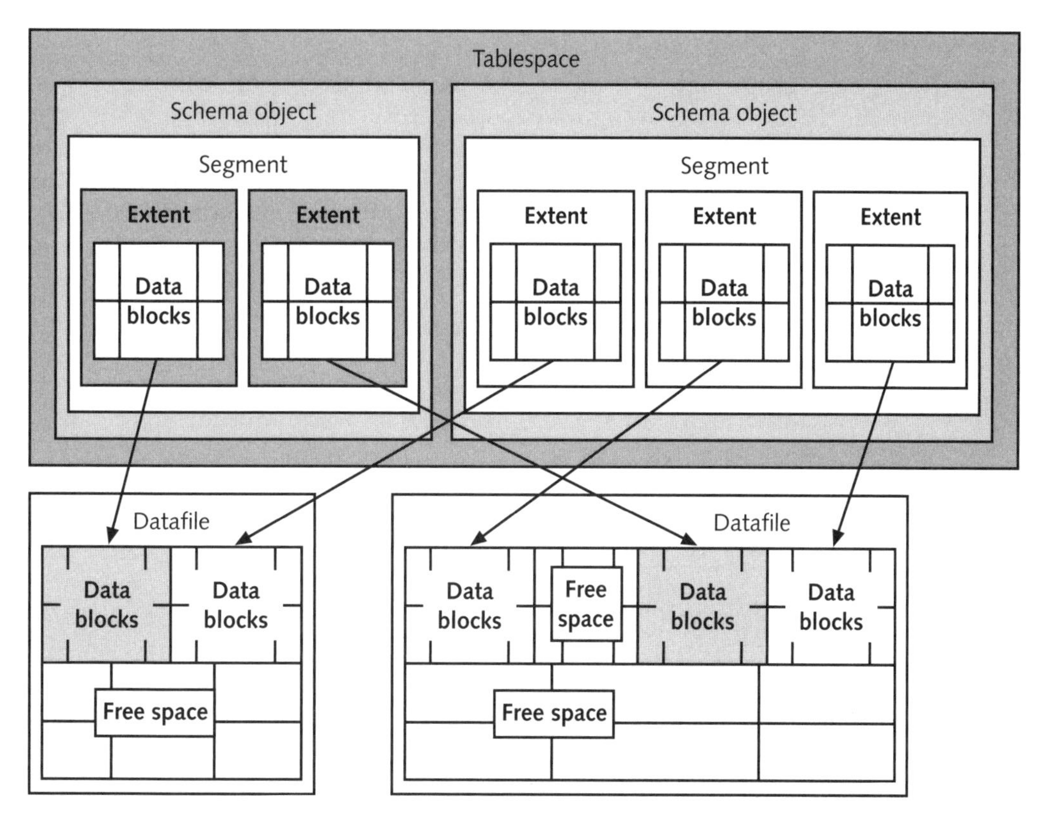

Figure 6-4 Ties between the logical and the physical components occur at several levels

The next section shows you how to create and configure tablespaces and datafiles.

Tablespaces and Datafiles

In this chapter, you learn how to create, manage, remove, and view metadata about tablespaces and datafiles. Tablespaces are always made up of at least one datafile. In fact, you cannot create a tablespace without also creating its initial datafile. Similarly, you cannot create a datafile without an associated tablespace. You can, however, create a new datafile to add to an existing tablespace.

Here is the syntax of the CREATE TABLESPACE command that includes clauses that are expanded later in this section.

```
CREATE TABLESPACE <tablespacename>
DATAFILE <filename> SIZE <nn> AUTOEXTEND ON|OFF
TEMPORARY|PERMANENT
EXTENT MANAGEMENT LOCAL|DICTIONARY
LOGGING|NOLOGGING
ONLINE|OFFLINE
SEGMENT SPACE MANAGEMENT MANUAL|AUTO
```

When you create a tablespace, you specify your choices for:

- **Temporary or permanent data:** **Temporary tablespaces** store objects during a session only. In addition, they are used for sorting data if you have assigned a temporary tablespace to a user. There are actually two formats for creating temporary tablespaces. See the section titled "Temporary Tablespaces" later in this chapter for the syntax and special considerations of this type of tablespace. **Permanent tablespaces** are the default type and store permanent objects, such as tables and indexes.
- **Local- or dictionary-management:** A **locally-managed tablespace** (the default) has a bitmap inside the tablespace that stores all the details about free space, used space, and location of extents. A **dictionary-managed tablespace** stores the details about its free space and other information inside the data dictionary tables in the SYSTEM tablespace. Locally-managed tablespaces are more efficient because less time is spent looking up locations of data.
- **Storage settings:** The storage settings include details on how Oracle9*i* allocates space within the tablespace, its initial size, and more.
- **Datafile information:** You can define one or more datafiles to be created for this tablespace. Specify the initial size of the datafile and whether it is allowed to automatically grow (AUTOEXTEND).
- **Online or offline:** A tablespace that is **offline** is not available for use, whereas one that is **online** is available (this is the default).
- **Logging or no logging:** This sets the default behavior for logging certain types of changes on objects in the tablespace. LOGGING is the default and specifies that all DDL commands and all mass INSERT commands (such as those issued by SQL*Loader) be recorded in the redo log buffer. NOLOGGING means that these two types of commands contain minimal logging. NOLOGGING speeds up the processing time, but in recovery, the changes must be redone manually by rerunning the original commands. A particular object can override the tablespace defaults. Other types of changes to objects, such as updates, inserts other than those done by SQL*Loader, and deletes, are always recorded in the redo log buffer, regardless of whether you choose LOGGING or NOLOGGING. The complete syntax for the CREATE TABLESPACE would fill a whole page. To view all the details, look at Oracle9*i* online documentation (choose **SQL and PL/SQL syntax and examples** from the main menu, and then look up CREATE TABLESPACE in the alphabetical list of commands).

The next sections go through examples of creating tablespaces using these parameters and clauses.

Implementing OMF with Tablespaces

When you use OMF, you allow the database to determine the actual names of datafiles, leave out the DATAFILE clause entirely, and minimize the details you need to supply for a tablespace. Complete the following steps to look at the existing tablespaces on the ORACLASS database and to create a locally-managed tablespace that is named USER_TBS.

1. Start up the Enterprise Manager console. In Windows, click **Start** on the Taskbar, and then click **Programs/Oracle - OraHome92/Enterprise Manager Console**. In UNIX, type **oemapp console** on the command line. The Enterprise Manager console login screen appears.
2. Select the **Launch standalone** radio button and click **OK**. The console appears.
3. Double-click the **Databases** folder, and then double-click the **ORACLASS** database icon. A list of tools displays below the database name.
4. Double-click the **Storage** icon to open the Storage Manager.
5. Click the **Tablespaces** folder. The right pane displays a list, as shown in Figure 6-5, showing the type of free space management (LOCAL or DICTIONARY), and the current amount of used and free space in each tablespace.

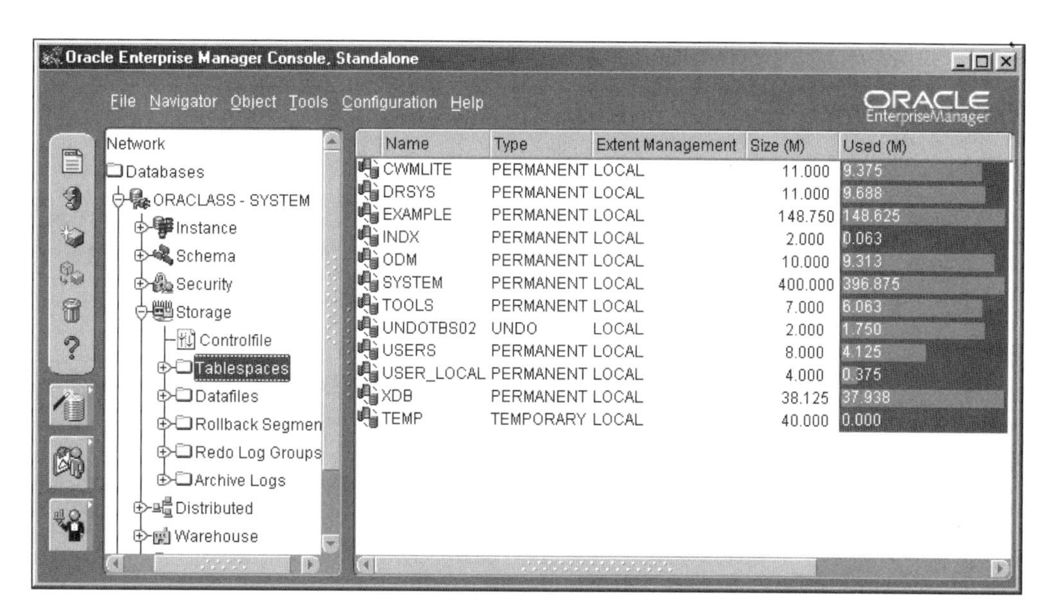

Figure 6-5 The used space of each tablespace appears as a bar graph

6. Double-click the **USERS** tablespace. The property sheet for the tablespace displays as shown in Figure 6-6. Here you can see that this is a permanent tablespace that is online and has one datafile.

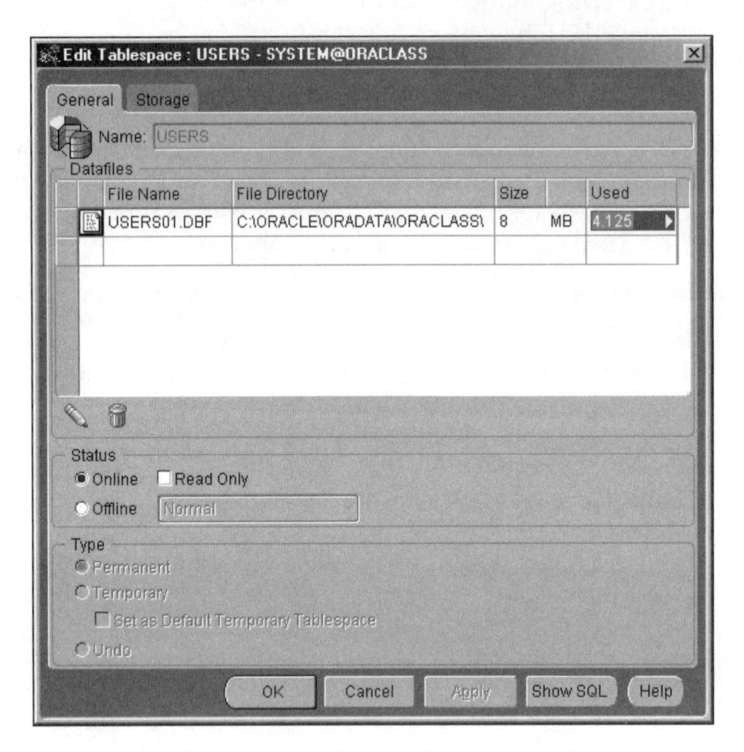

Figure 6-6 The USERS tablespace has one datafile

7. Click the **Storage** tab to display the storage settings for the tablespace as you see in Figure 6-7. The brief explanations of each setting are a useful feature of the console.

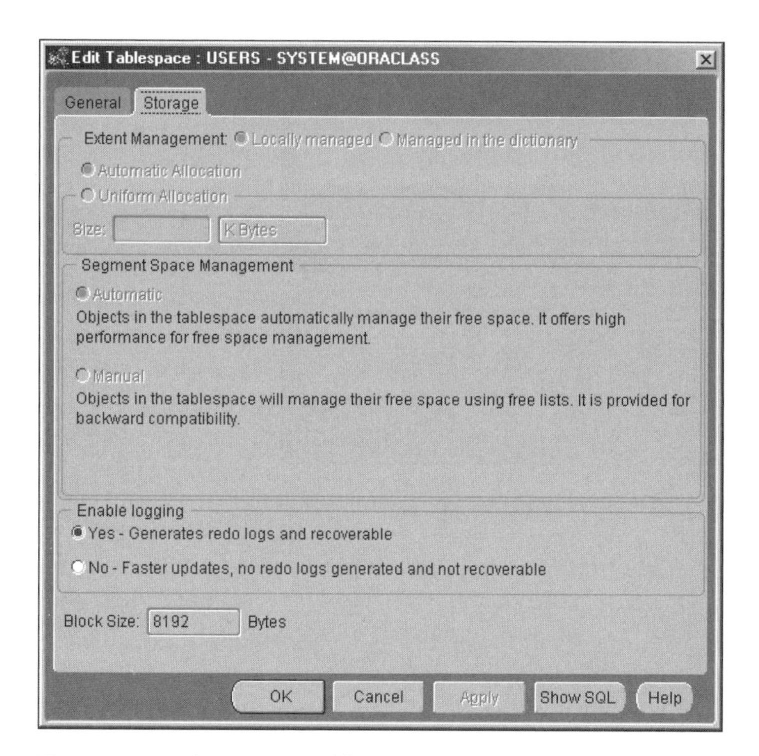

Figure 6-7 The USERS tablespace storage settings appear here

8. Click **Cancel** to return to the main console.
9. Start up the SQL*Plus Worksheet by clicking **Tools/Database Applications/SQL*Plus Worksheet** from the top menu in the console. A background SQL*Plus process starts, and then the SQL*Plus Worksheet window appears. If the background process appears in front, simply minimize it (click the **minus sign** in the top-right corner) so you can see the worksheet.
10. Type and execute the following command to check on the OMF parameters. The DB_CREATE_FILE_DEST must be defined before you can create an Oracle managed tablespace. Figure 6-8 shows the results.

    ```
    SHOW PARAMETERS DB_CREATE
    ```

11. Type and execute the following command to set the DB_CREATE_DEST for your current session. Replace <X:\xxx> with the actual directory where the new tablespace will be created. Your instructor will provide you with the exact directory name.

    ```
    ALTER SESSION SET DB_CREATE_FILE_DEST='<X:\xxx>';
    ```

    SQL*Plus Worksheet replies, "Session altered."

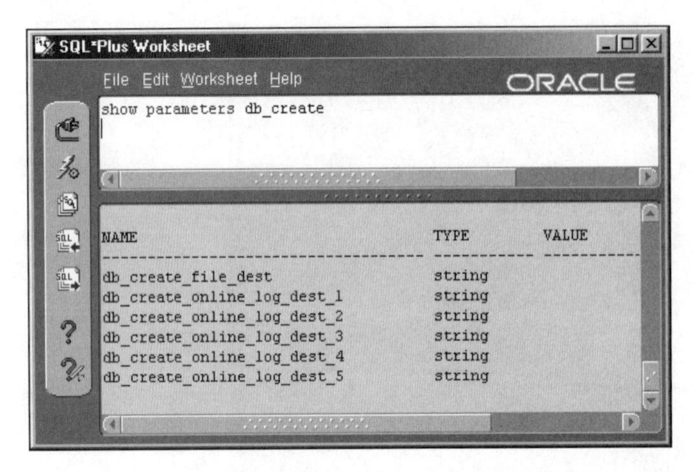

Figure 6-8 The DB_CREATE_FILE_DEST parameter is null

12. Type and execute the following command to create an Oracle managed tablespace. This command contains no additional parameters, so the default settings for OMF are used. The datafile is 100 M, named by the system, can automatically extend to an unlimited size.

    ```
    CREATE TABLESPACE USER_OMF1;
    ```

    SQL*Plus Worksheet replies, "Tablespace created."

13. Type and execute this command to create another tablespace using OMF. This time, your datafile is 20 M in size:

    ```
    CREATE TABLESPACE CUST_TBS
    DATAFILE SIZE 20M;
    ```

    SQL*Plus Worksheet replies, "Tablespace created."

14. Remain logged onto the console and SQL*Plus Worksheet for the next practice.

The next section shows how to create tablespaces with user managed datafiles.

The DATAFILE Clause

When creating a tablespace with a user managed file, specify a datafile name in the command. The syntax of the DATAFILE clause is:

```
DATAFILE '<datafilename>' SIZE <nn>|REUSE
AUTOEXTEND ON MAXSIZE <nn>|UNLIMITED|AUTOEXTEND OFF
```

- You can have more than one datafile, if you want, by simply listing more files in the DATAFILE clause.

- Each datafile must have a specified SIZE. The only exceptions are Oracle managed files and named files that already exist. OMF makes files 100 M by default. If the file already exists, you must also specify REUSE, and then either leave out the SIZE to keep the same size file, or include the SIZE to adjust the existing file's size. All data in an existing file is wiped out.
- By adding AUTOEXTEND ON, a datafile can be automatically extended up to MAXSIZE. For Oracle managed files, if you leave off this parameter and leave off the SIZE parameter, you get an AUTOEXTEND ON by default. You can specify that the datafile has no maximum size by specifying MAXSIZE UNLIMITED.

Follow these steps to create a user managed tablespace.

1. Type and execute the following command in your current SQL*Plus Worksheet session to create a tablespace named ACCOUNTING that has a user managed datafile. To make a user-managed datafile, simply include a name in the DATAFILE clause. Remember to replace <X:\xxx> with the actual directory provided by your instructor.

```
CREATE TABLESPACE ACCOUNTING
DATAFILE '<X:\xxx>\cust_tbs01.dbf' SIZE 10M;
```

   SQL*Plus Worksheet replies, "Tablespace created." Here are some other points about the DATAFILE clause:

2. Try out these options by typing and executing the following command. Here is an example of a tablespace with two datafiles. The first datafile is set at 20 M and can be automatically extended in 2 M increments to 200 M. The second datafile is initially 2 M and can expand to an unlimited size. Both files are user managed files.

```
CREATE TABLESPACE NEWS
DATAFILE '<X:\xxx>\news01.dbf'
     SIZE 20M AUTOEXTEND ON NEXT 2M MAXSIZE 200M,
     '<X:\xxx>\news02.dbf'
     SIZE 2M AUTOEXTEND ON MAXSIZE UNLIMITED;
```

3. Remain logged on for the next practice.

Changes to data are written to datafiles by the DBWn background process. To be more efficient, changes are accumulated in memory and then written later to the physical datafile. If the datafile needs more space, it is allocated when the memory buffers are written to the datafile. Without the AUTOEXTEND ON clause, transactions that cause the datafile to run out of space (such as an INSERT command) cause an error.

The EXTENT MANAGEMENT and SEGMENT SPACE MANAGEMENT Clauses

The EXTENT MANAGEMENT and SEGMENT SPACE MANAGEMENT clauses tell Oracle9*i* how to track the usage of blocks within each extent. Recall that the syntax for these two clauses in the CREATE TABLESPACE command is:

```
CREATE TABLESPACE <tablespacename>
...
EXTENT MANAGEMENT LOCAL|DICTIONARY
SEGMENT SPACE MANAGEMENT MANUAL|AUTO
...;
```

Blocks are either available for new inserts, available only for updates, or filled. An extent may have thousands of blocks, and every change to a block is tracked whenever data stored in the block changes. The method of tracking the blocks varies dramatically between the two types of extent management. LOCAL is the default for Oracle9*i*, although DICTIONARY was the default for earlier releases.

When the database was created, if the SYSTEM tablespace was created as a locally managed tablespace, no dictionary-managed tablespaces are allowed in the database.

Segment space management applies only to locally managed tablespaces. Specifying AUTO is recommended. This clause is discussed in more detail later in this section.

DICTIONARY was the default for earlier releases; however, Oracle plans to eliminate the ability to create new tablespaces with DICTIONARY management in the near future. Oracle recommends creating all new tablespaces with LOCAL management.

One of the biggest space wasters in a database is the management of extents within a tablespace. Dictionary-managed tablespaces make extent allocation more flexible, but the cost of flexibility is wasted space. Figure 6-9 illustrates a datafile that belongs to a dictionary-managed tablespace. As new objects are created and then grow, extents are allocated to each object's segment. These extents can be of various sizes. The diagram shows two extents, one 4 M and one 28 M. Extents are **deallocated** (returned to free status) when the object is dropped or truncated. (See Chapter 7 for more information on how to deallocate extents.) The diagram shows two deallocated extents, one 5 M and one 2 M. Deallocated extents return to the **free space list** in the Data Dictionary for the tablespace (the list of chunks of free space available in the datafile) as a contiguous chunk of data blocks. For example, the two deallocated extents shown in Figure 6-9 would be listed as two chunks of space in the freelist. For either of those chunks to be usable, the next object that needs an extent must be using that exact size (or smaller) extent. Otherwise, the deallocated data blocks are passed over, and more data blocks at the end of the datafile are used instead. Skipping over unused space in favor of free space at the end of the datafile causes the datafile to grow faster than necessary.

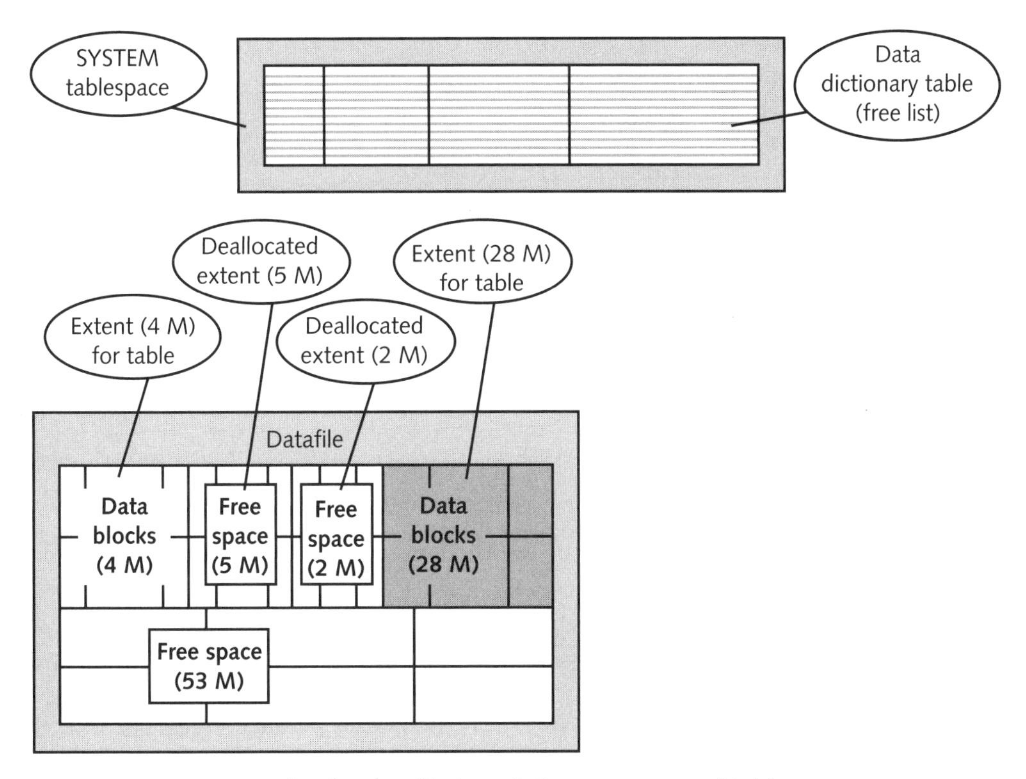

Figure 6-9 An example of a datafile in a dictionary-managed tablespace

If there are several deallocated extents next to one another, they are not seen as a single chunk of contiguous free space until they are **coalesced**, which is the term for combining multiple adjacent free extents into a single contiguous free extent. For example, in Figure 6-9, even though the two deallocated extents are adjacent to one another, they cannot be listed as a single chunk of 7 M of free space until they are coalesced. Coalescing occurs periodically through the SMON background process. The SMON process checks regularly for free extents that need coalescing. However, if you have set PCTFREE to 0 (the recommended setting), you must manually coalesce free extents by issuing the ALTER TABLESPACE command. The ALTER TABLESPACE command is discussed in detail later in this chapter.

In previous database releases, your only method of controlling the problem of unused free extents was diligent control over the storage settings for the tablespace and all objects within it. In this release, the DEFAULT STORAGE clause (see examples in the following bulleted list) sets the storage for any object created without specifying its own storage clause. If an object does specify its own storage clause, the settings must be monitored and carefully synchronized to assure that extent sizes stay in multiples of a common size. As the DBA, your job is to set the DEFAULT STORAGE clause for each tablespace and instruct all the developers who create tables to omit the STORAGE clause in their CREATE TABLE command, so that your default settings are used for all tables.

The syntax for dictionary-managed extents is:

```
EXTENT MANAGEMENT DICTIONARY
MINIMUM EXTENT <nn>
DEFAULT STORAGE (INITIAL <nn> NEXT <nn> PCTINCREASE <nn>
                 MINEXTENTS <nn> MAXEXTENTS <nn>)
```

The MINIMUM EXTENT clause helps keep objects that are created in the tablespace within similar parameters for extent size. If an object specifies a smaller extent size than the tablespace's MINIMUM EXTENT size, Oracle9*i* rounds the number up to the tablespace's MINIMUM EXTENT size.

Within the DEFAULT STORAGE clause, you can set the following parameters:

- **INITIAL:** The initial extent size that is allocated when the object is created.
- **NEXT:** The size of the next extent allocated when the initial extent runs out of room.
- **PCTINCREASE:** The percentage to increase any subsequent extents. For example, if INITIAL is set to 10 M, NEXT is set to 5 M, and PCTINCREASE is set to 10, then the first, second, and third extents will be 10 M, 5 M, and 5.5 M in size. Oracle recommends setting PCTINCREASE to zero at all times to keep extent sizes more uniform.
- **MINEXTENTS:** The minimum number of extents allocated when the object is created. This allows the object to use noncontiguous space for the initial extents. When creating large objects, this makes it easier to reuse the allocated storage as space is freed up by deletes.
- **MAXEXTENTS:** The maximum number of extents that an object can grow into.

Creating a Dictionary-Managed Tablespace

Because your SYSTEM tablespace is locally managed, you cannot practice actually creating a dictionary-managed tablespace. If you tried, you would get this error:

```
ORA-12913: Cannot create dictionary managed tablespace
```

You should, however, be familiar with the syntax of creating a dictionary-managed tablespace in case you work on an older version of the database. Here is an example to look over.

The following command creates a 250 M dictionary-managed tablespace with a MINIMUM EXTENT size of 15 M. Objects that use the default storage settings get an INITIAL EXTENT of 90 M, with all subsequent extents equal to 15 M, up to a maximum of 50 extents.

```
CREATE TABLESPACE USER_TEST
DATAFILE 'D:\oracle\data\user_test01.dbf' SIZE 250M
AUTOEXTEND ON
EXTENT MANAGEMENT DICTIONARY
MINIMUM EXTENT 15M
DEFAULT STORAGE (INITIAL 90M NEXT 15M PCTINCREASE 0
                 MINEXTENTS 1 MAXEXTENTS 50);
```


Dictionary-managed tablespaces have another problem that causes slower performance of DML commands. All the details of an extent and the data blocks within the extent are stored in data dictionary tables. A **freelist** is a list of individual blocks within an extent that have room available for inserting new rows. The freelist is not the same as the free space in a datafile. Free space is available for creating new extents. Blocks on the freelist, on the other hand, are blocks inside an existing extent with some extra room. Freelists are stored in the dictionary tables, and therefore the tablespaces are called "dictionary-managed." Updating the dictionary tables causes Oracle9*i* to create redo log entries. When, for example, you update rows in a table, redo log entries for those updates are recorded in the redo log buffers. In addition, more redo log entries for the updates to the data dictionary tables are also recorded. Undo records are also created for both your updates and for the updates to the data dictionary tables. The amount of work to record a change has doubled, because your update actually updates multiple tables behind the scenes.

Locally managed tablespaces reduce and even eliminate the problem of unused gaps of free space, by assuring that all the extents allocated to any object within the tablespace are of the same size or of multiples of that size as shown in Figure 6-10. In Figure 6-10, there are two extents, one 5 M and one 25 M. Both are multiples of 5 M. There are also two deallocated extents, both 5 M. Deallocated extents are automatically coalesced with any neighboring free space. In Figure 6-10, rather than two 5 M deallocated extents, the two would immediately be coalesced into one 10 M deallocated extent. This reduces the possibility that free space is passed over and left empty over time. No DEFAULT STORAGE clause is used for locally-managed tablespaces or the objects within them, so the work of keeping extent sizes uniform is left to the system. You can go further than this to homogenize the extent size. You can tell Oracle9*i* that all extents must be exactly the same size (the UNIFORM SIZE clause). If you used the UNIFORM SIZE clause when creating the datafile in Figure 6-10, the 25 M extent would have been created as five 5 M extents instead of one 25 M extent.

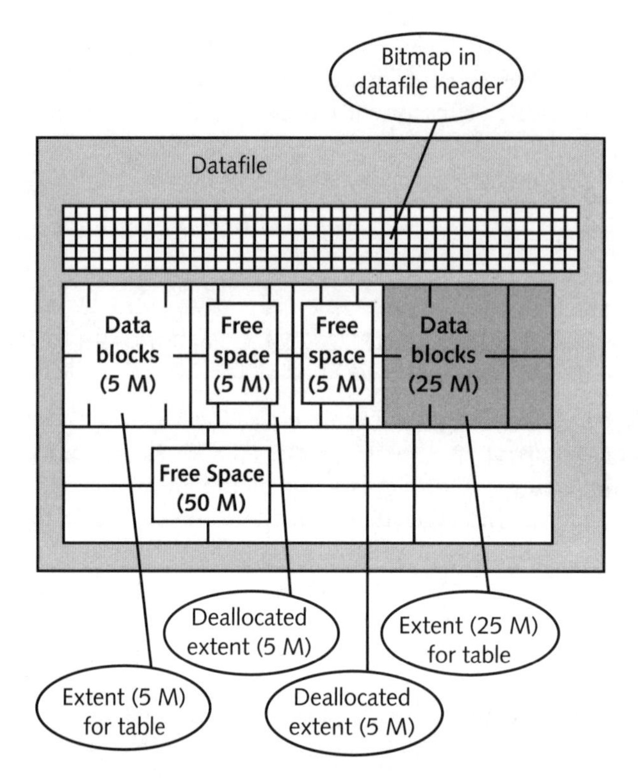

Figure 6-10 Locally-managed datafiles use a bitmap to track free space

Locally-managed tablespaces are more efficient than dictionary-managed tablespaces, because the information about storage is stored in a bitmap that is stored with the tablespace, as shown in Figure 6-10. There is a **bitmap** for each datafile in the tablespace. A bitmap is a small record in the datafile header with one bit for each data block number that marks the beginning of a used group or a free group of consecutive blocks in the datafile. Each bit in the bitmap stores information about a block or a group of blocks and tells whether they are used or free. As changes are made to rows stored in the extent, the bitmap updates its record of each changed block. Because the bitmap is stored locally within the tablespace, no redo or undo logs are generated for the bitmap. Therefore, the only redo and undo log records created are for the actual changes made to data. This reduction in redo and undo logging improves performance.

To further enhance the use of bitmaps and to eliminate the use of freelists, you can specify that the free space of each segment be stored locally as a bitmap as well as the free space of each extent. Do this by setting the SEGMENT SPACE MANAGEMENT to AUTO.

The syntax of the EXTENT MANAGEMENT and SEGMENT SPACE MANAGEMENT portions of the CREATE TABLESPACE command for a locally managed tablespace looks like this:

```
EXTENT MANAGEMENT LOCAL
AUTOALLOCATE|UNIFORM SIZE <nn>
SEGMENT SPACE MANAGEMENT MANUAL|AUTO
```

Here are the meanings of the settings:

- AUTOALLOCATE means that you allow the system to manage extent size allocation for objects in the tablespace. The MINIMUM EXTENT size the system can set is 64 K. The extents can vary in size.
- UNIFORM SIZE means that you set the size of all extents in the tablespace. The minimum size is 1 M.
- MANUAL means that a freelist tracks segment-free space.
- AUTO means that a bitmap tracks segment-free space.

Now, it is time to try out the command for yourself.

Creating a Locally Managed Tablespace

Use the following steps to create a locally managed tablespace.

1. Continuing with the SQL*Plus Worksheet session you began earlier in the chapter, enter the following command by typing each line as shown. Replace the variable with the correct path provided by your instructor. Execute the command by clicking the **Execute** icon.

   ```
   CREATE TABLESPACE USER_AUTO
   DATAFILE '<X:\xxx>\user_auto01.dbf'
   SIZE 20M AUTOEXTEND OFF
   EXTENT MANAGEMENT LOCAL AUTOALLOCATE
   SEGMENT SPACE MANAGEMENT AUTO;
   ```

   SQL*Plus Worksheet replies, "Tablespace created."

2. You have created a 20 M locally managed tablespace with system-managed extent allocation and automatic segment space management.
3. Remain logged on for the next practice.

As you know, the second largest logical structure is the segment. The next section describes the types of segments used in the Oracle9*i* database.

Segment Types and Their Uses

So far, the discussion in this chapter has focused on tablespaces and the segments and extents that are created when you create objects within the tablespaces. Tablespaces can contain other types of segments that store different data as well. The types of segments are shown in Table 6-1.

Table 6-1 Segment types

Segment type	Description and use	Allowed in which type of tablespace
Data segment	Used to store data for objects, such as tables, object tables, and triggers. Each object has one data segment. Exceptions to this rule are: partitioned tables, which have one data segment per partition, and clustered tables, which have one data segment per cluster.	Permanent tablespaces (locally- or dictionary-managed)
Index segment	Used to store all the index data. Each index has one index segment (except partitioned indexes, which have one index segment per partition).	Permanent tablespaces (locally- or dictionary-managed)
Temporary segment	Created during execution of SQL that needs space to perform sorting or other operations. After the execution is completed, the segment is dropped and its extents return to the free space in the tablespace. Also used for temporary tables and indexes created by a user.	Temporary tablespace (usually locally-managed but can be dictionary-managed) Exception: See the following note
Rollback segment	Created automatically by Oracle in an undo tablespace when using automatic undo management. See the section titled "Overview of Undo Data" later this chapter. If using manual undo in management, these segments are created manually.	Undo tablespace (if automatic undo management) or any permanent tablespace (if manual undo management)
LOB segment	Created when LOB data is stored out of line from the rest of the table's data. See Chapter 8 for more information.	Any tablespace except locally managed with a uniform extent size of two data blocks, or dictionary-managed with extent size of two data blocks. (LOB segment requires a minimum of three data blocks in its first extent.)

If you have not created any temporary tablespaces, the SYSTEM tablespace is used to store temporary segments. When creating a new database, if the SYSTEM tablespace is locally managed (standard for Oracle9*i*), a default temporary tablespace should be created during database creation. You can specify the temporary tablespace name, tempfile size, and location in the CREATE DATABASE command. If you do not supply a temporary tablespace name, the

SYSTEM tablespace is used, which can degrade performance. Oracle9*i* creates temporary segments as it needs them. Oracle9*i* allows you to create the temporary tablespace within the CREATE DATABASE command, or during database creation using the Oracle Database Configuration Assistant. You can add a temporary tablespace to an existing database as well, which you will do in the next section.

In Oracle9*i*, segments are created automatically when they are needed. For example, when you create a table, the corresponding data segment is created, and when you execute a query that sorts data, the needed temporary segment (for sorting) is created and managed totally behind the scenes. The only exception is the rollback segment, which you can specifically create. This is done to be compatible with older releases of the database that do not support automatic undo management. Oracle recommends that you use automatic undo management. See the section "Overview of the Undo Data" later in this chapter for more details.

Temporary Tablespace

As you saw in the previous section, temporary segments need a temporary tablespace. A temporary tablespace is almost identical to a permanent tablespace, except that when creating a temporary tablespace, fewer clauses are allowed in the CREATE command. Oracle recommends creating locally managed, temporary tablespaces.

In some cases, it is possible to create a dictionary-managed, temporary tablespace, although this capability will be unavailable in future releases, which will require locally managed, temporary databases. If you really want to create a dictionary-managed, temporary tablespace, use the CREATE TABLESPACE command shown in the previous section, and add the clause TEMPORARY into the command anywhere *after* the tablespace name. Here is an example:

```
CREATE TABLESPACE MYTEMP
TEMPORARY
DATAFILE 'C:\oracle\oratemp.dbf' SIZE 10M
AUTOEXTEND ON;
```

When creating a locally managed, temporary tablespace, the word TEMPORARY goes *before* the tablespace name instead of after it. The syntax for creating a locally managed, temporary tablespace is:

```
CREATE TEMPORARY TABLESPACE <tablespacename>
TEMPFILE <filename> SIZE <nn> AUTOEXTEND ON|OFF
EXTENT MANAGEMENT LOCAL UNIFORM SIZE <nn>
```

The TEMPFILE clause is just like the DATAFILE clause in the CREATE TABLESPACE command.

The EXTENT MANAGEMENT LOCAL clause is optional, because this command always creates locally managed, temporary tablespaces.

The UNIFORM SIZE clause is also optional. Use this clause only when you want to specify the size of the extents. The default size for uniform extents is 1 M.

Complete the following steps to create a locally managed, temporary tablespace that uses all the default settings and is not an Oracle managed file:

1. Type and execute the following command in your current SQL*Plus Worksheet session to create a tablespace named TEMP_STARTER that has a user-managed datafile. Replace <X:\xxx> with the actual directory provided by your instructor.

   ```
   CREATE TEMPORARY TABLESPACE TEMP_STARTER
   TEMPFILE '<X:\xxx>\temp_starter01.dbf'
   SIZE 2M;
   ```

   The resulting file is 2 M in size, with 1 M uniform extents that can extend to an unlimited number.

   AUTOEXTEND defaults to ON for temporary tablespaces.

2. Create another temporary tablespace. This one exemplifies using non default values. Type and execute this command in SQL*Plus Worksheet. Replace <X:\xxx> with the actual directory provided by your instructor.

   ```
   CREATE TEMPORARY TABLESPACE TEMP_STARTER2
   TEMPFILE '<X:\xxx>\temp_starter02.dbf' SIZE 6M
   AUTOEXTEND ON NEXT 2M MAXSIZE 50M;
   ```

   This command creates a 6 M file. The file extends automatically in 2 M increments to a maximum size of 50 M.

3. When a user runs a query or other transaction that needs temporary space, a temporary segment is either created or reused in the default temporary tablespace. The **default temporary tablespace** is the tablespace assigned to the user for all temporary segments. You can assign the default temporary tablespace in two ways:

   - Explicitly assign the **user** a default temporary tablespace using the CREATE USER or ALTER USER command. See Chapter 11 for the details.
   - Explicitly assign the **database** a default temporary tablespace and allow the user to use this tablespace by default. Use the CREATE DATABASE or ALTER DATABASE command, as described here.

4. If you created the database with the Oracle Database Configuration Assistant or included the DEFAULT TEMPORARY TABLESPACE clause in your CREATE DATABASE command, a default temporary tablespace has been created for you. Enter and execute the following query to determine the name (if any) of the current, default temporary tablespace.

   ```
   SELECT PROPERTY_VALUE
   FROM DATABASE_PROPERTIES
   WHERE PROPERTY_NAME = 'DEFAULT_TEMP_TABLESPACE';
   ```

The value returned is the current, default temporary tablespace for the database.

5. Occasionally, you may need to change the default that was originally set for the database. Your database may need a larger or smaller size tablespace, or you may want to change from a user managed tablespace to a locally managed tablespace. Oracle recommends always using locally managed, temporary tablespaces, because they are more efficient in handling frequent additions and deletions of segments (a frequent task in the default temporary tablespace). Switch the default temporary tablespace to the new tablespace you just created by typing and running the following command.

```
ALTER DATABASE DEFAULT TEMPORARY TABLESPACE
TEMP_STARTER;
```

6. Remain logged on for the next practice.

In some operating systems, the temporary tablespace's datafile is not even created until a temporary segment is created the first time.

As mentioned earlier, it is possible to create a tablespace with a data block size different than that set for the database with DB_DATA_BLOCK. The next section describes how this can be done.

Tablespaces with Nonstandard Data Block Size

You can create a special tablespace with its own data block size. The main reason to do this is to help you when transporting tablespaces from one database to another and the two data block sizes are different. Giving a tablespace a compatible data block size makes transporting the tablespace much more efficient.

Complete the following steps to find out what is required to create this type of tablespace.

1. First, it is important to know the database block size. Type and execute the following command into your SQL*Plus Worksheet session to find out.

```
SHOW PARAMETERS DB_BLOCK_SIZE
```

The block size is typically 8096 (8 K). Yours may be different. When you plan to create a tablespace of nonstandard data block size, you are allowed to use a block size of 2 K, 4 K, 8 K, 16 K, or 32 K. You cannot use 8 K if the database block size is 8 K. In addition, some operating systems don't allow 32 K block sizes. Check the specific manual for the operating system to determine any restrictions.

2. The second important piece of information is the current settings for cache memory. Before creating a tablespace with nonstandard block size, you are required to create a cache space with a matching block size. Recall that Oracle9*i* reads and writes data in data block sized chunks. All the memory cache is set up for the standard block size. You must allocate some memory

that can work with the new block size. Type and execute this command to view the current memory cache settings:

```
SHOW PARAMETERS CACHE
```

Figure 6-11 shows the results. Your results will be similar. Notice that there are zeros for the "db_2k_cache_size," "db_4k_cache_size," and so on. The only cache parameter that has space allocated is the "db_cache_size," which is calculated by Oracle and is the memory allocation for the standard data block size cache.

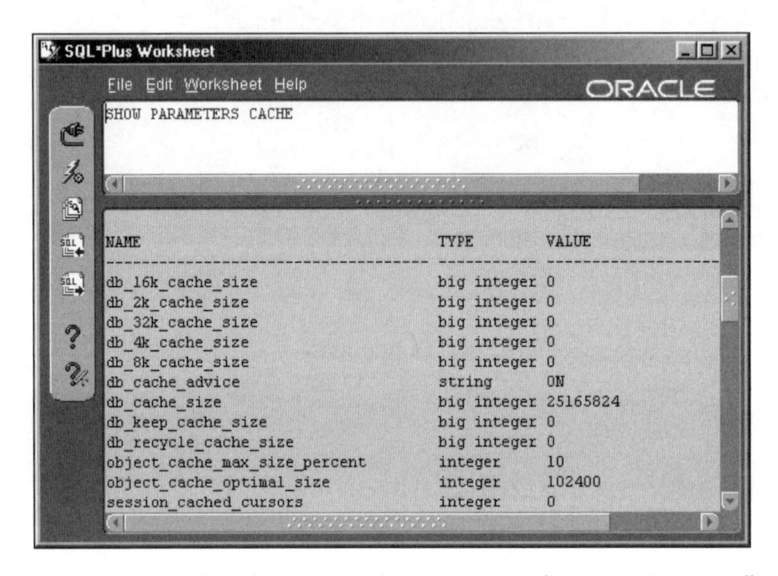

Figure 6-11 The alternate cache parameters have no storage allocation yet

3. For this example, you are creating a tablespace with a 2 K data block size. Therefore, you must allocate storage to the 2 K cache in Oracle's memory. Enter and execute this command to allocate 4 M to the 2 K cache.

   ```
   ALTER SYSTEM SET DB_2K_CACHE_SIZE=4M;
   ```

4. SQL*Plus Worksheet replies, "System altered." If you receive an error message, reduce the size of the db_cache_size (using ALTER SYSTEM), and then try again.

5. Now you have the buffer prepared, so you can create the tablespace. Execute the following command to create a tablespace with 2 K block size. Replace the variable with the directory path provided by your instructor.

   ```
   CREATE TABLESPACE TBS_2K
   DATAFILE '<X:\xxx>\tbs_2k.dbf'
   SIZE 4M BLOCKSIZE 2K;
   ```

SQL*Plus Worksheet replies, "Tablespace created."

6. Remain logged on for the next practice.

You may not create a temporary tablespace with nonstandard block size.

Now that you have created some tablespaces, how do you modify them? The next section discusses changes to tablespaces.

CONFIGURING AND VIEWING STORAGE

This section describes how to adjust the storage size of existing tablespaces and which data dictionary views can be queried to display information on storage.

Changing the Size, Storage Settings, and Status

After a tablespace has been created, you may want to change its settings to improve performance. For example, if your tablespace keeps running out of space, you may want to modify the AUTOEXTEND setting. Or, perhaps your tablespace has plenty of empty space, and you want to coalesce the extents. The ALTER TABLESPACE command handles these types of tasks as well as other tasks, such as taking the tablespace offline. Here is a list of the tasks you can handle using the ALTER TABLESPACE command:

- Change the DEFAULT STORAGE settings for any future objects created in the tablespace (dictionary-managed tablespaces only)
- Change the MINIMUM EXTENT size
- Change from LOGGING to NOLOGGING and vice versa
- Change from PERMANENT to TEMPORARY and vice versa
- Change from READ ONLY to READ WRITE and vice versa
- Coalesce contiguous storage space
- Add a new datafile or temporary file
- Rename a datafile or temporary file
- Begin and end an **open backup** (a backup of the tablespace while the tablespace is available for use)

The extent management mode (LOCAL or DICTIONARY) cannot be changed using the ALTER TABLESPACE command. If you want a different extent management mode, you have to create a new tablespace, migrate the data, and then drop the old tablespace.

The syntax for modifying the size and storage settings of the tablespace is:

```
ALTER TABLESPACE <tablespacename>
ADD|RENAME DATAFILE <filename>
       SIZE <nn> AUTOEXTEND ON|OFF REUSE
DEFAULT STORAGE (INITIAL <nn> NEXT <nn> PCTINCREASE <nn>
                 MINEXTENTS <nn> MAXEXTENTS <nn>)
MINIMUM EXTENT <nn>
COALESCE
```

Keep in mind that you may only perform one task per ALTER TABLESPACE command. For example, if you want to add a datafile and modify the DEFAULT STORAGE, you must issue two commands.

Let's say that you have created a tablespace that is locally managed and you want to add a new datafile to the tablespace to increase its storage space. Complete the following steps to accomplish this task.

1. Enter and execute the following command in SQL*Plus Worksheet by typing each line as shown. Replace the variable with the correct path provided by your instructor.

   ```
   ALTER TABLESPACE USER_AUTO
   ADD DATAFILE '<X:\xxx>\user_auto02.dbf' SIZE 5M;
   ```

2. SQL*Plus Worksheet replies, "Tablespace altered." You have added a 5 M datafile to the locally managed tablespace.
3. Remain logged on for the next practice.

Here is an example of changing the DEFAULT STORAGE clause of a dictionary-managed tablespace

```
ALTER TABLESPACE USER_TEST
DEFAULT STORAGE (INITIAL 120M MINEXTENTS 4 MAXEXTENTS 100);
```

When you modify the DEFAULT STORAGE from its original settings, the new settings only affect new objects. Old objects keep their current settings for INITIAL and MINEXTENTS, because these settings are only used during the creation of the object. Both old and new objects now extend to 100 extents instead of the original 50 extents.

Temporary tablespaces are changed in the same way, except that you use ADD TEMPFILE instead of ADD DATAFILE.

To change an existing datafile's storage, you must use the ALTER DATABASE command rather than the ALTER TABLESPACE command. Follow these steps to try this out.

1. In SQL*Plus Worksheet, change a datafile from AUTOEXTEND OFF to AUTOEXTEND ON, by entering and executing the following command.

As before, replace the variable with a real directory path provided by your instructor.

```
ALTER DATABASE DATAFILE '<X:\xxx>\user_auto02.dbf'
AUTOEXTEND ON;
```

2. Similarly, you can modify the size, extent settings, and online or offline status of a datafile. Enter and execute this command, replacing the variable as usual.

```
ALTER DATABASE DATAFILE '<X:\xxx>\user_auto02.dbf'
RESIZE 10M;
```

3. Remain logged on for the next practice.

To modify a tempfile, use the ALTER DATABASE TEMPFILE command.

The **status** of a tablespace defines its availability to end users and also defines how it is handled during backup and recovery. The two types of status are:

- **ONLINE:** This is the default status for new tablespaces. Online tablespaces are available to users when the database is open.
- **OFFLINE:** The tablespace is not available to users even though the database is open.

You might take a tablespace offline to make sure no users have access while you perform a backup of the tablespace, or while you update the application that uses that tablespace. You can perform an online backup of a tablespace; however, you may want to use the offline backup method because it is faster than the online method.

When you take a tablespace offline, there are three methods you can specify for Oracle9*i*'s execution of the command:

- **NORMAL:** This is the default, and Oracle9*i* performs a checkpoint on each datafile of the tablespace, assuring that all data in memory buffers is successfully written to all the datafiles before the tablespace goes offline. No recovery is needed to put the tablespace back online.
- **TEMPORARY:** Oracle9*i* performs a checkpoint on all the available datafiles in the tablespace. Use this mode when a datafile has been damaged, causing the checkpoint attempt while performing a NORMAL offline command to fail. You will have to recover unsaved changes in the bad datafile before putting the tablespace back online.
- **IMMEDIATE:** Oracle9*i* does not attempt to do a checkpoint, and you will be required to perform **media recovery** before the tablespace goes back online. Media recovery is the term used for recovery that requires restoring the database from a backup, rolling forward through archived redo logs, and finally rolling forward through online redo logs. It is called media recovery because the most common reason for this type of recovery is the failure of a disk drive or some other storage media.

Follow these steps to take a tablespace offline, and then restore it to online mode:

1. In your SQL*Plus Worksheet session, enter and execute the following command to take the tablespace offline.

```
ALTER TABLESPACE ACCOUNTING
OFFLINE NORMAL;
```

2. SQL*Plus Worksheet replies, "Tablespace altered." You have taken the tablespace offline. It is possible to take individual datafiles offline by using a similar command. Type this command to take one datafile of the USER_AUTO tablespace offline, replacing the variable with the correct directory path.

```
ALTER DATABASE DATAFILE '<X:\xxx>\user_auto02.dbf'
OFFLINE DROP;
```

The IMMEDIATE, TEMPORARY, and NORMAL options do not apply to taking a datafile offline. In addition, the DROP option is required, if you take a datafile offline when your database is in NOARCHIVELOG mode. This is convenient in some cases when you plan to drop the datafile anyway, because the command takes the datafile offline and drops the operating system file in one command. If, however, you don't want the datafile dropped, you must take the entire tablespace offline, or change the archiving mode of the database. When your database is in ARCHIVELOG mode, take a datafile offline without dropping it by removing the DROP keyword.

3. Place the ACCOUNTING tablespace back online by entering and executing the following command.

```
ALTER TABLESPACE ACCOUNTING
ONLINE;
```

4. Remain logged on for the next practice.

Because the operating system file was dropped when you took the datafile offline in the exercise, you cannot return the datafile to online status. If the database is in ARCHIVELOG mode, the syntax to place a datafile online is:

```
ALTER DATABASE DATAFILE '<X:\xxx>\<datafile>' ONLINE;
```

One of the options when creating or altering a tablespace is to make it a read-only tablespace.

Read-only Tablespaces

After you have loaded data into tables that don't change (or almost never change), such as a table of states in the United States, you might consider storing all this static data in a read-only tablespace.

The objects in a **read-only tablespace**, as its name implies, can be queried but not changed. No user can execute an INSERT, UPDATE, or DELETE command on the objects.

The primary purpose of a read-only tablespace is to make large amounts of static data available to the database without having to slow down the database backup and recovery. Back up the tablespace one time, and after that, it need never be backed up again.

If you need to alter data in the read-only tablespace, simply change it back to read-write status (the default) and make the changes. Then return it to read-only status.

Follow these steps to modify a tablespace to read-only status:

1. In SQL*Plus Worksheet, enter and execute this command to change the ACCOUNTING tablespace to read-only status:

   ```
   ALTER TABLESPACE ACCOUNTING READ ONLY;
   ```

2. Execute the following query to confirm the status of the tablespace.

   ```
   SELECT TABLESPACE_NAME, STATUS
   FROM DBA_TABLESPACES;
   ```

   Figure 6-12 shows part of the results.

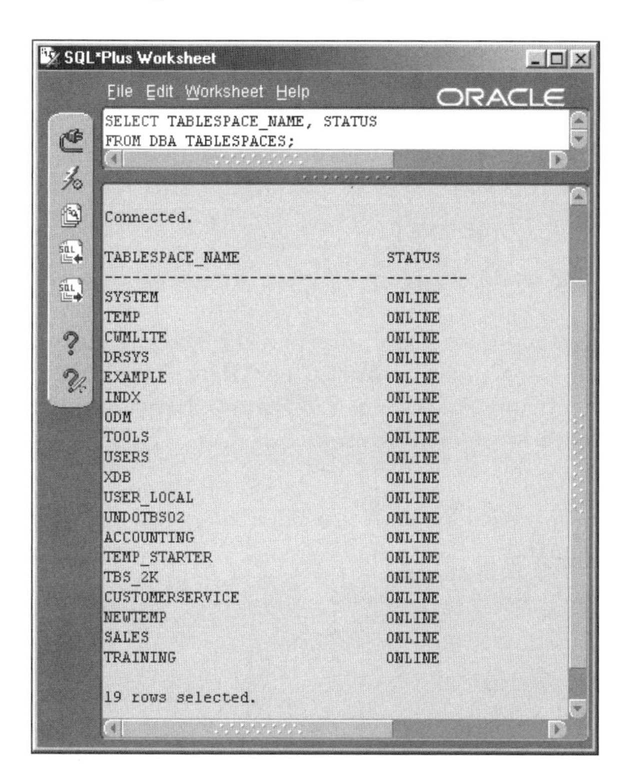

Figure 6-12 The DBA_TABLESPACES data dictionary view shows tablespace status

3. To return it to read-write status, use the same command, but specify READ WRITE. Enter and execute this command.

   ```
   ALTER TABLESPACE ACCOUNTING READ WRITE;
   ```

4. Run the query again and notice that the status changed from "READ ONLY" to "ONLINE."
5. Remain logged on for the next practice.

The last command in this chapter for working with tablespaces is for dropping them.

Dropping Tablespaces

A tablespace can be dropped. However, Oracle9*i* requires you to use specific clauses to drop tablespaces that contain data. Here is the syntax:

```
DROP TABLESPACE <tablespacename>
INCLUDING CONTENTS
AND DATAFILES
CASCADE CONSTRAINTS;
```

When you drop a tablespace, all references to the tablespace are removed from temporary tablespaces, the Data Dictionary, and the undo data. If there are transactions running that use data in the tablespace, the tablespace cannot be dropped.

Follow these steps to work with the command:

1. To drop a tablespace that contains no data, you use the basic DROP TABLESPACE command. Enter and execute this command in SQL*Plus Worksheet to drop the ACCOUNTING tablespace, which has no data.

   ```
   DROP TABLESPACE ACCOUNTING;
   ```

2. To experiment with dropping a tablespace containing data, first you will add some data to one of the tablespaces you created in this chapter. An easy way to do this is to run the prepared script found in the **Data\Chapter06** directory on your student disk. Click the **Worksheet/Run Local Script** on the SQL*Plus Worksheet menu. Navigate to the **Data\Chapter06** directory, and select the **add_data.sql** file. Click **Open** to run it in SQL*Plus Worksheet. The script creates tables in three tablespaces.
3. See how much space in each tablespace has been allocated to extents. Type and execute this command:

   ```
   SELECT TABLESPACE_NAME, SEGMENT_TYPE,
   EXTENT_ID, BLOCKS
   FROM DBA_EXTENTS
   WHERE TABLESPACE_NAME IN
   ('CUST_TBS', 'USER_OMF1', 'USER_AUTO');
   ```

   Figure 6-13 shows the results. As you can see, each table uses eight blocks, and the index also uses eight blocks.

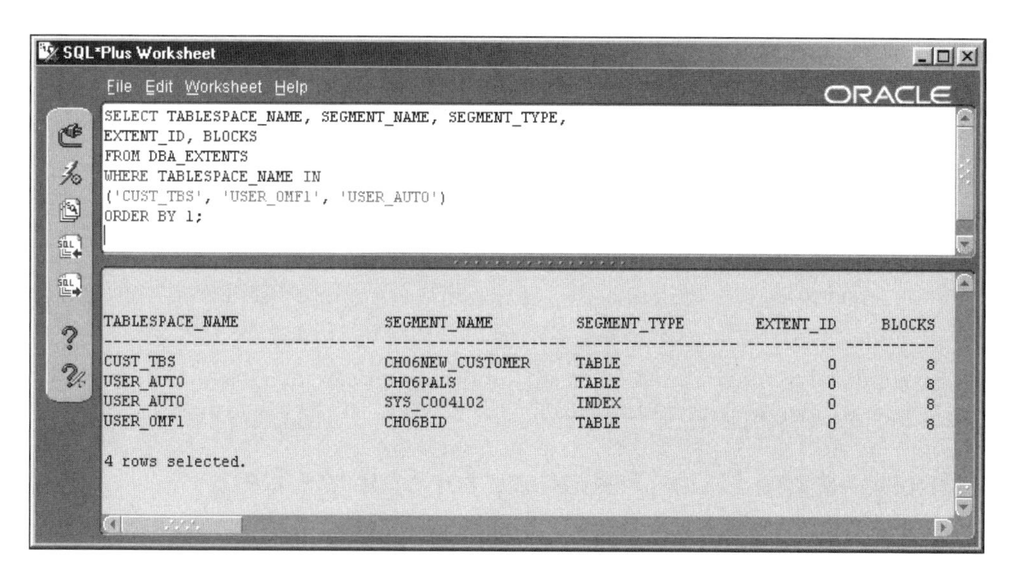

Figure 6-13 The first extent of a table has zero as its extent ID number

4. The CUST_TBS contains the CH06NEW_CUSTOMER table with one extent. Drop the CUST_TBS tablespace and all its data with the following command. The command also tells Oracle9*i* to drop the tablespace's datafiles with the AND DATAFILES clause.

   ```
   DROP TABLESPACE CUST_TBS INCLUDING CONTENTS
   AND DATAFILES;
   ```

5. The USER_AUTO tablespace contains the CH06PALS table and an index (created for the PRIMARY KEY constraint on CH06PALS). Another table, CH06BID, resides in the USER_OMF1 tablespace. This table has a FOREIGN KEY constraint referencing the CH06PALS table. (Read all about constraints in Chapter 10.) To drop the USER_AUTO tablespace, you must use the CASCADE CONSTRAINTS clause. The clause tells Oracle9*i* to remove any constraints that were created in other tablespaces referencing any table in the USER_AUTO tablespace. Type and execute this command to drop the tablespace and the constraint:

   ```
   DROP TABLESPACE USER_AUTO INCLUDING CONTENTS
   CASCADE CONSTRAINTS;
   ```

6. Oracle managed datafiles are automatically dropped without using the AND DATAFILES clause; however, you must still use the INCLUDING CONTENTS clause to drop this tablespace because it contains a table. Type and execute this command:

   ```
   DROP TABLESPACE USER_OMF1 INCLUDING CONTENTS;
   ```

7. Remain logged on for the next practice.

You can drop a tablespace whether it is online or offline. In fact, it is safer to take a tablespace offline before dropping it. This gives you time to discover any unexpected errors. For example, you may have assigned the tablespace to a developer as his default tablespace. When the tablespace is taken offline (or dropped), he suddenly is unable to create new tables.

Dropping a temporary tablespace is the same as dropping a permanent tablespace. If the tablespace is currently the default temporary tablespace, you must switch to another default tablespace, wait for any outstanding temporary segments remaining in the old temporary tablespace to clear, and then drop the tablespace.

You have examined how to create, modify, and drop tablespaces. The next section shows how to review your tablespaces by querying the Data Dictionary.

Querying the Data Dictionary for Storage Data

There are quite a few data dictionary views and dynamic performance views containing information about tablespaces. The most commonly used views are the DBA_TEMP_FILES and DBA_DATA_FILES views, because querying these two views shows you all the files connected with tablespaces in the database. Table 6-2 lists many of them.

Table 6-2 Views of tablespace information

View	Description
DBA_DATA_FILES	Details on datafiles
DBA_EXTENTS	Details on data extents in all tablespaces
DBA_FREE_SPACE	Free extents within all tablespaces
DBA_FREE_SPACE_COALESCED	Coalesced free space information
DBA_SEGMENTS	Information about all segments in all tablespaces
DBA_TABLESPACES	Descriptions of all tablespaces
DBA_TEMP_FILES	Details on tempfiles
DBA_TS_QUOTAS	Tablespace quotas for all users
DBA_USERS	Default and temporary tablespaces assigned to all users
V$DATAFILE	Datafiles and tablespaces that own them
V$TABLESPACE	All tablespaces from the control file
V$TEMP_EXTENT_MAP	All extents in all locally managed, temporary tablespaces
V$TEMP_SPACE_HEADER	Space used and space free for each tempfile
V$TEMPFILE	Tempfiles and tablespaces that own them

As with any other view, you can query these views.

Complete the following steps to query some data dictionary views.

1. As an example, imagine you want to know whether you should coalesce free extents in the USERS tablespace. Developers use the tablespace to test various database designs. Because of this, many extents would be created and then released as the developers create new tables and later drop the tables. Type and execute this query in SQL*Plus Worksheet to review the USERS tablespace:

   ```
   SELECT BLOCK_ID, BLOCK_ID+BLOCKS NEXT_BLOCK_ID, BLOCKS
   FROM DBA_FREE_SPACE
   WHERE TABLESPACE_NAME = 'USERS'
   ORDER BY BLOCK_ID;
   ```

   An example of the query results is shown next. Your results will look similar, but not identical. The goal of the query is to help you determine if any free extents can be coalesced. Recall that free extents must be adjacent to one another to be coalesced. Figure 6-14 illustrates how you interpret the results of the query to determine whether a tablespace needs coalescing. In Figure 6-14, there are two sets of blocks (deallocated extents) in which the NEXT_BLOCK_ID in the first is the same as the BLOCK_ID in the next set of blocks. These sets of blocks are adjacent and can be coalesced. As shown in the following query results and in Figure 6-14, the deallocated extents beginning with BLOCK_ID 2 and 6 can be coalesced and the extents beginning with BLOCK_ID 47 and 52 can also be coalesced.

BLOCK_ID	NEXT_BLOCK_ID	BLOCKS
2	6	4
6	12	6
21	25	4
31	32	1
40	44	4
47	52	5
52	58	6

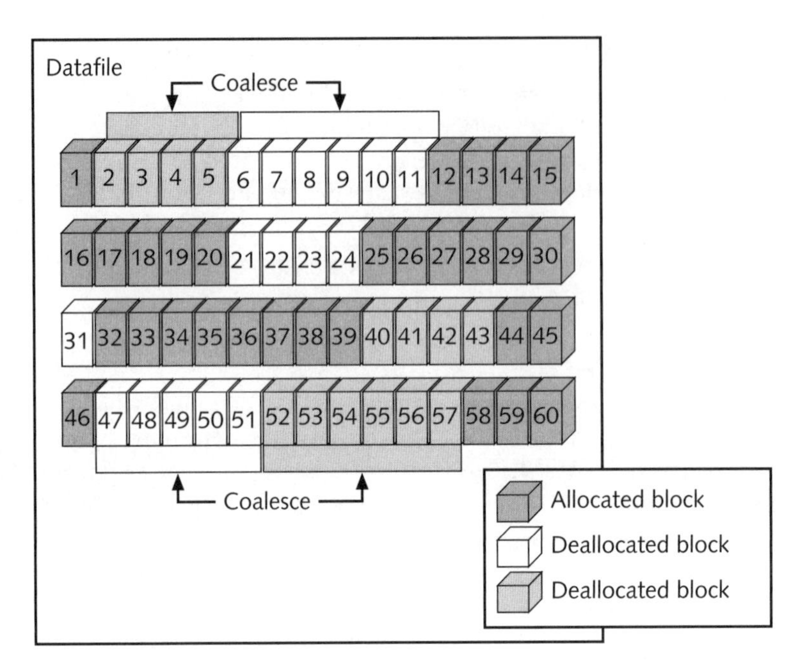

Figure 6-14 Adjacent free blocks can be combined into a single free space

2. If your own query results indicate that the USERS tablespace needs coalescing, enter and run this command:

```
ALTER TABLESPACE USERS COALESCE;
```

3. Remain logged on for the next practice.

Another type of data that can be stored in tablespaces is undo data. The next section discusses undo data.

OVERVIEW OF UNDO DATA

In Chapter 5, you read that one of the final steps in database recovery is to roll forward the online redo logs and then apply the undo data. What is undo data and where does it reside in the database?

Undo data is made up of undo blocks. Each **undo block** contains the before image of the data in the block. For example, when a row in the CUSTOMER table is updated to change Jane's last name to Smith, an exact copy of the block containing Jane's row is copied. This copy of the data is used to redo the original data if the user issues a ROLLBACK command. The undo data also provides read consistency for users accessing the table between the time the update is pending and the time the update has been committed. **Read consistency** means that only the user who issued the update command sees the changes to the data in queries, until that user commits the change. At that point, all users see the changed data.

The other use of undo data is during database recovery. As you recall from Chapter 5, the redo log files are occasionally updated before the data has officially been committed. This occurs when a checkpoint causes the buffers to be flushed. If the database fails and prevents the commit from happening, the data in the redo log is inaccurate. During recovery, the redo log updates the database with the uncommitted change. Then the undo data is applied and replaces the changed data with the original data.

You may want to keep undo data in the undo tablespace for a short time after the data has actually been rolled back or committed. This is called **undo data retention** and helps users to complete queries that were begun while the data was in limbo and continued after the data was either rolled back or committed.

There is a hard way and an easy way to manage undo data:

- **Manual undo management mode:** This is difficult, because you must set up and manage groups of files that store all the undo data in **user-managed redo segments**. Earlier versions of the database provided only this mode.
- **Automatic undo management mode:** This method is new and much easier to maintain. You need only create a special tablespace called the UNDO tablespace and Oracle9*i* manages all the space inside.

The next section describes how to set up automatic undo management.

Implementing Automatic Undo Management

To set up automatic undo management mode, you must perform these two tasks:

1. Set the UNDO_MANAGEMENT initialization parameter to AUTO.
2. Create an undo tablespace.

Optionally, you can set other related initialization parameters, such as UNDO_RETENTION and UNDO_TABLESPACE.

An undo tablespace is an entire tablespace reserved for undo data. Like data and index tablespaces, data in the undo tablespace is added in the form of extents. The extents are called **undo extents**.

You can create the undo tablespace when you create the database. If you set UNDO_MANAGEMENT to AUTO before database creation, you must include the UNDO_TABLESPACE clause in the CREATE DATABASE command. For example, the following command creates the NEWAUTO database with an OMF undo tablespace:

```
CREATE DATABASE newauto
UNDO TABLESPACE UNDO_TBS;
```

You can, of course, use user-managed files for the undo tablespace as well. In this case, supply the datafile name and storage information as you do for any tablespace.

Oracle9*i* uses only one undo tablespace for a database. If you created more than one, Oracle9*i* uses the first one it finds on startup. Or, you can specify which to use with the UNDO_TABLESPACE initialization parameter.

If you choose to change the undo management mode from user-managed to automatic, or if you want to switch to a new undo tablespace, you should create the undo tablespace with a CREATE UNDO TABLESPACE command.

If you start up a database in automatic undo management mode and no undo tablespace exists, Oracle9*i* creates a default undo tablespace, called SYS_UNDOTBS.

As with any other tablespace, drop an undo tablespace by using the DROP TABLESPACE command.

Follow these steps to switch the undo tablespace and adjust the related parameters.

1. Enter and run the follow SQL*Plus command in your SQL*Plus Worksheet session to find information about the undo mode and undo tablespace in your database.

```
SHOW PARAMETERS UNDO
```

   Figure 6-15 shows the results. The UNDO_MANAGEMENT parameter is set to AUTO, indicating automatic undo management. The current undo tablespace is UNDOTBS02. (Your results may show a different tablespace name.)

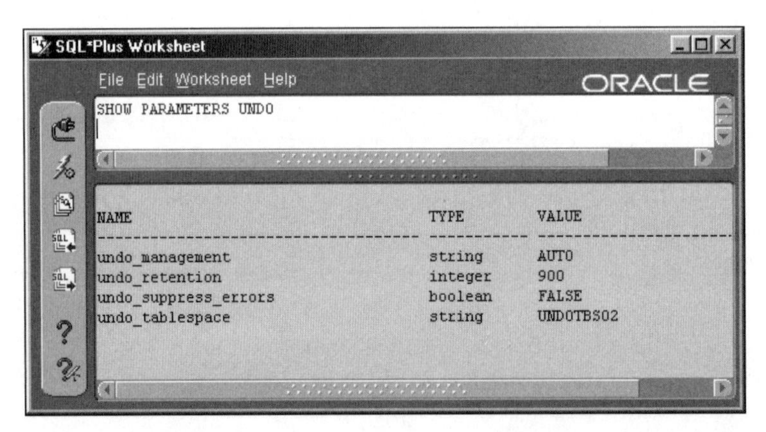

Figure 6-15 The four parameters listed define how undo data is managed

2. You want to create a new undo tablespace to replace the current one. The CREATE UNDO TABLESPACE command has fewer options than the CREATE TABLESPACE command. You can only specify the datafile name, size, and AUTOEXTEND mode. Oracle9*i* determines the rest. Enter and

execute the following command to create an undo tablespace named UNDO_IT with one 30 M datafile with AUTOEXTEND on. Replace the <X:\xxx> variable with the directory path provided by your instructor.

```
CREATE UNDO TABLESPACE UNDO_IT
DATAFILE '<X:\xxx>\undo_it.dbf'
SIZE 30M AUTOEXTEND ON;
```

3. Next, adjust the initialization parameters. Consult the following list of parameters to find out what they mean.
   - **UNDO_MANAGEMENT:** This tells Oracle9*i* which of the two modes to use for undo data. The two choices are AUTO and MANUAL. This parameter is static. Your database is already in automatic undo management mode. Manual undo management mode is provided for backward compatibility and requires manual creating of undo segments.
   - **UNDO_TABLESPACE:** This specifies the name of the undo tablespace to use for undo data, when the database is in automatic undo management mode.
   - **UNDO_RETENTION:** This is the time in seconds to save committed undo data. The default is 900 seconds (15 minutes). Retaining undo data provides support for a new feature: Flashback Query. This feature uses undo data to simulate going back in time and running a query on the old data.
   - **UNDO_SUPPRESS_ERRORS:** This tells Oracle9*i* how to handle errors caused by users or scripts issuing manual undo management mode commands, while the database is in automatic undo management mode. Set to TRUE to ignore errors, and FALSE to display errors. This is provided to smooth the transition from manual to auto by allowing scripts to continue, even if they use invalid manual undo commands.

   Switch the current undo tablespace by executing this command:

```
ALTER SYSTEM SET UNDO_TABLESPACE='UNDO_IT';
```

4. Lengthen the undo retention time to 30 minutes (1800 seconds) by executing this command:

```
ALTER SYSTEM SET UNDO_RETENTION=1800;
```

5. Adjust the parameter to suppress undo errors by executing this command:

```
ALTER SYSTEM SET UNDO_SUPPRESS_ERRORS=TRUE;
```

6. View the changes by executing this command to display the parameters:

```
SHOW PARAMETERS UNDO
```

7. Return the database to using the original undo tablespace by executing this command, replacing the variable with the previous undo tablespace name.

```
ALTER SYSTEM SET UNDO_TABLESPACE='<tablespacename>';
```

6

8. Prepare to drop the undo tablespace you created in this session by taking it offline. Execute this command:

```
ALTER TABLESPACE UNDO_IT OFFLINE;
```

   You must use caution when dropping an old undo tablespace because, unlike permanent tablespaces, the DROP TABLESPACE command does not return an error message if the tablespace contains data. The old undo tablespace may still contain undo data for some users.

9. Query the data dictionary view DBA_TABLESPACES to check the status of the tablespace. Execute this query:

```
SELECT TABLESPACE_NAME, STATUS
FROM DBA_TABLESPACES
WHERE CONTENTS='UNDO';
```

10. The status may be ONLINE, OFFLINE, or PENDING OFFLINE. If the status is PENDING OFFLINE, there are undo segments still in use on the old undo tablespace. Wait until the status is OFFLINE before dropping the tablespace.
11. If the status of the UNDO_IT undo tablespace is OFFLINE, execute this command to drop the tablespace:

```
DROP TABLESPACE UNDO_IT;
```

12. The SQL*Plus Worksheet replies, "Tablespace dropped."
13. Exit SQL*Plus Worksheet by typing **EXIT** and clicking the **Execute** icon. Click **OK** when the Enterprise Manager dialog box opens.
14. Exit the console by clicking the **X** in the upper-right corner of the window.

These are only a few of the many options of the ALTER TABLESPACE command that are allowed for undo tablespaces. You can also add more space by resizing or adding a datafile, rename a datafile, and start or stop an **open backup** (a backup done while the database is running) to an undo tablespace by using the ALTER TABLESPACE command. The other options of the ALTER TABLESPACE command are invalid for undo tablespaces.

Now that you have seen the methods of managing tablespaces, the next logical step is to look into creating the objects that go inside tablespaces: tables. The next chapter introduces table management.

Chapter Summary

- The internal structures of Oracle9*i* are either logical structures or physical structures.
- Physical structures have a physical presence in the operating system.
- Logical structures do not have a physical presence in the operating system.
- Logical structures cannot be viewed or modified outside the database.
- Physical structures include datafiles, redo log files, and control files.
- Logical structures include data blocks, extents, segments, schema objects, and tablespaces.
- A tablespace has one or more datafiles associated with it.
- Permanent tablespaces store objects, such as tables and indexes.
- Temporary tablespaces store temporary tables and data while the data is being sorted.
- Locally managed tablespaces use a bitmap to track used and unused space.
- Dictionary-managed tablespaces use the Data Dictionary to track used and unused space.
- The NOLOGGING setting does not log mass INSERT or DDL commands.
- To use OMF, omit the datafile name and (optionally) include the size of the file.
- The REUSE setting allows Oracle9*i* to reuse an existing file, erasing all its data.
- The AUTOEXTEND ON setting gives a datafile the ability to add to its size.
- Oracle recommends using locally managed tablespaces.
- Dictionary-managed files tend to waste storage space.
- Adjacent, free extents can be manually coalesced in a dictionary-managed tablespace.
- Free extents are automatically coalesced in a locally managed tablespace.
- The MINIMUM EXTENT setting overrides a smaller extent size specified by an object in the tablespace.
- Set PCTINCREASE to zero to keep extent sizes more uniform.
- Dictionary-managed tablespaces use a freelist in the dictionary to track blocks.
- Locally managed tablespaces keep all extents either the same size or a variable size that is controlled by the system.

- The SEGMENT SPACE MANAGEMENT setting is only used for locally managed tablespaces.
- The four types of segments are: data, index, temporary, and rollback.
- Temporary tablespaces should be locally managed, although it is possible to create a dictionary-managed temporary tablespace.
- Before creating a tablespace with a nonstandard data block size, you must create a cache with the corresponding data block size.
- Many of the initial settings of a tablespace can be changed using the ALTER TABLESPACE command.
- You cannot change a tablespace from LOCAL or DICTIONARY mode.
- Changing DEFAULT STORAGE settings affects future objects and future extents of existing objects.
- To block access, a tablespace can be changed from ONLINE to OFFLINE.
- Locally managed, temporary tablespaces cannot be taken offline.
- Taking a tablespace offline can be done in NORMAL, TEMPORARY, or IMMEDIATE mode.
- Media recovery is usually needed with any mode but NORMAL.
- A read-only tablespace is not included in regular backups or recoveries.
- Dropping a tablespace with the INCLUDING CONTENTS clause destroys all its data.
- DBA_FREE_EXTENTS tells how much free space is available in the tablespace.
- Undo data allows users to have read consistency, while other users make changes that are not yet committed.
- Manual undo management mode is from older versions and is not recommended.
- Automatic undo management mode is much easier to manage and requires an undo tablespace.
- Undo blocks reside in undo extents within either an undo tablespace or a rollback segment, depending on the undo management mode.

Review Questions

1. Place these in order from smallest to largest.
   1. Segment
   2. Data block
   3. Extent
   4. Tablespace

   a. 2, 1, 3, 4

   b. 2, 3, 1, 4

   c. 1, 2, 3, 4

   d. 4, 1, 3, 2
2. Extent management only pertains to locally managed tablespaces. True or False?
3. A logical structure always resides in one physical structure. True or False?
4. Explain the circumstance when the initial storage created for a table would *not* be made up of contiguous blocks.
5. One segment is created for each. Choose two.

   a. Table

   b. Partition

   c. Unpartitioned table

   d. Extent
6. A schema object spans multiple datafiles. Explain how this happens.
7. ______________ extents are automatically coalesced in ______________ managed tablespaces.
8. To create a locally managed tablespace with all extents 3 M in size, use the clause EXTENT MANAGEMENT LOCAL ______________.
9. Undo extents are stored in ______________ or ______________.
10. Review the following command, and fill in the missing words.

```
CREATE TEMPORARY TABLESPACE USERTEMP
______________ 'D:\oracle\data\usertemp01.dbf' SIZE 50M
EXTENT MANAGEMENT LOCAL ______________ SIZE 2M
```

11. Any temporary tablespace can be taken offline. True or False?
12. To find the names of the tempfiles in a temporary tablespace, query the ______________ view.

6

Exam Review Questions—Database Fundamentals I (#1Z0-031)

1. The two modes of extent management are:
   a. USER MANAGED and ORACLE MANAGED
   b. LOCAL and DICTIONARY
   c. MANUAL and AUTOMATIC
   d. USER MANAGED and SYSTEM MANAGED
2. Which of these statements about data blocks are true? Choose three.
   a. Data blocks are 2048 bytes in size.
   b. Data blocks are made up of multiples of operating system blocks.
   c. One data block can span more than one datafile.
   d. One database can have more than one data block size.
   e. Data blocks are tracked in freelists or bitmaps.
3. Look at the following command.

```
CREATE TABLESPACE USER_NEWDATA
DATAFILE SIZE 400M
EXTENT MANAGEMENT DICTIONARY
DEFAULT STORAGE (INITIAL 40M NEXT 20M PCTINCREASE 50);
```

   Which of the following statements are true regarding the previous command? Choose three.
   a. The tablespace has one Oracle managed datafile.
   b. An object created in this tablespace must have a 40 M initial size.
   c. An object created in this tablespace with no storage settings has 80 M after three extents.
   d. An object created in this tablespace with no storage settings has 90 M after three extents.
   e. The tablespace uses a bitmap to track free space.
4. Study the following command.

```
CREATE TABLESPACE ANYOBJECTS
DATAFILE 'D:\oracle\storage\anyobjects01.dbf' SIZE 500M
SEGMENT SPACE MANAGEMENT MANUAL;
```

   Which statement is true about this command?
   a. The datafile is an Oracle managed file.
   b. The statement will fail, because it is missing the DEFAULT STORAGE clause.
   c. The statement will fail because the EXTENT MANAGEMENT LOCAL clause is missing.
   d. The tablespace is locally managed.

5. The command ALTER TABLESPACE USER_TEST OFFLINE NORMAL has failed. What do you do?
   a. Attempt the command using TEMPORARY instead of NORMAL.
   b. Shut down and restart the database, and then take the tablespace offline.
   c. Drop the tablespace and re-create it.
   d. Back up the tablespace, and then drop and restore it.
6. The following command was issued on the tablespace containing the ZIP_CODE table.

   ```
   DROP TABLESPACE USER_REF;
   ```

   What happens?
   a. The tablespace and the ZIP_CODE table are dropped.
   b. The command fails because the tablespace contains data.
   c. The command fails because the ZIP_CODE table was not listed in the command.
   d. The tablespace is dropped, and all references to the ZIP_CODE table are removed.
7. Which statements are true about dictionary-managed tablespaces? Choose two.
   a. They store data with less wasted space than locally managed tablespaces.
   b. Changes cause dictionary tables to be updated.
   c. Extents are tracked in freelists.
   d. They are available in earlier versions of Oracle.
8. You want to switch a tablespace from dictionary-managed to locally managed. What do you do?
   a. Execute ALTER TABLESPACE ... LOCAL command.
   b. Drop the tablespace, and create a new one.
   c. Create a new tablespace, move the data, and then drop the old one.
   d. This task cannot be done.

HANDS-ON ASSIGNMENTS

1. Edit the file on your student disk, Data\Chapter06 directory, named **ho0601.sql**. *You cannot actually run the completed command because your database does not allow dictionary-managed tablespaces.* Replace all variables found in the file with valid values so that the command creates a 500 megabyte dictionary-managed tablespace. The tablespace has one datafile named **C:\oracle\oradata\newadmin01.dbf**. New objects in the tablespace that use the default storage settings will be created with five 5 M extents, and all subsequent extents will be 5 M, to a maximum of 100 extents. All objects, regardless of their storage settings, must have an extent size of

at least 5 M. Any other clauses listed in the command will have the default setting. Save the finished file, named **ho0601.sql**, in the **Solutions\Chapter06** directory on your student disk.

2. Create a SQL script that completes the following tasks successfully. Save the script in a file named **ho0602.sql** in the **Solutions\Chapter06** directory on your student disk.
    1. Create a locally-managed tablespace named HANDS_62 with one 10 M OMF datafile in the directory name provided by the instructor.
    2. Add another 25 M datafile (OMF) to the tablespace.
    3. Query the Data Dictionary, and display the total available space in the tablespace.
    4. Make the tablespace read only, and take it offline.
    5. Drop the tablespace.
3. One of your dictionary-managed tablespaces seems to run out of space long before you calculated that it would. There has been a lot of activity in which tables were created, then dropped, and re-created with different storage settings. Explain why your tablespace has a problem, and list two actions you can take to alleviate the problem. Save the answer in a file named **ho0603.txt** in the **Solutions\Chapter06** directory on your student disk.
4. Explain the difference between an extent, a segment, and a table. Give an example of how the three are related. Save the script in a file named **ho0604.txt** in the **Solutions\Chapter06** directory on your student disk.
5. Look up the CREATE TABLESPACE command in the Oracle9*i* SQL Reference Release 2 document found online in the Oracle 9*i* Database Online Documentation, Release 2 link. Find the description of the EXTENT MANAGEMENT clause. Figure 6-16 shows the diagram of this clause.

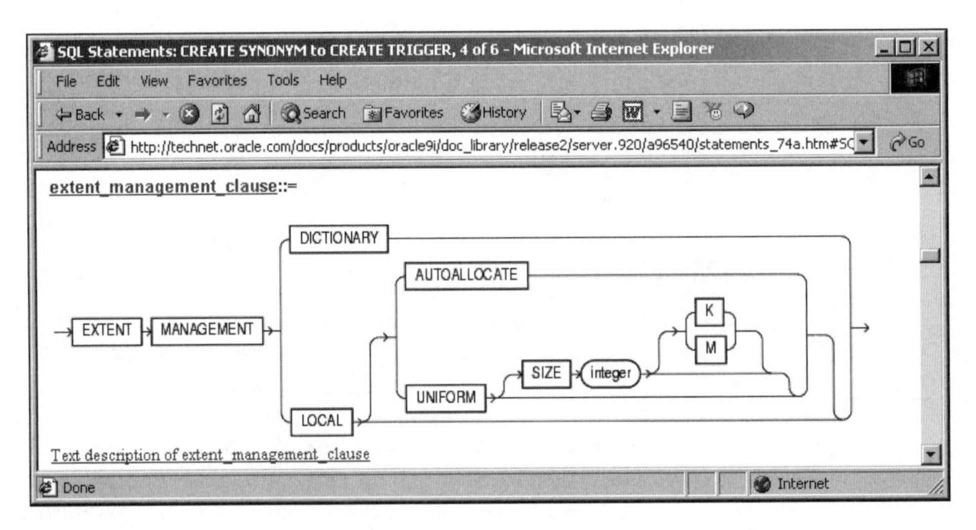

Figure 6-16 The EXTENT MANAGEMENT clause has many options

Write an EXTENT MANAGEMENT clause for each of the following three requirements. Save the script in a file named **ho0605.txt** in the **Solutions\Chapter06** directory on your student disk.

1. The extents are all 4096 bytes in size.
2. The extents are variable in size.
3. The extents have no bitmaps associated with them.

6. You have a tablespace that was created with this command:

```
CREATE TABLESPACE CUSTOMERSERVICE
DATAFILE 'D:\oracle\data\cust_serv01.dbf' SIZE 50M
AUTOEXTEND OFF;
```


Write a SQL script to accomplish these tasks:

1. Add another file that is identical (except the name) to the original one.
2. Make the first datafile 25 M in size.
3. Make the second datafile expandable forever.

After you have made the script, write a short explanation of what you believe are the pros and cons of using an unlimited datafile. Save the script and your explanation in a file named **ho0606.txt** in the **Solutions\Chapter06** directory on your student disk.

7. Write the command to create a temporary tablespace with 20 M storage space. The file must be an Oracle managed file. Modify the datafile so that it is 15 M. Write a query that shows the file name and size. Save the script and the query in a file named **ho0607.sql** in the **Solutions\Chapter06** directory on your student disk.
8. Write a query to determine your current, temporary tablespace, when you are logged onto SQL*Plus Worksheet as SYSTEM. Write another query to show the current space used and free space in all temporary tablespaces. Run the script in **ho0608.sql** in the Data/Chapter06 directory on your student disk. (*Hint*: Click **Worksheet/Run Local Script** from menu.) The script creates a temporary table and inserts one row. Rerun the query showing space used, and determine how much space was allocated for the script by subtracting the space used before and after the script. Save your queries and the amount of space used in a file called **ho0608a.sql** in the **Solutions\Chapter06** directory on your student disk.
9. Display the undo-related initialization parameters. Change the retention time to 10 minutes. Change the initialization parameter that suppresses errors to TRUE. Write a query using DBA_UNDO_EXTENTS to summarize the used blocks in the undo tablespace. Save your script in a file called **ho0609.sql** in the **Solutions\Chapter06** directory on your student disk.

CASE PROJECTS

1. Now that you have read all about locally managed tablespaces, you decide your Global Globe database should use exclusively locally managed tablespaces. Write a query to determine what management mode the SYSTEM tablespace uses. Save the query in your **Solutions\Chapter06** directory in a file named case0601.sql. Imagine that the SYSTEM tablespace is not locally managed. How will you be able to resolve this?

   Write a CREATE DATABASE command that creates the SYSTEM tablespace as a locally managed tablespace. Include a temporary tablespace in the CREATE DATABASE command. The database name is GLOBAL1. Assume that the CONTROL_FILES initialization parameter sets the control files, and the redo log files are Oracle managed files (therefore, the LOGFILE clause is not required).

   Your MIS manager may be dubious about your wish to drop and re-create the database, even though the programming staff has not begun using it yet. Write a paragraph justifying your choice to convince your boss and the Global Globe steering committee that you are right.

   Your request was turned down anyway, so you turn to the task of creating new tablespaces. You plan to use separate tablespaces for each department. Write SQL commands to create tablespaces with these parameters and execute the commands. (All *these* tablespaces will be locally managed, even if the SYSTEM tablespace is not.) The datafiles should be located in the directory provided by your instructor. Add your script to the file named **case0601.sql** in the **Solutions\Chapter06** directory on your student disk.

2. The Accounting Department will have tables with an initial volume of 50 M, a total volume of 500 M after one year. The Accounting Department's storage will grow slowly over the following two years to a volume of 1000 M. You want two datafiles initially and the storage space should be allowed to grow to a maximum of 1000 M combined. Provide uniform 5 M extents.

   The Sales Department will have tables that store frequently changing data with a total volume of 25 M to start and growing to an unknown size in two years. No one has been able to estimate the growth pattern accurately. You want to let them grow quickly without restraints, so you will set up their tablespace with one unlimited datafile that has variable system-managed constraints.

   The Training Department has a library of reference material that seldom changes. They want to store it in the database so it is easily searchable. You have estimated that the volume of data is about 100 M. The Training Department is concerned that others will make unauthorized changes to the material after it is loaded on the database and made widely available. You assure them that you can prevent any changes to the data. Save your scripts in a file called **case0602.sql** in the **Solutions\Chapter06** directory on your student disk.

CHAPTER

7

BASIC TABLE MANAGEMENT

In this chapter, you:

- Describe the different types of tables and their storage methods
- Create relational and temporary tables
- Create tables containing varrays and nested tables
- Create object and partitioned tables

In this chapter, you dig deeper into the database structure by studying how tables are stored and managed. You have a brief refresher on how to create a table, focusing on the storage options of the CREATE TABLE command. You find out how Oracle9*i* keeps track of rows, blocks, and extents for a table and how to monitor and adjust the settings that affect table storage.

Before beginning this chapter, you should install some additional objects into your database that are needed for exercises in this chapter. Follow these steps to run a script that creates the objects you need.

1. Start up the Enterprise Manager console. In Windows, click **Start/Programs/Oracle - OraHome92/Enterprise Manager Console**. In UNIX, type **oemapp console** on the command line. The Enterprise Manager Console login screen appears.
2. Select the **Launch standalone** radio button and click **OK**. The console appears.
3. Click **Tools/Database Applications/SQL*Plus Worksheet** from the top menu. A background SQL*Plus process starts, and then the SQL*Plus Worksheet window appears. If the background process appears in front, simply minimize it (click the **minus sign** in the top-right corner) so that you can see the worksheet.
4. Connect as the SYSTEM user. Click **File/Change Database Connection** in the menu bar. A logon window appears. Type **SYSTEM** in the Username box, *the SYSTEM password provided by your instructor* in the Password box, and **ORACLASS** in the Service box. Leave the connection type as "Normal." Click **OK** to continue.
5. Now that you are connected as the SYSTEM user, you can run the script prepared for this chapter. Click **Worksheet/Run Local Script** from the top menu. This opens a window in which you select a file. Navigate to the **Data\Chapter07** directory on your student disk, and select the **setup.sql** file. Click **Open** to run the script in the worksheet.
6. The SQL*Plus Worksheet runs the script, and replies, "Type created" several times.
7. Exit the worksheet by clicking the **X** in the upper-right corner, or typing **EXIT** clicking the **Execute** icon, and clicking **OK** in the Oracle Enterprise Manager dialog box that opens. This returns you to the console. Now, find the new object types and collection types created.
8. Exit the console by clicking the **X** in the upper-right corner.

The objects that were created are discussed later in this chapter.

Introduction to Table Structures

As you know, a table is the basic storage unit for data in the Oracle9*i* database. Tables are made up of rows and columns. When you create a table, you define the table's name as well as each column's name, data type, and size. In addition, you can define the storage settings of the table. After the table is defined, you use SQL to add data to the table. One row in a table contains data for one record. One column in a row contains one field of data for one record.

There are several different kinds of tables you can create, depending on what you need to store:

- **Relational table:** This is the standard table that is traditionally used in a relational database. In Oracle9*i*, it is referred to simply as a table. A **table** stores data of all types and is the most common form of storage in the database. Although it is not a strict requirement, most tables have a **primary key**, which is a column or set of columns that uniquely identify each row. A table resides in a single tablespace unless it is partitioned. Partitioned tables can store each partition in a separate tablespace. Tables are partitioned to store large quantities of data that is referenced frequently.
- **Index-organized table:** This type of table is a relational table with a primary key that specifies the location of the rows that physically store the values being sought. There is no separate index for the primary key, as there is with a relational table. Use an index-organized table when the data is retrieved by primary key values or ranges of primary key values. Retrieval by primary key is faster, because the database reads only the table, instead of reading the index and then reading the table. Index-organized tables can be partitioned just as relational tables can.
- **Object tables:** The **object table** is a table that holds objects and attributes. An object table is similar to a relational table, except that each row is a single unit of data defined by an object type. An **object type** is a set of column definitions that is defined ahead of time for use in objects. For example, before creating an object table that stores customer names and addresses, you create an object type that contains all the columns you plan to use. The object type's columns are called **attributes**. Some examples of attributes are customer name, street address, city, state, country, and zip code. After creating the object type, you use it to define an object table. The object table, like relational tables, holds the data in rows and columns. Object tables are used to store complex data that requires special handling whenever data is added or changed in the object table. The special handling can be programmed into **methods**, which make data maintenance simpler and more consistent. Methods are programmed processes stored with the object type's definition. You will not be working with methods in this book.
- **Temporary table:** A **temporary table** contains data like other tables; however, the data is private (not seen by other users) and disappears at the end of the user's session or when the user commits a transaction. The CREATE GLOBAL TEMPORARY TABLE command is used to create a temporary table. Temporary tables use temporary segments in either a temporary tablespace or a permanent tablespace.
- **External table:** An **external table** contains data in a file outside the database. You define the table and its columns as usual, and then define where the external data resides and how each column maps to the external file's records. External tables are new with Oracle9*i*. They are read-only.

- **Nested table:** A **nested table** contains data that is stored within a single column of another table. The data in a nested table is actually stored in its own table (called the **store table**), while the main table contains an identifier that locates the associated nested table for that row. Define the store table as you would define a relational table, with a name and storage settings.
- **XML Table:** A table created with one column of the XML type data type is an **XML table**. This is a new feature of Oracle9*i*. You now can store XML data in this new type of table. **XML** (Extensible Markup Language) is a programming language for the Internet that provides both the data and the display details to a Web page or to an application.
- **Cluster:** Technically, a **cluster** is not a table, but a group of tables stored together as if they were one table. Clusters store data of multiple tables in one segment. Clusters speed up the access to sets of tables that are frequently referenced together. For example, if your application always queries the CUSTOMER table joined with the CUSTOMER_ORDER table, creating a cluster of the two tables may speed up performance. A single I/O function can retrieve the combined data from both tables.

All these tables store data in tablespaces. As you saw in Chapter 6, tablespaces define how data is stored within them. In most cases, you can further define storage at the table level. To understand storage better, start by looking into the lowest level of logical data storage: the data block.

Setting Block Space Usage

As you saw in Chapter 6, with Oracle9*i*, a table's storage is contained in one segment. The segment contains one or more extents. Each extent contains a contiguous set of data blocks. What is inside a data block?

Each block represents a group of bytes on the physical file. The default block size is 2048 bytes. Oracle9*i* manages space down to the last byte in each block.

In locally managed tablespaces, you don't need to worry about the way the blocks are managed. However, in dictionary-managed tablespaces, you must understand how the STORAGE parameters work. This way, you can match the settings with the type of activity in the table.

The same STORAGE parameters discussed here for tables can be used in CREATE TABLE, CREATE INDEX, and other CREATE commands.

Recall that the DEFAULT STORAGE clause for a tablespace has this syntax:

```
DEFAULT STORAGE (INITIAL <nn> NEXT <nn> PCTINCREASE <nn>
MINEXTENTS <nn> MAXEXTENTS <nn>)
```

The STORAGE clause syntax for a table or index or other object looks like this:

```
TABLESPACE <tablespacename>
STORAGE (INITIAL <nn> NEXT <nn> PCTINCREASE <nn>
MINEXTENTS <nn> MAXEXTENTS <nn>
FREELISTS <nn> FREELIST GROUPS <nn>
BUFFER_POOL KEEP|RECYCLE|DEFAULT
PCTFREE <nn> PCTUSED <nn> MINTRANS <nn> MAXTRANS <nn>)
```

The INITIAL, NEXT, PCTINCREASE, MINEXTENTS, and MAXEXTENTS parameters all pertain to the size, number, and management of extents and were discussed in Chapter 6. All the tablespace parameters can be overridden by setting them when you create a table, index, and so on.

The PCTFREE, PCTUSED, MINTRANS, and MAXTRANS parameters help describe the way rows are stored and accessed inside each block in the extent. These four parameters are discussed later in this section.

The FREELIST and FREELIST GROUPS clauses tell Oracle9*i* if you want more than one freelist for each datafile. Usually, you do not need to change these parameters. By default, there is one freelist and one freelist group per extent. The freelist keeps track of all the data blocks that allow rows to be inserted into the block. You change the defaults mainly when you are running Real Application Clusters.

The BUFFER POOL clause identifies for Oracle9*i* the buffer for storing data blocks from the table when they are read from the disk and stored in the SGA. The default setting is BUFFER POOL DEFAULT and is appropriate for most tables. Specify BUFFER POOL KEEP for tables that are frequently used, so that their data blocks stay resident inside the buffer longer than normal. The data blocks stay in the KEEP buffer until it is full and more data blocks need to be added. Then, the data blocks are flushed from the buffer with the least recently used blocks (**LRU blocks**) being replaced first. You can adjust the size of the KEEP buffer by modifying the BUFFER_POOL_KEEP initialization parameter. Specify BUFFER POOL RECYCLE for tables that should be removed from the buffer as quickly as possible. This setting is best for infrequently used tables. Using this setting keeps the infrequently used table's data blocks from overwriting data blocks in the DEFAULT buffer. You can adjust the size of the RECYCLE buffer by modifying the BUFFER_POOL_RECYCLE initialization parameter.

There are three types of data blocks for storing the corresponding types of data: data, index, and undo. This chapter focuses on data blocks, which are used for tables. Index blocks and undo blocks behave similarly.

Tables store data in data blocks. Figure 7-1 shows the components that make up a single data block:

- **Common and variable header:** Identifying information, such as the type of block and block location.
- **Table directory:** Information about the table that has data in the block.
- **Row directory:** A list of row identifiers for rows stored in the block. This grows as more rows are inserted into the block. Even when a row is dropped, the row directory does not shrink.
- **Free space:** Bytes of storage space left unallocated. This shrinks as rows are inserted or updated with more data, causing both the row directory and the row data to take up more space. When a row is deleted or updated with less data, the free space grows, although the row directory never releases any of its space.
- **Row data:** Bytes of storage used for rows inserted or updated in the data block. Updates that make a row grow in size cause the row data to be shuffled so that the data for each row stays together. If a row is updated and requires more free space than the block contains, the entire row is **migrated**, that is, moved to another block. The original block has a pointer to the second block. If a row is too large to store in one block when it is inserted (for example, a row containing a 50 M recorded song), the row is started in one block and the overflow is stored in another block. The original block has a pointer to the second block. If it needs a third block, the second block has a pointer to the third block, and so on. A row that spans multiple blocks is called a **chained row**.

The first three data block components (common and variable header, table directory, and row directory) are called overhead, because they store information about the maintenance and use of the block rather than actual data. The average size of a data block overhead is 84 to 107 bytes. The size is affected by changes in the data (more rows mean more bytes in the row directory) and by the settings of MAXTRANS and INITRANS in the STORAGE clause of a table.

INITRANS and MAXTRANS tell Oracle9*i* how many concurrent transactions can access this block. The block header establishes a storage space inside the header for each transaction. The default value for INITRANS is one, and the default value for MAXTRANS depends on the block size. These parameters usually should not be modified. They affect the initial size and the growth of the block header. Additional concurrent transactions require more header space that does not return to free space, even after the transaction completes. If a table sets MAXTRANS at three and a fourth transaction attempts to access the block, the fourth transaction waits until one of the other transactions finishes.

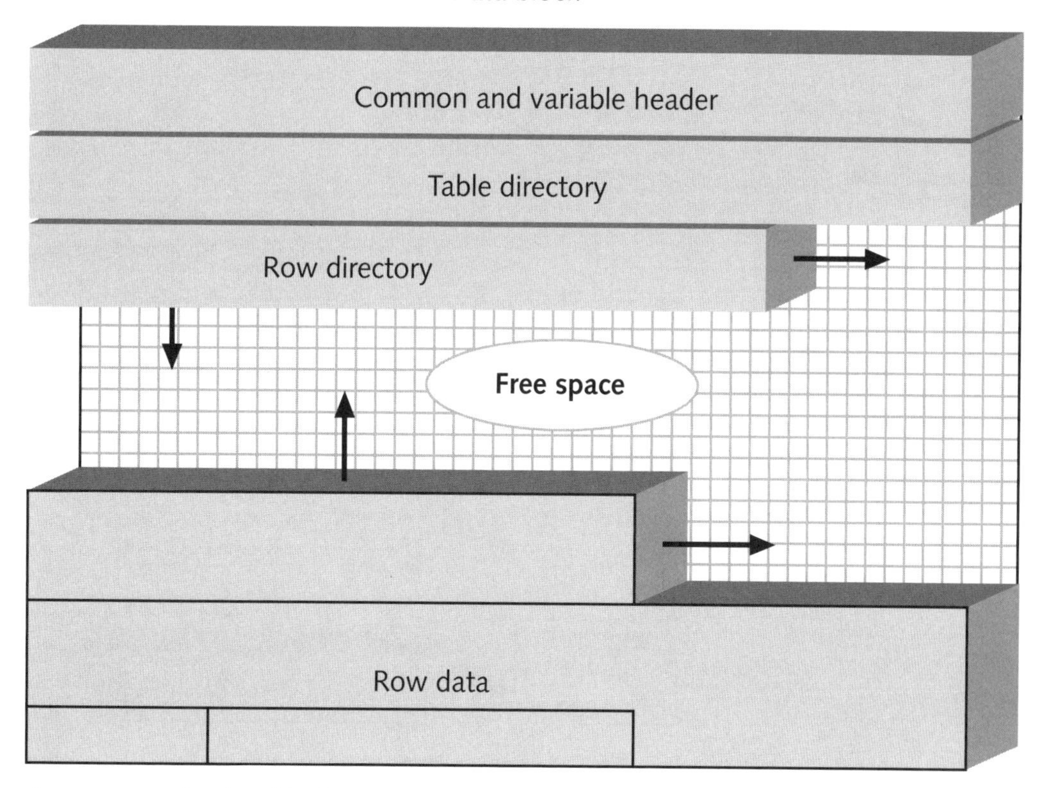

Figure 7-1 The data block's free space shrinks as the overhead grows and as the row data grows

One of the key issues in good performance is to avoid chaining and migrating of row data. A chained or migrated row requires additional work to retrieve. Oracle9*i* has no direct way to reach the data stored in the subsequent blocks. It must read the row header in the original block, locate the pointer to the second block, and read the second block. Figure 7-2 shows a normal row, a chained row, and a migrated row.

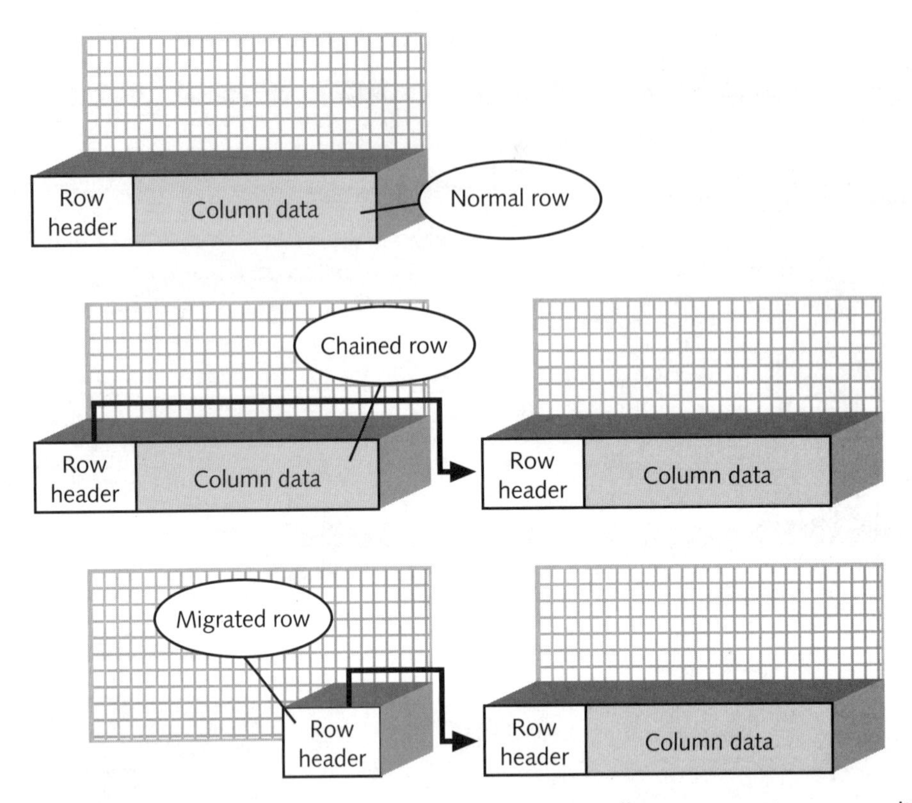

Figure 7-2 In some cases, chaining and migration allow one row to span multiple blocks

In the case of chained rows, the only solution is to increase the block size, and even that may not help in some cases. With large column data, such as those used by multimedia applications, some row chaining is inevitable.

To prevent row migration, you can adjust two storage parameters to suit the characteristics of your table's data. The two parameters are PCTFREE and PCTUSED. The two together must always add up to 100 or less.

PCTFREE tells Oracle9*i* how much free space to reserve for updates to existing rows. For example, if PCTFREE is 20, rows can be inserted until the free space shrinks to 20 percent of the block. At this point, Oracle9*i* allows no more inserting of rows. The remaining 20 percent of free space is reserved for the growth of the rows already in the block. For example, if you change the value in a column from "Hank" to "Henry-James," the size of that row grows, using some of the remaining free space. When a block stores rows that increase in size due to updates, a higher PCTFREE setting helps keep the rows from migrating.

PCTUSED tells Oracle9*i* when to allow more rows to be inserted into the block. If a row is deleted and the free space grows to more than 20 percent, the block still does not accept new rows (unless the PCTUSED is set at 80 percent). The free space has to increase to at least the number in PCTUSED, before new rows can be inserted. For example, if PCTUSED is 50 percent, the block initially is filled until only 20 percent is free (because of the PCTFREE of 20 percent). After this, if rows are deleted and the free space grows to 50 percent, rows can be inserted again. Rows can be inserted until the free space shrinks to 20 percent. This continues cycling as much as needed.

If your table data is not updated often, a lower PCTFREE allows you to use more space in the block. If your table data grows a great deal, a higher PCTFREE better prevents row migration. A PCTFREE of zero packs rows in all the free space and works well for inserting masses of static data, such as a table of telephone area codes and their city, county, and state locations.

A higher PCTUSED allows Oracle9*i* to reuse the space faster. The block is put back on the freelist that tracks block usage for the extent. However, because the row header keeps growing with every insert and does not shrink with deletes, it is conceivable that you could reach a point when the row header takes up more room than the data.

Refer to the next section for details on how to set INITRANS, MAXTRANS, PCTFREE, and PCTUSED for a table.

Storage Methods

Storage methods for tables depend on the type of table you are creating and the characteristics of the tablespace in which you create the table. You can define a table with no storage settings, and Oracle9*i* uses the default settings found in the tablespace, if they exist. If no storage defaults were set up for the tablespace, Oracle9*i* uses its own standard default settings which are:

- PCTFREE=10
- PCTUSED=40
- INITRANS=1
- MAXTRANS varies depending on the data block size

The most important difference in storage methods is between locally managed and dictionary-managed tablespaces.

How to Set Storage for Locally Managed Tables

As you learned in Chapter 6, locally managed tablespaces handle free space with a bitmap within the tablespace. The DEFAULT STORAGE clause is not used with locally managed tablespaces. Similarly, only a few of the parameters of the STORAGE clause should be used in the tables created in a locally managed tablespace.

What storage information do you specify for a table in a locally managed tablespace?

- **TABLESPACE:** Name the tablespace in which the table is created. If you omit this parameter, the table is created in your default tablespace. (The SYSTEM tablespace is used if the creator of the table does not have a default tablespace assigned.)
- **STORAGE (INITIAL <nn>):** Oracle9*i* looks at the tablespace attributes to determine extent size. If the tablespace specifies AUTOALLOCATE, Oracle can make the extent or extents any size that it determines is best. If the tablespace specifies UNIFORM, Oracle9*i* allocates enough uniform extents to reach the INITIAL size. For example, if UNIFORM 5M was set for the tablespace and INITIAL 50M was set for the table, Oracle9*i* allocates ten extents to the table initially. All the other parameters of the STORAGE clause are ignored, even if you include them.

Complete the following steps to create a table in a locally managed tablespace.

1. Start up SQL*Plus Worksheet. In Windows, select **Start/Programs/ Oracle - OraHome92/Application Development/SQLPlus Worksheet**. In UNIX, type **oemapp worksheet** on a command line. The standard logon window for all Enterprise Manager tools appears.
2. Type **CLASSMATE** in the Username box, **CLASSPASS** in the Password box, and **ORACLASS** in the Service box. Leave the Connect set to "Normal." The SQL*Plus Worksheet displays. If the background SQL box pops up in front of the worksheet, simply minimize it by clicking the **Minimize** button in the top-right corner.
3. Type the following CREATE TABLE command in the top pane of the SQL*Plus Worksheet window. You learn about the other components of the command later in this chapter.

```
CREATE TABLE CH07BICYCLE
(BIKE_ID NUMBER(10) PRIMARY KEY,
BIKE_MAKER VARCHAR2(50) NOT NULL,
STYLE VARCHAR2(15))
TABLESPACE USERS
STORAGE (INITIAL 25M);
```

4. Execute the command by clicking the **Execute** icon on the left side of the worksheet. Oracle9*i* creates the table and displays "Table created" in the SQL*Plus Worksheet window.
5. Exit SQL*Plus Worksheet by typing **exit**, clicking the **Execute** button, and clicking **OK** in the Oracle Enterprise Manager dialog box that opens.

You never need to issue a COMMIT command after executing DDL (Data Definition Language) commands, because they are automatically committed.

If you leave out the STORAGE clause altogether, Oracle9*i* looks for the tablespace settings. If the tablespace has no MINEXTENTS setting, Oracle9*i* creates the table with an initial extent of five data blocks (the default size) for AUTOALLOCATE, or one extent of whatever extent size is specified in UNIFORM.

Locally managed tables eliminate much work in managing storage space. However, it is still important to estimate the initial size of a table to keep its data in contiguous data blocks for faster data retrieval.

How to Set Storage for Dictionary-Managed Tables

This section digs into the details of specifying every possible storage parameter. INITIAL is the most important parameter, because it defines the table's initial size. The next most important parameters in dictionary-managed tables are the PCTFREE and PCTUSED parameters.

Do not execute the example code in this section. The code will cause errors, because your Oracle9*i* database has no dictionary-managed tablespaces.

Consider the tablespace's DEFAULT storage settings before writing the STORAGE settings for a table. It is generally better to keep extent sizes the same or multiples of the same size within the tablespace so free space that is deallocated can be more easily used. Here is an example of a table created in a dictionary-managed tablespace, in which all the default storage parameters are used, including the tablespace name. In this case, the tablespace name is derived from the user's default tablespace setting. That tablespace is dictionary-managed, and has default storage settings that are used to create the table:

```
CREATE TABLE BIKE_MAINTENANCE
(BIKE_ID NUMBER(10),
 REPAIR_DATE DATE,
 DESCRIPTION VARCHAR2(30));
```

In this example, imagine that the tablespace's default storage settings are like this:

```
DEFAULT STORAGE (INITIAL 10M NEXT 2M
PCTINCREASE 0 PCTFREE 10 PCTUSED 80)
```

The BIKE_MAINTENANCE table would inherit all these settings. By using the same settings as other tables in the tablespace, extents tend to be more uniform in size and, therefore, more easily reused when a table is dropped or storage is released by a TRUNCATE command.

Here is another table in which the initial size, extent size, and other parameters for the table override the tablespace settings:

```
CREATE TABLE TRUCK_MAINTENANCE
(TRUCK_ID NUMBER(10),
 REPAIR_DATE DATE,
 PROBLEM_DESCRIPTION VARCHAR2(2000),
 DIAGNOSIS VARCHAR2(2000),
 BILLING_DATE DATE,
 BILLING_AMT NUMBER (10,2))
TABLESPACE USER_DTAB
STORAGE (INITIAL 80M NEXT 40M PCT INCREASE 0
MINEXTENTS 2 MAXEXTENTS 25
PCTFREE 25 PCTUSED 50 MINTRANS 1 MAXTRANS 2)
```

The TRUCK_MAINTENANCE table sets its own storage settings, because it anticipates a higher volume required for storing its data. It still uses a multiple of the tablespace's default setting for INITIAL and NEXT so that any deallocated extents are more easily used by other tables.

The user can create a table in his default tablespace. In addition, if the user has the UNLIMITED TABLESPACE system privilege, he can create a table in any tablespace. Without the UNLIMITED TABLESPACE system privilege, a user must be given a quota of space on any tablespace he uses to store the tables that he creates. See Chapter 11 for more information.

The next section contains one final aspect of data storage: the makeup of row data. After working with row structure, you begin to dig into the details of what designers see as the building blocks of tables: columns.

Row Structure and the Rowid

Rows are stored in a data block in a compact structure, so a data block can accommodate the maximum data. Figure 7-3 shows the components of a row.

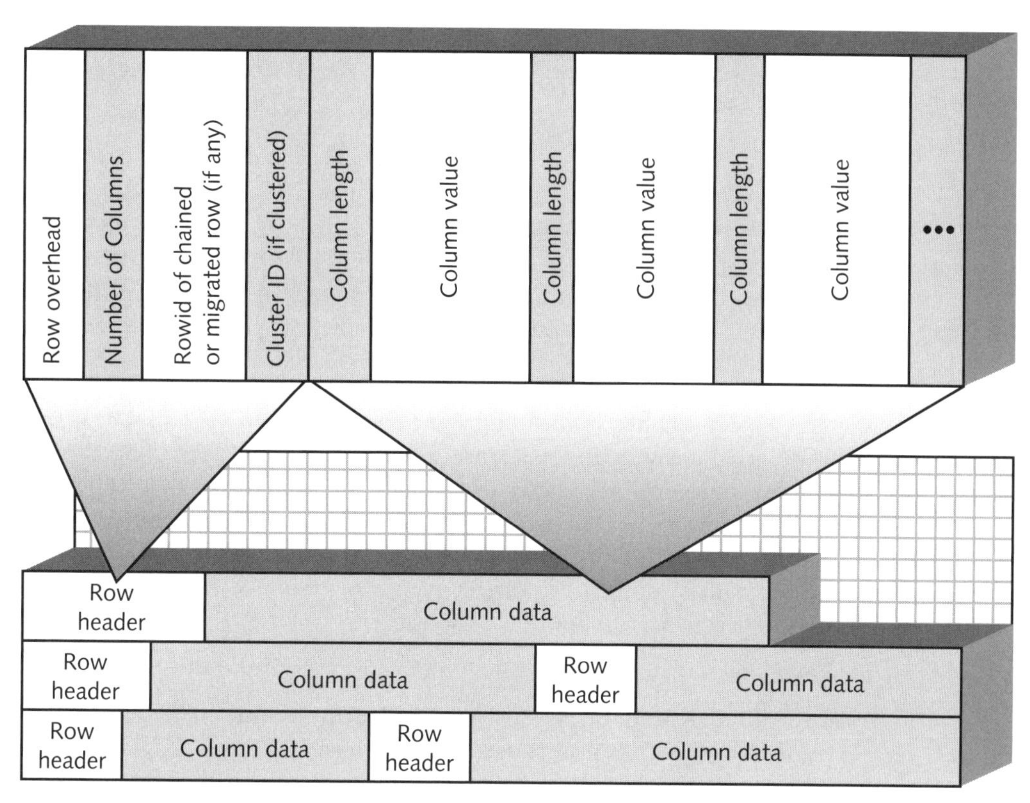

Figure 7-3 The row header points to a chained row if needed

The row is made up of two sections:

- **Row header:** The **row header** stores the number of columns contained in the column data area, some overhead, and the rowid pointing to a chained or migrated row (if any). A **rowid** contains the physical or logical address of the row. The row header is a minimum of three bytes long. If this is a row in a cluster, the cluster key ID is also included. A **cluster key ID** is a special type of rowid for clustered tables. Oracle9*i* stores only one **clustered row** (that is, the rows from each table in a cluster that correspond to one unique value in the cluster key) per block. Clustered tables are not covered in this book.
- **Column data:** The column data consists of a one- to three-byte section containing the length of the column's data followed by the actual data for the column. Notice that there is no column name or identifier in either the row header or the column data. Oracle9*i* stores column data in the exact order as defined when the table was created. If a row's column has null values *and is at the end of the columns*, nothing is stored. A null column *followed by one or more columns with data* must have a placeholder, because Oracle9*i* determines a column's name by

its order in the list of column data. Therefore, Oracle9*i* stores a column length of zero for the null column, followed immediately by the column length of the next column.

When creating the table, take advantage of space savings by placing the columns defined with the NOT NULL constraint first in the table, followed by columns that allow null values. (The NOT NULL versus NULL column constraint is covered in more detail in Chapter 10.) Placing columns that never contain null values first helps save space by improving the odds that the row won't have extra placeholders for null columns that appear between columns with data.

A chained or migrated row is located by the rowid stored in the row header. This is not the only use for the rowid. Indexes store the index column values and the rowid of the associated row in the table. The rowid is a special identifier that allows Oracle9*i* to retrieve a row quickly. Access by rowid is ranked the highest among methods of data retrieval. This is why indexes retrieve data more quickly than a scan of the table. See Chapter 8 for a discussion of optimizer plans, in which retrieval methods help determine the best method of finding data for a query.

Figure 7-4 shows a diagram of the two types of rowids found in Oracle9*i*. The types are:

- **Physical rowid:** A physical rowid identifies a row by its physical location in a datafile. The physical rowid of a row never changes, unless the table is reorganized (see Chapter 8). There are two formats for the physical rowid:
  - **Extended rowid:** The default format for Oracle8*i* and above. This contains four values, as shown in Figure 7-4.
  - **Restricted rowid:** This rowid format was standard for Oracle7 and below and is preserved for backward compatibility. It is missing the object identifier found in the extended rowid.
- **Logical rowid:** A logical rowid identifies a row by its primary key. It may also contain the probable location of the block in which the row resides. A logical rowid is used only for index-organized tables. The physical location of rows in an index-organized table changes as rows are added and deleted, so the physical rowid is not accurate.

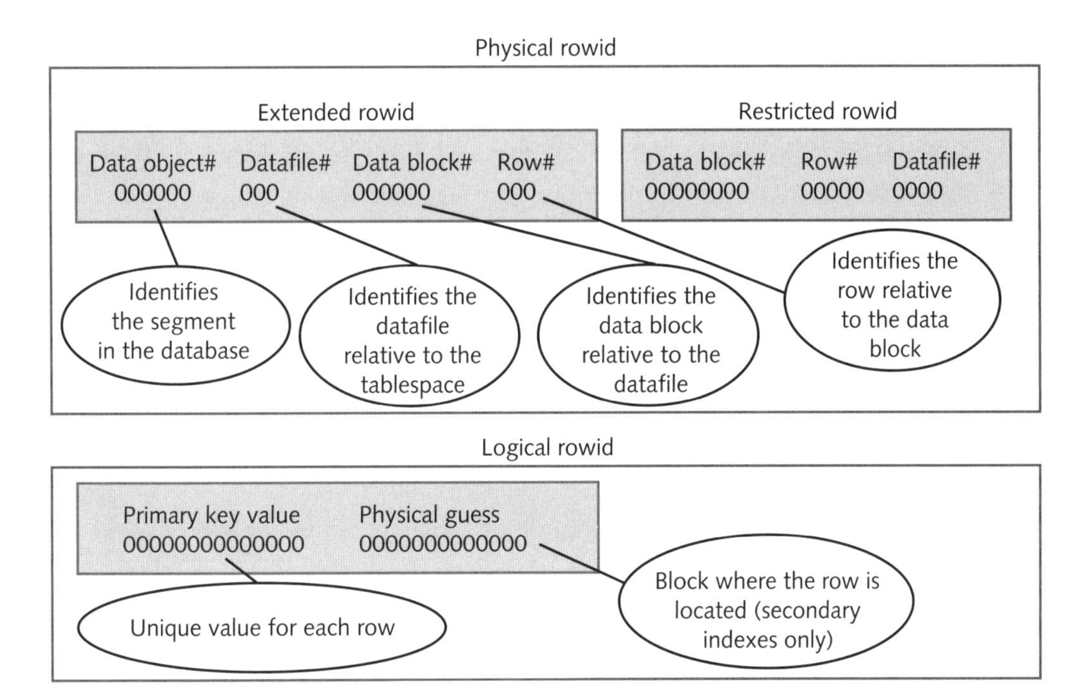

Figure 7-4 Physical and logical rowids have different internal formats

Rowids cannot be changed by any user. The rowid is actually not stored with the row at all. A rowid is only stored when an index on the table is created, or when you query the pseudocolumn ROWID, in which case the rowid is constructed for your viewing. A **pseudocolumn** acts like a column in a query, but is actually calculated by the database for the query. You can include a pseudocolumn when you query any table, and Oracle9*i* delivers the appropriate data in the pseudocolumn. Another pseudocolumn is SYSDATE, which returns the current date and time.

Indexes are automatically kept up-to-date as data in the table changes. An index is stored in the database as a table, and contains the values of the indexed columns and the rowid of the corresponding table row. The rowid is stored in a column with the UROWID data type (or ROWID data type for databases that are compatible with Oracle7 or lower). The **UROWID** data type stands for universal rowid data type, and it can hold any type of rowid, including a record identifier from a non-Oracle database.

You can create a table that contains the rowids of another table's rows to take advantage of the high-speed retrieval rate for rowids. For example, you have a CUSTOMER table with a primary key called CUST_ID. You build a CONTACT_LIST table that contains rows for each contact made to a customer. Ordinarily, you link the two tables together with a foreign key. However, you could create a column (with UROWID data type) in the CONTACT_LIST table that stores the rowid of the customer's row. Be cautious with this trick, because if a row in the CUSTOMER table is deleted, the corresponding rowid found in any rows in the CONTACT_LIST for that customer are not automatically removed. Later, a query that joins the CONTACT_LIST table and the CUSTOMER table by the rowid might fail because of the invalid rowid.

You can use the rowid to determine the physical location of a particular row in a table by looking at the portion of the rowid storing the datafile information.

Now that you are familiar with the data block, how rows are stored, and what the rowid contains, it is time to build some tables.

CREATING TABLES

You have already been exposed to the CREATE TABLE command earlier in this chapter. The command syntax varies according to what type of table you are creating. For relational tables (the most commonly used type of table), the syntax looks like this:

```
CREATE TABLE <schema>.<tablename>
(<column_name> <datatype> <size> NULL|NOT NULL
  DEFAULT <default_value> CHECK <check_constraint>,
... )
<constraints>
TABLESPACE <tablespacename>
STORAGE (INITIAL <nn> NEXT <nn> PCTINCREASE <nn>
MINEXTENTS <nn> MAXEXTENTS <nn>
FREELISTS <nn> FREELIST GROUPS <nn>
BUFFER_POOL KEEP|RECYCLE|DEFAULT
PCTFREE <nn> PCTUSED <nn> MINTRANS <nn> MAXTRANS <nn>)
```

The syntax for storage was discussed earlier in the chapter and in Chapter 6. This section primarily discusses the column definition section of the command.

The syntax for other types of tables and an alternative syntax for relational tables are discussed later in this chapter.

Columns and Data Types

A table must contain at least one column to be defined in the database. Columns have names and data types and can optionally have a default value and a restriction on null values. A **data type** is a predefined form of data. Oracle has 21standard data types. A **default value** is a value that is filled into the column when an inserted row does not specify the value for that column. To prevent inserted or updated rows from leaving a particular column null, add the NOT NULL constraint to the column definition. **Constraints** are rules that define data integrity for a column or group of columns. See Chapter 10 for a complete description of constraints.

Table 7-1 shows a categorized list of Oracle's predefined (built-in) data types. More details are listed following the table for some of the data types.

Table 7-1 Oracle9*i* built-in data types

Category	Data type	Range and Examples	Description
Character	CHAR (<length>)	1–2000 bytes CHAR(100)	Fixed length field; usually for smaller values. Data shorter than the specified size is padded with blank spaces.
	VARCHAR (<length>)	1–4000 bytes VARCHAR(45)	Variable length character data. Always use VARCHAR2 (used only for backward compatibility).
	VARCHAR2 (<length>)	1–4000 bytes VARCHAR2(30)	Variable length character data. Very commonly used because it adjusts to the size of the data to save storage space.
	NCHAR (<length>)	1–2000 bytes NCHAR(10)	Fixed length Unicode characters. Used only for national language support (see Chapter 14).
	NVARCHAR2 (<length>)	1–4000 bytes NVARCHAR2(1500)	Variable length Unicode characters. Used only for national language support.
	LONG	1 byte–2 gigabytes LONG	Variable length storage for large fields. Provided for backward compatibility. For new tables, use CLOB or NCLOB instead
Number	NUMBER (<s>,<p>)	1 x 10^{-130} to 9.99...99 x 10^{125}, positive or negative, to 38 significant digits	Storage for numbers. You can specify the total number of digits (scale) and the number of digits to the right of the decimal (precision).
DATE	DATE	January 1, 4712 BCE through December 31, 4712 CE (Common Era) DATE	Data type that stores both date and time. It stores a four-digit year and time in hours, minutes, and seconds.
	TIMESTAMP (<precision>)	Same range, but with fractional seconds (6 is default) TIMESTAMP(4)	Date with fractional seconds are needed.
	TIMESTAMP (<precision>) WITH TIME ZONE	Same range, but with fractional seconds and with the time zone in the data TIMESTAMP(5) WITH TIME ZONE	Date with fractional seconds and Time Zone information stored. Useful for geographically dispersed users sharing a central database.

Table 7-1 Oracle9*i* built-in data types (continued)

Category	Data type	Range and Examples	Description
	TIMESTAMP (<precision>) WITH LOCAL TIME ZONE	Same range, but with fractional seconds TIMESTAMP (3) WITH LOCAL TIMESTAMP	Date with fractional seconds. Time is adjusted to the database's local Time Zone. Time Zone information is not stored. Useful when you need all timestamps to be stored with a common Time Zone, even when users in a different Time Zone enter their local time.
	INTERVAL (<precision>) YEAR TO MONTH	Precision is the number of digits the year portion can contain. Default is 2. Range is 0–4. Interval (3) year to month	Time storage in years and months.
	INTERVAL DAY (<precision>) TO SECOND (<sec_precision>)	Precision is the number of digits (0–9, default of 2) in the day portion. Sec_precision is the number of digits to the right in the seconds portion. INTERVAL DAY (3) to SECOND (8)	Time storage spans of days and seconds.
LOB	BLOB	Large binary data up to 4 gigabytes BLOB	Unstructured binary data. Good for multimedia data, such as music, video, or images.
	CLOB	Large character data up to 4 gigabytes CLOB	Character data. Good for documents.
	NCLOB	Large Unicode data up to 4 gigabytes NCLOB	National language character data. Good for documents in languages, such as Chinese.
	BFILE	Large binary data up to 4 gigabytes, stored externally BFILE	Read-only data that is stored in a file. The database column contains a pointer to the datafile.

Table 7-1 Oracle9*i* built-in data types (continued)

Category	Data type	Range and Examples	Description
RAW	RAW (<length>)	Up to 4000 characters RAW(500)	Binary data that is not converted or interpreted by Oracle. Good for multimedia.
	LONG RAW	Up to 2 gigabytes LONG RAW	Used for binary data. Used for backward compatibility only. New columns should be defined as CLOB or BFILE instead.
ROWID	ROWID	Only for physical rowid in earlier versions ROWID	The rowid of data in standard form for databases from Oracle7 and lower. Always 10 bytes long.
	UROWID	Stores any type of rowid UROWID	Any type of row address.

Both DATE and TIMESTAMP contain hours, minutes, and seconds, as well as the year, month, and day. Only TIMESTAMP can contain fractions of a second (up to .000000001 second precision).

The following list contains a few more notes on some of the data types found in Table 7-1.

- **NUMBER:** If you use NUMBER without any scale or precision, the number can be any supported size. If you omit the precision, such as NUMBER(10), the number has 10 digits and no decimal places. Oracle9*i* rounds off numbers that are a different precision. For example, this column has a precision of two:

```
PAYRATE NUMBER(5,2)
```

  In this example, inserting the value 100.256 causes Oracle9*i* to round up and store 100.26. Another example of the format stores 15 digits, (5 to the right of the decimal, 10 to the left of the decimal): NUMBER(15, 5).

- **DATE:** The default display format for DATE is DD-MMM-YYYY (15-JAN-2003, for example). Use the TO_CHAR function to display times or other date formats. Use the TO_DATE function to insert any format of date and time you want. No matter what format you use to insert the date, the date displays in the default format unless you use the TO_CHAR function to specify how you want it displayed.

- **TIMESTAMP WITH LOCAL TIME ZONE:** This data type is the same as TIMESTAMP, except that the database converts the time to local time of the database. Oracle9*i* automatically converts the time to the user's local time when it is displayed. In addition, the time is adjusted to the local time zone of the database when inserted into the database.
- **INTERVAL YEAR TO MONTH:** Here is an example:

```
HOW_OLD_ARE_YOU INTERVAL(4) YEAR TO MONTH
```

  In this example, you can store a time span of up to 9999 years. You can subtract the current date from your birth date and store the results (in years and months) in the columns.
- **INTERVAL DAY TO SECOND:** Here is an example:

```
DAYLIGHT INTERVAL DAY(1) TO SECOND(4)
```

  In this example, subtract the sunrise date and time of today from the sunset date and time, and store the day and seconds (down to millionths of a second) of daylight in the column. The days would be zero in this example.
- **UROWID:** This stores rowids, including physical rowid, logical rowid, and external (for locating data in a non-Oracle database) rowids. It can store up to 4000 bytes.

  There are additional data types not listed previously, which are for other types of tables or complex data:
- **Collection data types: Collections** are repeating data contained within one column. They can be defined as an array (called **varray** in Oracle9*i*) or as a nested table. See the section titled "Creating Varrays and Nested Tables" later in this chapter.
- **Object data types:** Object tables can use any of the data types listed previously, and in addition, there are two more data types just for object tables: REF and user-defined. The REF data type is similar to a foreign key and links related object tables together. The **user-defined data type** can be any collection of the other data types you want assembled into its own data type. See the section titled "Creating Object Tables" later in this chapter.

The next sections guide you through creating various types of tables beginning with a relational table.

Creating Relational Tables

To create a relational table, you must design the table, deciding these factors:

- Name of the table
- Name and data type of all columns

- Estimated initial size and growth pattern
- Location of the table
- Constraints, relationships to other tables, default data (covered in Chapter 11)

Names of tables and columns follow naming rules of Oracle9*i*: They must be 1–30 characters long. If you enclose the name in double quotes, it is case sensitive and can begin with any letter, number, or symbol character, including spaces. If you do not enclose the name in double quotes, it is interpreted as uppercase. An unquoted column name must begin with a letter character, although it can contain any letter, number, and the symbols # (number sign), $ (dollar sign), and _ (underscore).

Estimating the initial size and growth pattern helps you specify the storage settings for the table. In essence, you determine the average length of each column's data, add some overhead space for each column, and add them up to determine the average row length. Then, you estimate the number of rows the table will have initially and multiply that by the average row length. This gives you a rough estimate of the initial size of the table (INITIAL). Growth patterns, such as how often rows are inserted, updated, and deleted, help you determine the size of the extents (NEXT) and how many extents to use (MAXEXTENTS).

Complete the following steps to create a relational table. During this exercise, you use the SQL*Plus Worksheet, which has a Windows-like format that makes editing long commands easier.

The tables you create in this chapter are the first tables you create in the Global Globe database project that you are developing throughout the book and, in particular, in the end of chapter Case Projects.

First, create a table to keep information about the Global Globe employees. You need to identify editors, so that you can assign them to specific sections of the classified ads. In addition, you need their names and phone numbers. The phone number can have an extension, so you want to allow both numbers and characters in the field.

1. Start up the Enterprise Manager console. In Windows, click **Start/Programs/ Oracle - OraHome92/Enterprise Manager Console**. In UNIX, type **oemapp console** on the command line. The Enterprise Manager Console login screen appears.
2. Select the **Launch standalone** radio button and click **OK**. The console appears.

3. Click **Tools/Database Applications/SQL*Plus Worksheet** from the top menu. A background SQL*Plus process starts, and then the SQL*Plus Worksheet window appears. If the background process appears in front, simply minimize it (click the **minus sign** in the top-right corner), so that you can see the worksheet, as shown in Figure 7-5.

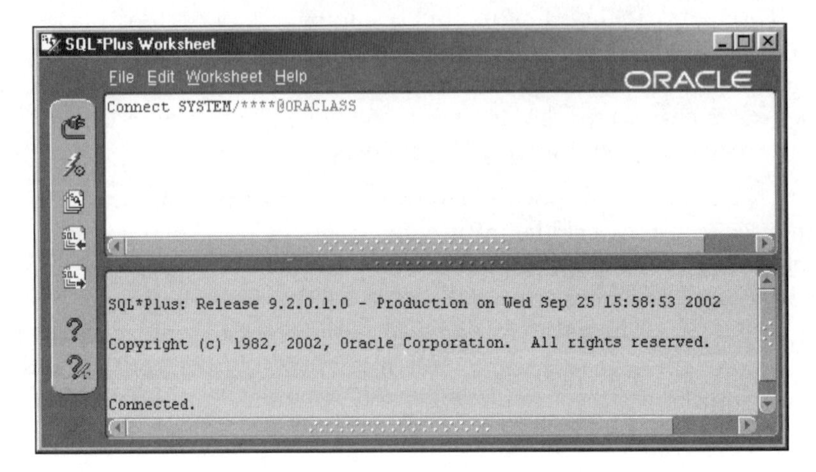

Figure 7-5 SQL*Plus Worksheet Starts from the Enterprise Manager console

4. Connect as the CLASSMATE user. Click **File/Change Database Connection** on the menu bar. A logon window appears. Type **CLASSMATE** in the Username box, **CLASSPASS** in the Password box, and **ORACLASS** in the Service box. Leave the connection type as "Normal." Click **OK** to continue.
5. Now that you are connected as the CLASSMATE user, you can create tables owned by CLASSMATE. (SYSTEM can also create objects owned by CLASSMATE; however, logging on as CLASSMATE uses this particular user's default tablespace and quotas.)
6. To begin creating a relational table, type this line in the top window of the worksheet:

   ```
   CREATE TABLE EMPLOYEE
   ```

   This line defines the table name as EMPLOYEE.

7. Now create the columns by typing these lines below the first line:

```
(EMPLOYEE_ID NUMBER (10),
JOB_TITLE VARCHAR2(45),
first_name varchar2(40),
Last_name varchar2(40),
phone_number Varchar2(20)
```

Our example shows a mixture of uppercase and lowercase letters simply to illustrate that you can use either uppercase or lowercase letters in the column names as well as in the data types. As long as you don't use double quotation marks around the column name, Oracle9*i* translates the name into all uppercase letters.

8. Complete the column list with a **closing parenthesis**. At this point, you can start defining storage settings. For this example, allow the tablespace defaults to be used by specifying no storage clause. Type a **semicolon (;)** to mark the end of the command. Figure 7-6 shows the entire command.

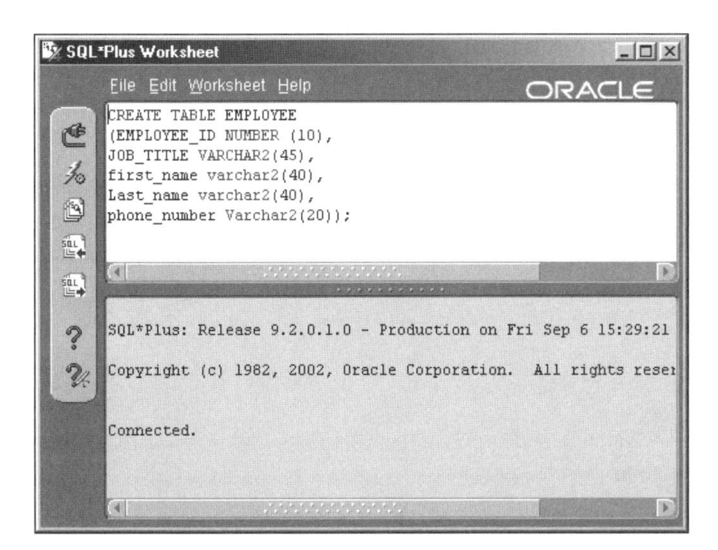

Figure 7-6 The completed command is ready to execute

9. Click the **Execute** icon on the left side of the window. The icon looks like a lightning bolt. SQL*Plus Worksheet creates the table and displays "Table created" in the lower portion of the Worksheet.

10. Save your work in the **Solutions\Chapter07** directory of your student disk by clicking the **File/Save Input As** in the top menu and by selecting the directory. Name the file **employee.sql**, and click **Save** to save the file.

11. Remain logged on for the next practice.

You can use the Enterprise Manager's Schema Manager to create tables. A Create Table Wizard guides you through all the steps, including calculating the storage requirements for your table. Alternatively, you can use the Create Table window and fill in a spreadsheet-like format for columns. Figure 7-7 shows the Create Table window of the Schema Manager. Use the Schema Manager when you want help calculating storage for a new table, or when you need help with the CREATE TABLE syntax. The Schema Manager has a feature with which you can view the DDL command that creates the table. This helps remind you of how to write a CREATE TABLE command.

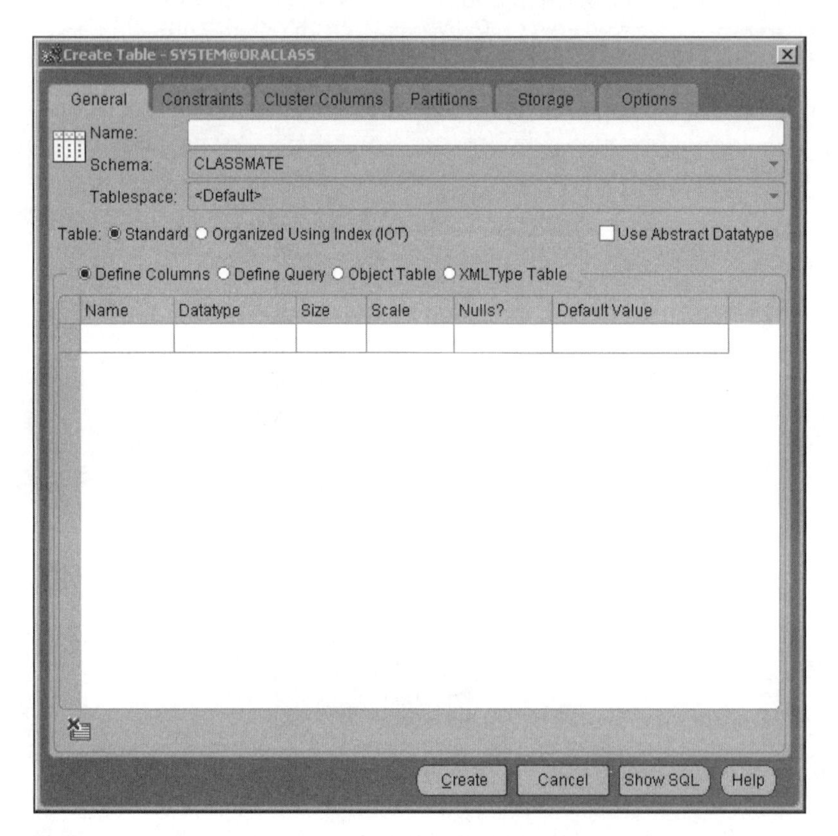

Figure 7-7 Creating a table in Schema Manager is fast and easy

For the next table, you add a table that stores all the classified ads running in the paper. Ads are not removed after they run, so customers can renew ads if they want. You want to keep track of the editor that worked with the customer and entered the ad, plus the final price of the ad and its start date, end date, and the number of days it runs. You also add storage settings for this table. The table resides in a locally managed tablespace. Follow along with the steps.

1. If you are not already there, log onto SQL*Plus Worksheet as the user CLASSMATE in the ORACLASS database.
2. If a previous command is displayed, clear it by highlighting the text and clicking the **Delete** button on your keyboard.
3. Type in the first three lines to start defining the table. The lines are:

```
CREATE TABLE CLASSIFIED_AD
(AD_NO NUMBER NOT NULL,
 SECTION_NO NUMBER NOT NULL,
```

   These first two columns must be filled in whenever a new row is inserted into the database, so the NOT_NULL integrity constraint follows the column's data type.

7

4. Type in the next four lines that define the next four columns.

```
AD_TEXT VARCHAR2(1000),
CUSTOMER_ID NUMBER,
INTAKE_EDITOR_ID NUMBER,
PRICE NUMBER (6,2),
```

   The price of the ad has a scale of six and a precision of two. If the number field is in units of dollars, the highest price allowed in the data is $9999.99.

5. Type in the next three lines that define three more columns.

```
PLACED_DATE DATE,
"Run Start Date" TIMESTAMP WITH LOCAL TIME ZONE,
"Run End Date" TIMESTAMP WITH LOCAL TIME ZONE,
```

   All three lines are variations on the DATE data type. All three can store date and time. If you do not insert a time, but only a date, the time is set to midnight of that day. Notice also, the double quotation marks around the column names. These names must always be surrounded by quotation marks when referenced in queries or other commands. After being created this way, the column names are case sensitive. Therefore, it is possible to add another column called "run start date" in this table. Use caution when creating column names this way, because it often leads to confusion. You are using it here simply for practice. In the real world, it is best to avoid case sensitive column names.

6. Type the final column definition. Notice the closing parenthesis that marks the end of the column list.

```
RUN_DAYS INTERVAL DAY(3) TO SECOND(0))
```

   The INTERVAL DAY TO SECOND data type is actually a DATE format. Because the ad runs for a number of days, you don't need any fractions of a second, so the precision of the second is zero, meaning whole seconds.

7. You add storage information after the column list. Type the tablespace and storage clauses and end with a semicolon to indicate that the command is complete.

```
TABLESPACE USERS
STORAGE (INITIAL 2M NEXT 2M MAXEXTENTS 10);
```

8. Click the **Execute** icon to create the table. SQL*Plus Worksheet replies "Table created" in the lower portion of the window.
9. Save the SQL command by selecting **File/Save Input As** in the menu bar and navigating to the **Solutions\Chapter07** directory. Type **classified_ad.sql** as the file name, and save by clicking the **Save** button.
10. Remain logged on for the next practice.

You have seen how to create a relational table with columns and storage settings. An alternative method of creating relational tables is illustrated in the next section.

Creating Temporary Tables

Temporary tables are tables created to store data for your session alone. The table remains after your session is finished; however, the data you added to the table is automatically deleted when you log off or commit. The timing of data removal (on log off or commit) depends on the ON COMMIT clause that you used to create the temporary table. The syntax for creating a temporary table is:

```
CREATE GLOBAL TEMPORARY TABLE <tablename>
ON COMMIT DELETE ROWS|PRESERVE ROWS
<table specifications>
TABLESPACE <tablespacename>
STORAGE (<storage settings>);
```

The <table specifications> and <storage settings> can be the same as those for permanent tables. If you omit the TABLESPACE clause, and you have a default temporary tablespace assigned to the user you are logged on as, all temporary tables are created in this tablespace. Otherwise, your temporary table is, by default, created in the SYSTEM tablespace. If you name the tablespace, it must be the current temporary tablespace for the database or any permanent, dictionary-managed tablespace you have authority to use.

Unlike working with permanent tables, adding or changing data in a temporary table does not generate redo log entries; however, it does generate undo log entries. Another difference between permanent tables and temporary tables is the allocation of segments. Temporary tables use temporary segments, and the segments are not allocated to the table until data is actually inserted into the table.

Follow along with this example to create a temporary table. In the example, you want to create a temporary table that contains data from both the CLASSIFIED_AD and the EMPLOYEE tables that you created earlier in the chapter. You use the subquery method to create the table that you used to create the other tables. A **subquery** is a query that is embedded in another SQL command, in this case, in the CREATE TABLE command.

1. If you are not already there, log onto the Enterprise Manager console, access the ORACLASS database, and start up the SQL*Plus Worksheet. Connect as the CLASSMATE user.
2. Clear any previous command by clicking **Edit/Clear All** in the menu bar.
3. Begin entering the temporary table command by typing this line:

   ```
   CREATE GLOBAL TEMPORARY TABLE EDITOR_REVENUE
   ```

4. Now specify how the rows are handled using the ON COMMIT clause. The default is ON COMMIT DELETE ROWS, which causes all the rows you inserted into the temporary table to be deleted when you issue a COMMIT command. In this case, you want to save the data for the duration of your session, even if you execute a COMMIT command, so you specify the ON COMMIT PRESERVE ROWS option:

   ```
   ON COMMIT PRESERVE ROWS
   ```

5. The next line is an alternative method (the subquery method) of creating a table. You can use this method for any type of table, not just temporary tables. One advantage of this method is that it not only handles the DDL (by generating column details from the select statement), but also it inserts data into the table (with the results of the query) at the same time. Type **AS** followed by the first line of the subquery:

   ```
   AS SELECT E.FIRST_NAME || ' ' || E.LAST_NAME EDITOR,
   ```

   The double vertical bars (||) are symbols meaning concatenate the left and right sides together. In this case, the FIRST_NAME column is concatenated with a single blank space and then concatenated to the LAST_NAME column. The column that results is given the name EDITOR (which will be its column name in the temporary table). EDITOR is a column alias. A **column alias** is a short name for a results column and is used when generating tables or views to give the results column a valid column name. In addition, a column alias is used to provide a more readable column heading for query results.

6. Type the second line of the subquery:

   ```
   SUM(CA.PRICE) ANNUAL_REVENUE
   ```

   This is the final column in the SELECT statement, so there is no comma following the column alias. The SUM function is a group function. **Group functions** act on sets of column data. In this case, as you see when you complete the command, you are adding up the price of every classified ad for the year 2003 for each employee. The employee's name and the sum of all that employee's ads are the two columns that are being created in the temporary table.

7. Type the FROM clause of the subquery:

```
FROM CLASSIFIED_AD CA JOIN EMPLOYEE E
```

As you can see, the subquery joins two tables. The tables have been assigned table aliases. A **table alias** is a shortcut name for a table used to prefix columns in an SQL command in place of using the entire table name. The table alias for CLASSIFIED_AD is CA and the table alias for EMPLOYEE is E. These aliases are used in the ON clause, which you type next.

8. Now, type in the criteria for joining the tables.

```
ON (CA.INTAKE_EDITOR_ID = E.EMPLOYEE_ID)
```

9. Besides the join criteria, you also include only the ads that were placed during the year 2003 by adding this WHERE clause.

```
WHERE TO_CHAR(CA.PLACED_DATE, 'YYYY') = '2003'
```

This line illustrates the TO_CHAR function, which you can use on DATE data types to extract any part of the date and time that is stored in the column.

10. The last line completes the subquery and the CREATE TABLE command.

```
GROUP BY FIRST_NAME || ' ' || LAST_NAME;
```

The GROUP BY clause is used only when you have a group function in the SELECT clause. It tells Oracle9*i* how to collect the data into groups.

11. Click the **Execute** icon to run the command. Figure 7-8 shows the entire command as you see it in the SQL*Plus Worksheet and the "Table created" reply in the lower half of the screen that appears after you execute the command.

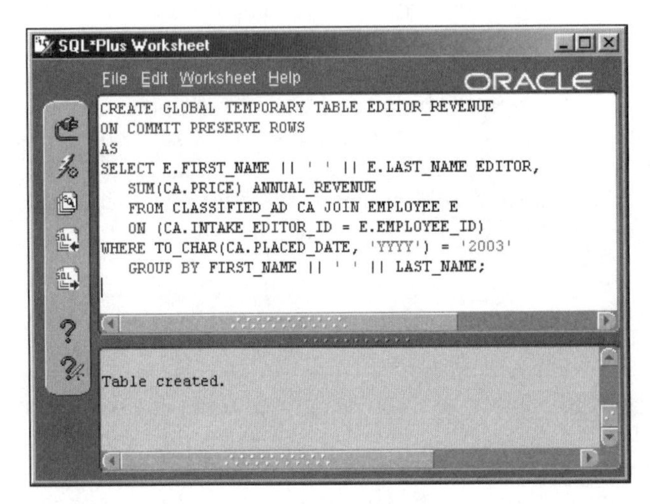

Figure 7-8 You must use column aliases when creating a table using a subquery, if the columns include functions or calculations

12. Save the command by clicking **File/Save Input As** from the menu bar. Navigate to the **Solutions\Chapter07** directory. Type **editor_revenue.sql** as the file name, and click **Save** to save the file.
13. Remain logged on for the next practice.

The additional spaces in a SQL command are ignored by Oracle9*i*, as are line breaks. The spaces and line breaks simply make the command more readable to you and other humans.

You can grant others access to a temporary table you create, but no two users ever share data. Every user session (or transaction, if you use the ON COMMIT DELETE ROWS option) has private rows. Two sessions can use the same temporary table at the same time; however, they do not see each other's data.

The next section describes tables within tables.

Creating Varrays and Nested Tables

When the capability to store a table within a column of another table was added to Oracle8, Oracle became an object-relational database. You can add a column that contains a table to a relational table. This column is most often used in object tables. The two types of embedded tables are called varrays and nested tables. Table 7-2 shows the similarities and differences between the two structures.

Table 7-2 Varrays versus nested tables

	Nested table	Varray
Column restrictions	Must contain a single column of a table type data type	Must contain a single column of either a built-in data type, or a collection type data type
Storage	Stored in a table apart from the column (**out of line**) and can have its own storage settings	Stored inside the column (**in line**) along with the rest of the row; if large enough, Oracle9*i* can move a varray (internally) to a RAW or CLOB column
Order	Unordered: Access each row by data values in the row in the same manner as a relational table; the order of the rows is not preserved	Ordered: Access each row in order by its sequence number, not by values in the row; the order of the rows is not preserved
Indexing	Can be indexed	Cannot be indexed
Number of rows	Unlimited	Limited when the varray is created

Use a varray when you have a small number of items to store, or you usually access the data by cycling through it. Use a nested table when you have a large or varying number of items, and you usually access the data by specific values.

Follow along with this example to create one relational table with a nested table column and one relational table with a varray column.

The first table, CLASSIFIED_SECTION, is used at the Global Globe to store information about each category of classified ads that the newspaper runs. Besides the number and the name of the section, you also want to store a list of editors who specialize in helping customers place this type of ad. You build a nested table with a list of editors and mark one of those editors as the primary contact. This way, the telephone operator can direct calls to the primary editor, and if that editor is unavailable, the operator can go down the list of additional editors, until a free editor is contacted for the customer.

The second table, CUSTOMER, stores details about a customer. Because many customers have several telephone numbers, you store the set of phone numbers in a varray with a maximum of ten phone numbers per customer.

1. If you are not already there, log onto the Enterprise Manager console, access the ORACLASS database, and start up the SQL*Plus Worksheet. Connect as the CLASSMATE user.
2. Go to the console window, double-click the **Databases** folder, and then double-click the **ORACLASS** database icon.
3. Double-click the **Schema** icon, and then double-click the **CLASSMATE** icon. The console displays objects owned by CLASSMATE in the right pane and categories of objects under the CLASSMATE icon in the left pane.
4. Double-click the **User Types** icon in the left pane, and then double-click the **Table Types** folder. The CONTACT_TABLE table type is displayed below the folder.
5. Double-click CONTACT_TABLE in the left pane. The property sheet for this table type is displayed as shown in Figure 7-9. As you can see, this table type is made up of rows of an object type named EDITOR_INFO.

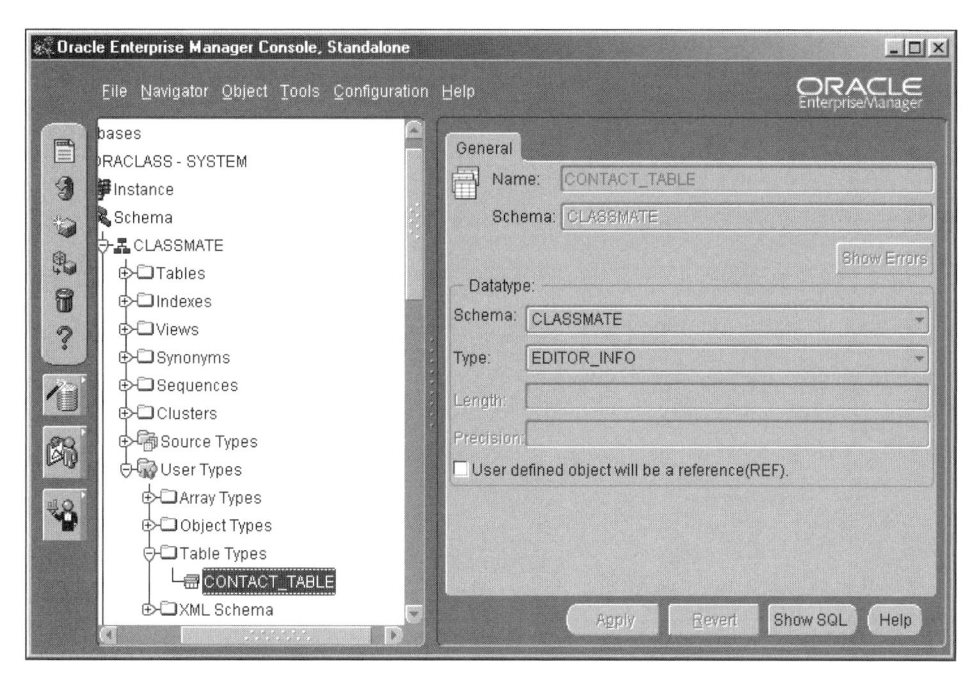

Figure 7-9 Object types, arrays, and table types used in this chapter appear under User Types

6. To view the attributes that will be stored in the CONTACT_TABLE table type, you must now look at the EDITOR_INFO object type. Double-click the **Object Types** folder to display a list of object types owned by CLASSMATE.
7. Click **EDITOR_INFO** to display the property sheet for this object type. As shown in Figure 7-10, the object type is made up of two attributes: PRIMARY_EDITOR and EDITOR_ID.

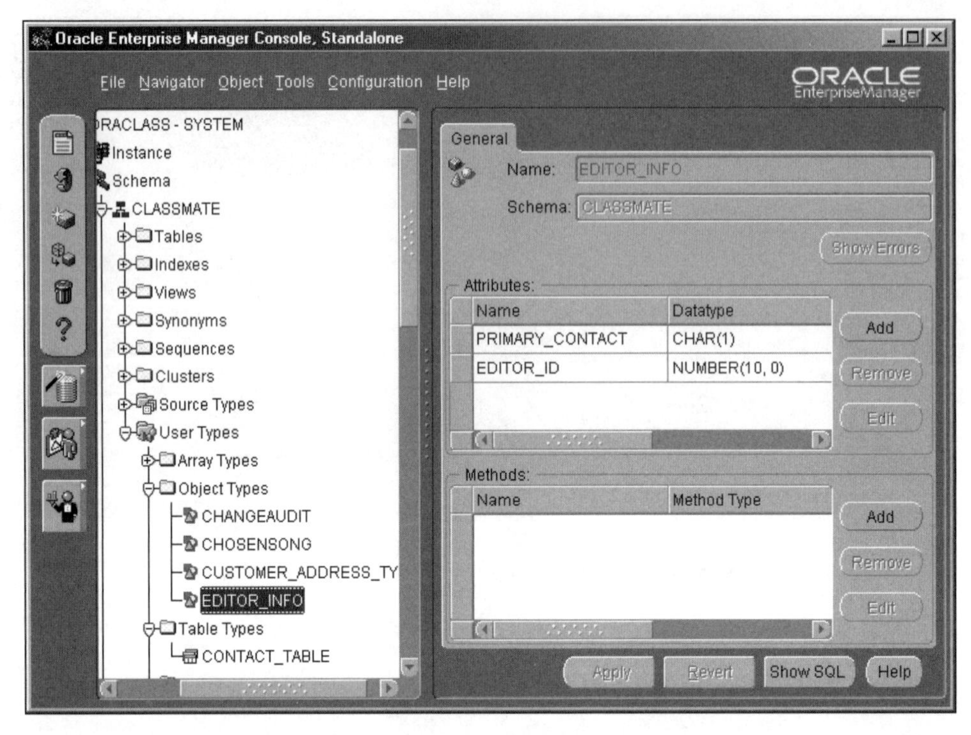

Figure 7-10 The EDITOR_INFO object type has two attributes

Now you know how to look up object types in the console. You may want to practice this during the chapter exercises, because it will help you complete the hands-on exercises at the end of the chapter.

8. Go to the SQL*Plus Worksheet window, and clear any previous command by clicking **Edit/Clear All** in the menu bar.

9. Begin entering the CLASSIFIED_SECTION table command by typing these lines:

```
CREATE TABLE CLASSIFIED_SECTION
(SECTION_NO NUMBER NOT NULL,
 SECTION_TITLE VARCHAR2(50),
 BASE_RATE_PER_WORD NUMBER(4,3),
```

   These first lines look like the beginnings of a typical relational table you have seen before with three columns of built-in data types.

10. Now type in the last column definition. This is the nested table column:

```
CONTACT_EDITOR CLASSMATE.CONTACT_TABLE)
```

The data type of the CONTACT_EDITOR column is a user-defined table type. It was created as part of the initial setup for the chapter. The CONTACT_TABLE type defines a table with two columns: EDITOR_ID and PRIMARY_CONTACT. The PRIMARY_CONTACT is a single character field and contains "Y" for the primary editor for that section and "N" for all other editors assigned to the section.

11. As described earlier, a nested table is stored in its own table. Although optional, it is advisable to define the name of the storage table, so you can identify it more easily. Type the following line (the last line for the CREATE TABLE command) to define NESTED_EDITORS as the storage table for the nested table column data.

```
NESTED TABLE CONTACT_EDITOR
STORE AS NESTED_EDITORS;
```


12. Click the **Execute** icon to run the command. Figure 7-11 shows the entire command as you see it in the SQL*Plus Worksheet and the "Table created" reply in the lower half that appears after you execute the command.

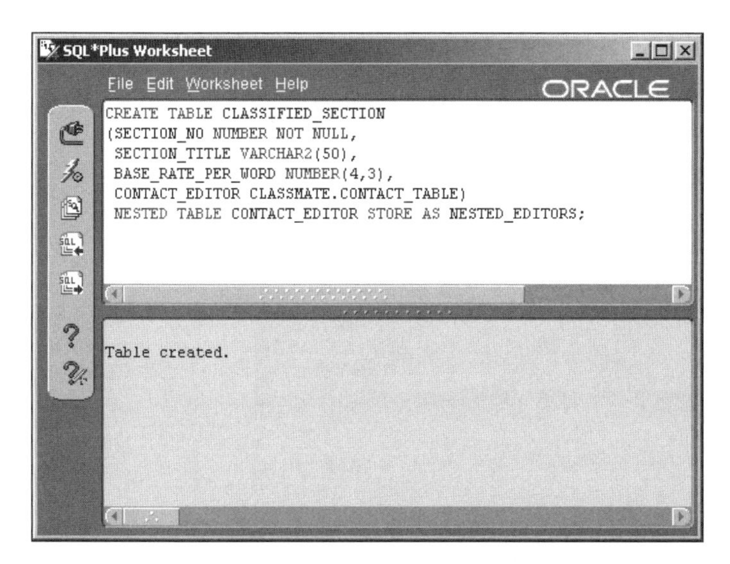

Figure 7-11 A nested table has its own name

13. Save the command by clicking **File/Save Input As** in the menu bar. Navigate to the **Solutions\Chapter07** directory. Type **classified_section.sql** as the file name and click **Save** to save the file.

14. Clear the command by clicking **Edit/Clear All** in the menu.

15. Next, begin entering the CUSTOMER table command by typing these lines:

```
CREATE TABLE CUSTOMER
(CUSTOMER_ID NUMBER NOT NULL,
 FULLNAME VARCHAR2(60),
 DISCOUNT_PERCENT NUMBER(3,1),
 EMAIL VARCHAR2(45),
```

    Again, these are typical columns. The FULLNAME column has no underscore; simply to illustrate that, although it is often used to make column names readable, it is not required.

16. Now add the varray column:

```
PHONE_LIST CLASSMATE.PHONE_ARRAY);
```

    The collection type PHONE_ARRAY is a varray that is predefined. It holds up to ten phone numbers in a repeating VARCHAR2(20) column.

17. Click the **Execute** icon to run the command. Figure 7-12 shows the entire command, as you see it in the SQL*Plus Worksheet, and the "Table created" reply in the lower half that appears after you execute the command.

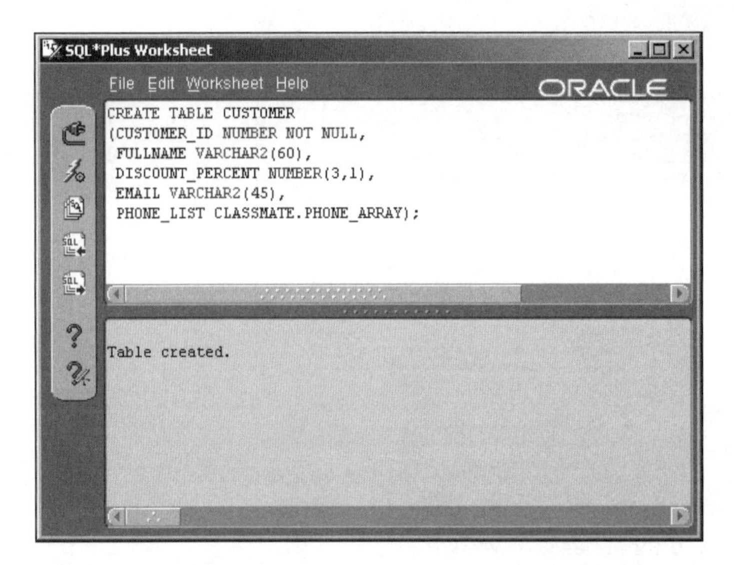

Figure 7-12 Varrays have a single column that is either a built-in data type or a user-defined data type

18. Save the command by clicking **File/Save Input As** in the menu bar. Navigate to the **Solutions\Chapter07** directory. Type **customer.sql** as the file name, and click **Save** to save the file.

19. Remain logged on for the final practice.

You now have two relational tables with collection data types in two of the columns. Figure 7-13 illustrates the layout of the CUSTOMER table that contains an array. Looking at the figure, the first row contains data for Mary Holman. She has three phone numbers, all recorded in the PHONE_LIST array. The second row has data for John Smith, who has five phone numbers in the PHONE_LIST array. A customer can have up to ten phone numbers.

CUSTOMER

CUSTOMER_ID	FULLNAME	DISCOUNT_PERCENT	EMAIL	PHONE_LIST
1	Mary Holman	5	mholman@ebb.com	544-322-1234 544-322-5490 544-322-4499
2	John Smith	10	jsmith@hhy.com	334-998-3837 ext. 12 334-487-3895 334-449-3344 800-339-0987 334-444-5533 ext. 4327
3	Edward Jones		jonesey@hiya.com	872-334-1234 872-230-8017 ext. 100 800-555-3049 888-203-1234 504-308-2343

Arrays stored inline with rows

Figure 7-13 A varray expands as new data is added to the array, but has a maximum number of entries

Nested table rows are associated with a row in the parent table by the nested table identifier. This identifier is an added column in the nested table. In the case of a relational table with a nested table column, all the nested table rows are stored in one single nested table. If the table is an object table containing a nested table, the nested table is still stored out of line with the parent table; however, there might be separate table instances created for each row of the parent table instead of storing all the nested table data in a single table.

Figure 7-14 shows the CLASSIFIED_SECTION table and its associated nested table named NESTED_EDITORS. The CONTACT_EDITOR column contains a pointer, called a **Nested table ID**, that directs you to the corresponding rows in the NESTED_EDITOR table. For example, the second row is for the Personal section. The CONTACT_EDITOR column contains a pointer with the value "9092af." Using this value, you can see that the Personal section has four contact editors and the editor with EDITOR_ID of 101 is the primary contact. The value shown is simply for illustration; the actual nested table ID value would not be displayed in the CONTACT_EDITOR column. The nested table ID value is stored internally and is not seen when you work with the table. When you use a nested table, an unlimited number of rows can be stored in the nested table that are associated with one row in the main table. In the example, one classified section could have an unlimited number of contact editors.

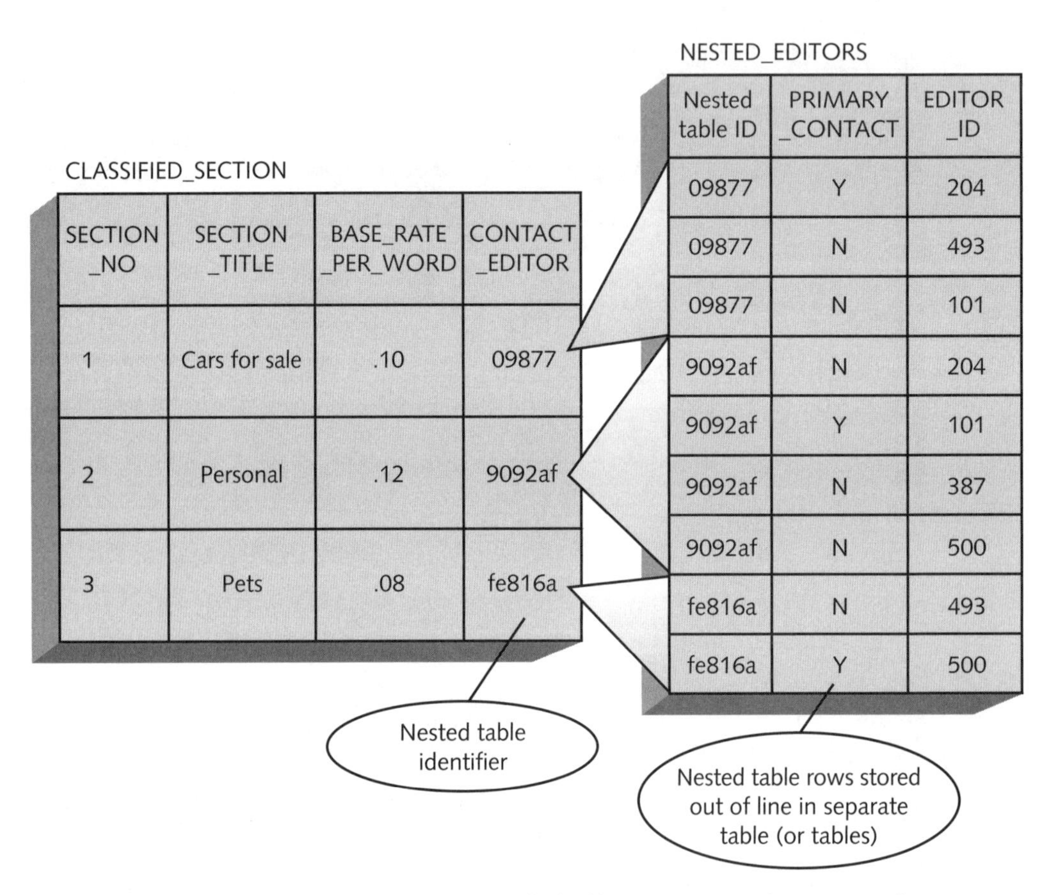

Figure 7-14 The nested table column actually holds a pointer to the rows in the separate nested table

The next section discusses how to create object tables.

Creating Object Tables

Object tables are always created with one column that always has a data type that is a user-defined object type. An object type contains one or more attributes, which can be similar to relational table columns or can be made up of another object type. Object tables, therefore, can be built with layers of object types in their definitions.

You must create an object type that has the design you want to use for your object table before actually creating the object table. Complete the following steps to view the predefined object type, and then create an object table.

In this example, you create an object table for storing customer addresses. A customer can have more than one address (such as a home and business address). The object type CUSTOMER_ADDRESS_TYPE was created for you to use when creating the object table.

1. If you are not already there, log onto the Enterprise Manager console, access the ORACLASS database, and start up the SQL*Plus Worksheet. Connect as the CLASSMATE user.
2. Clear any previous command by clicking **Edit/Clear All** in the menu bar.
3. Type and execute this command to view the attributes defined in the CUSTOMER_ADDRESS_TYPE object type:

   ```
   DESC CUSTOMER_ADDRESS_TYPE
   ```

   This object type contains attributes for the CUSTOMER_ID, POBOX_SUITE, STREET_ADDRESS, CITY, STATE, and COUNTRY.
4. Type and execute the following command to create an object table that contains these attributes.

   ```
   CREATE TABLE CUSTOMER_ADDRESS
   OF CUSTOMER_ADDRESS_TYPE;
   ```

   You can add storage settings to an object table in the same way as you would a relational table.
5. Save the command by clicking **File/Save Input As** in the menu bar. Navigate to the **Solutions\Chapter07** directory. Type **customer_address.sql** as the file name, and click **Save** to save the file.
6. Exit SQL*Plus Worksheet by clicking the **X** in the top-right corner or type **EXIT** at the prompt, press the **Execute** icon, and click **OK** in the Oracle Enterprise Manager dialog box that opens.

An object table is **instantiated** when you insert records into the table. Think of the object table you created as a template. Until the template is actually used (by inserting data), the object table is only a template and has no "real" existence. When you insert data (even if it is all null values) into the first row of an object table, the object table becomes instantiated. An object table row is also called an **object table instance**.

The next section describes how and why to create a partitioned table.

Creating Partitioned Tables

When a table grows very large, even an indexed search can become slow. To improve performance, you can break a large table up into separate sections, called **partitions**. You can partition any table except one that is part of a cluster. After a table is partitioned, queries that use the table access only the partition or partitions required to retrieve the data. Each partition is stored in a separate segment and can have its own storage settings. Partitions of the same table can even be stored on separate tablespaces, allowing you to divide the table across multiple disk drives if you want.

An application or SQL command does not have to be changed to access a partition. Oracle9*i* automatically handles the translation between the table and its partitions.

It is very important to understand how the table is queried and updated to best take advantage of partitioning. Some tables have no distinctive pattern to the queries, and so partitioning may not give you any performance improvements. An example is a table used by many diverse users to search for a variety of subjects or text strings in text columns that contain articles or descriptions. If, on the other hand, the table is consistently queried by primary key, then partitioning that divides the table according to the primary key improves performance. The idea is to determine the best partitioning method that enables queries (as well as updates, inserts, and deletes) to work within one partition rather than across multiple partitions.

Partitioning can be done in five ways:

- **Range:** To use range partitioning, define a set of dividing values so the table is stored in partitions according to a range of values in a set of columns (the **partitioning key**). You specify the lower boundary of each partition, aiming for an even distribution of the data volume in each partition. The last partition can be open-ended to allow for values higher than the current values in the table.
- **Hash:** With **hash partitioning**, you specify how many partitions to create and let Oracle9*i* use a hash value (calculated on the partitioning key) to divide the data evenly among the set of partitions. This can be useful when you are not sure of the best range boundaries, but you want to spread the table across multiple disk drives.

- **List:** To use **list partitioning**, set up a distinct list of partitioning key values, and define which values go into each partition. This is useful for dividing by values, such as state or district office name, when the alphabetical order is an appropriate method of dividing the data. For example, you might divide a customer table by the sales territory in which the customer lives.
- **Composite range-hash:** With this type of partitioning, you are actually partitioning the data and then partitioning the data within each partition (**subpartitioning**). Each subpartition is in a separate segment, so the subpartitions can reside in separate tablespaces. You specify range values (as with range partitioning), and then you specify the total number of subpartitions (as with hash partitioning) and the locations in which the subpartitions are to be stored.
- **Composite range-list:** This is another method for partitioning and subpartitioning, in which you again define a range of values for each partition. Then, you define subpartitions by lists of values on another set of columns in the table. For example, you may want to divide your historical sales data by ranges of sales dates, and then by sales territory within these ranges.

To create a partitioned table, you use the standard CREATE TABLE command as if you were creating a relational table, and add the PARTITION clause. Each type of partitioning has its own set of parameters. The following sections include examples for each type of partitioning.

Range Partitioning

The bank you work for has a massive transaction table. To manage it more effectively, you decide to create a table partitioned by ranges of account numbers.

```
CREATE TABLE TRANSACTION_RECORD
    ( ACCT_NO NUMBER NOT NULL,
      ACTION  VARCHAR2(5) NOT NULL,
      AMOUNT NUMBER(8,2) NOT NULL,
      SENDING_ACCT_NO NUMBER NOT NULL,
      ACTION_DATE DATE NOT NULL )
  PARTITION BY RANGE (ACCT_NO)
    ( PARTITION TRANS_P1 VALUES LESS THAN (2900999)
        TABLESPACE TBS1,
      PARTITION TRANS_P2 VALUES LESS THAN (5900999)
        TABLESPACE USER_DATA,
      PARTITION TRANS_P3 VALUES LESS THAN (9900999)
        TABLESPACE USER_ALT);
```

You can use multiple columns in the range by listing the columns and listing their range values within the parentheses, separated by commas. List ranges from lowest to highest.

Hash Partitioning

The bank has another large table that keeps track of mortgage loan history. You know that this historical table is nearly always accessed by date, and you want to allow the database to evenly divide the rows by hash value rather than determining a date range. Here is the CREATE TABLE command for a new, hash-partitioned history table.

```
CREATE TABLE MORTGAGE_HISTORY
     (LOAN_NO NUMBER,
      ACCT_NO NUMBER,
      DATE_CREATED DATE,
      MORTGAGE_AMOUNT NUMBER)
   PARTITION BY HASH (DATE_CREATED)
   PARTITIONS 3
   STORE IN (HISTORY_TBSP1, HISTORY2, HISTORYEXTENDED);
```

The tablespace to use for each partition is named in the STORE IN clause.

List Partitioning

The bank has a transaction history table that carries summarized transactions by branch. The table has grown so large that you need to divide it, and the BRANCH_REGION column provides a good basis for partitioning, because most reporting that is done with this table is by regional managers looking for their own historical data. Here is the CREATE TABLE command:

```
CREATE TABLE TRANS_BY_BRANCH_HISTORY
     (BRANCH_ID NUMBER(9,0),
      BRANCH_REGION VARCHAR2(10),
      TRANS_YEAR NUMBER(4,0),
      TRANS_MONTH NUMBER(2,0),
      TRANS_AMOUNT NUMBER(10,2))
   PARTITION BY LIST (BRANCH_REGION)
  (PARTITION WESTERN VALUES ('WESTCOAST', 'NORTHWEST')
       TABLESPACE HIST3,
   PARTITION MOUNTAIN VALUES ('ROCKIES', 'SOUTHWEST')
       TABLESPACE HIST3,
   PARTITION MIDWEST
      VALUES  ('MIDWEST', 'IL-METRO')
      STORAGE (INITIAL 20M NEXT 40M PCTINCREASE 20)
       TABLESPACE HIST1,
   PARTITION NORTHEAST VALUES ('NY-METRO', 'NE-STATES')
       TABLESPACE HIST5,
   PARTITION SOUTHEAST VALUES ('SOUTHEAST'))
        TABLESPACE HIST6;
```

You can only specify a single column for list partitioning.

The STORAGE clause was added for the MIDWEST partition as an example of how you add individual STORAGE clauses to each partition if needed. Add a STORAGE clause to any or all partitions, the way you would to a table.

Composite Range-hash Partitioning

The bank's range partitioned table has grown to such huge proportions that you decide to create a new table for the data that is partitioned by range as before, and in addition, subpartitioned by hash value, based on the transaction date. Here is the CREATE TABLE command.

```
CREATE TABLE TRANSACTION_RECORD
    ( ACCT_NO NUMBER NOT NULL,
      ACTION  VARCHAR2(5) NOT NULL,
      AMOUNT NUMBER(8,2) NOT NULL,
      SENDING_ACCT_NO NUMBER NOT NULL,
      ACTION_DATE DATE NOT NULL )
PARTITION BY RANGE (ACCT_NO)
SUBPARTITION BY HASH (ACTION_DATE)
    SUBPARTITIONS 8 STORE IN
    (TBS1, USER_DATA, USER_ALT, USER_ALT2)
PARTITION TRANS_P1 VALUES LESS THAN (2900999),
PARTITION TRANS_P2 VALUES LESS THAN (5900999),
PARTITION TRANS_P3 VALUES LESS THAN (9900999));
```

Because the subpartitions are divided with hash values, you do not specify any storage sizes.

Composite Range-list Partitioning

Another method of composite partitioning is the range-list partition. Here, the range partition is divided into subpartitions based on a list of column values. For example, instead of using hash values as in the previous example, let's say you decide that the TRANSACTION_RECORD table should be subpartitioned by values in the ACTION column. Here is the CREATE TABLE command that makes this type of table.

```
CREATE TABLE TRANSACTION_RECORD
    ( ACCT_NO NUMBER NOT NULL,
      ACTION  VARCHAR2(5) NOT NULL,
      AMOUNT NUMBER(8,2) NOT NULL,
      SENDING_ACCT_NO NUMBER NOT NULL,
      ACTION_DATE DATE NOT NULL )
PARTITION BY RANGE (ACCT_NO)
SUBPARTITION BY LIST (ACTION)
(PARTITION TRANS_P1 VALUES LESS THAN (9900999)
   TABLESPACE HIST3
  (SUBPARTITION WESTERN VALUES ('WESTCOAST', 'NORTHWEST'),
```

```
     SUBPARTITION MOUNTAIN VALUES ('ROCKIES', 'SOUTHWEST'),
     SUBPARTITION MIDWEST VALUES  ('MIDWEST', 'IL-METRO'),
     SUBPARTITION NORTHEAST VALUES ('NY-METRO', 'NE-STATES'),
     SUBPARTITION SOUTHEAST VALUES ('SOUTHEAST')),
  PARTITION TRANS_P2 VALUES LESS THAN (5900999)
     TABLESPACE HIST2
    (SUBPARTITION WESTERN VALUES ('WESTCOAST', 'NORTHWEST'),
     SUBPARTITION MOUNTAIN VALUES ('ROCKIES', 'SOUTHWEST'),
     SUBPARTITION MIDWEST VALUES  ('MIDWEST', 'IL-METRO'),
     SUBPARTITION NORTHEAST VALUES ('NY-METRO', 'NE-STATES'),
     SUBPARTITION SOUTHEAST VALUES ('SOUTHEAST')),
  PARTITION TRANS_P3 VALUES LESS THAN (2900999)
     TABLESPACE HIST1
    (SUBPARTITION WESTERN VALUES ('WESTCOAST', 'NORTHWEST'),
     SUBPARTITION MOUNTAIN VALUES ('ROCKIES', 'SOUTHWEST'),
     SUBPARTITION MIDWEST VALUES  ('MIDWEST', 'IL-METRO'),
     SUBPARTITION NORTHEAST VALUES ('NY-METRO', 'NE-STATES'),
     SUBPARTITION SOUTHEAST VALUES ('SOUTHEAST'));
```

As you can see, subpartitioning by list requires you to repeat the list criteria, as many times as there are partitions. It is possible to create a **subpartition template** instead of repeating the list criteria (or the hash criteria, for range-hash partitioned tables). The subpartition template describes all the subpartitions once and then all the partitions that use that template. The template is actually part of the CREATE TABLE command, not a specific object that is created. The template simply provides an alternative way to write the command for creating composite partitions. Here is the range-list partitioned table shown previously, using a template instead of repeating the subpartition criteria for each partition:

```
CREATE TABLE TRANSACTION_RECORD
    ( ACCT_NO NUMBER NOT NULL,
      ACTION  VARCHAR2(5) NOT NULL,
      AMOUNT NUMBER(8,2) NOT NULL,
      SENDING_ACCT_NO NUMBER NOT NULL,
      ACTION_DATE DATE NOT NULL )
PARTITION BY RANGE (ACCT_NO)
SUBPARTITION BY LIST (ACTION)
SUBPARTITION TEMPLATE
  (SUBPARTITION WESTERN VALUES ('WESTCOAST', 'NORTHWEST'),
   SUBPARTITION MOUNTAIN VALUES ('ROCKIES', 'SOUTHWEST'),
   SUBPARTITION MIDWEST VALUES  ('MIDWEST', 'IL-METRO'),
   SUBPARTITION NORTHEAST VALUES ('NY-METRO', 'NE-STATES'),
   SUBPARTITION SOUTHEAST VALUES ('SOUTHEAST'))
 (PARTITION TRANS_P1 VALUES LESS THAN (9900999)
    TABLESPACE HIST3,
```

```
    PARTITION TRANS_P2 VALUES LESS THAN (5900999)
       TABLESPACE HIST2,
    PARTITION TRANS_P3 VALUES LESS THAN (2900999)
       TABLESPACE HIST1);
```

Index-organized tables can also be partitioned with the following restrictions:

- The partition key must be all or a subset of the table's primary key.
- Only range and hash partitioning are allowed.
- Only range-partitioned, index-organized tables can contain large object (LOB) columns.

You have now examined and even created a wide variety of tables. The next chapter looks at how to make structural changes to existing tables.

Chapter Summary

- Relational tables are the most common form of table in an Oracle database.
- Relational tables reside in a single segment within a single tablespace, unless the table is partitioned.
- Indexed-organized tables always have a primary key and use the key to arrange the physical order of the rows.
- Object tables have rows made up of an object type, which in turn is made up of attributes.
- Besides data, object tables can have methods associated with them and can have other objects within them.
- A temporary table, which is seen only by one user, is created for private data.
- External tables are always read-only and are used to retrieve data from outside the database.
- A nested table is a table within a single column or within an attribute of an object table.
- An XML table has a single column of the XML type data type and contains XML formatted data.
- A cluster stores a group of related tables together in one segment.
- A table is made up of one segment, which contains one or more extents, which each contain a group of contiguous data blocks.
- Locally managed tablespaces use the storage settings of a table but actually control the extent sizes and how rows are inserted and updated automatically.

- Tables can use the default storage settings of the tablespace, add more to it, or override it entirely with their own storage settings.
- A data block has these components: common and variable header, table directory, row directory, free space, and row data.
- A row that spans multiple blocks because it is too large to fit into one block is a chained row.
- A row that has been moved from one block to another because it was updated and no longer fits within the original block is a migrated row.
- INITRANS and MAXTRANS involve the data block's header space and specify how many places should be held for concurrent transactions that access a data block.
- Adjusting the PCTFREE and PCTUSED parameters can help avoid migrated rows.
- Chained rows and migrated rows slow down performance speed.
- Tables in dictionary-managed tablespaces can use storage settings, such as PCTFREE and PCTUSED, plus settings that can also be set at the tablespace level.
- A row is made up of a row header and column data.
- A row header has row overhead, the number of columns, the rowid to chained rows, migrated row (if any), and a cluster key (if this is a cluster).
- Column data has the column data length, followed by data, then the next column data length followed by data, and so on.
- A rowid can be physical or logical.
- A physical rowid can be extended or restricted.
- A logical rowid is used to locate rows in an index-organized table.
- You can query any row's rowid value with the ROWID pseudocolumn.
- Each column in a table has a data type defining the type of data stored in the column.
- DATE data types contain time and date.
- To adjust times for time zones, use the TIMESTAMP WITH LOCAL TIME ZONE data type.
- Relational tables can be defined as a list of columns or as a subquery.
- Column names are case sensitive when enclosed in double quotes; otherwise, the names are interpreted as all uppercase.
- Temporary tables contain private data that can be viewed only by the user who inserted the data, and the data exists only for the duration of the user's session or until the user commits the transaction.
- Varrays and nested tables are like tables within one column of another table.

- A relational table or an object table can contain a varray or nested table in one of its columns.
- When creating a varray or nested table column, you must define a collection type (varray type or table type) that is used as the column's data type.
- Nested table data is stored out of line (in a separate location) from the primary table.
- A partition of a table resides in its own segment.
- If a table has partitions and subpartitions, each subpartition resides in its own segment.
- Partitioning can be done with range, list, hash, or composite methods.
- Composite partitioning can be range-hash or range-list.
- Range and hash partition keys can be made up of more than one column.
- List partition keys must contain only one column.
- To save typing, use composite range-list partitions with a subpartition template.
- Hash partitioning does not allow any storage settings other than what tablespace will contain the partition.

REVIEW QUESTIONS

1. Relational tables must have a primary key. True or False?
2. A nonpartitioned table has one ______________ and one or more ______________ for storage.
3. You want to speed up queries on a table. Queries are always looking up rows based on the values of the row's primary key. What type of table would improve query speed?
4. Which of the following statements are true about both object tables and relational tables?
   a. Rows are stored in extents.
   b. Collection types can be used in a column.
   c. The table can be partitioned.
   d. Rows can be made up of one column.
   e. All of the above.
5. The PCTFREE storage setting can only be used at the table level. True or False?

6. Examine this SQL command, and determine which line has an error.

```
1 CREATE TABLE BEARTRAIL
2 (TRAIL_NO NUMBER,
3 "BEAR_COUNT" NUMBER(10,0),
4 LASTCOUNT TIMESTAMP WITH LOCAL TIMEZONE);
```

   a. Line 1
   b. Line 2
   c. Line 3
   d. Line 4
   e. No errors exist in the SQL command.

7. Observe these two SQL commands.

```
CREATE TABLESPACE USER_SPACE
DATAFILE 'D:\oracle\data\user_space.dbf'
MINIMUM EXTENT 5M
DEFAULT STORAGE(INITIAL 25M);
CREATE TABLE ACCOUNT_LEDGER
(ACCOUNT_ID NUMBER, DEBIT NUMBER, CREDIT NUMBER)
TABLESPACE USER_SPACE
STORAGE (INITIAL 2M NEXT 10M);
```

   What will the size of the table's first extent be?

   a. 25 M
   b. 2 M
   c. 5 M
   d. 12 M
   e. 30 M

8. When you insert a row into a data block, and the free space drops below 20 percent, what happens?
   a. The data block is removed from the extent's freelist.
   b. The data block migrates the row to another block.
   c. The data block is not removed from the extent's freelist.
   d. There is not enough information to determine the action.

9. A ________________ or ________________ row spans multiple data blocks.

10. Look at the query and the first two rows of results below.

```
SELECT ROWID FROM STORE_INFO;
ROWID
---------------------
00000D45.0000.0021
00000D45.0001.0001
```

What type of rowid does the table contain?

a Universal rowid

b. Restricted rowid

c. Extended rowid

d. Logical rowid

11. A temporary table is dropped when your session ends. True or False?
12. To store repeating values within a column, use a ________________ or a ________________.
13. Which of the following is NOT a valid method of partitioning?

a. List-hash

b. Range

c. Hash

d. Range-hash

e. None of the above

14. You want to create a database table that stores songs recorded for a new music CD. Which data type is best suited for the songs?

a. CLOB

b. BFILE

c. LONGRAW

d. LOB

e. Any of the above

EXAM REVIEW QUESTIONS—DATABASE FUNDAMENTALS I (#1Z0-031)

1. Look at the following column definition.

```
COLUMN HATSIZE NUMBER(6,2)
```

Which of these statements are true?

a. The number 6.534 is rounded to 6.54 when inserted.

b. Nulls are not allowed.

c. The statement is invalid.

d. Numbers greater than 9999.99 return an error.

2. Which component is NOT part of the data block?
   a. Free space
   b. Column length
   c. Common and variable header
   d. Row directory
3. Look at this statement:

```
CREATE TABLE CAR_REPAIR (OWNER VARCHAR2(30),
REPAIR_DESC VARCHAR2(4000))
STORAGE (INITIAL 20M NEXT 5M MAXEXTENTS UNLIMITED
PCTFREE 20 PCTUSED 50 INITRANS 2 MAXTRANS 4)
TABLESPACE USERS;
```

   How much of each data block must be filled with either overhead or row data before the block is taken off the freelist?
   a. 20 percent
   b. 50 percent
   c. 80 percent
   d. 21 percent
4. Look at this SQL command.

```
CREATE TABLESPACE USER_NEW
EXTENT MANAGEMENT LOCAL UNIFORM SIZE 5M;
CREATE TABLE INVENTORY_UPDATES
(INVENTORY_NO NUMBER, LAST_CHANGED DATE)
STORAGE (INITIAL 20M NEXT 5M MAXEXTENTS UNLIMITED
PCTFREE 10)
TABLESPACE USERS_NEW;
```

   Which statements are true about the INVENTORY_UPDATES table? Choose three.
   a. The table's initial extent is 20 M.
   b. The INVENTORY_NUMBER column can store numbers with decimals.
   c. The table's next extent is 5 M.
   d. The PCTFREE setting is ignored.
5. Look at this table definition.

```
CREATE TABLE SALES_REPORT
(SALES_PERSON VARCHAR2(20), SALES_DATE DATE, SALES_AMOUNT
NUMBER(12,2), MANAGER_APPROVAL CHAR(3));
INSERT INTO SALES_PERSON VALUES ('Jerry Montell',
'15-JAN-2004');
```

Which statements are true about the inserted row? Choose two.

a. The rowid is stored in the row overhead.

b. The column length stored for the SALES_AMOUNT column is zero.

c. The column length for the MANAGER_APPROVAL column is three.

d. The SALES_AMOUNT is null.

e. The SALES_DATE has a time of midnight.

6. Which of the following are valid data types? Choose three.

   a. YEAR TO MONTH

   b. TIMESTAMP

   c. NVARCHAR

   d. BFILE

   e. NCLOB

7. You want to create a temporary table that stores rows for your entire session. Arrange the command lines in the correct order.

```
1 (TARGET VARCHAR2(30), HITS NUMBER(5), MISSES NUMBER(4,3))
2 ON COMMIT PRESERVE ROWS
3 STORAGE (INITIAL 5M)
4 CREATE GLOBAL TEMPORARY TABLE MYROWS
```

   a. 2, 4, 1, 3

   b. 4, 1, 2, 3

   c. 4, 2, 1, 3

   d. 2, 4, 3, 1

8. You want to store a repeating list of addresses in your CLIENT table. There can be a maximum of 20 addresses per client. You want to be able to list clients with one or more addresses in certain zip codes, so you plan to put an index on the zip code portion of the address. What is the best strategy for the table design?

   a. Create a nested table in the CLIENT table, and then create an index on the nested table.

   b. Create a varray in the CLIENT table, and then create an index on the varray.

   c. Create a new table for addresses, create a cluster, and then index it on the zip code.

   d. Create a partitioned table for the CLIENT table, and make zip code the partition key.

9. Which of these is NOT part of the CREATE TABLE syntax?
    a. MINIMUM EXTENT
    b. FREELIST GROUPS
    c. MINTRANS
    d. TABLESPACE
10. Look at the following command that creates a table for storing flights from 1996 through 1999.

```
CREATE TABLE AIRPORT_HISTORY
     (CODE CHAR(4),
      FLIGHT NUMBER,
      FLIGHT_DATE DATE)
   PARTITION BY HASH (FLIGHT_DATE)
   PARTITIONS 4
   STORE IN (HISTORY1, HISTORY2, HISTORY3, HISTORY4);
```

Which partition stores flights from the year 1998?
    a. The third partition
    b. All the partitions
    c. Depends on the dates in the data
    d. None of the above

HANDS-ON ASSIGNMENTS

1. You have a small project to complete. You have almost finished an application that reads data from one table, performs a set of calculations, and inserts data into two output tables and commits its work. It reports errors on the screen, which you capture in a spooled file. You have to repeat your test of the application several times, checking the error report each time. You want to create a set of temporary tables for the output tables. Then, you can run the application repeatedly without clearing out the tables before repeating. Write the SQL command to create two temporary tables and explain how you can use them for testing your application. The first table should be named CH07TEMPCUSTORDER and should have columns for customer name and total order amount. The second table should be named CH07TEMPCUSTORDERTOTALS and should have columns for customer order number, total state tax, and total shipping charges. Save the commands and the explanation in a file named **ho0701.sql** in the **Solutions\Chapter07** directory on your student disk.
2. Recall that you created this table for Global Globe's employees:

```
CREATE TABLE EMPLOYEE
(EMPLOYEE_ID NUMBER(10),
JOB_TITLE VARCHAR2(45),
first_name varchar2(40),
```

```
Last_name varchar2(40),
phone_number Varchar2(20));
```

Start with this CREATE TABLE command, and modify it to create a new range-partitioned table named CH07EMPLOYEE_RANGE that contains the same column as the employee table. The partition key is the EMPLOYEE_ID column, and there should be four partitions, all located in the USERS tablespace. Assume that the values in EMPLOYEE_ID start at one, reach 100,000, and are distributed evenly. Save the command in a file named **ho0702.sql** in the **Solutions\Chapter07** directory on your student disk.

3. Copy the range-partitioned table command from Assignment 2, and make these changes to it:
   - Name the table CH07EE_RANGEHASH
   - Keep the range partitions the same, and add hash subpartitioning on JOB_TITLE.
   - There are four subpartitions, all residing on the USERS tablespace.

   Run the command in your database so that the table is owned by CLASSMATE.

   Save your command in a file named **ho0703.sql** in the **Solutions/Chapter07** directory on your student disk.
4. You want a new table to create in the USER_DTAB tablespace. The table, named CH07GOLD, stores a history of the price of gold. A new row is added to the table approximately every two hours. Write the CREATE TABLE command for the table. Your command should address these points:
   - The columns are named PRICE, PRICE_DATETIME, and TIME_BETWEEN.
   - The PRICE can contain fractions of a penny as small as thousandths of a penny. The price is always under 1000 dollars.
   - The TIME_BETWEEN column stores the number of days (up to 99 days), hours, minutes, and seconds (to the hundredth of a second) between this record and the previous record.
   - Because the table is always receiving inserted data and is never updated, you want to minimize the storage space saved for updates.
   - Assume that the average row length is 20 bytes, the table gets an average of 12 inserted records per day, and you want the table to be created with enough storage space for approximately 6 months.

   Do not run the command in your database, because you do not have any dictionary-managed tablespaces, which this command requires. Save your command in a file name **ho0704.sql** in the **Solutions\Chapter07** directory on your student disk.

7

5. Your manager has just informed you that the GOLD table needs more columns. She asks you to create a new table called CH07GOLD_HISTORY that has the correct structure. Starting with the CREATE TABLE command you created in Assignment 4, modify the command to accommodate these changes:
   - Add a column called LOCATION that stores the city from which the price is quoted.
   - Add a column that contains multiple values for the price in five different currencies. The price is always added in this order: Francs, Yen, Pesos, Pounds, and Euro dollars. (*Hint*: Use the Enterprise Manager console to find a collection type in the CLASSMATE schema that you can use.)
   - The added columns add an average of 30 bytes to each row, including overhead.
   - Adjust the storage settings, so that the table resides in the locally managed tablespace, USERS.

   Run the command in your database as the CLASSMATE user. Save your command in a file named **ho0705.sql** in the **Solutions\Chapter07** directory on your student disk.

6. Once again, your manager says she wants a change to the GOLD table. It turns out that the table needs a few more columns and different storage settings. Instead of one price per record, each record holds ten prices. The date and time stamp columns are updated each time a new price is added to the table. The current record is updated until it is filled with ten prices. Then, a new record is inserted, and that becomes the current record to be updated.

   This time, she wants you to take the original table structure (from Assignment 4) and make these changes to the table:
   - Replace the price column with a column that holds up to 10 gold prices. (*Hint*: look for another collection data type in the CLASSMATE schema.)
   - Recalculate the initial space requirements: The new average row length is 30 when it is inserted, growing to 50 bytes after the gold prices are filled in. The table gets an average of two records inserted per day, and you want the table to be created with enough storage space for approximately six months.
   - Adjust the storage settings to minimize row migration.

   Name the table CH07GOLD_COLLECTION. Do not run the command, because it requires a dictionary-managed tablespace. Save your SQL in a file named **ho0706.sql** in the **Solutions\Chapter07** directory on your student disk.

7. Your new client has a collection of original music recordings in digital formats that he wants to store in his Oracle9*i* database. He plans to give online customers the ability to select songs and assemble a customized CD that is then printed and mailed to them. Your job is to create the table to store the music. The client wants to use software other than Oracle9*i* to update the music. Therefore, the music

must reside in AUD format in a file that can be viewed by the database. (AUD format is a format used especially for music that can be played on the Internet or printed on a CD and played on a stereo like any music CD.) Create a table named CH07SONG that stores the song' s unique ID number, artist name, song title, song length (in minutes and seconds), and the song itself. Create another table named CH07SONGLIST that is used to assemble the customer's song list. This table should have the customer's name and address, CD name (the customer fills this in online, and it can be up to 40 characters long), and a list of song ID numbers (up to 15 songs). Once again, you find a collection appropriate for the list of songs in the Schema Manager.

Create the tables in the CLASSMATE schema, and save the SQL commands in a file named **ho0707.sql** in the **Solutions\Chapter07** directory on your student disk.

8. Look at the CUSTOMER table you created earlier in the chapter. Create a temporary table that has all the columns except the varray and includes all the rows in the table. Name the table CH07TEMPCUSTOMER and rename all the columns with the prefix TEMP. The table should remove data rows after each commit.

   Execute the command as CLASSMATE in your database, and save your SQL command in a file named **ho0708.sql** in the **Solutions\Chapter07** directory on your student disk.

9. Create a table as described in Assignment 8, except make it a permanent table named ch07PERMCUSTOMER, and include the varray. Execute the command as CLASSMATE in your database, and save your SQL command in a file named **ho0709.sql** in the **Solutions\Chapter07** directory of your student disk.

10. Create an object table named CH07NEW_ADDRESSES using the CUSTOMER_ADDRESS_TYPE as its object type and with storage in the USER_DTAB tablespace. The table needs 10 M of storage and room for rows to double in size when they are updated.

    Do not create the table, because it requires a dictionary-managed tablespace. Save the SQL command in a file named **ho0710.sql** in the **Solutions\Chapter07** directory on your student disk.

CASE PROJECT

The Global Globe database now has tables for its classified ads (which were created in this chapter). Now, you turn your attention to the actual news articles. How will the articles be handled? You and your team of programmers and users decide that the articles should be moved from their current directory into the database, where they can be searched with Oracle's sophisticated text search tools. Here are the criteria for the table.

- The table is named NEWS_ARTICLE.
- One employee writes every article.

- Each article has a unique ID number, title, date it is run in the newspaper, and the article itself, which is a QuarkXPress document.
- The QuarkXPress document can be up to 500 M, because it sometimes contains digital images embedded within the document.
- The lead editor must approve an article before it is printed. Record the editor's employee ID and the date of approval.
- As an auditing and security feature, the table must store information about the last five times the article was modified. Store the employee's ID and the date and time of the modification. (Find an appropriate collection type owned by CLASSMATE.)
- Store the table in the USER_AUTO tablespace, and use the default storage settings.

After you have read the criteria, write and execute the command to create the table, and save your work in a file named **case0701.sql** in the **Solutions\Chapter07** directory on your student disk.

CHAPTER

8

ADVANCED TABLE MANAGEMENT

In this chapter, you will:

- Create tables with large object (LOB) columns and tables that are index-organized
- Understand the tasks involved in table management
- Use data dictionary views to find information about tables and their underlying structures

Table structures hold the vast majority of the data found in the database. You have practiced creating a variety of tables in the previous chapter. This chapter shows you how to create two more types of tables: tables with large object (LOB) columns and index-organized tables. You then learn how to make many kinds of changes to the table structure, such as adjusting the storage settings, removing a column, and removing all the data. The final section of the chapter shows you how to query the data dictionary views related to tables, segments, and extents.

ADVANCED TABLE STRUCTURES

Having seen so many ways to create tables, you are already familiar with the general syntax for creating a table. There are two more types of tables to learn to create. These have unusual methods of storing data: tables with LOB columns and index-organized tables.

A table with LOB columns can store huge amounts of data in a single column. For example, the digital audio file for one song on a CD, or an entire CD of songs, can be stored in one LOB record. Other uses for LOB columns are large documents in PDF or other formats that contain special formatting symbols that must be preserved. These include images, such as satellite multispectrum photographs, high-resolution digital photographs, and scanned images of artwork. These large-sized files can be loaded into a LOB record in the database, where they are protected from unauthorized use and possible theft or damage.

Index-organized tables help you query table rows more quickly by reducing the number of times a process must read either memory or disk to retrieve the row data. A good use for index-organized tables is a table in which most of the columns are indexed within the primary key of the table and the data that is not part of the primary key is relatively small and static. An example of this is a table used to look up the population of a city by its state, county, and municipal code. The state, county, and municipal code are already indexed, so moving the population into the index only increases the size of the index slightly. By not having to read the rowid from the index and then read the row from another block in the database, your queries perform better.

Tables with LOB Columns

As you know, a LOB column has one of four data types (BLOB, CLOB, NCLOB, and BFILE). These four LOB data types are divided into two groups, according to where they are stored:

- **Internal LOB:** The BLOB, CLOB, and NCLOB data types all have their data stored inside the database, so they are called internal LOBs. Internal LOBs use copy semantics, which means that if you copy a LOB from one row or column to another, the entire LOB—including its data—is copied to the new location.
- **External LOB:** The BFILE data type is the only external LOB, and its data is stored outside the database in an operating system file. External LOBs are read-only and can be stored on read-only media, such as CD-ROMs or DVDs. External LOBs use reference semantics, which means that when copying an external LOB, only the pointer to the location of the file is actually copied. The original file is not copied.

Oracle9*i*, Release 2, supports LOBs within locally managed tablespaces and in tablespaces in which segment management is set to AUTO.

Here is an example of a table with two LOB columns, one for a 90-minute music track stored in a file outside the database and one for a high-resolution image of the band for printing posters stored inside the database.

```
CREATE TABLE SOUNDBYTES
(ALBUM_ID VARCHAR2(20),
 ARTIST VARCHAR2(40),
 CD_MUSIC BFILE,
 POSTER_SHOT BLOB);
```

Oracle9*i* provides special PL/SQL packages to simplify the manipulation of data in LOB columns. Oracle9*i*, Release 2 adds two new procedures to the DBMS_LOB package:

- DBMS_LOB.LOADBLOBFROMFILE: A procedure to insert one file of binary data into one row's BLOB column
- DBMS_LOB.LOADCLOBFROMFILE: A procedure to insert one file of character data into one row's CLOB column

Oracle has consistently provided more tools for LOBs with each new release. Because LONG and LONGRAW are deprecated, all the capabilities and tools available for these old data types are being reintroduced and improved upon for the LOB data types.

LOB Storage

Because of their potentially huge size, Oracle9*i* has special storage methods for LOBs.

Figure 8-1 shows the components of an internal and an external LOB. The data stored inside both inline and out of line LOB types is called its **value**. Sometimes, Oracle9*i* stores an internal LOB's value within the row (inline). Unless you specify that all LOB values must be stored out of line, Oracle9*i* stores LOB values that are less than 4000 bytes long inline. Other times, Oracle9*i* stores the internal LOB's value elsewhere (out of line). Oracle9*i* stores LOB values out of line when they are more than 4000 bytes long. The pointer that directs the database to the actual location of the value is called the **LOB locator** for internal LOBs and the **BFILE locator** for external LOBs.

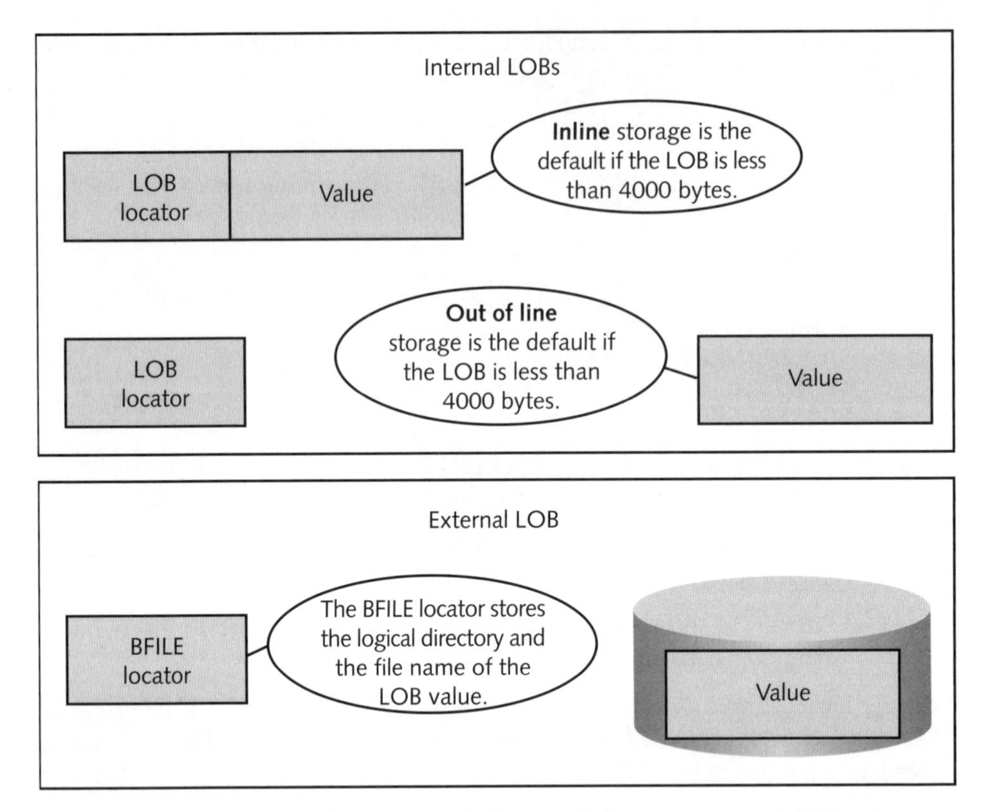

Figure 8-1 Even Large Objects stored inline with the row have a LOB locator

When an internal LOB is stored out of line, a separate LOB data segment is created. You can allow Oracle9*i* to handle this itself, or you can specify a location and size for the LOB data segment when you create the table. You can also specify that all LOB data (even data that is less than 4000 bytes) is to be stored out of line. Here is the syntax for creating a relational table with a BLOB column.

```
CREATE TABLE <tablename>
(<other_column_specs>,
 <LOBcolumnname> <LOBdatatype>)
LOB (<LOBcolumnname>) STORE AS <lobsegmentname>
 (TABLESPACE <tablespacename>
  ENABLE STORAGE IN ROW|DISABLE STORAGE IN ROW
  CHUNK <nn>
  STORAGE
     (INITIAL <nn> NEXT <nn> MAXEXTENTS UNLIMITED|<nn>)
  PCTVERSION <nn>|RETENTION
  LOGGING|NOLOGGING
   CACHE|NOCACHE);
```

Naming the LOB data segment (for example, STORE AS MOVIELOB) is optional; however, it makes queries on the data dictionary views more readable, because the name you choose appears rather than a system-generated name, such as SYS_103877.

The CHUNK parameter sets the number of bytes allocated for working with the LOB value. Oracle9*i* writes and reads one **chunk** of the LOB value at a time. In addition, when storing LOB values out of line, Oracle9*i* allocates space by chunks, rather than by data blocks. The chunk size must be a multiple of the database block size, and must be smaller than the INITIAL and the NEXT sizes (either in the tablespace or in the LOB storage clause). The maximum chunk size is 32 K. A larger chunk size is more efficient for the high volume I/O needed when manipulating very large LOB values. A smaller chunk size is better for LOBs, which have sizes that vary from row to row, because it reduces unused space allocated to the LOB storage. The default chunk size is one data block.

Specifying DISABLE STORAGE IN ROWS tells Oracle9*i* to always use out of line storage for the LOB value. This might be valuable if you have many rows with LOB values that are close to the default threshold (4000 bytes). Storing these portions of the data out of line may prevent row migration and allow faster access to the non-LOB columns. ENABLE STORAGE IN ROWS is the default, and it tells Oracle9*i* to store small LOB values (less than 4000 bytes) inline and to store larger LOB values out of line.

The PCTVERSION <nn> parameter tells Oracle9*i* to store old versions of the LOB value within the LOB itself, until the specified percent of storage is used up. After that, old versions are overwritten by newer versions. The RETENTION parameter is an alternative to the PCTVERSION and indicates that undo data should be generated as the method of retaining old versions of the LOB. Undo data is retained as long as it is specified in the UNDO_RETENTION initialization parameter. You can use one or the other, but not both of these parameters. The default depends on the database's undo mode. For automatic undo mode, RETENTION is the default. For manual undo mode, PCTVERSION is the default.

The CACHE parameter tells Oracle9*i* to place the LOB values into the data buffer for faster retrieval of frequently accessed data. NOCACHE is the default and indicates that the LOB values are placed at the end of the buffer, so they are first to be replaced when more space is needed in the buffer.

The LOGGING parameter assures that the creation of the LOB data segment and mass inserts (such as those from SQL*Loader) are recorded in the redo logs. The NOLOGGING parameter suppresses logging of these two types of activities. The default depends on the CACHE/NOCACHE setting. If you use CACHE, LOGGING is the default. If you use NOCACHE, the default is whatever the tablespace logging specifies. You cannot specify CACHE and NOLOGGING together.

In Oracle9*i*, Release 2, the LOB INDEXES clause is no longer supported and is ignored. Oracle9*i* automatically creates an internally managed index for each LOB column.

You can define attributes with LOB data types in object types for use in object tables or user-defined data type columns in relational tables. You can use LOB columns in partitioned tables and in index-organized tables.

Follow along with this example to create a relational table with LOB columns that are stored in their own storage segment.

1. Start up the Enterprise Manager console. In Windows, click **Start/Programs/Oracle - OraHome92/Enterprise Manager Console**. In UNIX, type **oemapp console** on the command line. The Enterprise Manager Console login screen appears.
2. Select the **Launch standalone** radio button and click **OK**. The console appears.
3. In the main window of the Enterprise Manager console, click **Tools/Database Applications/SQL*Plus Worksheet** from the top menu. A background SQL*Plus process starts, and then the SQL*Plus Worksheet window appears. If the background process appears in front, simply minimize it (click the **minus sign** in the top-right corner) so that you can see the worksheet.
4. Connect as the CLASSMATE user. Click **File/Change Database Connection** on the menu. A logon window appears. Type **CLASSMATE** in the Username box, **CLASSPASS** in the Password box, and **ORACLASS** in the Service box. Leave the connection type as "Normal." Click **OK** to continue.
5. For this example, start with a prepared SQL command. Click **File/Open**. A browse window opens. Navigate to the **Data\Chapter08** directory, and select the **movie.sql** file. Click **Open**. The file appears in the SQL*Plus Worksheet window, as shown in Figure 8-2.

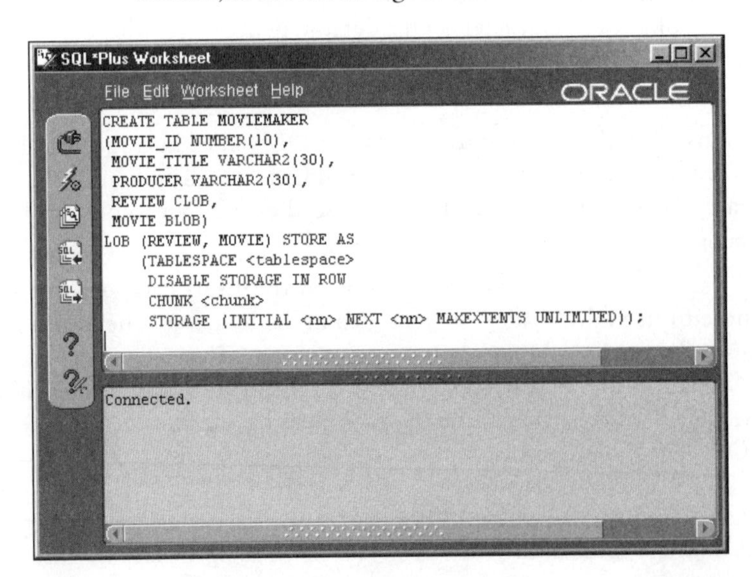

Figure 8-2 Load a file with File/Open and save a file with File/Save Input As

6. Replace the variables with actual names so that these characteristics are true:
   - The USER tablespace stores the LOB data segment.
   - The LOB is stored in 32768 size chunks.
   - The initial extent is 64 K, the next extent is 32 K, and the maximum extent is unlimited.

   There is no LOB data segment name in this example because when you specify the LOB storage for two LOB columns, you cannot name the LOB data segment. The extent sizes used here are low for LOB data and are used only to conserve space for these examples. Normally, the LOB extent is 50 M or more.
7. Run the command by clicking the **Execute** icon. The SQL*Plus Worksheet runs the command and responds "Table created."
8. Remain logged on for the next practice.

The file named **movie_final.sql** in the **Data\Chapter08** directory on your student disk contains the final version of the CREATE TABLE command.

LOB data segments can also be used to store a varray. The settings for the LOB storage are the same as those you saw here for storing a LOB column, except the varray's LOB cannot be in a different tablespace than the table. To specify that the varray should use a LOB data segment, add the STORE AS LOB clause immediately after the varray column name and data type in the list of columns. For example, the following table has a varray named RACE_LIST that is stored in a LOB data segment. (You create the varray type first.)

```
CREATE TYPE CLASSMATE.RACE_ARRAY AS VARRAY (1500) OF CHAR (25);
CREATE TABLE HORSERACE
(HORSE_NAME VARCHAR2(50),
 RACE_LIST CLASSMATE.RACE_ARRAY)
   VARRAY RACE_LIST STORE AS LOB RACEARRAYLOB
    (CHUNK 32768
     STORAGE (INITIAL 20M NEXT 40M MAXEXTENTS 100))
STORAGE (INITIAL 80M);
```

Use LOB storage for varrays, when the varray is intended to be very large.

Tables with LOBs are a type of table that requires special parameters in the CREATE TABLE command. The second type of table requiring special parameters is the index-organized table.

Index-Organized Tables

Tables with LOBs tend to have rows of extra large size. They are so large that part of the row is often separated into its own storage segment. In contrast, index-organized tables tend to have smaller sized rows that benefit from a consolidation of the table's index storage and the table's data storage. Whereas tables with LOBs are often used for multimedia applications, such as music, video, or images, index-organized tables are usually used in text-based or number-crunching applications, such as insurance rate estimates or airline reservations, in which speedy retrieval is key.

An index-organized table, as you saw in the previous chapter, is a relational table with a primary key, in which the rows are stored physically in order by the primary key Tables that are not stored in index order are called **heap-organized** tables. Heap-organized tables are simply relational tables. Most relational data is stored in heap-organized tables because it is the default format used by Oracle9*i* for relational tables. Figure 8-3 shows the difference between a heap-organized table with a primary key and an index-organized table.

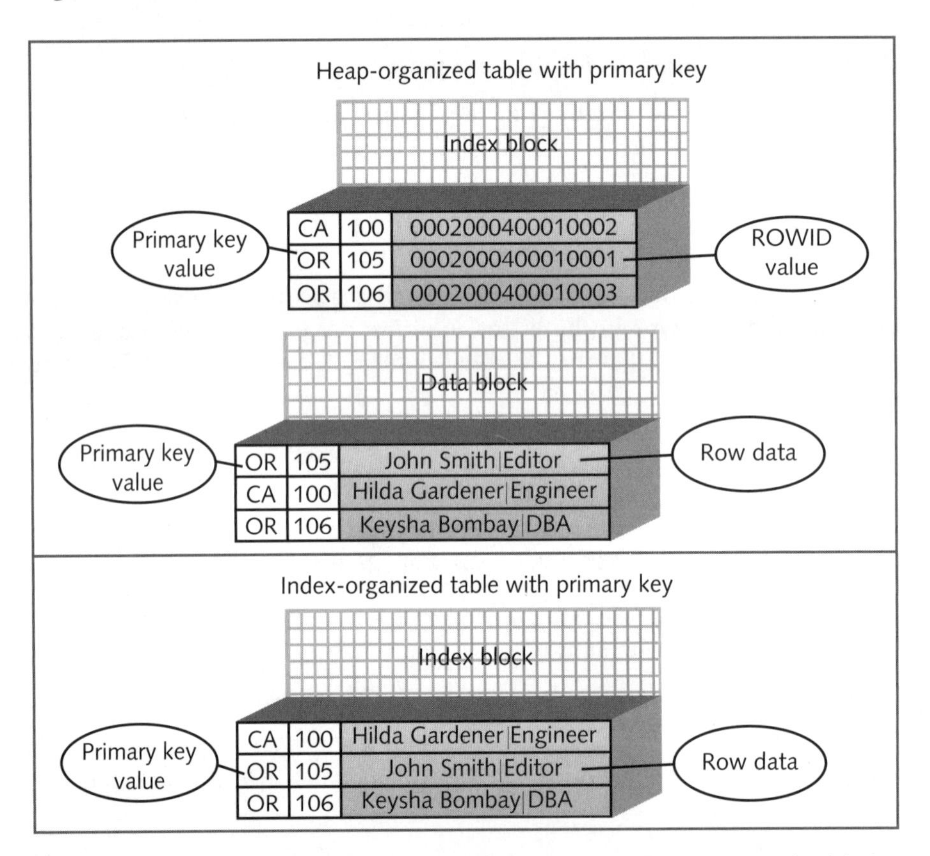

Figure 8-3 The index and the data are united in an index-organized table

A relational table with a primary key constraint automatically has a unique index associated with the primary key. That index is stored in primary key order in a b-tree index structure, making queries by primary key very fast. A **b-tree index** balances data in even branches that split into ever narrowing ranges of key values. See Chapter 9 for more details. The index stores the primary key columns and the rowid of the associated row. The actual data is stored in the data block. When you execute a query looking for a particular primary key value, Oracle9*i* searches the index, reads the rowid, locates the row in the table, and reads the row. Using Figure 8-3 as an example, imagine that this query was issued and finds the row for John Smith:

```
SELECT NAME, JOB
FROM EMPLOYEE
WHERE STATE='CA' AND ID='105';
```

If the table is heap-organized (as in the top of Figure 8-3), Oracle9*i* looks up the value of the ROWID for the matching primary key and then finds the row using the ROWID. It then retrieves the values of the NAME and JOB columns from the row.

An index-organized table does not have an index. Instead, it makes the entire table act like an index. Every row is stored in a b-tree index structure. The primary key values are used to arrange the data into branches. When you execute a query looking for a particular primary key value, Oracle9*i* searches the table, and retrieves the row. Continuing with the same example for Figure 8-3, Oracle9*i* looks up the primary key in the index and finds the matching row. Then it uses this same row to retrieve the values of the NAME and JOB columns. By avoiding the second read of a data block, the index-organized table saves I/O time. The index-organized table also saves space because it does not store the primary key values in both the index and the data blocks and because it does not store the ROWID values at all.

The primary advantage of an index-organized table is that queries based on the primary key are faster than queries in heap-organized tables. The primary disadvantage is that inserts, updates, and deletes are slower, because they may cause an imbalance in the b-tree structure, which requires Oracle9*i* to shuffle rows into different index blocks to rebalance the structure.

Follow along with this example to create an index-organized table.

1. If you are not already there, go to the SQL*Plus Worksheet, and log on as CLASSMATE/CLASSPASS in the ORACLASS database.
2. Start with a prepared SQL command found in the **Data\Chapter08** directory on your student disk. Click **File/Open**. A browse window opens. Navigate to the **Chapter08** directory, and select the **zip.sql** file. Click **Open**. The file appears in the SQL*Plus Worksheet window, as shown in Figure 8-4.

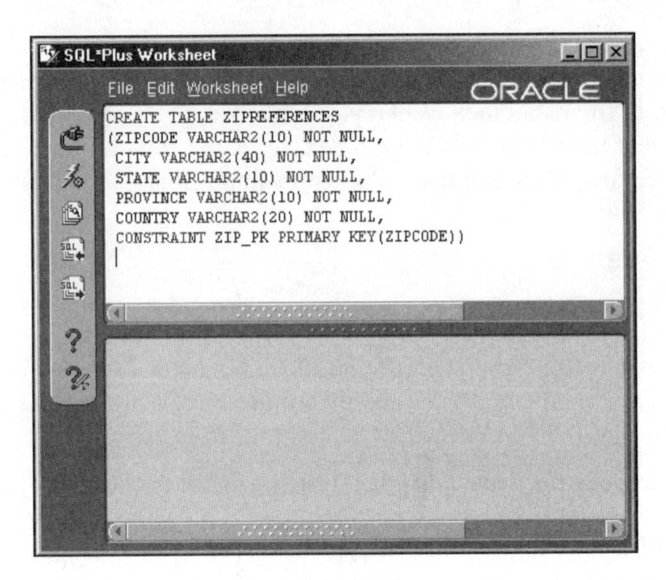

Figure 8-4 This relational table has a primary key constraint

3. As you can see, the only new feature in this relational table so far is a primary key constraint. (Primary key constraints are discussed in Chapter 10.) The first requirement of an index-organized table is that it must have a primary key on which the index is built. The second requirement is that you must include the ORGANIZATION INDEX clause. Type **ORGANIZATION INDEX** on the blank line below the command, and press **Enter**.

4. Optional additional phrases for index-organized tables are:
   - **OVERFLOW:** By restricting the length of each row within the table, adding this parameter helps keep the b-tree index tightly packed into index blocks. For any row that exceeds the specified length, the balance of the row is stored in a separate segment. You can add storage specifications, such as the tablespace, extent size, and so on, along with the OVERFLOW parameter.
   - **PCTTHRESHOLD:** This tells Oracle9*i* the maximum percentage of space in an index block that one row is allowed to use. If a row's total length exceeds this size, the row is divided, and the excess columns are stored in the overflow segment.
   - **INCLUDING <col>:** If you specify OVERFLOW, you specify the dividing line between which columns are stored in the main table and which are stored in the overflow segment. The primary key columns are always stored in the main table. When you specify INCLUDING, all columns after the named column are stored in the overflow segment. If you organize the list of columns in the table so that the most frequently accessed columns come first (after the primary key columns), this parameter can help reduce I/O by keeping the commonly accessed columns together.

- **TABLESPACE and STORAGE:** Add storage and tablespace information for the overflow segment by specifying them after the OVERFLOW parameter.

Type the following line to define the table's tablespace. Then press **Enter**.

```
TABLESPACE USERS
```

5. Type the following two lines to indicate that a threshold of 20 percent per row will restrict the row size, and the remaining portion of the row (if any) will be stored in an overflow segment located in the USER_LOCAL tablespace. Press **Enter** after typing the lines.

```
PCTTHRESHOLD 20
OVERFLOW TABLESPACE USERS
```

6. Add storage parameters for the overflow segment by typing these two lines and pressing **Enter**.

```
STORAGE (INITIAL 64K NEXT 32K
MAXEXTENTS 50 PCTINCREASE 0);
```

7. Execute the command to create the table by clicking the **Execute** icon (the lightning bolt). The worksheet replies, "Table created" in the lower half of the screen.

8. Remain logged on for the next practice.

The complete command for the index-organized table named ZIPREFERENCES is in the **zip_final.sql** file in the **Data\Chapter08** directory on your student disk.

Here are some additional points about index-organized tables.

- You can include LOB columns.
- You can partition index-organized tables, but only using HASH or RANGE partitioning. In addition, the partition key must be all or a subset of the primary key.
- You can add secondary indexes; however, they should be rebuilt if the table is updated often, so that the physical guess stored in the index is kept accurate. Inaccurate physical guesses cause extra I/O.

This completes the details of procedures for creating complex table structures. You have practiced creating several tables in this chapter and in Chapter 7. What happens when you want to change something about a table you have created? For example, you may want to add a new column, change the data type of a column, or add a default value to a column. You might want to give a table more storage space or reduce the storage it has been allocated but is not using. The next section shows you how to handle these kinds of tasks.

OVERVIEW OF TABLE MANAGEMENT

After you create a table, you begin loading data into it by using an application, a SQL*Loader script, or SQL. As time goes by, you may find your table structure needs adjustment. Here are the types of changes you can make to a table's structure:

- **Change the storage settings:** You can adjust the size of the next extent, maximum extents, percent free, and most of the other storage settings.
- **Reorganize the table online:** As a new feature of Oracle9*i*, Release 2, you can now rearrange column order, add or remove columns, and change column names or data types while the table remains online.
- **Drop columns:** You can mark a column so it is unavailable until it is dropped, or simply drop the column immediately.
- **Truncate or drop the table:** You can remove all the rows in a table without generating redo log entries by using the TRUNCATE command. Dropping the table removes the rows and the table structure as well.

Before making changes to a table's structure, it is always a good idea to analyze the data, so that you understand the table's makeup better. In fact, you may discover that your table needs adjusting based on the outcome of the analysis. For example, you know a table needs better storage settings, because, after analyzing the table, you find it has ten extents and 25 percent of the rows are migrated.

The next section describes how to analyze a table or an entire schema. The following sections describe each of the table structure changes mentioned in the previous list.

Analyzing a Table

You should analyze the tables in a schema periodically to give the optimizer up-to-date information for optimizing queries and other SQL commands. If the table data changes (new rows, updated rows, or deleted rows) every day, you may want to analyze the table once a week. If the table data changes more slowly, you could analyze the table on a monthly or quarterly basis. Analyzing a table also gives you up-to-date information so that you can make valid decisions regarding changes to the storage parameters of the table.

To **analyze** a table, you issue a command that causes Oracle9*i* to read the table's structure and update the table's metadata with current information about the size of the table, average row length, total number of rows, free space remaining in extents, and so on. There are two reasons to analyze a table. First, analysis provides accurate statistics for the cost-based optimizer. Second, analysis gives you, the DBA, information to help you decide which storage or column settings to change (if any).

The **Optimizer** is a process used within Oracle9*i* that decides the most efficient plan. A **plan** is a list of steps taken to retrieve data for a query. For example, one query might require a full table scan followed by a sort operation. Another query might require an index lookup, followed by a merge with another table's data, followed by another index lookup on a third table. Queries are executed according to the plan selected by the Optimizer.

There are two types of optimizers available to Oracle9*i*: cost-based and rule-based. The **rule-based Optimizer** is available for backward compatibility. It uses static rules to rank possible access paths and select the best path. The **cost-based Optimizer** is the default Optimizer. It uses statistics on the actual volume and distribution of table data to determine the best path to retrieve the data, taking into account the relative costs of I/O, CPU time, execution time, and other factors. The statistics used by the cost-based Optimizer are stored in the table's metadata when you analyze the table.

To analyze a table, you can use either the ANALYZE command or the DBMS_STATS predefined package. Follow along to run examples of each on the ORACLASS database.

1. If you are not already there, go to the SQL*Plus Worksheet, and log on as CLASSMATE/CLASSPASS in the ORACLASS database.
2. Type the following line to enter the command to analyze the CUSTOMER table.

   ```
   ANALYZE TABLE CUSTOMER COMPUTE STATISTICS;
   ```

3. Click the **Execute** icon to run the command. The worksheet replies, "Table analyzed" when it completes its work. The command automatically gathers statistics on the table, all its columns, and all its indexes.
4. For very large tables, computing statistics takes more time. Therefore, Oracle9*i* provides the ESTIMATE STATISTICS SAMPLE clause. You can specify a number of rows for the sample or a percentage of the table's rows. Two other clauses you can add to the ANALYZE TABLE command are the VALIDATE STRUCTURE (to validate the integrity of every data block and every row) and LIST CHAINED ROWS (to detect chained and migrated rows) options. Type the following command to experiment with the command.

   ```
   ANALYZE TABLE CUSTOMER ESTIMATE STATISTICS
   SAMPLE 1000 ROWS;
   ```

5. Click the **Execute** icon to run the command. The worksheet replies, "Table analyzed" when it completes its work.
6. Oracle recommends that you use the DBMS_STATS package instead of the ANALYZE command for gathering statistics. Oracle requires that you gather statistics on at least one table in a query to enable the cost-based Optimizer. If no tables have statistics, the rule-based Optimizer is used. Using the rule-based Optimizer is a disadvantage and should be avoided because the cost-based Optimizer is the newer and better optimizer of the two. Type in the following command to analyze the table with the DBMS_STATS package.

   ```
   EXECUTE DBMS_STATS.GATHER_TABLE_STATS -
   ('CLASSMATE','CUSTOMER');
   ```

7. Run the command by clicking the **Execute** icon (the lightning bolt). The worksheet replies, "PL/SQL procedure successfully completed" in the lower half of the screen.

8. An alternate method of running the procedure is to place it into a PL/SQL block and execute it. The advantage is that you don't have to type it all on one line, which makes the command more readable. Type and execute the following command, which has the same effect as the command in Step 7.

```
BEGIN
DBMS_STATS.GATHER_TABLE_STATS
('CLASSMATE','CUSTOMER');
END;
```

9. You can also gather statistics on an entire schema at one time using the GATHER_SCHEMA_STATS procedure. To gather statistics for CLASSMATE, type and execute the following command:

```
EXECUTE DBMS_STATS.GATHER_SCHEMA_STATS('CLASSMATE');
```

10. This command takes longer to complete. After it is finished, the worksheet replies again with, "PL/SQL procedure successfully completed" in the lower half of the screen.

11. Remain logged on for the next practice.

Now that you have gathered statistics, you can query those statistics in the USER_, ALL_, and DBA_TABLES, in the USER_, ALL_, and DBA_INDEXES, and in the DBA_TABLES data dictionary views, or you can see them in the Schema Manager within the Enterprise Manager console.

Adjusting Table Storage Structure

Many portions of the storage structure of a table can be modified after the table is created. For example, you can increase or decrease the size of the next extent.

All these changes are made by using various clauses within the ALTER TABLE command. The ALTER TABLE syntax, with each clause listed, looks like this:

```
ALTER TABLE <schema>.<tablename>
PCTFREE <nn> PCTUSED <nn>
INITTRANS <nn> MAXTRANS <nn>
STORAGE (NEXT <nn> PCTINCREASE <nn>
         MAXEXTENTS <nn>|UNLIMITED)
ALLOCATE EXTENT SIZE <nn> DATAFILE <filename>
DEALLOCATE UNUSED KEEP <nn>
COMPRESS|NOCOMPRESS
MOVE TABLESPACE <tablespacename>
      STORAGE (INITIAL <nn> NEXT <nn> PCTINCREASE <nn>
               MAXEXTENTS <nn>|UNLIMITED)
      COMPRESS|NOCOMPRESS
      ONLINE
```

You do not have to specify the table schema when you are logged on as the owner of the table.

The following list describes each clause. At the end of the section, you will try out the MOVE command.

- **PCTFREE <nn> PCTUSED <nn>:** Modifies the settings for these two parameters. The new settings affect all future rows inserted or updated. You cannot specify these parameters when the table is in a locally managed tablespace.
- **INITTRANS <nn> MAXTRANS <nn>:** Changes the number of initial and maximum transactions allowed to concurrently access the data block. You cannot specify these for tables in locally managed tablespaces.
- **STORAGE (...):** Adjusts anything except the INITIAL extent of the table. Changes affect future extents, not current extents.
- **ALLOCATE EXTENT:** Explicitly adds an extent of a designated size and location to the table.
- **DEALLOCATE UNUSED:** Releases unused data blocks above the high watermark of the table. The **high watermark (HWM)** is the boundary between used data blocks and unused data blocks in a table. There is often room in some of the blocks below the HWM due to deleted rows or rows that were migrated. The boundary marks the last data block that is formatted for data. When deallocating, the data blocks that are above the high watermark are released back to the database. Adding the KEEP <nn> clause tells Oracle9*i* to keep that many bytes of storage above the HWM within the table.
- **COMPRESS:** Changes a table from decompressed (the default) to compressed data storage. This saves space and memory; however, it is usually applied to tables that are primarily static. Updates and inserts take more time on compressed tables. You can reverse the process by specifying the NOCOMPRESS parameter for a table that is compressed.
- **MOVE:** Moves the table to another tablespace, although MOVE can also be used to change storage settings while keeping the table in the same tablespace. In the move, you can adjust any or all storage settings, including INITIAL. You can also compress or decompress the data by adding the COMPRESS or NOCOMPRESS parameter.
- **ONLINE:** Allows users access to the table during the move process. This is only currently available for index-organized, nonpartitioned tables.

You cannot use all the clauses at once. Some clauses, such as DEALLOCATE and ALLOCATE, cannot be used together. Others, such as MOVE, must be used alone.

Here is an example of modifying a table to release all unused space except 50 K above the HWM:

```
ALTER TABLE HORSERACE DEALLOCATE UNUSED KEEP 50K;
```

Follow along to modify the CUSTOMER table by moving it to a different tablespace and adjusting part of the storage settings.

1. If you are not already there, go to the SQL*Plus Worksheet, and log on as CLASSMATE/CLASSPASS in the ORACLASS database.
2. Let's assume that you have looked at the current storage values for the CLASSIFIED_AD table and have decided that the table will not need the 2 M extents that you set when creating the table. The table appears to be using much less space. You cannot change the INITIAL storage allocation of 2 M; however, you can change the size of subsequent extents by changing the NEXT parameter to a smaller size. Type and execute the following two lines to modify the CLASSIFIED_AD table.

   ```
   ALTER TABLE CLASSIFIED_AD MOVE TABLESPACE USERS
   STORAGE (NEXT 56K);
   ```

3. The worksheet replies "Table altered" when it completes its work.
4. Remain logged on for the next practice.

The next section describes a new feature of Oracle9*i*: the ability to reorganize a table while users query and modify rows in the table.

Reorganizing a Table

Oracle9*i* has a new feature called the **online table redefinition** that allows you to make just about any change to a table you need, while keeping the table available for inserts and updates nearly the entire time. The feature is implemented in a new PL/SQL package called DBMS_REDEFINITION.

DBMS_REDEFINITION was created for high-availability applications, such as online airline reservations or online banking services. The package is intended for DBAs, and, therefore, you must have several DBA-level privileges to use the package. The package cannot handle all types of tables. For example, object tables and relational tables with object type columns are not supported.

Here is an outline of the steps you follow to redefine a table online. See the Oracle 9i, Release 2 document named *Oracle Administrator's Guide* in the section titled "Managing Tables" for more details.

1. Let's say you want to use the package to modify the table structure of the CLASSIFIED_AD table. First of all, the CLASSMATE user does not have the

privileges needed to run the package, so connect to the SYSTEM user by typing and executing the following command, replacing *<password>* with the actual password for SYSTEM.

```
CONNECT SYSTEM/<password>@ORACLASS
```

2. Another minor detail to handle before you begin working with the package is that you must grant a privilege that is required for the package. Type and execute this command.

```
GRANT EXECUTE_CATALOG_ROLE TO SYSTEM;
```

3. Verify that the table is a candidate for online redefinition by typing and executing these commands:

```
BEGIN
DBMS_REDEFINITION.CAN_REDEF_TABLE
('CLASSMATE','CLASSIFIED_AD');
END;
```


This error message displays: ORA-12089: cannot online redefine table "CLASSMATE"."CLASSIFIED_AD" with no primary key. Because the statement failed, you cannot use the package to restructure the CLASSIFIED_AD table.

4. Try another table by typing and executing this command.

```
BEGIN
DBMS_REDEFINITION.CAN_REDEF_TABLE
('CLASSMATE','CLIENT');
END;
```

The statement succeeds, so the table can be redefined using the package.

5. Create an interim table in the same schema as the original table with all the changes to columns and storage finished. For example, let's add a new column FULLNAME just after the LAST_NAME column that contains the FIRST_NAME and LAST_NAME combined. Type and execute this command to create the new table.

```
CREATE TABLE CLASSMATE.CLIENT1
(CLIENT_ID NUMBER(10) NOT NULL,
 FIRST_NAME VARCHAR2(10) NOT NULL,
 LAST_NAME VARCHAR2(20) NOT NULL,
 FULLNAME VARCHAR2(32),
 CONTACT_PHONE VARCHAR2(15),
 CONTACT_EMAIL VARCHAR2(30),
 CONSTRAINT CLIENT_PK1 PRIMARY KEY(CLIENT_ID))
TABLESPACE USERS PCTFREE 0
    STORAGE (INITIAL 64K NEXT 8K
    MINEXTENTS 1 MAXEXTENTS 10);
```

You can also include whatever storage parameters you want, regardless of what they are in the old table, including the INITIAL clause. Notice that the table and primary key names are different from the original table and primary key names. This is needed because you cannot have two tables with the same name in one schema, and you cannot have two constraints with the same name in one schema. You can rename the table afterward. To rename the primary key constraint, you have to drop and re-create it, so you might want to simply leave it as it is.

6. Start the redefinition process by running this command:

```
BEGIN
DBMS_REDEFINITION.START_REDEF_TABLE
  ('CLASSMATE','CLIENT', 'CLIENT1',
   'CLIENT_ID,
    FIRST_NAME,
    LAST_NAME,
   SUBSTR(CONCAT(RPAD(FIRST_NAME,LENGTH(FIRST_NAME)+1)
                 ,LAST_NAME),1,32) FULLNAME,
    CONTACT_PHONE,
    CONTACT_EMAIL');
END;
```

7. The column mapping tells Oracle9i which column in the old table should map to each column in the new table. List the old column names in the order of the new table's columns.You can use functions or literals as you do in a query. At this point, you can create any constraints, grants, triggers, or indexes you need on the interim table.These are transferred to the redefined table later. In this example, there is nothing that we need to create, so continue by running this command to complete the redefinition of the table:

```
BEGIN
DBMS_REDEFINITION.FINISH_REDEF_TABLE
  ('CLASSMATE','CLIENT', 'CLIENT1');
END;
```

   The procedure re-creates the table in the image of the interim table (including indexes, grants, and constraints), pours all the data in according to the column mapping, and then applies any data changes that took place between the beginning and end of the process to the redefined table. Constraints, indexes, and grants from the old table are transferred to the interim table and are dropped when you drop the interim table.

8. To view the results, query the restructured CLIENT table by typing and executing the following command. Figure 8-5 shows the results.

```
SELECT * FROM CLASSMATE.CLIENT;
```

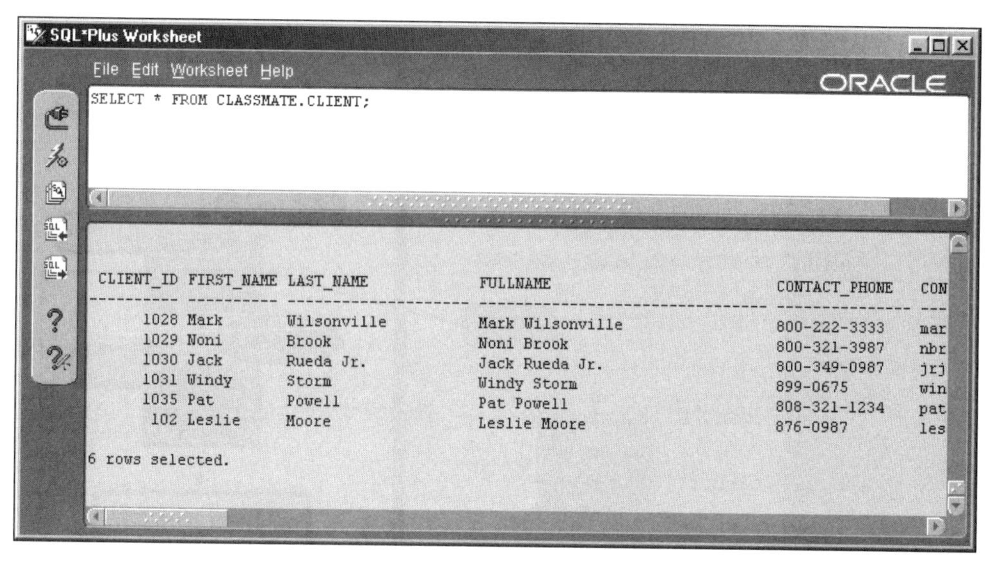

Figure 8-5 The new column, FULLNAME, was populated by the procedure

9. Remain logged on for the next practice.

If errors are encountered, there are a few other procedures included in the DBMS_REDEFINITION package for clean up and restoration. Refer to the Oracle 9*i*, Release 2, documentation in the book titled *Oracle9i Supplied PL/SQL Packages and Types Reference*.

When little or no down time is available for database maintenance, use the package for tables in the high-demand system.

The next command is a form of the ALTER TABLE command that is strictly for object tables and relational tables with object type columns. Object tables and object type columns are based on the definition of an object type. If the definition changes, the tables in which the object type is used do not change automatically. You have to upgrade the table structure. **Upgrading** a table structure causes any object types used in the table to be updated with the most recent version of the object type definition. Here is an example of the command:

```
ALTER TABLE CUSTOMER UPGRADE INCLUDING DATA;
```

The INCLUDING DATA clause is the default, and data is modified to match the new type definition. You can use NOT INCLUDING DATA to leave data unchanged; however, this could cause problems in queries or other operations on the table's data.

You have seen how to make changes to a table's columns and to its storage settings. There are some other changes, such as changing the row movement mode, that you can make on a table's structure. The next section describes these miscellaneous table changes.

Making Other Table Changes

As with most table changes, all the changes described in this section are made using the ALTER TABLE command. Here is the syntax you examine in this section:

```
ALTER TABLE <schema>.<tablename>
RENAME TO <newname>
LOGGING|NOLOGGING
MONITORING|NOMONITORING
ENABLE|DISABLE ROW MOVEMENT
CACHE|NOCACHE
```

These parameters have the following effects:

- **RENAME TO:** As is very apparent, this changes the name of the table. RENAME must be used alone as the only clause in the ALTER TABLE command.
- **LOGGING|NOLOGGING:** These change the logging mode of the table. Recall from Chapter 7 that NOLOGGING suppresses the creation of redo log entries for DDL commands on the table and for mass insert commands, such as those run by SQL*Loader.
- **MONITORING|NOMONITORING:** These clauses turn on or off statistics gathering for the table. When monitoring is on, the system keeps track of whether the table is used while monitoring is turned on. Monitoring is used by the DBA to find old, unused tables or indexes. Unused indexes cause unnecessary overhead, whereas unused tables take up extra space in the database. Data is collected in the V$OBJECT_USAGE dynamic performance view.
- **ENABLE|DISABLE ROW MOVEMENT:** These allow or prevent Oracle9*i* from moving rows during operations, such as data compression. Moving a row changes its rowid, which could cause a problem, if you stored rowids in a reference table for looking up data. It is usually not a problem, and the default setting is ENABLE ROW MOVEMENT.
- **CACHE|NOCACHE:** When you set CACHE for a table, data blocks read are placed in the "most recently used" section of the buffer. This keeps them stored in the buffer longer than the default NOCACHE setting. Use CACHE for frequently accessed tables.

Follow along with an example of how to modify the cache and monitoring of a table.

1. If you are not already there, go to the SQL*Plus Worksheet, and log on as CLASSMATE/CLASSPASS in the ORACLASS database.
2. Type and execute the following line to enter the command to modify the EMPLOYEE table.

```
ALTER TABLE EMPLOYEE MONITORING CACHE;
```

3. Note that the worksheet replies, "Table altered" when it completes its work.
4. Remain logged on for the next practice.

There may be a time when you need to add or change a column in a table. The next section describes how to accomplish these types of changes.

Dropping, Adding, or Modifying a Column in a Table

The previous section described a way to totally rework a table's structure using the DBMS_REDEFINITION package. Many times, you will not have to go through a total redefinition of a table. More often, you need one more column added or a column's length increased. These kinds of changes can be accomplished using the ALTER TABLE command and the clauses listed in the following syntax:

```
ALTER TABLE <schema>.<tablename>
RENAME COLUMN <oldcolname> TO <newcolname>
ADD (<colname> <datatype>, ... )
MODIFY (<colname> <datatype>, ... )
DROP (<colname>, <colname>,...)|COLUMN <colname>
      CASCADE CONSTRAINTS
SET UNUSED (<colname>,<colname>,...)|COLUMN <colname>
     CASCADE CONSTRAINTS
DROP UNUSED COLUMNS
```

Each of these clauses must be used alone in a single ALTER TABLE command. No other clauses can be added to the command. Notice, however, that the ADD, DROP, MODIFY, and SET_UNUSED clauses allow you to list multiple columns.

The CASCADE_CONSTRAINTS parameter must be added if there are any constraints in other tables that reference the column or columns you drop or set to unused status. Otherwise, Oracle9*i* does not change the table and issues an error message. See Chapter 10 for more details about constraints and dependencies.

Follow along with these steps to use examples of each of these column clauses.

1. If you are not already there, go to the SQL*Plus Worksheet, and log on as CLASSMATE/CLASSPASS in the ORACLASS database.
2. Click **Worksheet/Run Local Script** in the Worksheet menu. Navigate to the file named **create_table.sql** in the **Data\Chapter08** directory on your student disk, and click **Open**. The commands are executed in the lower window as shown in Figure 8-6. This creates a table called CH08SURGERY and adds rows so you can experiment with various changes to the columns.

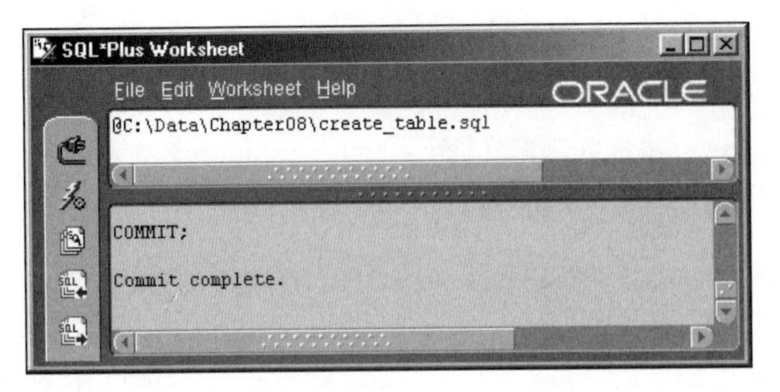

Figure 8-6 The commands run automatically when you select Run Local Script

3. Type **DESC CH08SURGERY** in the top window, and click the **Execute** icon. You will see a list of column names in the lower window. Notice that the third column name is misspelled. It is named PATIENT_FISRT_NAME.
4. Type and execute the following lines to enter the command to correct the column name.

   ```
   ALTER TABLE CH08SURGERY
   RENAME COLUMN PATIENT_FISRT_NAME TO PATIENT_FIRST_NAME;
   ```

   The worksheet replies, "Table altered" when it completes its work.
5. This table tracks surgeries in a hospital. The doctors need information about the outcome of the surgery and the operating room used for the surgery. Type and execute the following lines to enter the command to add these two columns.

   ```
   ALTER TABLE CH08SURGERY
   ADD (OUTCOME VARCHAR2(40), OPERATING_ROOM_NO CHAR(4));
   ```

   The worksheet replies, "Table altered" when it completes its work.
6. You decide that the DOCTOR_NAME column is too long. It is currently defined as VARCHAR2(40). You cannot think of any doctor whose name is longer than 20 characters. Type and execute the following command to reduce the size of the column.

   ```
   ALTER TABLE CH08SURGERY
   MODIFY (DOCTOR_NAME VARCHAR2(20));
   ```

   The worksheet displays an error in the lower part of the window: "ORA-01441: cannot decrease column length because some value is too big." This is caused by the modification of the data in the column. At least one row has data that is longer than 20 characters, so the command failed.
7. Change the command so that the column is 25 rather than 20 characters long. Then click the **Execute** icon to run the command. The worksheet replies "Table altered" indicating that all the data fits within 25 characters or less.

8. Imagine that time has gone by, and the doctors report to you that they are not using the PROCEDURES column or the OUTCOME column because the formats are incorrect. You decide the best course of action is to first render the two columns unusable and later, during an upcoming maintenance period, add replacement columns and drop the old columns. Type and execute the following command for making the two columns unusable.

   ```
   ALTER TABLE CH08SURGERY
   SET UNUSED (PROCEDURES, OUTCOME);
   ```

   The worksheet replies, "Table altered" when it completes its work. The SET UNUSED clause does not fully remove the columns or the data in those columns. However, it does remove the columns from the Data Dictionary's information about the table and makes it impossible to update the data. The SET UNUSED clause is faster than the DROP COLUMN clause, because the data is not actually removed from each row in the table. You cannot undo an unused column.

9. Now it is time for maintenance, and you have conferred with the doctors and determined the proper data types for the PROCEDURES and OUTCOME columns. Type and execute the following command to add the new definitions of the two columns.

   ```
   ALTER TABLE CH08SURGERY
   ADD (PROCEDURES CLASSMATE.PROCEDURES_ARRAY,
        OUTCOME NUMBER(2,0));
   ```

   The worksheet replies, "Table altered" when it completes its work. The added columns have the same names as the two columns you set to unused status. This is valid in Oracle9*i*.

10. Now it is time to drop the two unused columns. Type and execute the following command to drop all unused columns.

    ```
    ALTER TABLE CH08SURGERY
    DROP UNUSED COLUMNS;
    ```

    The worksheet replies, "Table altered." Any column that was marked unused is dropped at this point.

11. Remain logged on for the next practice.

When you execute an ALTER TABLE DROP COLUMN command, all columns that are set to unused status are dropped along with the column or columns you name in your DROP command.

There are a few restrictions when changing the data type or length of columns. Basically, if all the existing data cannot be translated into the intended data type or length, Oracle9*i* cannot make the change. If you have special conversions that must be done to translate the data, you may be better off creating a new column and writing an UPDATE command to translate the column data from the old column into the new column. Then, you can drop the old column, and rename the new column, if you want.

Picture this scenario: You are running a program that inserts ten thousand rows into a table, committing its work each time it inserts one hundred rows. You change the program, and then you want to run the program again. You issue a DELETE command and then wait for a long time while Oracle9*i* deletes ten thousand rows. Then, you issue a commit and wait again. How can you speed up this process? The next section shows how to drop the entire table, or remove all its contents quickly.

The commands you ran in this section are all listed in the **alter_table.sql** file in the **Data\Chapter08** directory on your student disk.

Truncating and Dropping a Table

It is possible to remove all the rows from a table by typing:

```
DELETE FROM <tablename>;
```

However, this method of removing all the rows is slow, but can be reversed. Executing a DELETE command that removes all the table's rows causes Oracle9*i* background processes to write undo records for every row, and to write redo log entries for every row. This takes time, which might be a disadvantage. On the other hand, it also allows you to reverse the command (before committing) by executing the ROLLBACK command, which might be an advantage.

If you are sure you do not need the data and want to remove thousands or millions of rows quickly, you are better off using the TRUNCATE command. The TRUNCATE command does not generate any undo records or any redo log entries. This can improve the response time for removing all the rows in a table dramatically. The more rows you have to delete, the bigger the performance gain by using the TRUNCATE command. The syntax is:

```
TRUNCATE TABLE  <schema>.<tablename>
        DROP STORAGE|REUSE STORAGE
```

DROP STORAGE is the default setting and frees up all but the space allocated to the table except space required by the MINEXTENTS setting of the table or tablespace. In addition to the table's storage space, all indexes on the table are dropped, and all their space is released as well. This space returns to the tablespace and is available for creating new segments in the future.

Specify REUSE STORAGE to keep the space allocated to the table for future inserts and updates. This is appropriate if you are purging all the existing rows in preparation for inserting an equal number of new rows.

You can also remove all the table's rows *along with its structure* by using the DROP TABLE command.

The primary difference between dropping a table and truncating the table is that only truncating preserves the table's structure. In addition, you cannot truncate a table that has constraints from other tables connected to any of its columns, whereas you can drop a table that has constraints from other tables by adding the CASCADE_CONSTRAINTS clause to the DROP TABLE command.

For example, the table COPCAR is related to the COP table with a foreign key in the COP table. To drop the COPCAR table, you must run this command:

```
DROP TABLE COPCAR CASCADE CONSTRAINTS;
```

In this chapter and the previous chapter, you have seen how to create tables, modify their structure, and drop them. The next section describes some of the data dictionary views available to view information about your table's structure and contents.

QUERYING TABLE-RELATED DATA DICTIONARY VIEWS

Because tables are so important in a database, there are many data dictionary views that contain information about them. The obvious views have either TAB or TABLE in their names, such as DBA_TABLES and USER_TAB_COLUMNS. The lesser known views report on segments, extents, and columns. Table 8-1 lists data dictionary views and dynamic performance views that contain information about tables. When the view begins with DBA_, remember there is almost always another view prefixed with USER_ and ALL_ with the same characteristics.

Table 8-1 Table-related data dictionary views

Name	Description
DBA_EXTERNAL_TABLES	List names of external tables and access methods used for retrieving data
DBA_LOBS	Lists LOB statistics, LOB storage settings, and so on for all but BFILE LOBS
DBA_NESTED_TABLES	List names and storage settings of nested tables
DBA_OBJECT_TABLES	List names and storage settings of object tables
DBA_SEGMENTS	List information about segments, such as the extents, blocks, freelists, and buffers used
DBA_SEGMENTS	Lists segment identification, type, owner, and location
DBA_TAB_COLUMNS	Lists all columns in the table, their data types, default values, and so on; also contains statistics gathered by ANALYZE and DBMS_STATS
DBA_TAB_PARTITIONS	List information about partition key, storage settings, and statistics on partitions
DBA_TAB_PRIVS	List table privileges granted on any table to any user

Table 8-1 Table-related data dictionary views (continued)

Name	Description
DBA_TABLES	Lists the name, storage, and other settings assigned when the table was created; also contains statistics gathered by ANALYZE and DBMS_STATS
DBA_UNUSED_COL_TABS	Lists tables that contain unused columns
USER_TAB_PRIVS	List table privileges granted **to** the current user on any table
USER_TAB_PRIVS_MADE	List table privileges granted **by** the current user on tables owned by the current user
V$OBJECT_USAGE	Lists results of monitoring tables

To try out one of these data dictionary views, follow these steps.

1. If you are not already there, go to the SQL*Plus Worksheet, and log on as CLASSMATE/CLASSPASS in the ORACLASS database.
2. Earlier in the chapter, you ran the DBMS_STATS package for the entire CLASSMATE schema. Look at the statistics now with this query. Type the query into the SQL*Plus Worksheet.

```
SELECT TABLE_NAME, TABLESPACE_NAME, MONITORING,
BLOCKS, EMPTY_BLOCKS
FROM USER_TABLES;
```

3. Click the **Execute** icon to run the command. The worksheet displays the list of tables and their information in the lower half of the window.
4. Log off of SQL*Plus Worksheet by clicking the **X** in the top-right corner.

You should run ANALYZE or DBMS_STATS on a regular basis. The time between runs depends on the volume of activity on your database. Keeping statistics current enables the Optimizer to function more efficiently and predict more accurately which plan offers the fastest performance. In addition, the statistics are available to you through the data dictionary views to help determine whether you need to add storage space, move tables to different disk drives, and so on.

Chapter Summary

- LOB data types are either internal LOBS, stored within the database, or external LOBS, stored outside the database.
- Oracle9*i* added procedures to the DBMS_LOB package to simplify loading LOB data into the database.
- By default, internal LOB values are stored inline, if they are less than 4000 bytes long.

- Internal LOB values larger than 4000 bytes are stored out of line in a LOB data segment.
- Oracle9*i* works with LOBs by reading and writing one chunk at a time (a chunk can be from one data block that is up to 32 K in size).
- LOB segment storage settings are similar to settings for table storage.
- LOB data types can be part of partitioned tables, index-organized tables, and object tables.
- LOB data segments can also be used to store varray data.
- Index-organized tables require a primary key and store data in order, as if the whole table were an index.
- You can split the column data in an index-organized table into the main segment and an overflow segment.
- The ANALYZE command gathers statistics on a table's size, number of rows, column distribution, and free space.
- You can use DBMS_STATS.GATHER_TABLE_STATS instead of ANALYZE.
- Changing a table's storage parameters involves using the ALTER TABLE command.
- ALTER TABLE ... DEALLOCATE UNUSED releases unused space in a table's extents above the high watermark.
- ALTER TABLE ... MOVE ... can move a table to a different tablespace.
- DBMS_REDEFINITION is a package that can help you restructure a table while keeping the table online most of the time.
- Upgrading an object table or a relational table with object columns redefines their object types with the latest definition.
- Rename a table with the ALTER TABLE ... RENAME command.
- Modifying columns within a table also uses the ALTER TABLE command.
- To drop a column that is involved in another table's foreign key constraint, you must specify the CASCADE_CONSTRAINTS clause.
- The SET_UNUSED clause marks a column for dropping later.
- If you truncate a table, you cannot roll back the transaction.
- TRUNCATE ... DROP STORAGE releases space allocated to the table.
- TRUNCATE ... REUSE STORAGE keeps the space with the table.
- Use DROP TABLE ... CASCADE CONSTRAINTS, if the table is named in foreign keys in other tables.
- Data dictionary views store table, segment, extent, column, and LOB segment information.

REVIEW QUESTIONS

1. List all the internal LOB data types.
2. When a BFILE data type column's value is copied into another row, the external file is copied into a new location. True or False?
3. ________________ LOBs can be stored on read-only devices such as a DVD.
4. A row of data has a BLOB column. The column's value is 4200 bytes long. Where is the value stored?
   a. Inline with the row data
   b. Out of line in the LOB data segment
   c. Depends on the LOB storage setting
   d. In an external file
5. Examine this LOB storage clause.

   ```
   LOB (COMMERCIAL) STORE AS COMM_LOB
      (DISABLE STORAGE IN ROW
       CHUNK 32768
       STORAGE (INITIAL 5K NEXT 32K MAXEXTENTS UNLIMITED)
   ```

   Which statement is true about this clause?
   a. The clause is invalid, because you cannot name the LOB data segment.
   b. The clause is valid and will create a 5 K LOB data segment.
   c. The clause is invalid, because the initial extent must be a multiple of the chunk size.
   d. The clause is invalid, because the chunk size is not a multiple of the database block size.
6. Why is inserting a row in an index-organized table usually slower than inserting a row in a heap-organized table?
7. Index-organized tables can be partitioned using ________________ or ________________ partitioning.
8. Which of the following reasons are valid cause for analyzing a table? Choose all that apply.
   a. To improve the accuracy of the rule-based Optimizer
   b. To help you decide if storage settings fit the actual data
   c. To speed up inserts and updates
   d. To improve query performance

9. To release unused data blocks in a table, which of these commands can you use? Choose all that apply.
    a. TRUNCATE TABLE
    b. ALTER TABLE
    c. DROP TABLE
    d. MOVE TABLE
10. An object type has changed its definition. What should you do to a table containing a column with the changed object type?
11. Is the following command valid or invalid? Why?

```
ALTER TABLE CAR
DROP COLUMN CAR_OWNER
RENAME COLUMN CAR_MANUFACTURER TO CAR_MANUF;
```

12. If a table specifies MINEXTENTS 3 and INITIAL and NEXT are 5 K, what size is the table after this command is run?

```
TRUNCATE TABLE BIKERS KEEP 10K;
```

    a. 10 K
    b. 15 K
    c. 25 K
    d. 5 K

Exam Review Questions—Database Fundamentals I (#1Z0-031)

1. The CAR table has a foreign key referencing the MANUFACTURER table. Which statement will remove all rows in the CAR table? Choose two.
    a. DROP TABLE CAR CASCADE CONSTRAINTS;
    b. TRUNCATE TABLE CAR;
    c. TRUNCATE TABLE CAR CASCADE CONSTRAINTS;
    d. DELETE FROM CAR;
2. You issue the following command:

```
ALTER TABLE CAR MONITORING;
```

Which view stores the information gathered?
    a. V&TABLE_STATS
    b. V$OBJECT_STATS
    c. V$MONITOR
    d. DBA_TABLES

3. You have a column called SNOW_DEPTH that you want to change from CHAR(4) to NUMBER(10, 2). The column contains numerical characters. What should you do? (Choose the best response.)
   a. Take the table offline, alter the column, update the data, and put the table online.
   b. Use DBMS_REDEFINITION to restructure the table with minimal down time.
   c. Execute an ALTER TABLE MODIFY command, and let the data automatically be converted.
   d. Create a new column, use an UPDATE command to load the data, and then drop the old column.
4. You are creating a table to keep digital copies of old movie clips. The table must meet the following requirements:
   - The table must use an internal LOB for the movie clips.
   - All LOB values must be stored out of line.
   - The LOB should store old versions of the value until 25 percent of the space is full.

   Which of the following CREATE TABLE commands meets all these criteria?

   a.
   ```
   CREATE TABLE OLDMOVIE
   (MOVIE_ID NUMBER(10),
    MOVIE_TITLE VARCHAR2(30),
    MOVIECLIP BLOB)
   LOB (MOVIECLIP)  STORE  AS  MOVIELOB
       (TABLESPACE USER_LOCAL
        ENABLE STORAGE IN ROW
        PCTVERSION 25
        STORAGE (INITIAL 64M NEXT 32M));
   ```
   b.
   ```
   CREATE TABLE OLDMOVIE
   (MOVIE_ID NUMBER(10),
    MOVIE_TITLE VARCHAR2(30),
    MOVIECLIP BLOB)
    LOB (MOVIECLIP)  STORE  AS  MOVIELOB
        (TABLESPACE USER_LOCAL
         DISABLE STORAGE IN ROW
         PCTVERSION 25
         STORAGE (INITIAL 64M NEXT 32M));
   ```
   c.
   ```
   CREATE TABLE OLDMOVIE
   (MOVIE_ID NUMBER(10),
    MOVIE_TITLE VARCHAR2(30),
    MOVIECLIP BFILE)
   LOB (MOVIECLIP)  STORE  AS  MOVIELOB
   ```

```
      (TABLESPACE USER_LOCAL
       DISABLE STORAGE IN ROW
       PCTVERSION 25
       STORAGE (INITIAL 64M NEXT 32M));
```

d.

```
CREATE TABLE OLDMOVIE
(MOVIE_ID NUMBER(10),
 MOVIE_TITLE VARCHAR2(30),
 MOVIECLIP BLOB)
LOB (MOVIECLIP)  STORE  AS  MOVIELOB
    (TABLESPACE USER_LOCAL
     DISABLE STORAGE IN ROW
     RETENTION 25
     STORAGE (INITIAL 64M NEXT 32M));
```

5. Which of the following statements are true about an index-organized table? Choose two.
   a. An index-organized table can contain LOB columns.
   b. Add the index-organized parameter when creating an index-organized table.
   c. Overflow segments always contain the same columns.
   d. Index-organized tables can be partitioned using HASH or RANGE partitioning.
6. Which of these dictionary views should you query to list all the columns in a table?
   a. USER_TABLES only
   b. USER_TABLES and USER_TAB_COLUMNS
   c. USER_TAB_COLUMNS and USER_UNUSED_COL_TABS
   d. USER_TAB_COLUMNS only
7. You want to move a table to a new tablespace and resize the initial extent, next extent, and turn on monitoring. Choose the best sequence of steps from this list.
   1. Create a new table with new extent sizes in the new tablespace.
   2. Drop the old table.
   3. Alter the table for new extent sizes.
   4. Alter the new table to turn on monitoring.
   5. Copy all data to the new table.
   6. Move the old table to the new tablespace.
   7. Move the old table to the new tablespace, and change extent sizes at the same time.

   a. 1, 3, 4, 5, 2 (6 and 7 not needed)
   b. 6, 3, 4 (1, 2, 5, 7 not needed)
   c. 7, 4 (1, 2, 3, 5, 6 not needed)
   d. 7, 4, 5, 2 (1, 3, 6 not needed)

8

8. Which of the following statements change(s) the size of the table's segment? Choose all that apply.
   a. ALTER TABLE CUSTOMER ALLOCATE EXTENT SIZE 55M
   b. ALTER TABLE CUSTOMER STORAGE (NEXT 55M)
   c. ALTER TABLE CUSTOMER DEALLOCATE UNUSED
   d. ALTER TABLE CUSTOMER COMPRESS
9. Which of the following statements are true concerning the DBMS_REDEFINITION package? Choose three.
   a. Use the package instead of ALTER TABLE for all table modifications.
   b. The package minimizes the amount of time a table is unavailable to users.
   c. The package cannot be used for all types of tables.
   d. The package is intended for use by DBAs.
   e. The package can be used while the database is mounted but not open.
10. Examine the following SQL commands:

```
ALTER TABLE NEWSPAPER SET UNUSED COLUMN VOLUME_SOLD;
ALTER TABLE NEWSPAPER ADD (VOLUME_SOLD NUMBER(10,0));
ALTER TABLE NEWSPAPER DROP UNUSED COLUMNS;
```

    Which of the following statements is true?
    a. The second command will fail, and one column will be dropped.
    b. One column will be dropped, and one column will be added.
    c. The second command will fail, and no columns will be dropped.
    d. None of the above.

Hands-on Assignments

1. Write the SQL command to create a table with the following traits:
   - The table is named BEARS.
   - The table has nine columns: BEAR_TAG, BEAR_NAME, TAGGED_DATE, WEIGHT, HEIGHT, BIRTH_DATE, LAST_KNOWN_LOCATION (allow 2000 bytes), PHOTO (a BLOB), and MAP (a BFILE).
   - The table resides in the USERS tablespace and uses the default storage settings for the tablespace.
   - The table is index-organized on the BEAR_TAG column.
   - Overflow should begin after the BIRTH_DATE column.

Create the table in the CLASSMATE schema in the ORACLASS database. Save the SQL command in the **Solutions\Chapter08** directory in a file named **ho0801.sql** on your student disk.

2. Copy the CREATE BEARS command you created in Hands-on Assignment 1. Change the name of the table to BEARS2 and define a LOB storage clause that places the BLOB column in a separate segment and requires all BLOB column values be stored in the LOB data segment. The chunk size is 8 K, and there should be an initial and next extent size of 16 K (small because you are working on a test database), with unlimited extents. Allow 10 percent of the space to be used for old versions. Write the SQL command, execute it to create the table, and save the SQL command in the **Solutions\Chapter08** directory in a file named **ho0802.sql** on your student disk.
3. This assignment uses the CH08SURGERY table. If you did not create the table while working in this chapter, create it now by using SQL*Plus Worksheet to run the **create_table.sql** file in the **Data\Chapter08** directory on your student disk.

   Analyze the CH08SURGERY table.

   Write one or more queries on data dictionary view(s) to answer these questions:

   - How many rows are in the table?
   - What is the average length of a row in the table?
   - How many columns have no values in any rows?
   - What is the average length of the DOCTOR_NAME column?
   - How many distinct values are there in the PATIENT_FIRST_NAME and PATIENT_LAST_NAME columns?
   - What is the segment name and extent ID of the table's segment and extent?

   Save your SQL commands in a file named **ho0803.sql** in the **Solutions\Chapter08** directory on your student disk.
4. Copy the CREATE TABLE statement from the **create_table.sql** file in the **Data\Chapter08** directory on your student disk. Modify the statement to create a new table named CH08IOTSURGERY. Make it an index-organized table in the USERS tablespace. Insert the data from the CH08SURGERY table into the CH08IOTSURGERY table, except put a zero in every PROCEDURES column. Write a query to display the extent name and number of bytes for all segments associated with the CH08SURGERY table and the CH08IOTSURGERY table. (*Hint*: The primary key indexes have segments.) Which table uses more space? Save your SQL script and answer to the question in a file named **ho0804.sql** in the **Solutions\Chapter08** directory on your student disk.
5. Revise the BEARS table created in Hands-on Assignment 1. Make the following changes:

   - You have found that the bears' names are too long to fit into the column. The names sometimes are as much as 10 characters longer than the current maximum size.

- In addition, the hand-held computer that loads data into the mainframe's database stores the TAGGED_DATE with a precision of hours, minutes, seconds, and hundredths of a second. You want your data to keep this level of accuracy.
- The bear pictures are sometimes photos, but other times they are videos, so you want a different column name.
- Save your script in a file named **ho0805.sql** in the **Solutions\Chapter08** directory on your student disk.

6. For this assignment, you should log onto SQL*Plus Worksheet as CLASSMATE and run the script called **ho0806setup.sql** in the **Data\Chapter08** directory. This script creates two related tables: CH08REPAIR_TYPE and CH08HOUSE_REPAIR. The CH08HOUSE_REPAIR table has a foreign key referencing the CH08REPAIR_TYPE table. You have discovered that the table CH08REPAIR_TYPE is loaded with invalid data, and that data has been carried over into the CH08HOUSE_REPAIR table. You want to remove all the data from CH08REPAIR_TYPE while leaving the structure intact. You also want to remove the foreign key data in the CH08HOUSE_REPAIR table, so it can be reloaded later. Write two different SQL scripts to accomplish these tasks. One script should use TRUNCATE, and one should use DELETE. Save the script that uses TRUNCATE in a file named **ho0806a.sql** in the **Solutions\Chapter08** directory on your student disk. Save the script that uses DELETE in a file named **ho0806b.sql** in the **Solutions\Chapter08** directory on your student disk.
7. There are several tables owned by the CLASSMATE schema. Copy the file named **ho0807setup.sql** in the **Data\Chapter08** directory on your student disk. Use the file to build a file that tests all the tables owned by CLASSMATE. Test each one to see if it is eligible for the DBMS_REDEFINITION package. Which tables can use the package? Which ones cannot? Save the SQL commands in a file named **ho0807a.sql** in the **Solutions\Chapter08** directory on your student disk. Save the outcome of the SQL commands (showing the success or failure of the package) in a file named **ho0807b.txt** in the **Solutions\Chapter08** directory on your student disk.
8. For this assignment, you should log onto SQL*Plus Worksheet as CLASSMATE and run the script called **ho0808setup.sql** in the **Data\Chapter08** directory on your student disk. This creates a table called CH08FAMILYTIES and an object type called CH08CHILDINFO that is used in the table in a column called CHILD_DATA. After creating the table, the script modifies the CH08CHILDINFO object type so that it contains an additional attribute. Write the SQL command needed to modify the attributes in the CHILD_DATA column in the CH08FAMILYTIES table. Save your command in a file named **ho0808.sql** in the **Solutions\Chapter08** directory on your student disk.

9. Using the table that you created in hands-on assignment 8, you now find that more changes are needed to the table's columns. Write the SQL commands to make these changes, and run them using SQL*Plus Worksheet.
   - The PARENT_ONE column needs to be split into two columns: PARENT1_FIRST_NAME and PARENT1_LAST_NAME. (*Hint*: use an UPDATE command with a SUBSTR function to load the data from the old column into the new columns.)

   The PARENT_TWO column should be split the same way that PARENT_ONE was split.

   Remove the old columns after the data has been transferred to the new columns.

   Save your script in a file named **ho0809.sql** in the **Solutions\Chapter08** directory on your student disk.

CASE PROJECT

The Global Globe's database system is developing well. You have a few changes to make as you learn more about the way the office runs. Here are the revisions you want to incorporate to improve the database. Use these to create SQL scripts, then run the scripts, and save your work in a file named **case0801.sql** in the **Solutions\Chapter08** directory.

- There are many interactive applications that need the employee's ID and other data from the EMPLOYEE table. You have determined that the majority of queries read the table by primary key, so you decide this table should be index organized. Make the appropriate changes to the EMPLOYEE table. (*Hint*: You will have to re-create the table.)
- You discovered that many of the customers that place ads in the newspaper are businesses. You want the CUSTOMER table to have additional fields to track the business name, business Web site address, and business type. In addition, you want to rename the FULLNAME column to CONTACT_NAME. Make these changes to the CUSTOMER table.
- The newspaper has decided to add a new feature to its classified ads. For an additional fee, a customer's ad can be placed at the front of the section with a red star in front of it. You must add a new column to the CLASSIFIED_AD table to mark the ads that have priority placement. Also add a column to the CLASSIFIED_SECTION table, so the editors can set the priority placement pricing for each section. In addition, because it is so frequently used, you want the CLASSIFIED_AD table to stay cached in memory.

CHAPTER

9

INDEX MANAGEMENT

In this chapter, you will:

- Learn the types of indexes Oracle offers and when to use each type
- Understand how to create each type of index
- Determine which data dictionary views contain information on indexes
- Find out how to monitor index usage and when to drop an index
- Learn how to modify, rebuild, and coalesce an index

In the previous two chapters, you learned how to create many types of tables. You can add indexes to most tables. This chapter explores the various types of indexes and how they store data and help speed up performance. You create some indexes, after examining the parameters and what they mean. You learn to determine when each type of index is warranted and how to monitor the use of an index. Finally, you find out how to modify an existing index and how to drop an index.

INTRODUCTION TO INDEXES

An **index** is a database structure that is associated with a table or a cluster and speeds up data retrieval when the table or cluster is used in a query. An index has its own storage and does not have to be in the same tablespace as the associated table. You can create indexes on relational tables, object tables, nested tables, index-organized tables, and partitioned tables. When creating an index, you choose the type of index, the column or columns to be included in the index, and the location of the index. Oracle recommends that you separate the index and the table into separate tablespaces for better performance.

Oracle9*i* automatically adds new rows to the index as new rows are added to the table. In addition, it updates or deletes index entries as the table rows are updated or deleted. Figure 9-1 shows a logical view diagram of a table and its index. In the figure, the rows in the CUSTOMER table are stored in two blocks within one table extent Block #000 and Block #001. The rows are not ordered by the primary key, CUST_ID. The index is based on the primary key and is named CUST_PK. The index resides in its own segment and contains one block of index entries.

Entries, on the other hand, are stored in order by primary key. Each entry stores the primary key value of one row and the rowid for that row. For example, Marsha Kalani is in the sixth row in Block #000 and has a primary key of 222. In the index, the third row is hers and the last two numbers in the rowid (representing the block and row numbers) are "000" and "006".

An index does not have to be based on a primary key. Any combination and order of columns can be used to create an index for a table. An index can have from one to 32 columns (only 30 columns if it is a bitmap index). An index containing multiple columns is called a **composite index**. Columns can be used in more than one index, although no two indexes on a table can contain the same combination and order of columns. For example, one index might use the LAST_NAME, FIRST_NAME, and ZIPCODE columns, whereas a second index might use the same three columns in a different order, and a third index might include only the ZIPCODE column.

In addition to providing faster access to data, certain indexes also have a second purpose: enforcing a PRIMARY KEY constraint or UNIQUE key constraint. Chapter 10 discusses these and other constraints. The index for the PRIMARY KEY or UNIQUE key constraint is automatically created and has the same name as the constraint. Both constraints cause a unique index to be created. A **unique index** requires every row inserted into the table to have a unique value in the indexed column or columns.

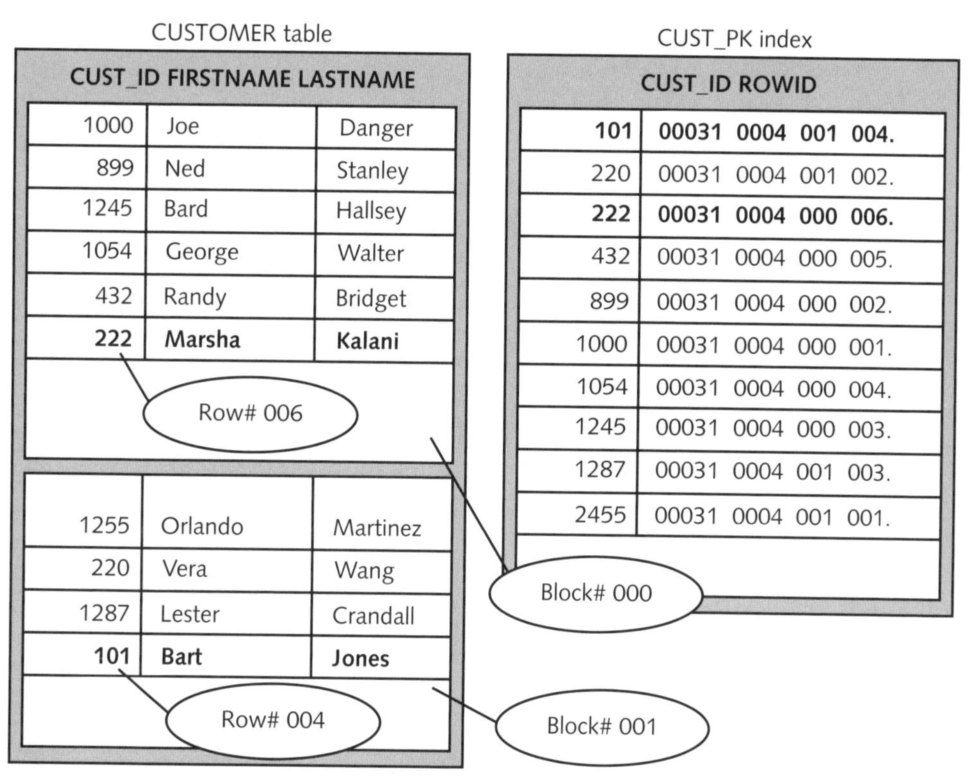

Figure 9-1 An example of a table and its index

An index does not store an entry for a row when all the indexed columns are null except if the index is a bitmap index. Bitmap indexes do store an entry for a row with all null values in the indexed columns. An index created to support a PRIMARY KEY constraint does not allow any null values in any of the indexed columns, because there is an automatic NOT NULL constraint added to the primary key columns. However, an index to support a UNIQUE key constraint does allow null values in the indexed columns and lists an entry in the index, unless, like most other indexes, all the indexed columns are null. Figure 9-2 shows the behavior of an index when rows have null values in the indexed columns. The figure shows a table named DOG and an index named DOGX that indexes the BREED, NAME, and OWNER columns. The second row in the table has null values in all three columns and is not included in the index. All the other rows are included in the index, even though they have null values in one or two of the indexed columns.

9

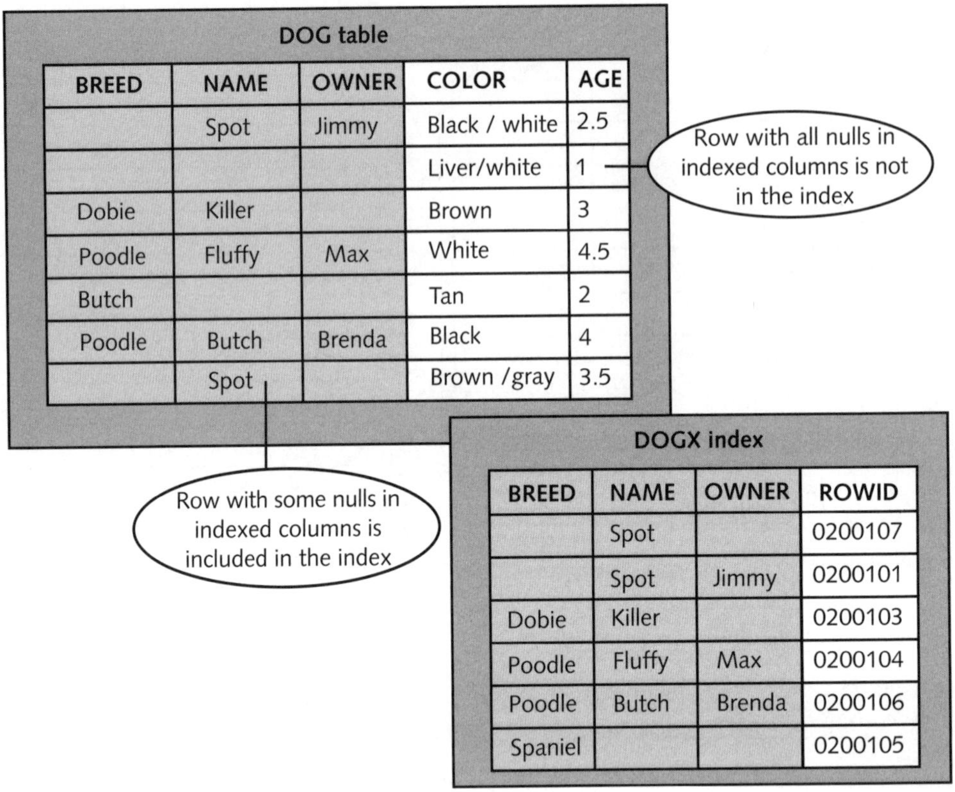

DOG table

BREED	NAME	OWNER	COLOR	AGE
	Spot	Jimmy	Black / white	2.5
			Liver/white	1
Dobie	Killer		Brown	3
Poodle	Fluffy	Max	White	4.5
Butch			Tan	2
Poodle	Butch	Brenda	Black	4
	Spot		Brown /gray	3.5

Row with all nulls in indexed columns is not in the index

Row with some nulls in indexed columns is included in the index

DOGX index

BREED	NAME	OWNER	ROWID
	Spot		0200107
	Spot	Jimmy	0200101
Dobie	Killer		0200103
Poodle	Fluffy	Max	0200104
Poodle	Butch	Brenda	0200106
Spaniel			0200105

Figure 9-2 Indexes can include null values in some cases

Armed with this general overview of how indexes work, you are ready to learn about the types of indexes supported by Oracle9*i* and when to use each type.

Types and Uses of Indexes

There are nine types of indexes you can create in Oracle9*i*:

- **B-tree index:** Used as the default type of index Oracle9*i* builds for tables
- **Bitmap index:** Used to store index entries in a compact format and is used primarily for very large tables with a small number of distinct index values
- **Local partitioned index:** Used on partitioned tables in which the index and the table are partitioned identically
- **Global partitioned index and global nonpartitioned index:** Used on partitioned tables in which the index is either partitioned differently from the table or not partitioned at all

- **Reverse key index:** Used to improve efficiency of Oracle Real Application Clusters
- **Function-based index:** Used to store precomputed expression values based on table columns
- **B-tree cluster index:** Used specifically for creating indexes on clusters
- **Hash cluster index:** Used specifically for hash clusters
- **Domain index:** Used for application- or cartridge-specific indexes

All these index types except the last three are described in this chapter. The last three (cluster and domain indexes) are beyond the scope of this book.

Before moving into the b-tree index section, take a look at the syntax for creating an index. This syntax is used for all types of indexes.

```
CREATE UNIQUE|BITMAP INDEX <schema>.<indexname>
ON <schema>.<tablename>
(<colname>|<expression> ASC|DESC,
 <colname>|<expression> ASC|DESC, ..)
TABLESPACE <tablespacename>
STORAGE (<storage_settings>)
LOGGING|NOLOGGING
ONLINE
COMPUTE STATISTICS
NOCOMPRESS|COMPRESS <nn>
NOSORT|REVERSE
NOPARALLEL|PARALLEL <nn>
PARTITION|GLOBAL PARTITION <partition_settings>
```


Many of these parameters have meanings that are identical to those described for the CREATE TABLE command. Refer to Chapter 7 for a detailed description of the following parameters:

- TABLESPACE <tablespacename>
- STORAGE (<storage_settings>)
- LOGGING|NOLOGGING
- ONLINE

Other parameters are probably new to you because they are not used when creating tables. They are:

- **UNIQUE|BITMAP:** Specify UNIQUE to create an index that requires every row in a table to contain a unique value in the indexed column(s). Specify BITMAP to create a nonunique bitmap index (an index stored as a bitmap within the index segment). Omit them both to create a nonunique b-tree index. As of Oracle9*i*, Release 2, a bitmap index cannot be a unique index.

- **(<colname>|<expression> ASC|DESC, ...):** Use this phrase to list all the columns that are to be indexed. ASC is the default order, indicating that the column values should be arranged in ascending order. DESC signifies descending order. Separate columns by commas. An alternative to a column name is to list an expression. When an expression, such as UPPER(LAST_NAME), is listed within the list of columns included in the index, the index is called a **function-based index**.
- **COMPUTE STATISTICS:** Use COMPUTE STATISTICS to gather statistics while building the new index.
- **NOCOMPRESS|COMPRESS <nn>:** Use **key compression** to cause repeating values of a key column to be eliminated, saving space. Specify how many of the columns in the key you want compressed (1, 2, and so on). Key compression is used primarily for unique keys that have multiple columns. You cannot compress a bitmap index.
- **NOSORT|REVERSE:** Specify NOSORT to tell Oracle9*i* to assume that the rows in the table are already in the same order as the index, and, therefore, no sorting needs to be done. The purpose of specifying NOSORT is to save time and temporary storage space when building the index. If the rows are not in order as expected, the CREATE INDEX command fails. The REVERSE parameter tells Oracle9*i* to store the bytes of the index values in reverse order. This is primarily used with Real Application Clusters to improve performance on a distributed database. The NOSORT and REVERSE parameters are not allowed on bitmap indexes or on index-organized tables. In addition, REVERSE and NOSORT cannot be specified together.
- **NOPARALLEL|PARALLEL <nn>:** Use PARALLEL to indicate that <nn> number of threads is to be used in parallel to create the index. The database instance must be running in parallel to use the PARALLEL parameter. Note that a thread is a connection between the database server and the user's process. Unless the database instance runs in parallel with one or more other database instances, a user's process uses one thread. NOPARALLEL is the default.
- **PARTITION|NOPARTITION:** Partitioning an index can occur on partitioned tables as well as nonpartitioned tables and uses many of the same settings as those needed for partitioning a table.

This chapter contains examples of the parameters just listed. The next sections explain each type of index, beginning with the b-tree index.

B-tree Index

The **b-tree index** is an index structure in which data is divided and subdivided based on the index key values to minimize the look-up time when searching for a key value. The b-tree appears to be so named, because it is a more complex variation on the binary tree. To understand the b-tree, it is easiest to begin by learning about the binary tree.

A **binary tree** uses a branching formation that always divides the list of data into two halves. Each half is then divided in half again, and so on. Figure 9-3 shows two binary tree structures.

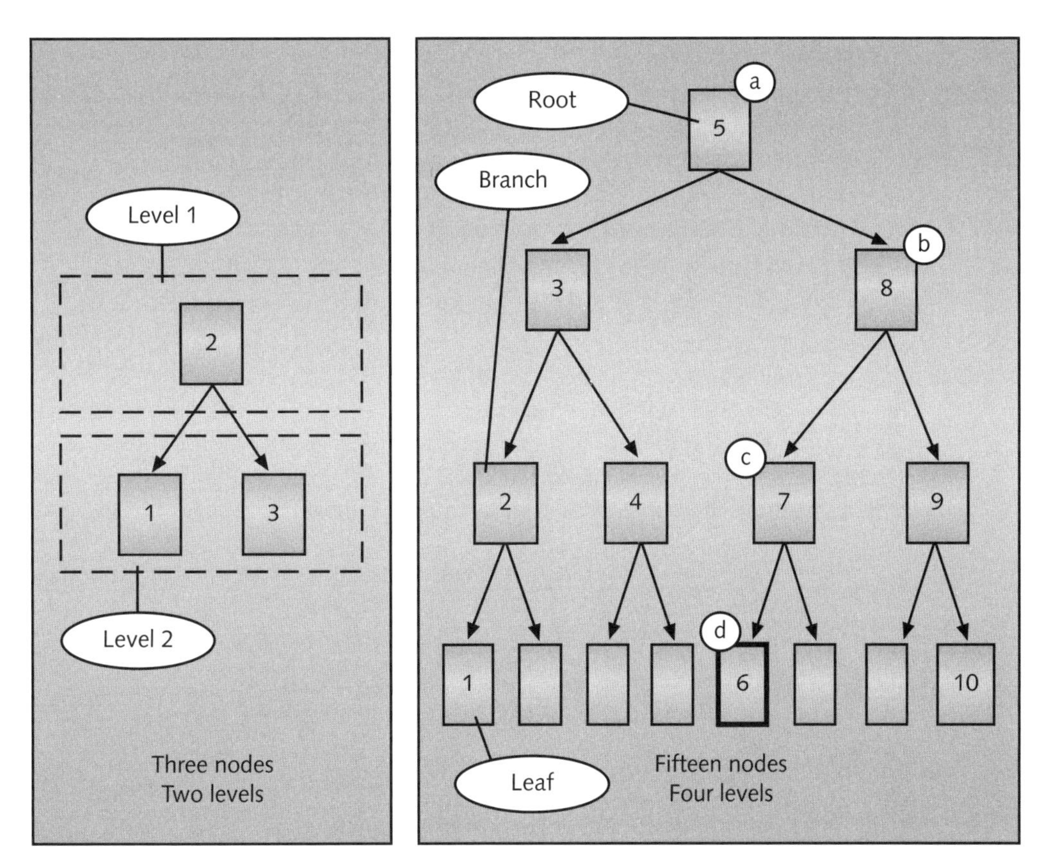

Figure 9-3 To find a value, follow the branches to the leaves

The tree on the left is the simplest example; one root has two leaves. The tree on the right has one root, six branches, and eight leaves. A **root** is the starting point for searching a binary tree or a b-tree. A **node** is any point on the tree. A **branch** is a node that has more nodes below it. A **leaf** is the bottom level and is a node with no nodes below it. A leaf on the left has a value less than its parent branch, and a leaf on the right has a value greater than its parent branch. In the figure, five of the leaves are blank and available for additional data. Three of the leaves contain data. To find a value on the binary tree, you always start at the root. An example of a binary search is illustrated with the letters "a" through "d" in Figure 9-3, which shows how to search for the number six. Here are the steps:

a. Start with the root (5). Six is greater than the node (5), so go down the right branch to the next node (8).

b. Six is less than the node (8), so go down the left branch to the next node (7).

c. Six is less than the node (7), so go again down the left branch to the next node (6).

d. Six is equal to the node (6), so you have located the node.

You have found the number in four steps. When every node has two branches, as in a binary tree, the number of levels increases quickly as you add more nodes. Each time you add a level, the search for a node takes more steps. For example, a binary tree with 1000 nodes requires ten levels. Imagine what the tree would look like with 100,000 nodes. If this type of structure were used for indexing a table with, for example, 100,000 rows, the number of levels to search through would be so large that the index would probably be slower than a full table scan.

To alleviate the problem of increasing the steps for a search as the number of nodes increases, the b-tree was invented. A b-tree can divide the data into two, three, or more parts each time it branches. And each branch node can contain two, three, or more values from which to branch. The leaf node can also contain two, three, or more values. Figure 9-4 shows an example of a b-tree index, in which each branch node contains two values, each node branches in three directions, and each leaf node contains up to four values.

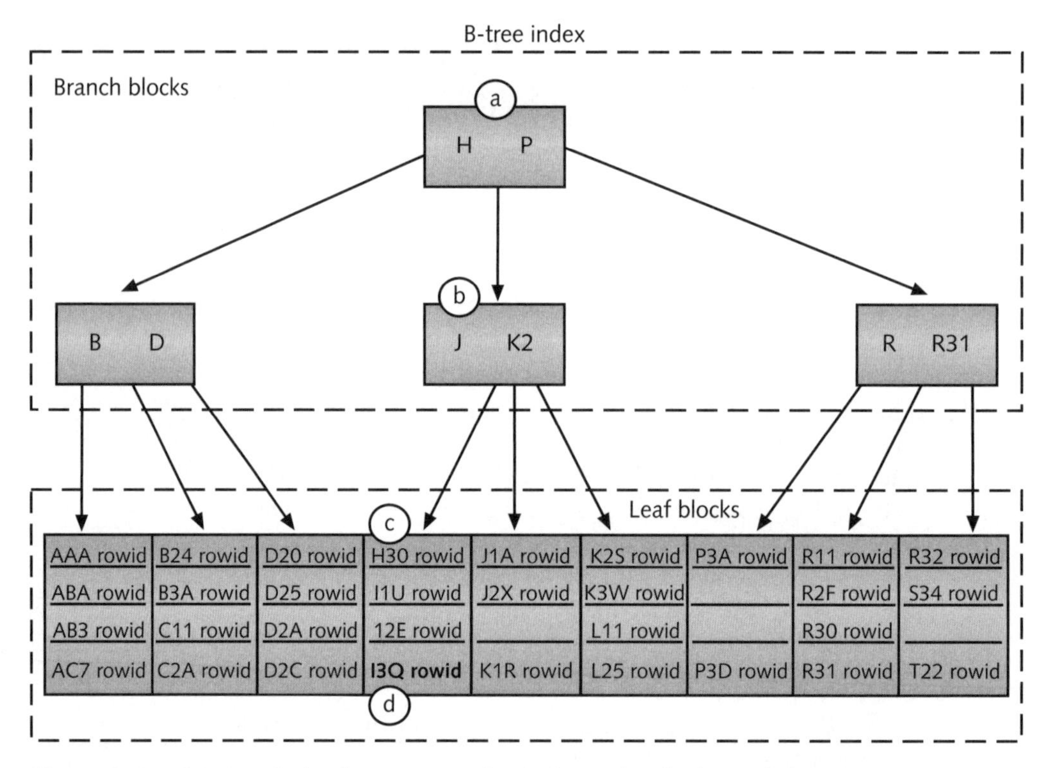

Figure 9-4 A b-tree index has more nodes in fewer levels than a binary tree

One big difference between the binary tree and the b-tree is that the final values are always found in a leaf node, never in a branch or root node. Figure 9-4 shows the structure of a b-tree index for a table. The indexed column contains a three-character code with numbers and letters.

The example in Figure 9-4 depicts the storage technique used for b-tree indexes. The approximate midpoint of the data is at the top of the tree. The key values or partial key values are listed in the branch nodes, and the complete key values with their associated rowids are listed in the leaf nodes. With a b-tree using three-branch nodes and four-value leaves, as in the example, your search algorithm is more complex than with a binary tree. An **algorithm** is a formula for solving a problem. For example, when guessing a number from 1 to 100, you might start at 50 and ask, "Is 50 higher or lower than the number?" If the answer is "higher," you divide 50 in half and ask, "Is 25 higher or lower than the number?" The pattern of dividing in half, asking a question, and repeating is an algorithm. In the case of an index, the algorithm tells you what to do at each branch node and at each leaf node. The goal of the b-tree is to keep the leaf nodes from getting too far down the tree levels. The b-tree accomplishes this by adding more values in each leaf node, adding more values in each branch node, and allowing each branch node to branch in more than two directions. All these changes help minimize the levels, which in turn shorten the number of steps needed to find any indexed value. For example, a four level b-tree with fifty values on each node can handle ten million values.

To see how a b-tree search works, follow along with these steps to find the rowid of the row with an index key value of "I3Q". Each step letter is noted in Figure 9-4.

a. Start with the root (H P). "I3Q" is between H and P, so you take the center branch. (If the value were smaller than H, you would take the left branch, and if it were larger than P, you would take the right branch.)

b. The next branch node is "J K2". "I3Q" is less than "J", so you take the left branch.

c. You are now in the leaf node that contains "I3Q". Search through the values until you find "I3Q".

d. The fourth key is equal to "I3Q", so you read the rowid and using this rowid, you can quickly retrieve the actual data row.

Notice that the values in the branch nodes are sometimes a partial key value. This is a way for the b-tree to save time. It stores only the minimum characters needed to make a decision on which branch to take. For example, look at the center branch node in Figure 9-4. The lower value is "J", whereas the upper value is "K2". There are leaf values that begin with "K" in both the center leaf and the right leaf below this branch node; therefore, a decision on which leaf to go to cannot be made using just the letter "K". By adding one more character, a valid decision point is made. "K2" is less than the first value in the right leaf ("K2S") and greater than or equal to the last value in the middle leaf ("K1R").

It is time to create a b-tree index. Follow along with these steps.

1. Start up the Enterprise Manager console. In Windows, click **Start** on the Taskbar, and then click **Programs/Oracle - OraHome92/Enterprise Manager Console**. In UNIX, type **oemapp console** on the command line. The Enterprise Manager Console login screen appears.
2. Select the **Launch standalone** radio button and click **OK**. The console appears.
3. In the main window of the Enterprise Manager console, navigate starting at the **Databases** icon to **ORACLASS/Schema/CLASSMATE/Indexes**. Your screen should look similar to Figure 9-5.

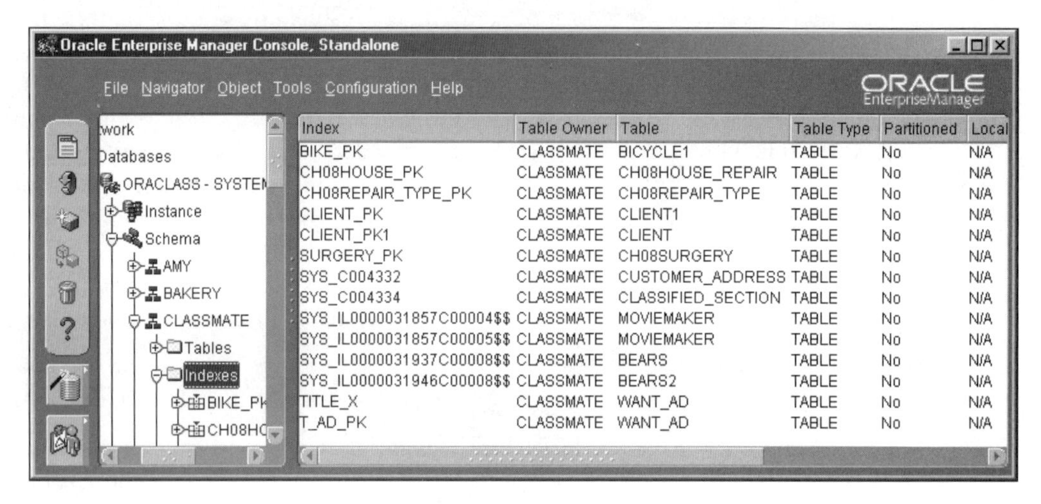

Figure 9-5 Use the console for a quick look at all the indexes owned by CLASSMATE

4. Notice that there are several indexes already in existence even though you have not specifically created any up to this point in the book. All these indexes were automatically created to support PRIMARY KEY constraints on tables created in earlier chapters. Some have readable names, such as SURGERY_PK, because the constraint was named when the table was created. Others have names that are system-generated such as "SYS_IL0000031857C00005$$" for constraints that were not named when the table was created. All these indexes are unique b-tree indexes.
5. To create a nonunique b-tree index, start up the SQL*Plus Worksheet by clicking **Tools/Database Applications/SQL*Plus Worksheet** from the top menu in the console. A background SQL*Plus process starts, and then the SQL*Plus Worksheet window appears. If the background process appears in front, simply minimize it (click the **minus sign** in the top-right corner) so you can see the worksheet.

6. Connect as the CLASSMATE user. Click **File/Change Database Connection** on the menu. A logon window appears. Type **CLASSMATE** in the Username box, **CLASSPASS** in the Password box, and **ORACLASS** in the Service box. Leave the connection type as "Normal". Click **OK** to continue.

7. For this chapter, start with a prepared SQL command found in the directory named **Data\Chapter09** on your student disk. Click **Worksheet/Run Local Script**. A browse window opens. Navigate to the **Data\Chapter09** directory, and select the **ch09setup.sql** file. Click **Open**. The file runs in the SQL*Plus Worksheet window and creates tables needed for the examples in this chapter.

8. Clear the top part of the worksheet by selecting **Edit/Clear All** from the menu. Then type the first two lines of the command to create a unique index on the LIBRARYBOOK table.

   ```
   CREATE UNIQUE INDEX CLASSMATE.DEWEY_IX
       ON CLASSMATE.CH09LIBRARYBOOK
   ```

   The UNIQUE parameter is optional. Add it for unique indexes and omit it for nonunique indexes. If you are logged on as the owner of the index you plan to create, you can also omit the schema. It is possible to create an index owned by one user on a table owned by another user, provided that the creating user has either the CREATE ANY INDEX privilege or the INDEX privilege on the target table, or the user has been granted specific privileges for creating an index on that table.

9. Type the list of columns to be included in the index. For this example, one column, called DEWEY_DECIMAL, is included in the index.

   ```
   (DEWEY_DECIMAL)
   ```

10. Add other parameters, such as LOGGING, TABLESPACE, STORAGE, and so on. For this example, include the INITRANS, PCTFREE, and LOGGING parameters by typing these lines:

    ```
    INITRANS 2 PCTFREE 20 LOGGING
    ```

    These are all default settings. Although INITRANS is seldom changed for tables or indexes, Oracle recommends that you set the index's INITRANS to double the value of the table's INITRANS. The default MAXTRANS for tables is 1, so the default MAXTRANS for indexes is 2. You cannot set PCTUSED on an index, because when you update a column that is indexed, the index entry usually moves and is re-created anyway, rather than staying in the same location, as a table row does.

11. Add the parameter to have Oracle9*i* gather statistics while building the index, and mark the end of the command by typing this line.

    ```
    COMPUTE STATISTICS;
    ```

12. Run the command by clicking the **Execute** icon. SQL*Plus Worksheet runs the command and responds, "Index created." Figure 9-6 shows the complete command in SQL*Plus Worksheet.

13. Remain logged on for the next practice.

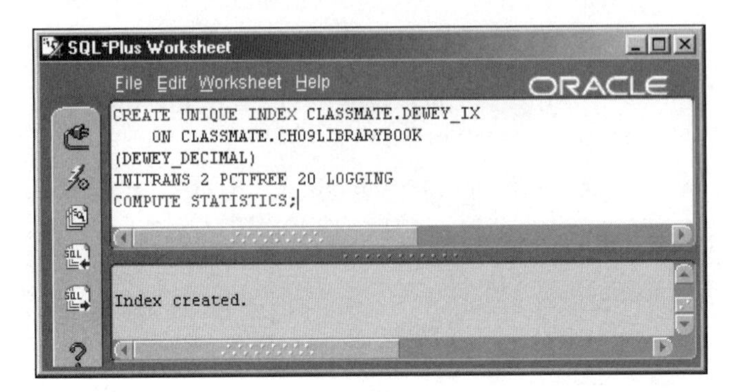

Figure 9-6 A unique index is created on the CH09LIBRARYBOOK table

Unless you specifically instruct Oracle9*i* to create a bitmap index, it always creates a b-tree index. Bitmap indexes have a completely different search algorithm and storage method than the b-tree index. You find out how they work in the next section.

Bitmap Index

A **bitmap index** is an index that does not use the b-tree method to store index key values and rowids. Instead, it uses a bitmap to store index key values and row locations. A bitmap index takes up a very small amount of storage space compared to a b-tree index; however, the bitmap index is appropriate only for certain special circumstances:

- The indexed columns should have low cardinality. **Low cardinality** means that the number of distinct values in an indexed column should be low compared to the number of rows in the table. For example, the column MARITAL_STATUS might have three or four values over the entire table. Another example is a table with 100,000 rows and a column with 1000 distinct values. This column still qualifies as having low cardinality, because the ratio of distinct values to rows is 1 percent. Oracle advises that if the ratio of distinct values of a column to the total number of rows is less than 1 percent, or if the column has values that repeat more than 100 times, the column is a candidate for a bitmap index.

- The table should be used primarily for queries rather than updates, such as those in a data warehouse system. A data warehouse is a database used to collect and store information from many areas of a business. Usually, a data warehouse is used to perform queries for statistical applications, such as business trends and customer profiles.

- The majority of queries should use AND, OR, NOT and "equal to" in the WHERE clause referencing the table's indexed columns. Queries that contain "greater than" or "less than" in the WHERE clause cannot use a bitmap index.
- The majority of queries include complex conditions in the WHERE clause, such as those used in statistical analysis of a data warehouse table.
- The table should not have concurrent updates, such as when a table is involved in an OLTP (online transaction processing) system.
- A bitmapped index cannot be a unique index.
- A bitmapped index cannot contain any DESC columns (columns in descending order in the index).

Bitmap indexes are only available in the Enterprise Edition of Oracle9*i*.

With all these qualifiers, you can see that the bitmap index has limited uses. However, for very large tables in a data warehouse system for which the types of queries described previously are used frequently, a bitmap index can save a significant amount of space and perform faster than an equivalent b-tree index.

A bitmap index does not store rowid values, nor does it store key values. Figure 9-7 illustrates a conceptual view of a bitmap index and the corresponding table rows. In this example, the column BLOOD_TYPE and GENDER are indexed using this command:

```
CREATE BITMAP INDEX PATIENT_BITMAP_X
ON PATIENT
(BLOOD_TYPE, GENDER);
```

The BITMAP parameter appears immediately after the word "CREATE" in the command. You cannot create a bitmap index that is also a UNIQUE index.

As a bitmap index is built, Oracle9*i* builds a chart of each distinct value in each indexed column. There are four blood types and two genders in the PATIENT table in our example, so the bitmap has six values listed across the top. Oracle9*i* walks through the table, from the first row to the last row building the bitmap index records. Each row is listed, in order of its appearance in the table, as 1a row in the bitmap chart. Under each value column, a zero or a one is marked. A zero means, "No, this value is not in this row." A one means, "Yes, this value is in this row." You can see now why a small number of distinct values are important to a bitmap's efficiency, because every row gets a tick mark (0 or 1) for every value.

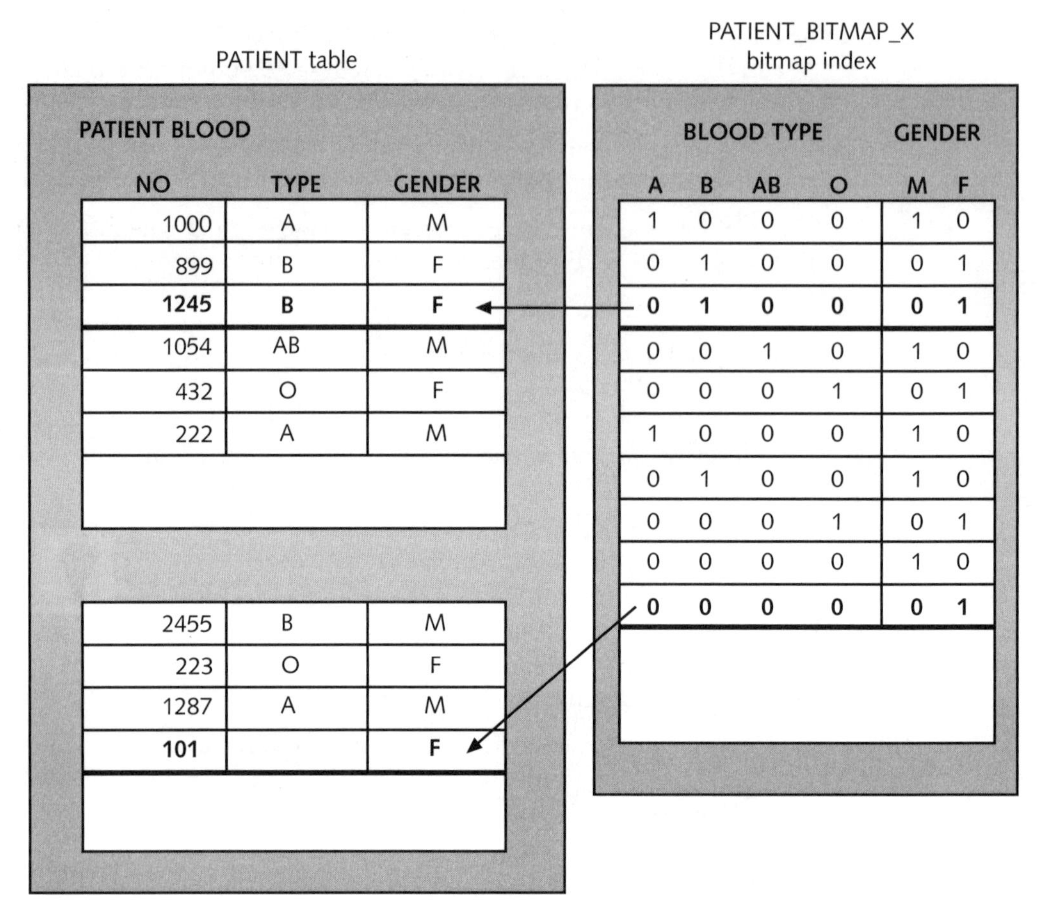

Figure 9-7 A bitmap index stores bits (ones and zeros) which take up very little space

Look at Figure 9-7, and find the third row in the bitmap index. This row has a blood type of "B" and a gender of "F." Look at the third row in the table to see that it corresponds to these values. Similarly, the last row in the table has a null value in the BLOOD_TYPE column and "F" in the GENDER column. Looking at the last row in the bitmap index, you can see that all the blood type values are zero and the gender "F" has a 1 in the column.

Unlike b-tree indexes, bitmap indexes include table rows that have null values in all the indexed columns.

Imagine that four new rows are added to the PATIENT table in Figure 9-7. Recall from Chapter 7 that inserted 1rows are sometimes placed in data blocks that already have rows. If that happened here, then a new row might be added in the first data block. The bitmap index would require an inserted row in the same order as the table. Because each record in a bitmap index is so compact, inserting one record between the others causes much rewriting of data within a single index block. You can see how this could cause a problem and slow down performance. Slow performance on inserts (and updates) is why Oracle recommends using bitmaps for tables that are used almost entirely for queries.

To illustrate why bitmap indexes are faster than b-tree indexes for certain types of queries and low-cardinality columns, imagine that you run the following query on the PATIENT table:

```
SELECT COUNT(*) FROM PATIENT
WHERE BLOOD_TYPE='A' and GENDER='M';
```

Looking again at Figure 9-7, you can quickly find the results by looking for the entries with "1" under the "A" column and the "M" column. If you were using a b-tree index, you would be walking through several branches, finding the leaf nodes, and then counting the number of rows in each leaf node.

On the other hand, the following query simply cannot use a bitmap index, because there is not enough information stored in the bitmap index to determine an answer:


```
SELECT COUNT(*) FROM PATIENT
WHERE BLOOD_TYPE >'A' and GENDER='M';
```

This is why bitmap indexes are not appropriate for tables that are primarily queried with "greater than" or "less than" in the WHERE clause.

Local Partitioned Index

A **local partitioned index** is an index on a partitioned table, in which the index is partitioned in the same way and on the same columns as the table. Each partition of the local partitioned index has data from the associated table partition, which makes queries using the index faster than an index that contains values from the entire table. A local partitioned index is automatically updated if you update the partitioning of the table. For example, if you use the ALTER TABLE command to add a new partition, the local index automatically receives a new partition. Oracle9*i* handles naming and locating the index partitions, unless you specifically name or locate the index partitions in the CREATE INDEX command.

Follow along with a few examples of partitioned indexes.

1. If you are not already there, start up the Enterprise Manager console and SQL*Plus Worksheet, logging on as CLASSMATE.
2. If you have already run the **ch09setup.sql** file earlier in the chapter, skip this step. Otherwise, click **Worksheet/Run Local Script**. A browse window opens. Navigate to the **Data\Chapter09** directory on your student disk, and select the **ch09setup.sql** file. Click **Open**. The file runs in the SQL*Plus Worksheet window and creates tables needed for the examples in this chapter.

3. Clear the top part of the worksheet by selecting **Edit/Clear All** from the menu. The CH09MORTGAGE_HISTORY table is a HASH-partitioned table with two partitions, one in the USERS tablespace and one in the USER_AUTO tablespace. You want to add a local partitioned index to speed up query-retrieval time. The table is partitioned on the DATE_CREATED column. To begin the command, type the following two lines.

```
CREATE INDEX LOCAL_X ON CH09MORTGAGE_HISTORY(DATE_CREATED)
LOCAL
```

   If you wanted Oracle9*i* to do the rest (name the partitions and locate them in the corresponding tablespaces), you could run the command at this point. Instead, you will add details as to where each index partition is to be stored.

4. To specify the tablespaces for each index partition while allowing Oracle9*i* to name the partitions, type the following line.

```
STORE IN (USERS, USER_AUTO);
```

5. Run the command by clicking the **Execute** icon. SQL*Plus Worksheet replies, "Index created." Other types of partitioned tables are handled the same way. Here is another example.

6. In this example, you are working with CH09MORTGAGE_CLIENT, which is RANGE partitioned by the LOAN_DATE. You are going to create a local partitioned index, in which you specify the partition names and the table-space locations. In addition, this index is a bitmap index.

```
CREATE BITMAP INDEX MCLIENT_LOCAL_X
ON CH09MORTGAGE_CLIENT(LOAN_DATE)
LOCAL
```

7. Add details on each of the two partitions, including the index partition name, tablespace, and storage settings by typing these lines:

```
(PARTITION OLDER_X TABLESPACE USERS
           STORAGE (INITIAL 50K NEXT 10K),
 PARTITION NEWER_X TABLESPACE USER_AUTO
           STORAGE (INITIAL 40K NEXT 15K));
```

8. Run the command by clicking the **Execute icon**. SQL*Plus Worksheet replies, "Index created."

9. View the index partitions in the data dictionary view by querying the USER_IND_PARTITIONS. Type the following query to see the work you have done so far:

```
SELECT INDEX_NAME, PARTITION_POSITION POS,
PARTITION_NAME, TABLESPACE_NAME
FROM USER_IND_PARTITIONS
ORDER BY 1,2;
```

10. Run the query by clicking the **Execute** icon. SQL*Plus Worksheet displays the results, as shown in Figure 9-8. The system-generated index partition names may be different from the ones in the figure.
11. Remain logged on for the next practice.

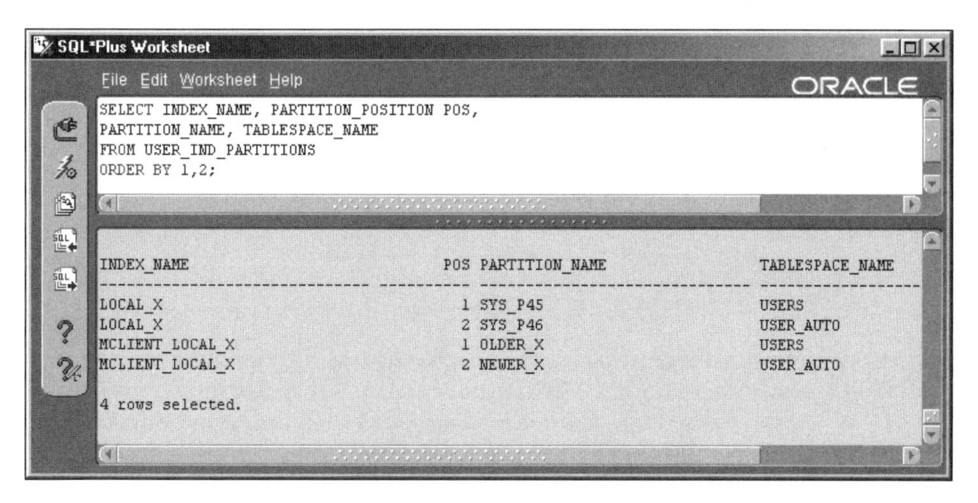

Figure 9-8 Oracle9*i* named the SYS_P45 and SYS_P46 index partitions

If you accidentally specify the wrong number of partitions for a local partitioned index, Oracle9*i* returns an error and does not create the index.

You cannot create a local partitioned index on a nonpartitioned table; however, you can create a different type of partitioned index—a global partitioned index—on a non-partitioned table.

Global Partitioned Index and Global Nonpartitioned Index

A **global partitioned index** is an index that is partitioned when either the table is not partitioned or the table is partitioned in a different way than the index. Use a global partitioned index when you have queries on a table that are not using the partitions efficiently. For example, you have a list partitioned EMPLOYEE table that is partitioned by DEPARTMENT. You install a new online application that retrieves data by EMPLOYEE_NAME instead. So, you create a global partitioned index that is RANGE partitioned by EMPLOYEE_NAME to support efficient data retrieval for the new application.

Global partitioned indexes can only be RANGE partitioned.

A **global nonpartitioned index** is a normal index created on a partitioned table. The index contains data from all the partitions in the table. Use a global nonpartitioned index for partitioned tables in which queries are posed on columns that do not lend themselves to partitioning. For example, the list partitioned EMPLOYEE table (partitioned by DEPARTMENT) also has a column (RETIREMENT_ACCT_NO) that is unique. You want to create a unique global nonpartitioned index on this column.

Global indexes cannot be bitmap indexes; however, they can be unique indexes.

Follow along with these examples to add a global partitioned index to a table.

1. If you are not already there, start up the Enterprise Manager console and SQL*Plus Worksheet, logging on as CLASSMATE.
2. If you have already run the **ch09setup.sql** file earlier in the chapter, skip this step. Otherwise, click **Worksheet/Run Local Script**. A browse window opens. Navigate to the **Data\Chapter09** directory on your student disk, and select the **ch09setup.sql** file. Click **Open**. The file runs in the SQL*Plus Worksheet window and creates tables needed for the examples in this chapter.
3. Clear the top part of the worksheet by selecting **Edit/Clear All** from the menu. The CH09MORTGAGE_HISTORY table is a HASH-partitioned table partitioned on the DATE_CREATED column. You want to add a global partitioned index to speed up queries that use the ACCT_NO column. To begin the command, type the following two lines.

```
CREATE INDEX G_ACCT_X ON CH09MORTGAGE_HISTORY (ACCT_NO)
GLOBAL PARTITION BY RANGE (ACCT_NO)
```

4. The next part of the command looks just like the code to create a RANGE-partitioned table, in which you specify the partition details. In this example, you are dividing the table data into three partitions. Type the following lines to define the three partitions.

```
(PARTITION LOWEST_ACCT VALUES LESS THAN (5000),
 PARTITION MIDDLE_ACCT VALUES LESS THAN (10000),
 PARTITION HIGHEST_ACCT VALUES LESS THAN (MAXVALUE));
```

5. The final partition of a global partitioned index must contain MAXVALUE as the value. This ensures that all rows in the table fit into one of the partitions. (You can use MAXVALUE as the value in the highest range of a RANGE-partitioned table as well.)
6. Run the command by clicking the **Execute** icon. SQL*Plus Worksheet replies, "Index created."
7. Remain logged on for the next practice.

When you create a nonpartitioned global index on a partitioned table, the command looks just like any other nonpartitioned index, except you add the GLOBAL parameter to the command. For example, a unique index on the LOAN_NO column of the CH09MORTGAGE_HISTORY partitioned table is created with this command:

```
CREATE UNIQUE INDEX G_LOAN_X
ON CH09MORTGAGE_HISTORY (LOAN_NO)
GLOBAL;
```

The global index can be useful in large tables. The next section describes an index type that can be used for large tables on a clustered database.

Reverse Key Index

A reverse key index creates an unusual form of b-tree index. Every byte in the indexed column is reversed. The leaf blocks still include the rowid (not reversed); however, the distribution of the data is changed.

A reverse key index is not the same as specifying DESC (for descending order) on an indexed column. The order is not reversed; the data is turned backwards and then put in order. For example, a nonreversed index on LAST_NAME would contain "Huxley," "Lopez," and "Madison" in alphabetical order. However, a reversed index on LAST_NAME would place them in this order: "Madison," "Huxley," and "Lopez," because the REVERSE function essentially spells each name backwards ("nosidaM", "yelxuH", and "zepoL"), and then alphabetizes the list.

What good is a list in reverse order? You cannot run a range search on the index, such as "LAST_NAME greater than 'Martin.'" In fact, the Optimizer only uses a reverse key index when the query contains a single record search or a full index scan.

The real usefulness of the reverse key index is found when you have a clustered database using Real Application Clusters, in which many users are accessing the same section of data from a large table, causing I/O contention. I/O bottlenecks caused by too many users attempting to read or update the same group of data blocks can slow performance, because some users must wait for others to complete their work before accessing the data. For example, imagine that you have a large table called CURRENT_EVENTS that is indexed by CREATE_DATE. You have a Real Application Clusters database with an online application, from which all your clients want the last three days' events. This has caused a bottleneck in the database. Fix the problem by reversing the index. The reversed index distributes the three days across many 1have dropped the old index and are creating a new reverse key index with this command (*do not actually execute this example*):

```
CREATE INDEX EVENT_REVERSE ON CURRENT_EVENTS(CREATE_DATE)
REVERSE;
```

You can change an index from normal to reverse key and vice versa using the ALTER INDEX command.

The final type of index, which is described in the next section, is a new feature of Oracle9*i*: function-based indexes.

Function-Based Index

Traditionally, both the cost-based and rule-based Optimizer took full advantage of indexes when evaluating a query. However, only the cost-based Optimizer is able to use the new function-based index type. A **function-based index** is an index with one or more columns indexed that are actually expressions or functions instead of columns.

Allowing functions and expressions within an index gives the Optimizer the ability to search an index for values that otherwise would require a full table scan. For example, the CH09LIBRARYBOOK table has a column called PUB_DATE that contains the date a book was first published. Imagine that the online library search application has a box where the user types in a cutoff date for publication. For example, the user wants to search for books published after 1950, so he types '1950' in the box. The application generates a query that looks like this:

```
SELECT TITLE, AUTHOR, DEWEY_DECIMAL,
TO_CHAR(PUB_DATE,'YYYY') PUB_YEAR
FROM CH09LIBRARYBOOK
WHERE TO_CHAR(PUB_DATE,'YYYY') >= '1950';
```

Before Oracle9*i*, neither the cost-based nor the rule-based Optimizer was able to use an index to resolve this query, *even when there was an index on the PUB_DATE column*. The index would be ignored, because there is a function on the indexed column in the WHERE clause. Today, a function-based index solves this problem.

The user who creates a function-based index must have the usual privileges granted to a user (the RESOURCE role), plus the QUERY REWRITE system privilege. Steps 1 through 4 in the following list grant this privilege to the CLASSMATE user.

Follow along to create a function-based index for this table.

1. If you are not already there, start up the Enterprise Manager console and SQL*Plus Worksheet.
2. Connect as SYSTEM by clicking **File/Change Database Connection** on the menu and filling in the Database Connection Information window with **SYSTEM** in the Username box, the current password for SYSTEM in the Password box, and **ORACLASS** in the Service box. Leave the Connect as box set to "Normal". Click **OK**.

3. Clear the top part of the worksheet by selecting **Edit/Clear All** from the menu. To create a function-based index, the user must have the QUERY REWRITE system privilege, because in some cases the Optimizer handles a function-based index by rewriting the query. Type the following command to grant the privilege to the CLASSMATE user.

   ```
   GRANT QUERY REWRITE TO CLASSMATE;
   ```

4. Run the command by clicking the **Execute** icon. SQL*Plus Worksheet replies, "Grant succeeded."

5. Connect to CLASSMATE again by either clicking the **File/Change Database Connection** or by typing this line and pressing the **Execute** icon.

   ```
   CONNECT CLASSMATE/CLASSPASS@ORACLASS
   ```

6. Now you are ready to create the function-based index. For this example, you will create an index on the PUB_DATE column so that the query you saw earlier can use the index. Type the command as follows, and click the **Execute** icon to run it:

   ```
   CREATE INDEX PUB_YEAR_IX
   ON CH09LIBRARYBOOK(TO_CHAR(PUB_DATE,'YYYY'));
   ```

   SQL*Plus Worksheet replies, "Index created."

7. Type the following query to look at the indexes owned by CLASSMATE. Click the **Execute** icon to run the query.

   ```
   SELECT INDEX_NAME, INDEX_TYPE FROM USER_INDEXES;
   ```

8. Figure 9-9 shows the results in the lower half of SQL*Plus Worksheet. Your results should look similar. Notice that the index type for the function-based index you just created is "FUNCTION-BASED NORMAL."

9. You can use expressions in function-based indexes as well. For example, a complex mathematical calculation might work well for a function-based index, because the value of the calculation is stored, already calculated, in the index. This makes searches faster by shifting the computation time from the query to the building of the index (or to the insertion of new rows after the index is built). Type and execute this SQL command:

   ```
   CREATE INDEX TOTAL_PAIDX
   ON CH09MORTGAGE_CLIENT
   ((MORTGAGE_AMOUNT*NUMBER_OF_YEARS*MORTGAGE_RATE)
         +MORTGAGE_AMOUNT)
   TABLESPACE USERS
   STORAGE (INITIAL 20K NEXT 10K) LOGGING COMPUTE STATISTICS;
   ```

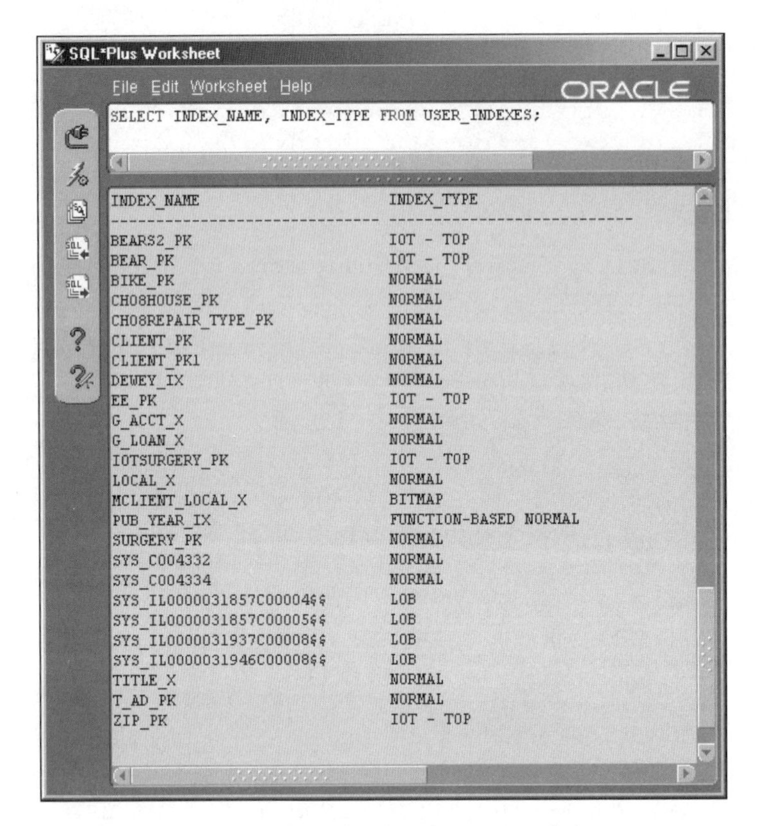

Figure 9-9 Function-based indexes are stored the same way as other indexes

10. Remain logged on for the next practice.

As with other indexes, you can add storage settings, logging, and other parameters to the CREATE INDEX command of a function-based index.

You have queried some of the data dictionary views that contain index information as you went through the practices in this chapter. The next section describes more data dictionary views that you can review when looking for information about indexes.

Data Dictionary Information on Indexes

Before looking at any specific data dictionary views for indexes, you should be sure to periodically run the DBMS_STATS command to analyze all your indexes. This supplies you with up-to-date statistics about the indexes and provides the cost-based Optimizer with the information it needs to best utilize the indexes. Follow these steps to analyze all indexes in the CLASSMATE schema.

1. If you are not already there, start up the Enterprise Manager console and SQL*Plus Worksheet, logging on as CLASSMATE.
2. Clear the top part of the worksheet by selecting **Edit/Clear All** from the menu. Type and execute the following command to analyze all the tables and indexes in the CLASSMATE schema. Oracle recommends using the DBMS_STATS package to analyze indexes (rather than the ANALYZE command), because it gathers some unique statistics for the cost-based Optimizer.

```
BEGIN
 DBMS_STATS.GATHER_SCHEMA_STATS
 (ownname=>'CLASSMATE',cascade=>TRUE);
END;
```

The process takes several seconds, and for schemas with more data or more tables, the process could take several minutes. When it is done, SQL*Plus Worksheet replies, "PL/SQL procedure successfully completed."

When executing a package or a procedure, you must either wrap the command in a PL/SQL block by using the BEGIN and END; clauses on either side of the command, or type EXECUTE and the command all on a single line. Using the BEGIN and END; clauses allows you to type the command on more than one line.

3. One data dictionary view, INDEX_STATS, is updated with statistics that currently are only gathered by the ANALYZE command. Type and execute the following command to gather these statistics for the function-based index you just created. The SQL*Plus Worksheet replies "Index analyzed."

```
ANALYZE INDEX TOTAL_PAIDX VALIDATE STRUCTURE;
```

4. Now type and execute this query to view INDEX_STATS data.

```
SELECT NAME, BR_ROWS, BR_BLKS, LF_ROWS, DEL_LF_ROWS
FROM INDEX_STATS;
```

   Figure 9-10 shows the results. This view shows the number of branches (BR_ROWS), leaves (LF_ROWS), and deleted leaves that still remain in the structure (DEL_LF_ROWS). If 30 percent or more of the leaves are deleted leaves, the index should be rebuilt. The next section describes how to rebuild an index. As Figure 9-10 shows, we don't need to rebuild the index at this time.

5. Remain logged on for the next practice.

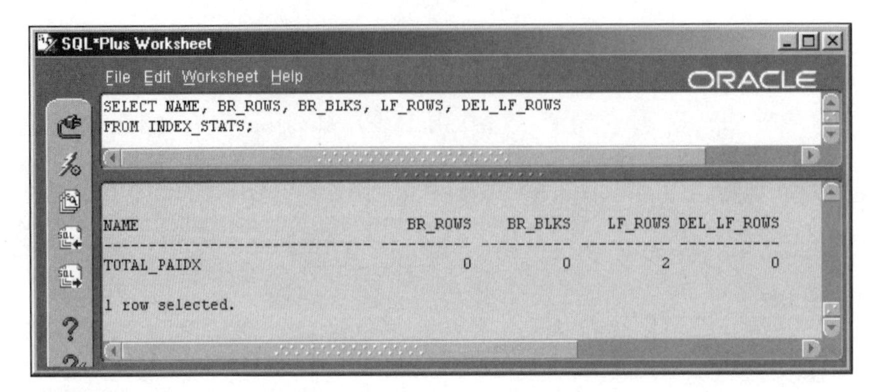

Figure 9-10 The leaves and branches of an index may need pruning

Table 9-1 has a list of data dictionary views that display information about indexes. The views that begin with "DBA_" have similar views that begin with "USER_" and "ALL_". For example, the first view listed is the DBA_IND_COLUMNS that lists indexed column names and statistics for all indexes in the database. A second view named USER_IND_COLUMNS lists indexed column names and statistics owned by the user. A third view named ALL_IND_COLUMNS lists indexed column names and statistics either owned by the user or accessible to the user.

Table 9-1 Data dictionary views on indexes

View name	Description
DBA_IND_COLUMNS	Indexed column names and statistics
DBA_IND_EXPRESSIONS	Expressions found in function-based indexes
DBA_IND_PARTITIONS	Index partition names and statistics
DBA_IND_SUBPARTITIONS	Index subpartition names and statistics
DBA_INDEXES	Index name, owner, type, and storage statistics
DBA_PART_INDEXES	High-level information on partitioned indexes, such as name, type of partitioning, storage settings, and statistics
INDEX_STATS	Statistics on leaf, branch, and deleted leaves gathered with ANALYZE ... VALIDATE STRUCTURE
V$OBJECT_USAGE	Details on index usage when you set MONITORING on for an index

Managing Indexes

Like tables, your indexes may require some maintenance now and then. The ALTER INDEX command has many parameters for helping you maintain your index. In addition, there is the DROP INDEX command for removing an index from the database. The next few sections describe how to monitor, modify, rebuild, and remove your indexes.

Monitoring Indexes and Dropping Indexes

You can monitor an index the same way that you monitor a table:

```
ALTER INDEX <schema>.<indexname> MONITORING USAGE;
```

After you turn monitoring on, you see what, if any, activity has been using the index in question. This is a good way to determine whether the index is actually being used for queries. Be sure that you monitor the index during a time frame when you know that typical work is being done. For example, monitor the index on Monday through Friday for one week instead of over the weekend, if your users generally access the database during weekdays.

It is important to review the usage of indexes, because indexes create overhead that slows down insert, update, and delete tasks on a table. Although indexes usually make up for the slight lag in changing data by speeding up queries on data, an index that is not used has no benefits and should be removed.

To query the monitoring information, type the following query (*do not actually execute this code because we have not initiated any monitoring*):

```
SELECT * FROM V$OBJECT_USAGE;
```

If the index is being used for queries, the USED column contains "YES." If not, the USED column contains "NO." Before dropping the index that is not being used, you should review the queries and the index columns to see if you can modify one or the other, so the index can be used. For example, perhaps the queries add a function to the column in the WHERE clause. In this case, you might consider a function-based index, or a revision of the WHERE clause, so the function is not used on the indexed column.

If you determine that an index should be removed, because it is no longer being used, use the DROP INDEX command:

```
DROP INDEX <schema>.<indexname>;
```

Often, the change you want to make to cause a query to use an index involves reordering the columns used in the index. You must create a new index to accomplish this. You might also want to drop the old index, if it is not being used.

You cannot change the column order of an existing index.

Reorganizing and Modifying Indexes

When reviewing an index's statistics, you may find that the index has much unused storage allocated or that the index has an incorrect setting for PCTINCREASE, NEXT, or other storage settings. In these cases, you can modify the index with the ALTER INDEX ... REBUILD command.

Even indexes created by Oracle9*i* to enforce a PRIMARY KEY constraint or UNIQUE key constraint can be modified to have different storage settings or to free up unused storage space.

Changing storage settings on an index is similar to changing the settings on a table. Like tables, you cannot modify the INITIAL and INITRANS settings of an existing index, but you can modify other storage settings, such as TABLESPACE, NEXT, MAXEXTENTS, and so on. The basic syntax for modifying storage and basic index settings is:

```
ALTER INDEX <schema>.<indexname>
REBUILD PARTITION|SUBPARTITION
REVERSE|NOREVERSE
TABLESPACE <tablespacename>
STORAGE (NEXT <nn> MAXEXTENTS <nn>)
PCTFREE <nn>
COMPUTE STATISTICS
COMPRESS|NOCOMPRESS
LOGGING|NOLOGGING
ONLINE;
```

All these settings have been discussed either earlier in this chapter (where the CREATE INDEX command is described) or in Chapter 7 (where the CREATE TABLE command is described). Refer to those areas for more information.

The ALTER INDEX ... REBUILD command also has these features:

- It automatically **rebuilds** the b-tree structure of a normal index, which adjusts levels and leaves as needed.
- If successful, an index rebuild automatically corrects an index that has been marked "UNUSABLE" because a change was made to the structure of the underlying table or partition.
- An index rebuild can be performed on only one partition at a time for partitioned indexes.
- It can change a reverse key index to a normal index or vice versa.

Complete the following steps to experiment with an index rebuild.

1. If you are not already there, start up the Enterprise Manager console and SQL*Plus Worksheet, logging on as CLASSMATE.

2. Clear the top part of the worksheet by selecting **Edit/Clear All** from the menu. You want to move the PUB_YEAR_IX index to a different tablespace. This is the function-based index created earlier in the chapter. Type and execute the following command in SQL*Plus Worksheet.

   ```
   ALTER INDEX PUB_YEAR_IX REBUILD
   TABLESPACE USER_AUTO ONLINE;
   ```

   SQL*Plus Worksheet replies, "Index altered." The ONLINE parameter tells Oracle9*i* to keep the table and index available for queries during the operation.

3. Exit SQL*Plus Worksheet by clicking the **X** in the top-right corner.

4. Exit the console by clicking the **X** in the top-right corner.

You can change other index setting with the ALTER INDEX command as well. The following syntax shows additional settings that you can modify on indexes.

```
ALTER INDEX <schema>.<indexname>
COALESCE
UPDATE BLOCK REFERENCES
UNUSABLE
ONLINE
RENAME <oldindexname> TO <newindexname>
RENAME PARTITION <oldname> TO <newname>
DEALLOCATE UNUSED KEEP <nn>
LOGGING|NOLOGGING
NOPARALLEL|PARALLEL <nn>
MONITORING USAGE|NOMONITORING USAGE;
```


The first three parameters are unique to indexes, whereas the last four parameters should be familiar to you by now (refer to an earlier section of this chapter or to Chapter 7). Here is a description of the three new parameters.

- **COALESCE:** You **coalesce** an index to consolidate fragmented storage space in the leaf blocks. The excess space is kept unless you add DEALLOCATE UNUSED to the command. Coalescing an index is faster and takes less temporary storage space than rebuilding an index. Coalescing an index is useful to quickly compact the index without completely rebuilding it. You should coalesce or rebuild an index when the leaves contain more than 30 percent unused space due to deleted entries. Query the INDEX_STATS view to check this condition. The COALESCE parameter must be used alone on either an entire index or on a partition.
- **UPDATE BLOCK REFERENCES:** The UPDATE BLOCK REFERENCES clause must be used alone. It is only for normal indexes on index-organized tables and is used to update the physical guesses of the indexed row's location stored in the index. Recall from Chapter 7 that the logical rowid used for index-organized tables is a best physical guess of the location of a row.

Refresh these guesses to aid performance of an index on the index-organized table after you have added or updated many rows of data on the table.

- **UNUSABLE:** The UNUSABLE clause tells Oracle9*i* to mark the index unusable, which causes the Optimizer to ignore the index when determining execution plans for queries. You might do this to experiment with performance time.

Here is an example using the COALESCE parameter:

```
ALTER INDEX DEWEY_IX COALESCE;
```

Partitioned indexes can be modified in most of the same ways that partitioned tables can be modified: You can add new partitions, split partitions, remove partitions, and rename partitions. There are a few restrictions on certain types of partitioned indexes. For example, you cannot add a new partition to a global partitioned index, because the highest partition must be set to MAXVALUE, so no additional partitions can fit above it. See the Administrator's Guide for details on maintenance of partitioned indexes and tables.

CHAPTER SUMMARY

- An index is a structure that holds data and has its own storage in the database.
- An index contains columns and rowids.
- A normal index stores rows in order by the index key values.
- An index can be based on a table's primary key or any combination and order of columns.
- An index can have up to 32 columns (30 for bitmap indexes).
- There are a number of types of indexes: b-tree, bitmap, local partitioned, global partitioned, reverse key, function-based, b-tree cluster, hash cluster, and domain indexes.
- Indexes are created using the CREATE INDEX command.
- B-tree is Oracle's default type of index.
- B-tree is a more complex variation of a binary tree structure.
- B-tree has a root, branches, and leaves.
- Searches of b-trees take longer as more levels are added to the tree.
- B-tree indexes are efficient even for very large amounts of data.
- Create a unique index with the CREATE UNIQUE INDEX command.
- Bitmap indexes take up far less storage space than b-tree indexes.

- Bitmap indexes are best for columns with low cardinality and complex WHERE clauses involving AND, OR, or NOT, but not involving "greater than" or "less than."
- Bitmap indexes cannot be unique and are best suited for data warehouse systems.
- Local partitioned indexes mirror the partitioning of the partitioned table.
- The USER_IND_PARTITIONS data dictionary view lists partitioned indexes.
- Global partitioned indexes can be created on partitioned or nonpartitioned tables.
- Global partitioned indexes can only be RANGE partitioned.
- Global nonpartitioned indexes can be used on partitioned tables.
- Reverse key indexes contain columns that have been stored in reverse byte order.
- Reverse key indexes are usually only for Real Application Cluster systems.
- Function-based indexes substitute a function or expression for a column.
- You must run ANALYZE or DBMS_STATS to give the Optimizer information about an index.
- The ALTER INDEX command allows you to make changes to the index.
- The ALTER INDEX REBUILD command lets you change storage parameters, change a reverse index to a normal index, and move the index to a different tablespace.
- The ALTER INDEX ... COALESCE command performs faster than REBUILD.
- REBUILD can consolidate free space, but not release it, whereas COALESCE consolidates and releases free space, if you specify the DEALLOCATE UNUSED parameter.
- You can only rebuild or coalesce one partition at a time of a partitioned index.

Review Questions

1. When you create an index, it is automatically stored in the same tablespace as the table. True or False?
2. What is the difference between a unique index that supports a PRIMARY KEY constraint and a unique index that supports a UNIQUE key constraint?
3. What is the difference between specifying REVERSE on an index and specifying DESC on an indexed column?
4. To create a b-tree index named BOOK_X, the CREATE command should start like which of the following:
   a. CREATE INDEX BOOK_X
   b. CREATE BTREE INDEX BOOK_X
   c. CREATE UNIQUE BTREE INDEX BOOK_X
   d. CREATE BITMAP INDEX BOOK_X

5. Look at the following command:

```
1 CREATE INDEX CUST_IX ON
2 (CUST_ID ASC, BIRTH_DATE, AGE-18)
3 TABLESPACE USERS
4 LOGGING ONLINE;
```

   Which line has an error?

   a. Line 1

   b. Line 2

   c. Line 3

   d. Line 4

6. You have a table that has been loaded into the database in order by the CREATE_DATE column. You plan to create an index on the CREATE_DATE column. What parameter helps save time in creating the index?

   a. REVERSE

   b. NOSORT

   c. BITMAP

   d. NOCOMPRESS

7. You have a b-tree index on a table with 1000 rows. How many levels will the index have?

   a. Four levels

   b. Ten levels

   c. Depends on the indexed columns

   d. A maximum of twenty levels

   e. None of the above

8. The CUSTOMER table has 100,000 rows and the FAVORITE_COLOR column has low cardinality. Queries usually look for an exact match on the color value. What kind of index would you use for the FAVORITE_COLOR column?

   a. Partitioned

   b. B-tree

   c. Bitmap

   d. Function-based

9. A ________________ partitioned index is partitioned just as a table is, but a ________________ partitioned index is partitioned differently from the table.

10. A global index must be partitioned. True or False?

11. A function-based index cannot be used by the ________________ Optimizer.

12. The ________________ data dictionary view can tell you the name of a column in an index.

Exam Review Questions — Database Fundamentals I (#1Z0-031)

1. Your index, called ACCT.PROCESS_X, has just been created. You run a query and look at the plan that the Optimizer will use to execute the query. You expected to see the index used in the plan. Which command will help you get the results you want?
   a.
   ```
   BEGIN
      DBMS_STATS.GATHER_SCHEMA_STATS
     (ownname=>'ACCT',cascade=>TRUE);
   END;
   ```
   b. `ANALYZE INDEX PROCESS_X VALIDATE STRUCTURE;`
   c. `GRANT REWRITE QUERY TO ACCT;`
   d. `ALTER SESSION SET OPTIMIZER_MODE='COST';`
2. You can gather usage information by setting the index for MONITORING USAGE. Which view shows the details of the monitoring activity?
   a. DBA_INDEXES
   b. DBA_INDEX_STATS
   c. V$OBJECT_USAGE
   d. V$INDEX_USAGE
3. When would it be wise to partition an index?
   a. When the underlying table is partitioned on the same columns as the index
   b. When the table contains low cardinality of the column to be indexed
   c. When a new application in a remote system needs the index
   d. When the table has a primary key on the columns to be indexed
4. Which query shows you all your indexes in the USERS tablespace?
   a.
   ```
   SELECT INDEX_NAME FROM DBA_INDEXES
   WHERE TABLESPACE_NAME = 'USERS';
   ```
   b.
   ```
   SELECT INDEX_NAME FROM INDEX_STATS
   WHERE TABLESPACE_NAME = 'USERS'
   ```
   c.
   ```
   SELECT INDEX_NAME FROM USER_INDEXES
   WHERE TABLESPACE_NAME = 'USERS';
   ```
   d.
   ```
   SELECT INDEX_NAME FROM USER_IND_COLUMNS
   WHERE TABLESPACE_NAME = 'USERS';
   ```

5. Which of these tasks can be performed with the ALTER INDEX ... REBUILD command? Choose three.
   a. Changing a reverse index to a normal index
   b. Moving an index to another tablespace
   c. Adding a new column to the index
   d. Moving an index partition to another tablespace
   e. Changing a bitmap index to a normal index
6. Examine the command below.
   ```
   ALTER INDEX BOOK_IX COALESCE ONLINE DEALLOCATE UNUSED;
   ```
   What effect does this command have?
   a. Rebuilds the branches and leaves of the b-tree
   b. Releases free space in leaves of the b-tree
   c. Keeps users from using the index in the future
   d. Combines two partitions into one
   e. None of the above
7. The LIBRARY_FINES table needs an index. The majority of queries search for a range of the sum of FINE_TOTAL plus INTEREST_TOTAL. The table has 100,000 rows and is in the USERS tablespace. What type of index should you create?
   a. Reverse key index
   b. Bitmap index
   c. Function-based index
   d. B-tree index
   e. None of the above
8. Examine the following command.
   ```
   CREATE INDEX CLASSMATE.READER_IX
   ON CLASSMATE.LIBRARYBOOK(READER_ID, BOOK_ID)
   STORAGE (INITIAL 10M NEXT 5K)
   TABLESPACE USERS NOLOGGING PARALLEL 3;
   ```
   What type of index is created?
   a. Partitioned index
   b. Bitmap index
   c. Function-based index
   d. Composite index
   e. None of the above

9. Which of these situations calls for a bitmap index?
   a. Several columns with low cardinality need indexing; the table has 50 million rows and infrequent updates.
   b. The columns have low cardinality; the table has 25 million rows and has frequent OLTP activity and few queries.
   c. The low cardinality columns need a unique index; the table has 50 million rows and is rarely updated.
   d. Several high cardinality columns need indexing; the table has 8 million rows and is updated frequently.
   e. None of the above
10. You need to create an index on the CUST_ID and ORDER_NO columns in the BILLING table. The table is in CUST_ID order. The index must be unique and have 5 M extents. Which of these commands will create the index?
    a.
    ```
    CREATE INDEX BILLX ON BILLING
      (CUST_ID, ORDER_NO)
       UNIQUE NOSORT STORAGE (INITIAL 5M NEXT 5M)
       TABLESPACE INDEXES ONLINE;
    ```
    b.
    ```
    CREATE INDEX BILLX ON BILLING
      (CUST_ID, ORDER_NO)
       UNIQUE STORAGE (INITIAL 5M NEXT 5M)
       TABLESPACE INDEXES;
    ```
    c.
    ```
    CREATE UNIQUE INDEX BILLX ON BILLING
      (CUST_ID, ORDER_NO)
       ONLINE STORAGE (INITIAL 5M NEXT 5M)
       TABLESPACE INDEXES;
    ```
    d.
    ```
    CREATE BITMAP INDEX BILLX ON BILLING
      (CUST_ID, ORDER_NO)
       UNIQUE STORAGE (INITIAL 5M NEXT 5M)
       TABLESPACE INDEXES;
    ```
11. Examine the following command.
    ```
    CREATE INDEX LOANX
    ON ACTMGR.MORTGAGELOAN (LOANID, CUST_CODE)
    GLOBAL PARTITION BY RANGE (LOANID, CUST_CODE)
    (PARTITION L1 VALUES LESS THAN ('L005',001)
     TABLESPACE USERS STORAGE (INITIAL 100M NEXT 10M),
     PARTITION L2 VALUES LESS THAN ('L499',025),
     PARTITION L3 VALUES LESS THAN (MAXVALUE, MAXVALUE));
    ```

Which of the following statements are true? Choose two.

a. All the partitions are located in the USERS tablespace.

b. A row with LOANID='M001' and CUST_CODE=004 will fall into the L3 partition.

c. The MORTGAGELOAN table is partitioned.

d. The L2 partition will be located in ACCTMGR's default tablespace.

12. Examine the following command.

```
ALTER INDEX LOANX
COALESCE PARTITION L1
RELEASE UNUSED KEEP 10K ONLINE;
```

Which of the following statements are true? Choose two.

a. The index's leaves and branches are rebalanced to improve performance.

b. Unused space is released except 10 K above the HWM.

c. Users can query the table while the command runs.

d. The index's leaves are adjusted to release unused space.

e. The partition's leaves are adjusted to release unused space.

HANDS-ON ASSIGNMENTS

Prior to working these assignments, run the setup script named **ch09_handson_setup.sql** found in the **Data\Chapter09** directory on your student disk. To run the script, log onto SQL*Plus Worksheet as CLASSMATE on the ORACLASS database and run the script.

1. Write and execute the command to create a b-tree index on the INTAKE_EDITOR_ID of the CLASSIFIED_AD table. The index should contain a 5 K first extent and 2 K subsequent extents. Create the index with a compressed key and in descending order. Save your SQL in a file named **ho0901.sql** in the **Solutions\Chapter09** directory on your student disk.

2. Move the index you created in Hands-on Assignment 1 to the USERS tablespace. In the same command, change the PCTINCREASE to 0 and decompress the key. Save your SQL in a file named **ho0902.sql** in the **Solutions\Chapter09** directory on your student disk.

3. The newspaper has added a special deal for the classified ads online. For a small fee, the words "BEST BUY" are added to the text of the ad, and these ads are listed on a special "Best Buys" page online. The online classified section has a query to bring up all the ads that qualify, but it runs very slowly. Here is the query.

   ```
   SELECT AD_TEXT, AD_NO
   FROM CLASSMATE.CLASSIFIED_AD
   WHERE INSTR(AD_TEXT,'BEST BUY') > 0
   ```

   Create an index that will be used for this query and, therefore, speed up its response time. Save your SQL in a file named **ho0903.sql** in the **Solutions\Chapter09** directory on your student disk.

4. There is an object table named CUSTOMER_ADDRESS in the CLASSMATE database. Add a nonunique index that sorts on the CUSTOMER_ID attribute followed by the STATE and CITY attributes of the table. Save your SQL in a file named **ho0904.sql** in the **Solutions\Chapter09** directory on your student disk.

5. Modify the index created in Hands-on Assignment 4. Consolidate the index leaf blocks and release unused free space. Save your SQL in a file named **ho0905.sql** in the **Solutions\Chapter09** directory on your student disk.

6. The CUSTOMER table has a column called DISCOUNT_PERCENT that gives a number for the usual discount each customer receives (if any). You have found that the column has only ten distinct values. The table has 20,000 rows and grows slowly with few updates to the discount for each customer. Create a bitmap index on the column. Save your SQL in a file named **ho0906.sql** in the **Solutions\Chapter09** directory on your student disk.

7. Modify the bitmap index created in Hands-on Assignment 6 by changing the PCTFREE to 0 and moving the index to the USER_AUTO tablespace. The DDL command should run without logging. Save your SQL in a file named **ho0907.sql** in the **Solutions\Chapter09** directory on your student disk.

8. Modify the index you created in Hands-on Assignment 3, changing it to a reverse key index. Then coalesce the index. Save your SQL in a file named **ho0908.sql** in the **Solutions\Chapter09** directory on your student disk.

9. Write a query on the data dictionary views that shows these components of all the indexes owned by CLASSMATE:
   - Index name
   - Index type
   - Table name

- Expression (if any)
- Column name

Save your SQL in a file named **ho0909.sql** in the **Solutions\Chapter09** directory on your student disk.

10. Write a query that generates a set of ANALYZE INDEX commands for every index owned by CLASSMATE. For a hint on getting started, look at this example of a query that generates a sentence made up of a combination of literals and column values.

```
SELECT 'I own the ' || UT.TABLE_NAME ||
' table in the '||UT.TABLESPACE_NAME ||
' tablespace.'
FROM USER_TABLES UT
ORDER BY TABLE_NAME;
```

You will find the previous query in the **Data\Chapter09** data directory on your student disk in a file named **ho0910setup.sql**. Run the previous query to see how it works. Then create your own query. Save your SQL in a file named **ho0910.sql** in the **Solutions\Chapter09** directory on your student disk.

Case Project

Global Globe has implemented an on-line classified ad Web Site. You have found that nearly every query on the CLASSIFIED_AD table for the online ad Web site contains this clause in the WHERE clause:

```
AND PLACED_DATE > SYSDATE-7
```

You are sure an index would help, especially if the index could quickly eliminate older ads that have expired and are just stored for historical purposes. Create an index that is partitioned on the PLACED_DATE column. Decide what type of partitioning is best for the column. Although your test database has only a few sample rows, assume that the table contains 50,000 rows and the table grows by 2500 rows per month. The oldest want ad was run on March 15, 1998. The PLACED_DATE column is 7 bytes long. Estimate the INITIAL size to handle the current number of rows and NEXT extent size to handle one month of data. Make three partitions. Store two of the partitions in the USERS tablespace and one in the USER_AUTO tablespace. Use the same storage parameters for all three. Run the command to create the index. Save your SQL command and also include a few sentences on why you chose a particular type of partitioning for the index. (*Hint:* You must use the TO_DATE function to specify a literal date in a partition range.) Save your SQL and comments in a file named **case0901.sql** in the **Solutions\Chapter09** directory on your student disk.

CHAPTER

10

DATA INTEGRITY CONSTRAINTS

In this chapter, you will:

- Learn the types and the uses of constraints
- Examine the syntax and options for creating constraints
- Work with practical examples of creating, modifying, and dropping constraints
- Query database dictionary views to monitor constraints

Defining the relationship of tables on paper or in a design tool is a good start, but the database system must incorporate the design and prevent invalid data from corrupting the design. **Business rules** are the statements defined during the design of a database system that inform both the database designer and the application programmer how data is used to support the business. For example, a business rule might state that every customer is assigned a unique customer identification number when he makes his first purchase. Applications were designed to handle all the business rules, including those at the database level (such as a primary key in a table) early in the development of relational database systems. Today, many of the business rules that apply to the database can be defined and enforced within the database itself, removing the burden from applications. One of the easiest and most direct methods of defining business rules is the use of integrity constraints.

Introduction to Constraints

Imagine that you are in charge of reservations for a theater. You and two assistants answer the telephone making reservations as quickly as possible. You sell seats until the theater is filled. What would happen if two of you sold the same seat? An argument, a fight, perhaps even a refund. You could avoid this situation by using a method in which each of you knows when the other sells a seat. When ticket sellers share the same office, selling duplicates is less of a problem. However, if the reservations can be made online across the country, you need a database. The database can handle concurrent updates to theater seat reservations. You must tell the database not to sell a seat more than once. The rule, "Only one customer per seat" is called a constraint. **Constraints** are rules or restrictions that guide database inserts, updates, and deletions. Constraints keep invalid or erroneous data out of your database.

You can enforce integrity constraints in several ways. This chapter deals with the first method, declaring integrity constraints.

- **Declare an integrity constraint:** **Integrity constraints** are constraints defined at the database level on either a column or a table.
- **Write a database trigger:** A **trigger** is a program that runs when a certain event, such as inserting a row in the THEATER table, occurs. Like constraints, triggers are at the database level and provide more flexibility in designing complex constraints than integrity constraints, while keeping the constraint within the database.
- **Include constraints in an application:** This type of constraint is outside the database. Although application constraints are flexible and can be fine-tuned for each application, they enforce only those changes that are made within the application. Modifications to the database, which are made outside the application, are not affected by the constraint.

The integrity constraint method of enforcing constraints has several advantages over the other methods. First, integrity constraints are simple to create and maintain. Second, integrity constraints are always enforced, no matter what application or tool is used to modify a table's data. Finally, integrity constraint checking performs faster than other methods of constraint enforcement, because the Optimizer has been programmed to handle them efficiently.

Types of Integrity Constraints

Oracle9*i* supports five types of integrity constraints. The constraint types handle all the basic requirements for relational database design and include the following:

- **PRIMARY KEY:** A **primary key** is the column or set of columns that define a unique identifying value for every row in the table. For example, the CLIENT table has a unique identifying column called CLIENT_ID. A primary key can contain a single column or multiple columns. When multiple columns make up the key, it is called a **compound key** or **composite key**. A primary key is often a system-generated number, guaranteeing uniqueness and stability. A PRIMARY KEY constraint has two rules underlying its enforcement: The column or columns must contain unique values, and the column or columns must not be null. The PRIMARY KEY constraint is important in relational databases. Most tables have a primary key to comply with the standards of normalization of tables. A table can have no more than one PRIMARY KEY constraint. **Normalizing** tables is part of the design process, and one of the normalizing rules is that every table should have a key that contains a unique value for every row.

- **UNIQUE:** This constraint is similar to the PRIMARY KEY constraint because both enforce unique values. The difference is that the UNIQUE key constraint allows null values in the column or columns that are named in the constraint. Columns with null values are ignored by the constraint. This constraint is useful for defining an alternate unique key for a table. A table can have more than one UNIQUE constraint.
- **FOREIGN KEY:** A **foreign key** is the name of a relationship between two tables, in which one is the parent and the other is the child. The FOREIGN KEY constraint is placed in the child table, referencing the primary key of the parent table. Figure 10-1 shows two tables with a FOREIGN KEY constraint and two PRIMARY KEY constraints. The FOREIGN KEY constraint requires that every row in the child table (the ORDER table in Figure 10-1) contain either a value that matches a primary key value in the parent table or a null value in the CUST_ID column. The PRIMARY KEY constraint in each table requires that every row in the table should contain a unique value in the primary key column. A FOREIGN KEY constraint can also reference a UNIQUE constraint rather than a PRIMARY KEY constraint in the parent table.
- **NOT NULL:** Placing the NOT NULL constraint on a column indicates that all rows in the table must contain a value in the column. A null value is an empty column within a table row and usually means that the column value for that row is unknown or not applicable. For example, the column OCCUPATION can contain a null value for a row containing information about an unemployed person or a child. Use the NOT NULL constraint for columns holding essential information, such as the LAST_NAME column in a CUSTOMER table.
- **CHECK:** The CHECK constraint enforces a predefined list of values for a column. For example, the APPROVED column can have a CHECK constraint that requires the column to contain <YES>, <NO>, or a null value. Unless you also specify the NOT NULL constraint, a column with a CHECK constraint may contain a null value.

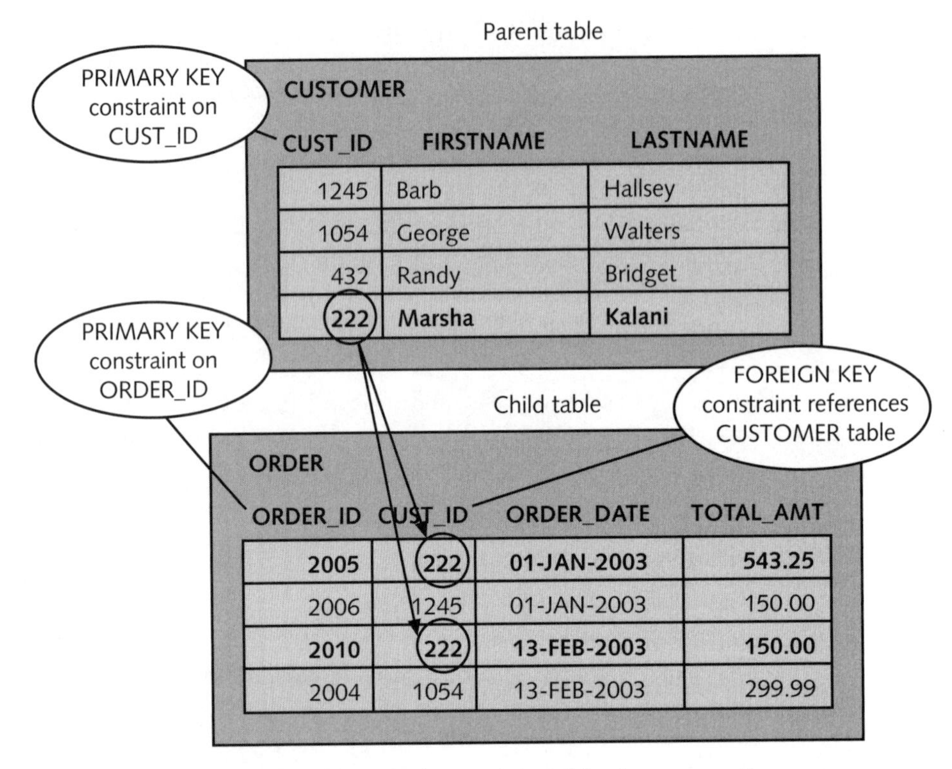

Figure 10-1 Primary and foreign keys relate tables to one another

A unique index that is automatically created enforces the PRIMARY KEY and UNIQUE key constraints. The index has the same name as the constraint, although you can specify the index name and location if you want.

The next section describes how to create integrity constraints.

How To Create and Maintain Integrity Constraints

There are two ways to create integrity constraints on columns and tables.

1. Use the CREATE TABLE command to create the constraint while creating the table. This is the most common method for PRIMARY KEY and NOT NULL constraints, because these are essential parts of the table design and are usually known when the table is created.

2. Use the ALTER TABLE command later to create the constraint. Use this to add constraints that were either missed or added to the table design after table creation. Typically, all rows in the table must conform to the new constraint, although you can modify this default action if needed.

The next section describes how to add constraints to the CREATE TABLE command.

Creating Constraints Using the CREATE TABLE Command

Figure 10-2 shows the syntax of the CREATE TABLE command with notation pointing out the placement of constraints. Constraints that apply to a single column, such as the NOT NULL constraint, are called **column constraints** and appear inline with the column. An **inline constraint** appears immediately next to the column to which it applies. Constraints that apply to multiple columns, such as a constraint for a compound foreign key, are called **table constraints** and are placed immediately after the list of columns in the CREATE TABLE command. An **out of line constraint** appears after the full list of columns in the CREATE TABLE command and usually applies to multiple columns. You can place single column constraints out of line, if you choose. The NOT NULL constraint must always be defined inline.

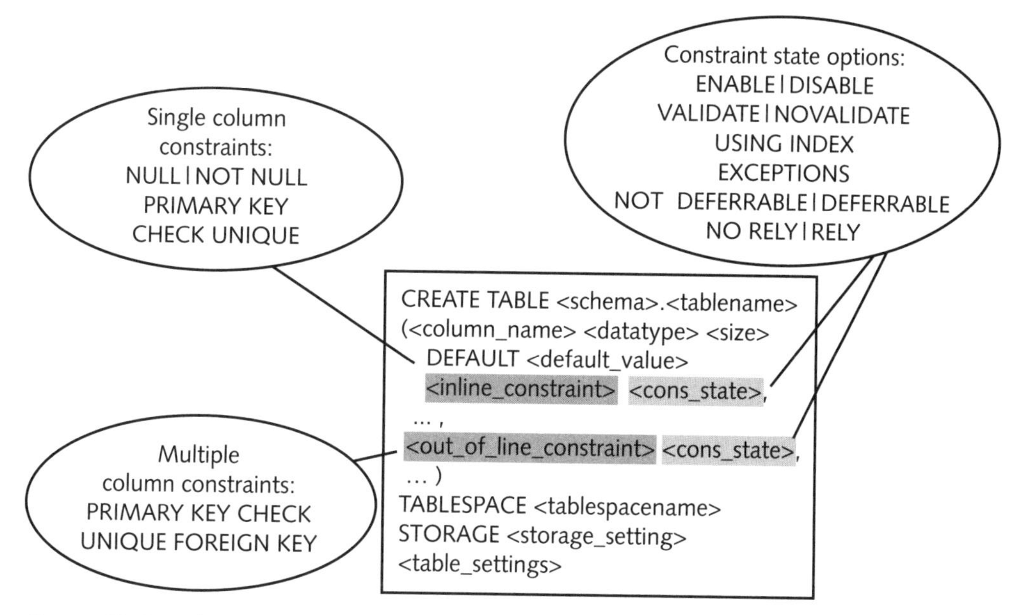

Figure 10-2 There are two valid locations for constraints in the CREATE TABLE command

Some constraints can be defined either inline or out of line. Table 10-1 shows the ways each constraint can be defined.

Table 10-1 Constraints: inline, out of line, or both?

Constraint	Inline in CREATE TABLE	Out of line in CREATE TABLE	Inline in ALTER TABLE	Out of line in ALTER TABLE
PRIMARY KEY	Yes, if single column	Yes	No	Yes
FOREIGN KEY	Yes, if single column and not named	Yes	No	Yes
NOT NULL	Yes	No	Yes	No
CHECK	Yes, if single column	Yes	No	Yes
UNIQUE KEY	Yes, if single column	Yes	No	Yes

Figure 10-3 shows an example of a table with several constraints in the CREATE TABLE command.

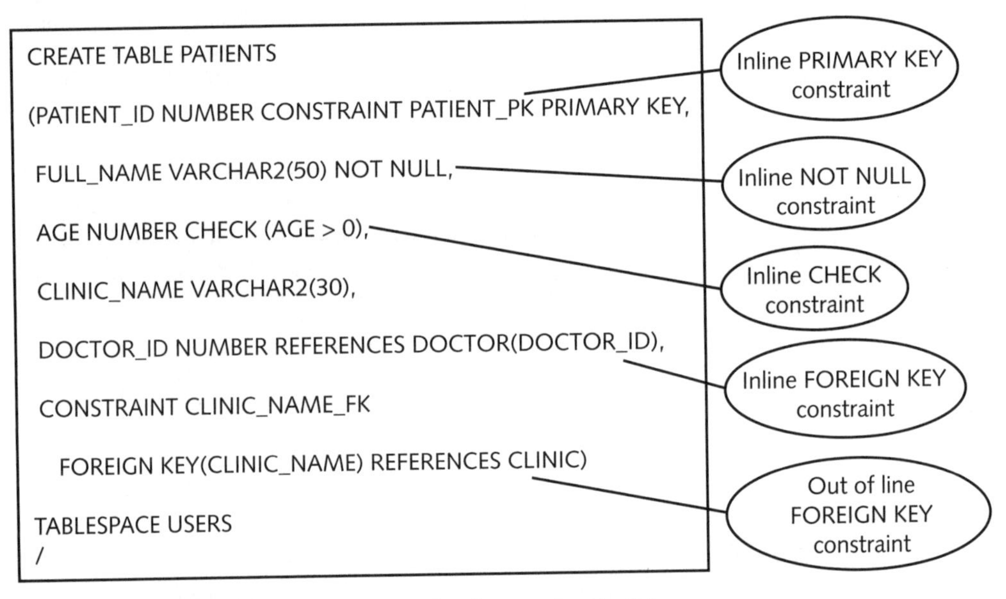

Figure 10-3 Constraints created both inline and out of line

In the code example, you can see these constraints:

- **PRIMARY KEY:** The constraint is named PATIENT_PK; the PATIENT_ID column is the primary key.
- **NOT NULL:** The constraint is inline on the FULL_NAME column and has no name.
- **CHECK:** The constraint is inline on the AGE column and has no name.

- **FOREIGN KEY:** The constraint is inline on the DOCTOR_ID column and has no name.
- **FOREIGN KEY:** The constraint is named CLINIC_NAME_FK and is out of line.

Frequently, NOT NULL and CHECK constraints are not named, which is an option for all constraints. When no name is specified, omit the CONSTRAINT keyword and the constraint name. A name can be useful, however, when you are rapidly reviewing constraints later to confirm that you have not missed anything in your schema. In general, you should name all PRIMARY KEY and FOREIGN KEY constraints.

Although CHECK constraints can check multiple conditions, it is usually better to use multiple CHECK constraints instead of one complicated constraint. Name the CHECK constraints with names that describe their function, such as CHK_GRTR_THN_ZERO. The error message includes the check constraint name when the constraint is violated, and the user may have a better idea of how to correct the invalid data when she sees a meaningful constraint name.

A FOREIGN KEY constraint that is defined *inline* is not allowed to have a name. You must omit the CONSTRAINT keyword and the FOREIGN KEY keywords as well as the constraint name. Figure 10-3 shows both an inline FOREIGN KEY constraint and an out of line FOREIGN KEY constraint. A FOREIGN KEY constraint that is defined *out of line* can have a name, but is not required to have a name.

You can create or modify constraints in several different states. Figure 10-2 shows a partial list of **constraint states**. A constraint state is the attribute that tells Oracle9*i* how to use the constraint when data is added to the table. If the state changes from DISABLE to ENABLE, the constraint state also tells Oracle9*i* how to use the constraint on existing data. Here is a list of the constraint states and how they are used:

- **ENABLE | DISABLE:** Constraints are enabled by default when you create them. You can create a disabled constraint (or change an existing constraint) by adding DISABLE to the constraint. Disabling an existing constraint does nothing to the data, although if the constraint is a PRIMARY KEY or UNIQUE key constraint, the associated unique index is dropped. As an example, you might disable a FOREIGN KEY constraint to allow data to be loaded into a child table prior to loading data into a parent table. After the load is complete, you enable the FOREIGN KEY constraint again. You can change a disabled constraint to enabled by using the ENABLE keyword. If you enable a constraint on a table with existing rows, all the rows are checked for compliance, unless you specify ENABLE NOVALIDATE.
- **VALIDATE | NOVALIDATE:** Constraints are validated (that is, all rows in the table must comply with the constraint) by default when they are created. If you are creating a constraint on a table that already has rows and use the

NOVALIDATE keyword, the existing rows are not checked against the constraint, but new rows must comply (so long as the constraint is enabled). A typical use for the NOVALIDATE keyword is when you enable a constraint that you disabled temporarily to add several thousand rows. You know the rows comply with the disabled constraint, so you use ENABLE NOVALIDATE to switch the constraint to enabled because this saves time: Oracle9*i* does not run through all the rows looking for rows that fail to comply with the constraint.

- **INITIALLY IMMEDIATE | INITIALLY DEFERRED:** Prior to Oracle8*i*, all constraints were in the immediate state. A constraint in **immediate state** is validated as soon as a statement is executed. For example, you have a DOG_OWNER table and a DOG table. The DOG table has a FOREIGN KEY constraint referencing the DOG_OWNER table. You run an INSERT statement that has an invalid foreign key value. The INSERT statement fails immediately with an error caused by the failed constraint validation. A constraint in **deferred state** is validated only when you execute a COMMIT command. Continuing with the example, imagine that the FOREIGN KEY constraint on the DOG table is deferred. Now, you can insert a row in the DOG table first (without an error) and insert the dog's owner row into the DOG_OWNER table second. As long as you commit the rows after both are inserted, the COMMIT succeeds. The INITIALLY IMMEDIATE state is the default state for constraints. Use INITIALLY DEFERRED to change the current state of the constraint to deferred state.
- **DEFERRABLE | NOT DEFERRABLE:** The default for all constraints is NOT DEFERRABLE. You must specify DEFERRABLE when you create the constraint to allow it to be deferred. *Furthermore, Oracle9i does not allow you to use ALTER TABLE to change this constraint state.* Only constraints that are created as deferrable can be later modified to deferred state. You must drop and re-create the constraint to change this setting.
- **RELY | NORELY:** This state tells the Optimizer to either use (RELY) or ignore (NORELY) disabled constraints when rewriting queries to better optimize the performance of complex queries. The RELY state is useful for data warehouses using materialized views and query rewrites. Otherwise, it is rarely used. A materialized view is a view that has been parsed and executed and its results stored in a table so that subsequent queries that use the materialized view do not have to repeat the parsing and executing of the view's SQL. Materialized views are efficient when the data in the underlying tables seldom changes. A **query rewrite** is the ability of Oracle9*i* to modify the query you execute, changing it into an equivalent but more efficient statement before actually running the query. The query rewrite is especially useful when implementing materialized views, because the Optimizer can rewrite your query so that a materialized view can be used to retrieve the data, making the query more efficient. Deferred constraints are normally ignored during the Optimizer's rewrite activity. Therefore, the default state is NORELY,

which tells the Optimizer to ignore constraints that are disabled when rewriting queries or evaluating materialized views. RELY tells the Optimizer to use disabled constraints.

- **USING INDEX <index> | <storage>:** You can specify this state only for the PRIMARY KEY and UNIQUE constraints. This tells Oracle9*i* that you have made one of two choices: 1) You have an existing index and, therefore, no index needs to be created when creating or enabling the constraint, or 2) you are specifying the storage setting for the index that Oracle9*i* creates for the constraint. For example, you can specify the tablespace in which to create the associated index when you add this state to a CREATE TABLE command. As another example, when you have a disabled PRIMARY KEY or UNIQUE constraint, you can create a unique index yourself. Then, when you enable the constraint, you specify that Oracle9*i* uses the existing index to enforce the constraint. You try out the USING INDEX state when you create a UNIQUE key constraint later in the chapter.
- **EXCEPTIONS | EXCEPTIONS INTO <tablename>:** This state is used only when enabling an existing constraint or adding a constraint to an existing table. The EXCEPTIONS keyword instructs Oracle9*i* to list the rowids of rows that fail validation in a table. If you don't specify a table, Oracle9*i* inserts the exception information into a table, named EXCEPTIONS, that you own or to which you have access. Create this table by running the script: **utlexcpt.sql** (for supporting physical rowids) or **utlexpt1.sql** (for supporting universal rowids, such as those in index-organized tables). The script is found in **ORACLE_HOME\rdbms\admin**. If you want to use a table with a different name, copy the script and use it to create your own table. Then use the EXCEPTIONS INTO <tablename> format, specifying the table you created.

10

Creating or Changing Constraints Using the ALTER TABLE Command

The syntax for constraints within the ALTER TABLE command varies according to the type of constraint. The NOT NULL constraint has one variation, adding constraints to existing columns has a second variation, and changing existing constraints has a third variation. The next sections describe each variation.

Adding or Removing NOT NULL on an Existing Column

The NOT NULL constraint is always associated with a single column. Use the following syntax to either add or remove the NOT NULL constraint on a column.

```
ALTER TABLE <tablename>
MODIFY(<columnname> NULL|NOT NULL);
```

Use NULL to remove the NOT NULL constraint. Use NOT NULL to add the constraint.

To add a NOT NULL constraint successfully, all rows in the table must contain values for the column.

Adding a New Constraint to an Existing Table

Regardless of whether the constraint is for a single column or multiple columns, use the out of line format you saw in the CREATE TABLE command to add a new constraint to an existing table. Here is the syntax:

```
ALTER TABLE <tablename>
ADD CONSTRAINT <constraintname>
  PRIMARY KEY (<colname>, ...) |
  FOREIGN KEY (<colname>, ...)
      REFERENCES <schema>.<tablename> (<colname>, ...) |
  UNIQUE (<colname>, ...) |
  CHECK (<colname>, ...) (<check_list>);
```

Specific examples of this format appear in many of the sections in this chapter to show you how to create and modify each type of constraint. If you omit the CONSTRAINT keyword, you must omit the constraint name as well.

Changing or Removing a Constraint

There are only two things you can change on an existing constraint: the *name* and the *state* of the constraint. Here is an example in which you change the name of the FOREIGN KEY constraint:

```
ALTER TABLE CUSTOMER
RENAME CONSTRAINT CUST_FK TO CUST_ORDER_FK;
```

The general syntax for changing the constraint state is:

```
ALTER TABLE <tablename>
MODIFY CONSTRAINT <constraintname>
  <constraint_state> <constraint_state> ...;
```

Here is an example in which you change the state of a UNIQUE constraint. You enable the constraint, report invalid rows to the BADCUSTOMERS table, and use the existing index, CUST_UNQ_INDEX, rather than creating a new unique index.

```
ALTER TABLE CUSTOMER
ENABLE CONSTRAINT CUST_UNQ
EXCEPTIONS TO BADCUSTOMERS
USING CUST_UNQ_INDEX;
```

You cannot change the state of a NOT NULL constraint; a NOT NULL constraint is always enabled. Removing a NOT NULL constraint is the only way to "disable" it.

PRACTICAL EXAMPLES OF WORKING WITH CONSTRAINTS

The next five sections provide you with step-by-step examples in which you create, modify, and drop constraints of all types. New details about constraints are explained as you work through the examples.

The first constraint you examine is the easiest one: the NOT NULL constraint.

Adding or Removing a NOT NULL Constraint

The NOT NULL constraint is often used when creating tables. Use the NOT NULL constraint to require a value in a column when a row is inserted or updated. Follow along with the example to see how the NOT NULL constraint is handled.

1. Start up the Enterprise Manager console. In Windows, on the Taskbar, click **Start/Programs/Oracle - OraHome92/Enterprise Manager Console**. In UNIX, type **oemapp console** on the command line. The Enterprise Manager console login screen appears.
2. Select the **Launch standalone** radio button and click **OK**. The console appears.
3. Start up the SQL*Plus Worksheet by clicking **Tools/Database Applications/SQL*Plus Worksheet** from the top menu in the console. A background SQL*Plus process starts, and then the SQL*Plus Worksheet window appears. If the background process appears in front, simply minimize it (click the **minus sign** in the top-right corner), so you can see the worksheet.
4. Connect as the CLASSMATE user. Click **File/Change Database Connection** on the menu. A logon window appears. Type **CLASSMATE** in the Username box, **CLASSPASS** in the Password box, and **ORACLASS** in the Service box. Leave the connection type as "Normal." Click **OK** to continue.
5. You are creating a sample table to practice adding and modifying the NOT NULL constraint. This table is used later for other constraint examples. Type this command to create a small table in which the three columns have NOT NULL constraints and one has a DEFAULT value. Execute the command by clicking the **Execute** icon. The SQL*Plus Worksheet replies, "Table created."

```
CREATE TABLE CH10DOGSHOW
(DOGSHOWID NUMBER NOT NULL,
 SHOW_NAME VARCHAR2(40) NOT NULL,
 DATE_ADDED DATE DEFAULT SYSDATE NOT NULL);
```

   SYSDATE is a **pseudocolumn** that represents the current date and time. A pseudocolumn is available for use within SQL*Plus anywhere you can use a column value or (as in this case) a static value. When a new row is inserted with null values in the DATE_ADDED column, Oracle9*i* places the current date and time in the column as the default value.

6. Try inserting a row with null in the SHOW_NAME column by typing this command and executing it. When you leave out a column, it defaults to a null value unless the column was defined with a DEFAULT value. In that case, the column omitted is assigned the DEFAULT value. In this example, there is no DEFAULT value, so the SHOW_NAME column value is null.

```
INSERT INTO CH10DOGSHOW
(DOGSHOWID, DATE_ADDED) VALUES
(1, '11-MAY-03');
```

Figure 10-4 shows the error message that is returned stating that the SHOW_NAME column cannot be null.

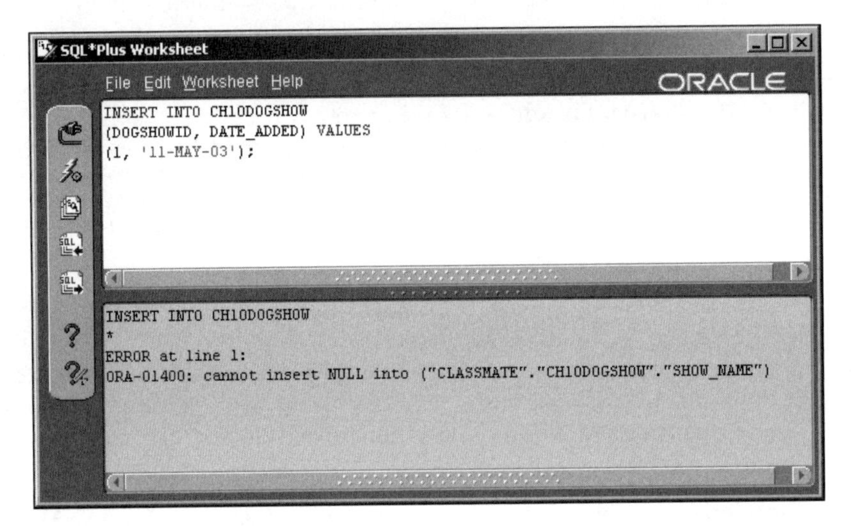

Figure 10-4 Leave out the column to allow it to default to null (or its default value)

7. Now insert a row with null in the DATE_ADDED column. This is accepted, because the default value of SYSDATE is used. Type the following command and execute it.

```
INSERT INTO CH10DOGSHOW
(DOGSHOWID, SHOW_NAME) VALUES
(1, 'AKC Portland');
```

The SQL*Plus Worksheet replies, "1 row created."

8. Remove the constraint from SHOW_NAME using the ALTER TABLE command. Type and execute this command.

```
ALTER TABLE CH10DOGSHOW
MODIFY (SHOW_NAME NULL);
```

The SQL*Plus Worksheet replies, "Table altered." Note that the row inserted in Step 7 was automatically committed when you ran the ALTER TABLE command. Any DDL command causes Oracle9*i* to commit previous changes that have not yet been committed.

9. To place the NOT NULL constraint back on the SHOW_NAME column, use the same format you used to remove it. Type the following command and execute it.

   The SQL*Plus Worksheet replies, "Table altered."

```
ALTER TABLE CH10DOGSHOW
MODIFY (SHOW_NAME NOT NULL);
```

10. Remain logged on for the next practice.

Including a DEFAULT setting on the column provides a fail-safe method of populating a NOT NULL column. Any row inserted with a null value in the column is automatically modified so that the column contains the default value instead.

Adding and Modifying a PRIMARY KEY Constraint

The PRIMARY KEY constraint is important for the integrity of the table. The primary key can be one column, such as the PATIENT_ID in the previous example, or it can be more than one column. The columns are usually placed at the beginning of the table, although this is not a requirement. Follow along with this example to create a new table with a compound primary key.

1. If you are not already logged on, start up the Enterprise Manager console and log onto the SQL*Plus Worksheet as CLASSMATE.
2. You are creating a sample table to practice adding and modifying the primary key. In this example, the primary key is the OWNER_ID column. Type the following two lines into SQL*Plus Worksheet to begin defining the table and to define the PRIMARY KEY constraint.

```
CREATE TABLE CH10DOGOWNER
(OWNER_ID NUMBER CONSTRAINT CH10_PK PRIMARY KEY,
```

   As you can see, the constraint is defined inline with the OWNER_ID column and is named CH10_PK.

3. Type the remaining lines and press the **Execute** icon to create the table.

```
OWNER_NAME VARCHAR2(50),
MEMBER_OF_AKC CHAR(3) DEFAULT 'NO',
YEARS_EXPERIENCE NUMBER(2,0));
```

   The SQL*Plus Worksheet replies, "Table created."

4. The primary key has been created along with the table. In addition, an index named CH10_PK has been created in CLASSMATE's default tablespace. You decide the constraint should be named CH10_DOG_OWNER_PK, so you must modify the constraint. To modify a constraint, you must use the ALTER TABLE command. The NOT NULL constraint is the only constraint that can be modified using an inline constraint. All other constraints must be modified using an out of line constraint format. Type the following command to rename the constraint.

```
ALTER TABLE CH10DOGOWNER
RENAME CONSTRAINT CH10_PK TO CH10_DOG_OWNER_PK;
```

5. Run the command by clicking the **Execute** icon. SQL*Plus Worksheet replies, "Table altered."

6. Imagine that you decide you don't want the key after all. To remove the primary key, you must drop the constraint. Type the following command, and execute it by clicking the **Execute** icon.

```
ALTER TABLE CH10DOGOWNER
DROP CONSTRAINT CH10_DOG_OWNER_PK;
```

   When you drop a PRIMARY KEY constraint, the associated index is also dropped.

Any constraint other than NOT NULL can be dropped using the ALTER TABLE DROP CONSTRAINT command.

   The SQL*Plus Worksheet replies, "Table altered."

7. As with all Oracle DDL commands, you cannot roll back (undo) the command after it has been executed. In this example, you discover you really need the PRIMARY KEY constraint that you just dropped. Place it back onto the table, this time using an out of line constraint and the ALTER TABLE command. Type the following command and execute it.

```
ALTER TABLE CH10DOGOWNER
ADD CONSTRAINT CH10_DOG_OWNER_PK PRIMARY KEY (OWNER_ID)
DISABLE;
```

   The SQL*PLUS Worksheet replies, "Table altered."

   The final line contains the DISABLE keyword. This means the constraint was created but is not currently enforced. A complete discussion of constraint states is presented earlier in this chapter.

8. Enforce the constraint by typing and executing this command.

```
ALTER TABLE CH10DOGOWNER
MODIFY CONSTRAINT CH10_DOG_OWNER_PK ENABLE;
```

   The SQL*Plus Worksheet replies, "Table altered."

9. Remain logged on for the next practice.

As you go through more examples, you practice using constraint states as well as other types of constraints.

Adding and Modifying a UNIQUE Key Constraint

As you read earlier, the UNIQUE key constraint and the PRIMARY KEY constraint are very similar. Both enforce a rule stating that every row in the table must contain a unique value within the column or columns in the constraint. The difference is that the PRIMARY KEY also enforces a NOT NULL constraint on the column or columns.

A UNIQUE constraint is often used in addition to a PRIMARY KEY rather than in place of it. For example, assume you are converting an old database system in which the primary key on the old table must be carried into the new system for reference. The old primary key does not conform to modern normalization rules because it is actually three distinct values combined into a single column. You want to break these distinct values into three columns and use the three new columns as the primary key. New rows inserted into the table do not carry the old key, so it must allow null values. You create a PRIMARY KEY constraint on the new key and a UNIQUE key constraint on the old key.

Follow these steps to create and modify a UNIQUE key constraint.

1. If you are not already logged on, start up the Enterprise Manager console, and log onto the SQL*Plus Worksheet as CLASSMATE.
2. You are creating a sample table to practice adding and modifying a UNIQUE key constraint. In this example, the table stores names of people from around the world. The primary key is made up of the COUNTRY and the PERSON_ID columns. The unique key is the US_TAX_ID column. Not every person has a U.S. tax number; however, every U.S. tax number is different. Type the following lines into SQL*Plus Worksheet to begin defining the table and to define the UNIQUE key constraint.

```
CREATE TABLE CH10WORLD
(COUNTRY VARCHAR2(10),
 PERSON_ID NUMBER,
 US_TAX_ID NUMBER(10) CONSTRAINT US_TAX_UNIQUE UNIQUE,
```

3. Finish the command by adding two more columns and the compound PRIMARY KEY constraint. Type these lines and execute the command.

```
FIRST_NAME VARCHAR2(10),
LAST_NAME VARCHAR2(20),
CONSTRAINT CH10WORLD_PK PRIMARY KEY (COUNTRY, PERSON_ID));
```

The SQL*Plus Worksheet replies, "Table created."

4. A common event in the life of a database system is the loading of data from one source, such as an old file system, to a new table. Constraints are often disabled to prevent the data load from failing, because some rows do not comply with the new constraints. Therefore, to simulate a data load event, you first disable the UNIQUE key constraint. Type the following command and execute it.

```
ALTER TABLE CH10WORLD
MODIFY CONSTRAINT US_TAX_UNIQUE DISABLE;
```

The SQL*Plus Worksheet replies, "Table altered."

5. Next, run a script that inserts rows into the table. Select **Worksheet/Run Local Script** and locate the file named **ch10data.sql** in the **Data\Chapter10** directory. Click **Open** to start the script. SQL*Plus Worksheet runs the script. You see that the phrase "1 row created" appears in the lower section of the screen several times, as shown in Figure 10-5.

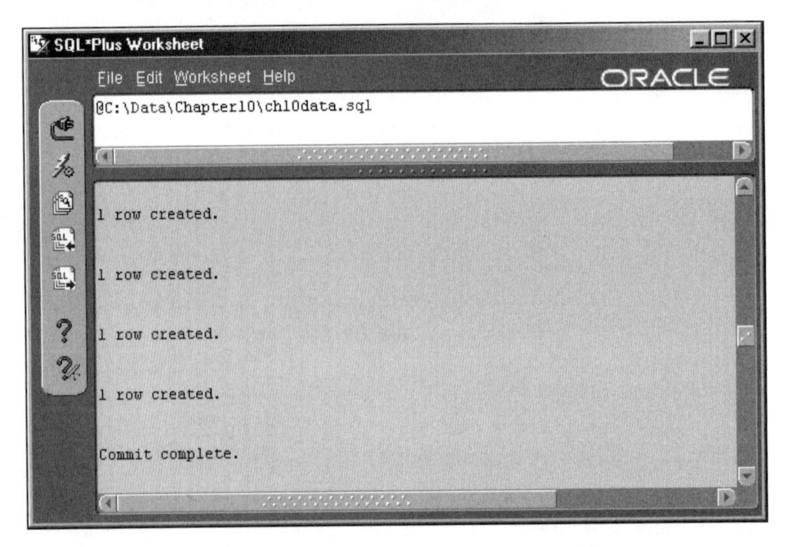

Figure 10-5 Running a script displays feedback as commands run

6. Before enabling the constraint, you decide that if the constraint fails, you want to get a list of invalid rows. Using the EXCEPTIONS constraint state handles this for you. However, you must create the EXCEPTIONS table before enabling the constraint. To do this, run the script that Oracle provides to create the table. Select **Worksheet/Run Local Script** and locate the file named **utlexcpt.sql** in the **ORACLE_HOME/rdbms/admin** directory. (Remember that ORACLE_HOME represents the full path of the directory that holds the Oracle9*i* software. For example, on a Windows machine, ORACLE_HOME might be **C:\oracle\ora92**.) Click **Open** to start the

script. SQL*Plus Worksheet runs the script. You see the response, "Table created" in the lower part of the window. The table named EXCEPTIONS is created and owned by CLASSMATE (because you are logged on as CLASSMATE). The EXCEPTIONS table has these four columns:

- **ROW_ID:** The rowid of the row that failed the constraint rule
- **OWNER:** The owner of the table where the row resides
- **TABLE_NAME:** The name of the table where the row resides
- **CONSTRAINT:** The name of the constraint violated by the row

7. Now, enable the constraint by typing this command and executing it:

```
ALTER TABLE CH10WORLD
MODIFY CONSTRAINT US_TAX_UNIQUE ENABLE VALIDATE
EXCEPTIONS INTO EXCEPTIONS;
```

When you run the command, you see this error message: "ORA-02299: cannot validate (CLASSMATE.US_TAX_UNIQUE) - duplicate keys found." The constraint is not enabled.

The ENABLE VALIDATE tells Oracle9*i* to enable the constraint and validate all existing rows. (VALIDATE is the default when you specify ENABLE, so you can leave it out of the command and get the same results.) The last line tells Oracle9*i* to load the rowid of any rows that failed the validation into the table named EXCEPTIONS.

8. View the invalid records logged in the EXCEPTION table by typing and executing this query. The query joins the EXCEPTIONS table (alias of "E") with the CH10WORLD table (alias of "W") by matching the rowid found in the EXCEPTIONS table with the actual rowid of the rows in the CH10WORLD table. Matching rows are the ones that violated the constraint.

```
SELECT E.CONSTRAINT, W.*
FROM CH10WORLD W JOIN EXCEPTIONS E
ON (W.ROWID = E.ROW_ID)
ORDER BY W.US_TAX_ID;
```

Figure 10-6 shows the results. As you can see, there are two rows for Beth Blanks and two rows for Sean Watson. To successfully enable the UNIQUE constraint, the duplicate rows must be removed from the CH10WORLD table. For this exercise, you leave the constraints disabled.

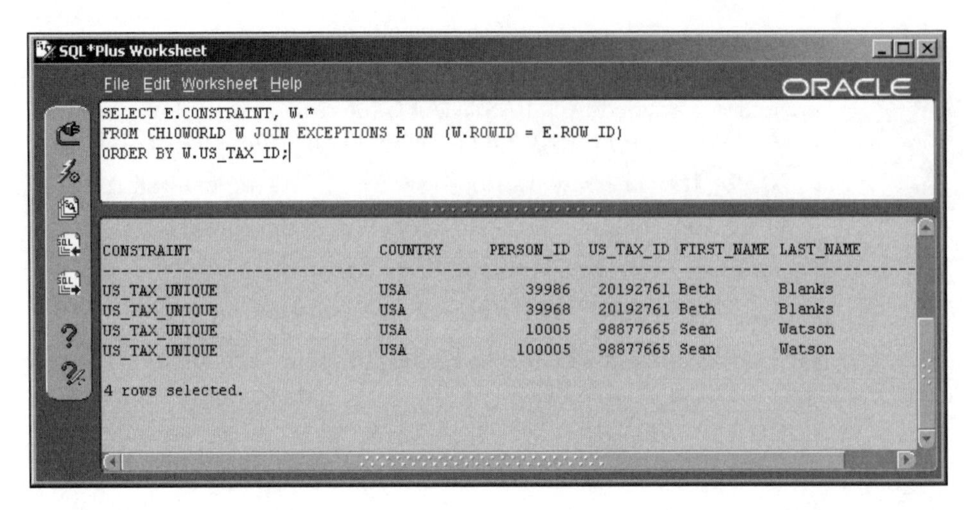

Figure 10-6 The constraint was violated four times

9. Remain logged on for the next practice.

The query executed in this section is formatted in the new ANSI-standard JOIN syntax introduced in Oracle9*i*. For information on this syntax, see the *Oracle9i SQL Reference* document in the chapter titled *SQL Queries and Subqueries*. The equivalent query using the proprietary Oracle9*i* format is:

```
SELECT E.CONSTRAINT, W.*
FROM CH10WORLD W, EXCEPTIONS E
WHERE W.ROWID = E.ROW_ID
ORDER BY W.US_TAX_ID;
```

When you join tables together, you are nearly always using a column or set of columns in one table that corresponds to a column or set of columns in another table. The corresponding columns are connected with a constraint called the FOREIGN KEY constraint.

Working with a FOREIGN KEY Constraint

The FOREIGN KEY constraint requires a bit more coding than the previous constraints. As you saw earlier, a foreign key references another table's primary key or unique key. Follow these examples to create and work with a FOREIGN KEY constraint.

1. If you are not already logged on, start up the Enterprise Manager console and log onto the SQL*Plus Worksheet as CLASSMATE.
2. You are creating a sample table to practice adding and modifying a FOREIGN KEY constraint. In this example, you are creating a table to register dogs for a dog show. You already created a table for dog owners earlier in this chapter. The CH10DOG table uses a foreign key to connect the dogs to their owners. Type the following lines into SQL*Plus Worksheet to begin defining the table.

```
CREATE TABLE CH10DOG
(DOG_ID NUMBER,
 OWNER_ID NUMBER(10) ,
 DOG_NAME VARCHAR2(20),
 BIRTH_DATE DATE,
```

3. Type these final lines defining the table and execute the command.

```
CONSTRAINT CH10DOGOWNER_FK
     FOREIGN KEY (OWNER_ID) REFERENCES CH10DOGOWNER
     DEFERRABLE INITIALLY IMMEDIATE);
```

   The SQL*Plus Worksheet replies, "Table created."

   The FOREIGN KEY constraint is defined out of line (because to name the FOREIGN KEY constraint, you must define it out of line), so you must name the column in the table that contains the foreign key, in this case, the OWNER_ID column. The last line defines the constraint's state. DEFERRABLE means that at a later date, this constraint can be modified so that the constraint is validated only when inserts and updates are *committed* instead of immediately when the inserts or updates are *executed*. INITIALLY IMMEDIATE means that for now, the constraint is executed immediately when a row is inserted or updated.

4. Type and execute the following command to insert a row into the CH10DOG table. Because there are no rows in the CH10DOGOWNER table, the foreign key constraint will fail validation.

```
INSERT INTO CH10DOG VALUES
(1,2,'Chow Moon','15-JAN-02');
```

   Let's say you have a new owner to enter into the CH10DOGOWNER table along with this entry to the CH10DOG table. You have several choices to successfully insert both records:

   - Insert the owner, then the dog, and commit the transaction. The FOREIGN KEY constraint is not violated when you insert the owner row first.
   - Alter the CH10DOG table to disable the FOREIGN KEY constraint. Then, you can enter the rows in either order and commit the transaction. Then modify the table again to enable the constraint. This works only if you are the owner of the table, a DBA, or you have been granted the object privilege required to modify the table's constraints.
   - Use the SET CONSTRAINT command to adjust your *session* to defer all deferrable constraints. Now, you can enter the dog first, then the owner, then commit the transaction. This option provides a simplified method of deferring constraints without issuing any DDL commands. To defer a constraint using the SET CONSTRAINT command, you must have the SELECT privilege on the table, or be the owner. To reverse the setting, issue another SET CONSTRAINT command, or log off.

The final selection (adjusting your session to defer constraints) is used for this example.

5. The SET CONSTRAINT command provides a way to change a named constraint or all constraints to a deferred state during your session. Of course, the only constraints that are affected were created with the DEFERRABLE setting. (Recall that by default, a constraint is not deferrable.) Type and execute the following command to defer all deferrable constraints. Remember, using this command does not change the constraint's state in the table; it only changes the constraint's state for your current session.

```
SET CONSTRAINTS ALL DEFERRED;
```

   The SQL*Plus Worksheet replies, "Constraint set."

6. Repeat the insert command from Step 4. SQL*Plus Worksheet replies, "1 row created."

7. Insert a row for the dog's owner by typing and executing the following command.

```
INSERT INTO CH10DOGOWNER VALUES
(2, 'Jack Maylew','YES', 3.5);
```

   SQL*Plus Worksheet replies, "1 row created."

8. Finally, commit both rows. When you execute the command, the constraint is validated. The FOREIGN KEY constraint is valid at this point. Type **COMMIT;** and click the **Execute** icon. SQL*Plus Worksheet replies, "Commit complete."

9. Reset your session by typing and executing this command.

```
SET CONSTRAINTS ALL IMMEDIATE;
```

   The SQL*Plus Worksheet replies, "Constraint set."

10. After a FOREIGN KEY constraint has been added, the parent table's PRIMARY KEY constraint cannot be removed without adding the CASCADE keyword that tells Oracle9*i* to drop the PRIMARY KEY constraint and all FOREIGN KEY constraints that reference the table. Try this out with the CH10DOGOWNER table by typing and executing the following command.

```
ALTER TABLE CH10DOGOWNER
DROP CONSTRAINT CH10_DOG_OWNER_PK CASCADE;
```

   The SQL*Plus Worksheet replies, "Table altered."

   The CH10DOGOWNER no longer has a PRIMARY KEY constraint, and the CH10DOG table no longer has a FOREIGN KEY constraint.

11. Add the PRIMARY KEY constraint again by typing and executing this command.

```
ALTER TABLE CH10DOGOWNER
ADD CONSTRAINT CH10_DOG_OWNER_PK PRIMARY KEY(OWNER_ID);
```

The SQL*Plus Worksheet replies, "Table altered."

12. Add the FOREIGN KEY constraint back to the CH10DOG table, but this time, use a new parameter. Type and execute the following command:

```
ALTER TABLE CH10DOG
ADD CONSTRAINT CH10DOGOWNER_FK FOREIGN KEY (OWNER_ID)
REFERENCES CH10DOGOWNER
ON DELETE CASCADE;
```

The SQL*Plus Worksheet replies, "Table altered."

The ON DELETE CASCADE parameter tells Oracle9*i* that when a parent row is deleted, all the related child rows are also deleted. Another alternative to this is ON DELETE SET NULL, which instructs Oracle9*i* to set the value of the foreign key column in the related child rows to null instead of deleting the rows. If you omit the ON DELETE parameter, you cannot delete a row from the parent table, if it has related children in other tables. Try this out by deleting the owner of the dog row now in the CH10DOG table.

13. Delete the row of the dog's owner by typing and executing this command.

```
DELETE FROM CH10DOGOWNER WHERE OWNER_ID = 2;
```

SQL*Plus Worksheet replies, "1 row deleted." But in fact, two rows were deleted. One row was deleted from the CH10DOGOWNER table because you executed the DELETE statement. Due to the cascading effect of deleting a parent row, one child row was also deleted from the CH10DOG table.

14. Execute this query to prove that the row was deleted from the CH10DOG table.

```
SELECT * FROM CH10DOG;
```

The SQL*Plus Worksheet replies, "no rows selected."

As you can see, no rows were retrieved. The one row that was inserted earlier has been deleted. This is useful when you know that the child rows are not needed after the parent row is deleted. For example, a CALENDAR_DAY table has a child table called DAY_APPOINTMENTS. When a day is deleted from the CALENDAR_DAY table, all the appointments for that day can be deleted automatically without issuing another DELETE command.

15. Roll back the deleted rows by typing **ROLLBACK;** and clicking the **Execute** icon.

16. The SQL*Plus Worksheet replies, "Rollback complete." Remain logged on for the next practice.

When using SET CONSTRAINTS, you can list one or more constraints by name, separated with commas, to defer only certain constraints. For example, to defer two constraints named DOG_FK and SHOW_NAME_FK, type and execute this command.

```
SET CONSTRAINTS DOG_FK, SHOW_NAME_FK DEFERRED;
```

The final constraint to examine helps validate data in one or more columns within the same row.

Creating and Changing a CHECK Constraint

The CHECK constraint helps validate the value within a column or a set of columns within one row. This is useful for columns with a fixed number of values, such as the days of the week or calendar months. You can use a few expressions to check against a range of values as well. For example, the number of wheels on a bus must be between four and sixteen and divisible by two.

You have already created several tables, so you are using the ALTER TABLE command to add CHECK constraints to columns that already exist. The syntax for adding a CHECK constraint with CREATE TABLE and ALTER TABLE is identical. Like the other constraints, add the constraint inline with a new column if the constraint only refers to that column; add the constraint out of line when it refers to more than one column or when the column already exists.

Follow these steps to add CHECK constraints to some of the tables created in this chapter.

1. If you are not already logged on, start up the Enterprise Manager console, and log onto the SQL*Plus Worksheet as CLASSMATE.
2. Type and execute this command to create a CHECK constraint that requires the MEMBER_OF_AKC to have the values <YES>, <NO>, or null. (Null values automatically pass the constraint unless you specifically state the column cannot be null in the constraint.)

   ```
   ALTER TABLE CH10DOGOWNER ADD CONSTRAINT AKC_YN
   CHECK (MEMBER_OF_AKC IN ('YES','NO'));
   ```

   The SQL*Plus Worksheet replies, "Table altered."
3. Add another constraint that requires that the names of all dog shows must be all uppercase letters. Because there is one row that violates the constraint, create the constraint in a disabled state. This gives you a chance to correct the data and enable the constraint. Type and execute this command:

   ```
   ALTER TABLE CH10DOGSHOW ADD CONSTRAINT ALL_CAPS
   CHECK (SHOW_NAME = UPPER(SHOW_NAME)) DISABLE;
   ```

   The SQL*Plus Worksheet replies, "Table altered."

4. Correct the data in the CH10DOGSHOW table to make all names uppercase by typing and executing this command.

   ```
   UPDATE CH10DOGSHOW SET SHOW_NAME = UPPER(SHOW_NAME);
   ```

   The SQL*Plus Worksheet replies, "1 row updated."

5. Enable the constraint now by typing and executing the following command:

   ```
   ALTER TABLE CH10DOGSHOW MODIFY CONSTRAINT ALL_CAPS ENABLE;
   ```

   The SQL*Plus Worksheet replies, "Table altered."

6. You can also use the CHECK constraint to compare one column against another column. Remember that the constraint can only use values within the current row. Add a constraint that rejects a row if both the first and last name of a person in the CH10WORLD table contain null values, or if the two names are identical. Type and execute the following command to accomplish this.

   ```
   ALTER TABLE CH10WORLD ADD CONSTRAINT CHK_NAMES
   CHECK ((FIRST_NAME IS NOT NULL OR LAST_NAME IS NOT NULL)
           AND(FIRST_NAME <> LAST_NAME));
   ```

   The SQL*Plus Worksheet replies, "Table altered."

7. Test the constraint by trying to insert a row where the first and last names match.

   Type and execute the following command:

   ```
   INSERT INTO CH10WORLD VALUES
   ('USA', 1995, 99877689, 'Jeremy', 'Jeremy');
   ```

   SQL*Plus Worksheet replies, "ORA-02290: check constraint (CLASSMATE.CHK_NAMES) violated."

8. Test again by modifying the INSERT command so that the first and last names are both null values. The row is rejected with the same error message.

   Type and execute the following command.

   ```
   INSERT INTO CH10WORLD VALUES
   ('USA', 1995, 99877689, NULL, NULL);
   ```

   SQL*Plus Worksheet replies, "ORA-02290: check constraint (CLASSMATE.CHK_NAMES) violated."

9. For the last test, revise the INSERT command so that only the first name has a value. This row should pass the constraint. Type and execute the following command:

   ```
   INSERT INTO CH10WORLD VALUES
   ('USA',1995, 99877689, 'Jeremy', NULL);
   ```

   The SQL*Plus Worksheet replies, "1 row created."

10. Commit your work by typing **COMMIT;** and clicking the **Execute** icon.

    The SQL*Plus Worksheet replies, "Commit complete."

11. Remain logged on for the next practice.

CHECK constraints cannot contain queries or any references to other tables. They also cannot use pseudocolumns, such as SYSDATE and USER.

Now that you have created many constraints, the next section takes a look at the data dictionary views that display constraint information.

DATA DICTIONARY INFORMATION ON CONSTRAINTS

There are only two constraint-related data dictionary views. They are:

- **ALL_CONSTRAINTS:** This view (and its corresponding views, DBA_CONSTRAINTS and USER_CONSTRAINTS) contains the definition of a constraint. The columns in this view include:
  - CONSTRAINT_NAME
  - CONSTRAINT_TYPE (letters identifying each type, such as "P" for PRIMARY KEY, "C" for CHECK, and "R" for FOREIGN KEY (also called referential constraint)
  - SEARCH_CONDITION (lists the condition for CHECK constraint)
  - STATUS (enabled or disabled)
- **ALL_CONS_COLUMNS:** This view lists the columns associated with each constraint. There is one row for each column in the constraint. For example, a primary key made up of two columns has two rows in this view.

Complete the following steps to examine the contents of the ALL_CONSTRAINTS view.

1. If you are not already logged on, start up the Enterprise Manager console, and log onto the SQL*Plus Worksheet as CLASSMATE.

2. Type and execute this command to see the constraints created by CLASSMATE.

```
SELECT CONSTRAINT_NAME, CONSTRAINT_TYPE, TABLE_NAME,
STATUS, SEARCH_CONDITION
FROM ALL_CONSTRAINTS
WHERE OWNER='CLASSMATE' ORDER BY TABLE_NAME;
```

   Figure 10-7 shows the results of the query. Notice that the NOT NULL constraints are listed as CHECK constraints and have system-assigned names. (You can name a NOT NULL constraint like any other constraint if you choose.)

3. Log off SQL*Plus Worksheet by clicking the **X** in the top-right corner.
4. Log off the Enterprise Manager console by clicking the **X** in the top-right corner.

You have seen how to create your own tables, indexes, and constraints. The next chapter shows you how to allow other users to view or modify data in your tables.

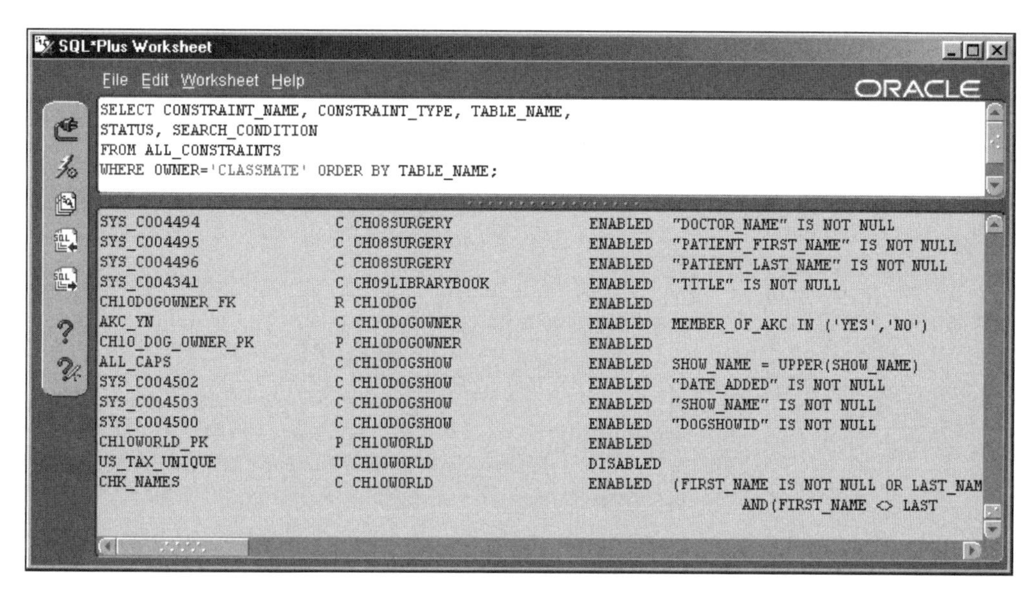

Figure 10-7 FOREIGN KEY constraints are indicated with the letter R (for "referential constraint")

CHAPTER SUMMARY

- Integrity constraints can be enforced using declared constraints, triggers, or application programming.
- A PRIMARY KEY constraint can cover one or more columns in a table and requires unique, non-null values in the primary key columns for every row.
- A UNIQUE constraint requires unique values or null values in the unique key columns.
- A FOREIGN KEY constraint identifies a parent/child relationship between two tables and is defined on the child table.
- A NOT NULL constraint must always be on a single column and prevents null values in the column.
- A CHECK constraint compares the column to a list or range of values.

- Constraints can be created with the CREATE TABLE and the ALTER TABLE commands.
- Use the ALTER TABLE statement to rename or drop a constraint.
- Use the ALTER TABLE statement to change the constraint state.
- To remove the NOT NULL constraint, use ALTER TABLE MODIFY (column...) statement.
- The PRIMARY KEY constraint is crucial to a relational database design.
- When a PRIMARY KEY constraint is created (and not disabled), a unique index is created to help enforce the constraint.
- Dropping or disabling a PRIMARY KEY constraint causes the underlying index to be dropped.
- Use the NOVALIDATE constraint state when you do not want existing rows to be checked for compliance with a constraint.
- The default states of a constraint are ENABLE, VALIDATE, INITIALLY IMMEDIATE, NOT DEFERRABLE, and NORELY.
- USING INDEX applies only to PRIMARY KEY and UNIQUE key constraints in which you want to use a preexisting index or you want to specify storage settings for the index to be created when the constraint is created.
- EXCEPTIONS can be used after creating a table (usually called EXCEPTIONS) to hold the rowid of rows that violate a constraint.
- Specifying DEFERRABLE INITIALLY IMMEDIATE specifies that the constraint is enforced after every statement (immediate), but can be deferred later if the user either alters the constraint or alters his session.
- SET CONSTRAINTS ALL DEFERRED switches all deferrable constraints to a deferred state during your current session only.
- A FOREIGN KEY constraint can define the behavior of the database when a parent row is deleted using the ON DELETE CASCADE or ON DELETE SET NULL parameter.
- The CHECK constraint can look for a specified list of values or other simple expressions.
- The ALL_CONSTRAINTS data dictionary view contains details on all five types of constraints.
- The ALL_CONS_COLUMNS data dictionary view lists all the columns involved in a constraint.

REVIEW QUESTIONS

1. In which of these situations would you enforce integrity constraints using database triggers?
   a. Your database contains many parent/child relationships.
   b. Your integrity constraints are complex and involve multiple table lookups.
   c. Your tables often use compound primary keys.
   d. All of the integrity constraints of your database system are enforced at the application level.
2. List the five types of integrity constraints.
3. Define an inline integrity constraint. When is it used?
4. Which constraints can be compound constraints?
5. You cannot specify storage information when creating a table that has integrity constraints. True or False?
6. The ________________ constraint allows nulls but not duplicate values, whereas the ________________ constraint does not allow nulls or duplicate values.
7. The following statement fails with a syntax error. True or False?

```
ALTER TABLE CLIENT
MODIFY(FIRST_NAME NOT NULLS);
```

8. A FOREIGN KEY constraint can reference columns in a UNIQUE key constraint. True or False?
9. The ________________ command changes a deferrable constraint to deferred state during your session.
10. Write the SQL statement to change the name of the KEY1 constraint (a PRIMARY KEY constraint) to PK1 on the MOUSETRAP table.
11. List the letters and what they stand for in the CONSTRAINT_TYPE column of the USER_CONSTRAINTS view.
12. The ________________ state tells Oracle9*i* to check the constraint only when a transaction commits.

EXAM REVIEW QUESTIONS — DATABASE FUNDAMENTALS I (#1Z0-031)

1. The following constraints CANNOT be defined inline. Choose three.
   a. A compound primary key defined with ALTER TABLE
   b. A CHECK constraint on an existing column
   c. A NOT NULL constraint on an existing column
   d. A FOREIGN KEY constraint on an existing column
   e. A single column primary key defined with CREATE TABLE

2. The following statements were executed. What are the results?

```
CREATE TABLE MYCUSTOMER
  (ID NUMBER PRIMARY KEY, FIRSTNAME VARCHAR2(10),
   AGE NUMBER(4,0) CONSTRAINT CHK_AGE
   CHECK (AGE > 18) DEFERRABLE);
SET CONSTRAINT ALL DEFERRED;
INSERT INTO MYCUSTOMER VALUES (1, 'JANE',12);
INSERT INTO MYCUSTOMER VALUES (1, 'MARY',21);
UPDATE MYCUSTOMER SET AGE = 19 WHERE FIRSTNAME = 'JANE';
COMMIT;
```

a. The first insert succeeds, the second insert succeeds, the update succeeds, and the commit fails.

b. The first insert succeeds, the second insert fails, the update succeeds, and the commit fails.

c. The first insert succeeds, the second insert succeeds, the update succeeds, and the commit succeeds.

d. The first insert succeeds, the second insert fails, the update succeeds, and the commit succeeds.

3. The DEFERRABLE INITIALLY IMMEDIATE clause has what effect?

a. Existing rows are validated, and then the constraint is deferred.

b. Existing rows are not validated, and then the constraint is immediate.

c. Existing rows are validated, and then the constraint is immediate unless later deferred.

d. Existing rows are not validated, and the constraint is deferred unless later made immediate.

e. None of the above.

4. Which command would you use to defer a constraint? Choose two.

a. ALTER CONSTRAINT

b. DEFER CONSTRAINT

c. DROP CONSTRAINT

d. ALTER TABLE

e. SET CONSTRAINT

5. What does it mean when a constraint is created with ON DELETE SET TO NULL?

a. All parent rows are deleted when a child row is deleted.

b. All child rows have the column(s) set to null when a parent row is deleted.

c. All child rows are deleted when a parent row is deleted.

d. All parent rows have the column(s) set to null when a child is deleted.

e. A single column primary key is defined with CREATE TABLE.

6. Which query displays whether a constraint is deferred?
   a. SELECT CONSTRAINT_NAME, STATUS FROM USER_CONS_COLUMNS;
   b. SELECT CONSTRAINT_NAME, DEFERRED FROM DBA_CONSTRAINTS;
   c. SELECT CONSTRAINT_NAME, DEFERRABLE FROM ALL_CONSTRAINTS;
   d. SELECT CONSTRAINT_NAME, DEFERRED FROM DBA_TAB_CONSTRAINTS;
7. You query USER_CONSTRAINTS and find three constraints on the CLIENT table with the constraint type 'R'. What can be said about these constraints? Choose two.
   a. Each constraint contains a different set or order of columns.
   b. This is invalid because there cannot be three constraints of this type on one table.
   c. All three constraints are FOREIGN KEY constraints.
   d. When enabled, the constraints created unique indexes.
8. What type of constraint can enforce this business rule: "An employee's salary may be no more than 10 percent higher than his boss's salary." Choose two.
   a. CHECK constraint
   b. NOT NULL constraint
   c. Application code
   d. Database trigger
9. Examine the following statement. Which statements are true regarding the constraint? Choose two.

```
ALTER TABLE FRIENDS
ADD CONSTRAINT FRIEND_ALT_KEY UNIQUE (PHONENO)
USING INDEX PHONEX;
```

   a. The statement fails if the PHONEX index does not exist.
   b. A new index named PHONEX is created.
   c. The PHONENO column cannot contain null values.
   d. Existing rows are validated immediately.
10. Examine the following statement. What might cause the statement to fail? Choose the best two.

```
ALTER TABLE CLIENT
ADD CONSTRAINT CLIENT_CO FOREIGN KEY (COMPANY_ID)
REFERENCES COMPANY (COMPANY_ID) ON DELETE CASCADE;
```

   a. The COMPANY table has a UNIQUE key constraint on COMPANY_ID.
   b. The COMPANY table has no rows.
   c. The CLIENT table has duplicate values in the COMPANY_ID column.
   d. The COMPANY table has no PRIMARY KEY constraint.
   e. The CLIENT table has all null values in the COMPANY_ID column.

HANDS-ON ASSIGNMENTS

The assignments in this section require that you have completed all the practices in the chapter and that you also run another script to add rows to some of the tables. Follow these steps to prepare for the hands-on assignments.

1. Start up the SQL*Plus Worksheet. In Windows, on the Taskbar, click **Start/ Programs/Oracle - OraHome92/Application Development/SQLPlus Worksheet**. In UNIX, type **oemapp worksheet** on a command line. A logon window displays.
2. Type **CLASSMATE** in the Username box, **CLASSPASS** in the Password box, and **ORACLASS** in the Service box. Leave the connection type as "Normal". Click **OK** to continue.
3. If you have not completed all the chapter exercises, on the SQL*Plus Worksheet menu, click **Worksheet/Run Local Script**. Then navigate to the **Data\ Chapter10** directory and select the file named **ch10catchup.sql** on your student disk. Click **Open** to execute the script.
4. Run a script to add data to some of the tables. On the SQL*Plus Worksheet menu, click **Worksheet/Run Local Script**. Then navigate to the **Data\ Chapter10** directory, and select the file named **ch10_handson_setup.sql** on your student disk. Click **Open** to execute the script.
5. Remain logged on for the hands-on assignments.
6. Delete the two invalid rows in the CH10WORLD table and enable the US_TAX_UNIQUE constraint. Assume that the row with lower value in the PERSON_ID column is the correct one for each duplicate record. Save the SQL you use in your **Solutions\Chapter10** directory in a file named **h1001.sql**.
7. Add a PRIMARY KEY constraint to the CH10DOG table. The primary key column is named DOG_ID. The constraint should be nondeferrable, immediate, and enabled. Include the parameters for the constraint states even though they are the default settings and not technically required. Save the SQL you use in your **Solutions\Chapter10** directory in a file named **h1002.sql**.
8. Add a PRIMARY KEY constraint to the CH10DOGSHOW table. The primary key column is named DOGSHOWID. The constraint should be named by the system, and the index should reside in the USER_LOCAL tablespace. Save the SQL you use in your **Solutions\Chapter10** directory in a file named **h1003.sql**. (*Hint*: Omit the word CONSTRAINT and the constraint name to allow the system to name the constraint.)
9. Create a new table that keeps track of which dogs attend which dog shows. The table should contain the foreign key columns needed to reference the CH10DOG and CH10DOGSHOW tables, plus columns for the placement category (such as Best of Show or Best of Breed) and rank (first through fourth place). The table name should be CH10DOGATTENDANCE. Include a primary key that is made up of the columns from the two foreign keys. Include a CHECK constraint on

the rank column so that only "First", "Second", "Third", "Fourth," and a null value are allowed. Save the SQL you use in your **Solutions\Chapter10** directory in a file named **h1004.sql**.

10. Add a new column to the CH10DOGSHOW table. The column tracks the show date. It does not allow null values; however, because there is one record in the dog show table, do not add a NOT NULL constraint when you create the column.

    Next, update the row with a show date. Finally, add the NOT NULL constraint to the new column. Save the SQL you use in your **Solutions\Chapter10** directory in a file named **h1005.sql**.
11. Create a unique key on the CH10DOGSHOW table. The key should contain the SHOW_NAME and the SHOW_DATE columns. The constraint is deferrable but currently set to immediate state. Save the SQL you use in your **Solutions\Chapter10** directory in a file named **h1006.sql**.
12. Modify a constraint so that when a parent row in the CH10DOGSHOW table is deleted, all the related rows in the CH10DOGATTENDANCE table are deleted. Save the SQL you use in your **Solutions\Chapter10** directory in a file named **h1007.sql**.
13. Change the CH10DOGOWNER table so that these business rules are enforced:
    - A dog owner must have at least one year of experience.
    - If the actual number of years experience is not known, the dog owner is assigned One year of experience until the actual number is known.

    Save the SQL you use in your **Solutions\Chapter10** directory in a file named **h1008.sql**.
14. Add new columns to the CH10DOGSHOW table as follows:
    - The location of the dog show needs to be tracked, and you need one column for the city and one column for the state of the show. The state must be a two character state code and the only values allowed are: NV, CA, OR, WA, and TX. These can be left out if the location has not been determined yet.
    - The dog show must have a sponsor, such as PETCO or Purina. Allow up to 50 characters in the sponsor name. Sometimes, a tentative sponsor is recorded, so add another column that is a "T" for "tentative" or an "F" for "finalized." This column must be filled in, and should default to "T" if no value is specified.

    Save the SQL you use in your **Solutions\Chapter10** directory in a file named **h1009.sql**.
15. Modify the CH10WORLD table so that if a person is in the United States, he or she must have a US tax ID and if the person is in any other country, he or she must NOT have a US tax ID. Preexisting data should not be checked. Save the SQL you use in your **Solutions\Chapter10** directory in a file named **h1010.sql**. (*Hint*: Use "IS NULL" and "IS NOT NULL" to check a column for null or non-null values in a CHECK constraint.)

CASE PROJECT

The Global Globe newspaper database now has several tables. Your mission for this project is to add integrity constraints to all the tables you have created to enforce the business rules. Documentation is important in any project, so in addition to the SQL commands, your boss wants a document describing the connection between the constraints you create and the business rules. The business rules are listed in a text file named **case1001.txt** in the **Data\Chapter10** directory. Open the file, complete the entries in the text file as you do your SQL work, and save the modified text file in a file named **case1001.txt** in the **Solutions\Chapter10** directory on your student disk. Save all your SQL work in a file named **case1001.sql** in the **Solutions\Chapter10** directory on your student disk. (*Hint*: to create a NOT NULL constraint with a name, use the following syntax.

```
ALTER TABLE <tablename>
MODIFY(<columnname> CONSTRAINT <constraintname> NOT NULL);
```

CHAPTER

11

USERS AND RESOURCE CONTROL

In this chapter, you will:

- Create, modify, and remove users
- Discover when and how to create, use, and drop profiles
- Manage passwords
- View information about users, profiles, passwords, and resources

You have already been exposed to several users during previous exercises in this book. For example, the CLASSMATE user owns the sample data used in the book. You generally logged on as the CLASSMATE user to perform your exercises. Making changes to the database settings required you to log on as the SYSTEM or SYS users.

This chapter gives you in-depth coverage of how to create, modify, and remove database users. In addition, you learn how to set limits on the amount of space and other resources users may consume and query the data dictionary views for information about users.

OVERVIEW OF DATABASE USERS

When a new database instance is created, at least two users are created so that these users can create the tables that form the backbone of the database. The SYS user owns most of the tables needed to run the database, as well as the data dictionary views, and a host of packages and procedures built into the database for your use and for internal use. The SYS user can start up and shut down a database instance and perform recovery and backup tasks. You should log on as the SYS user to do these tasks. However, because the SYS user can also drop its own tables (which could crash the entire system), you should log on as the SYS user only for these tasks, and not for other, more routine tasks.

The SYSTEM user owns some tables, packages, and procedures. It is given the DBA role (see Chapter 12), which allows it to perform routine database administration tasks, such as creating new users, monitoring the database activities, and regulating database resources. You should log on as the SYSTEM user when you want to perform these routine tasks.

During database creation, Oracle9*i* creates other users to help it install some of the features of the database. For example, the MDSYS user is created to own functions, packages, views, and objects related to Oracle Spatial, a feature for storing dimensional data in the database. After database creation, these users are disabled to prevent anyone from logging onto the database with their accounts. (You'll learn about disabling users later in the chapter.)

After the database instance is up and running, you create users that own tables and other objects needed to implement whatever business system you plan to implement. Creating new users keeps system tables of the database and your tables in distinct logical groups. You can limit each user's abilities to create objects by setting storage limits on tablespaces. You can also create a special group of resource limits, called a profile, and assign the profile to any user.

After creating users to own the business tables, you need to create users who access these tables but don't need to create any tables of their own. These are the users that the word "user" fits best: the people who "use" the tables created to support their business functions by adding, changing, removing, and viewing data. You can control how often the user's password expires and also what resources are available to these kinds of users.

Begin exploring users by learning how to create new users in the next section.

Creating New Users

The two types of users (those who own tables and other objects and those who use tables and objects owned by other users) are created with the same command. Figure 11-1 shows the syntax of the CREATE USER command and outlines the commands that are explained in the first section of this chapter.

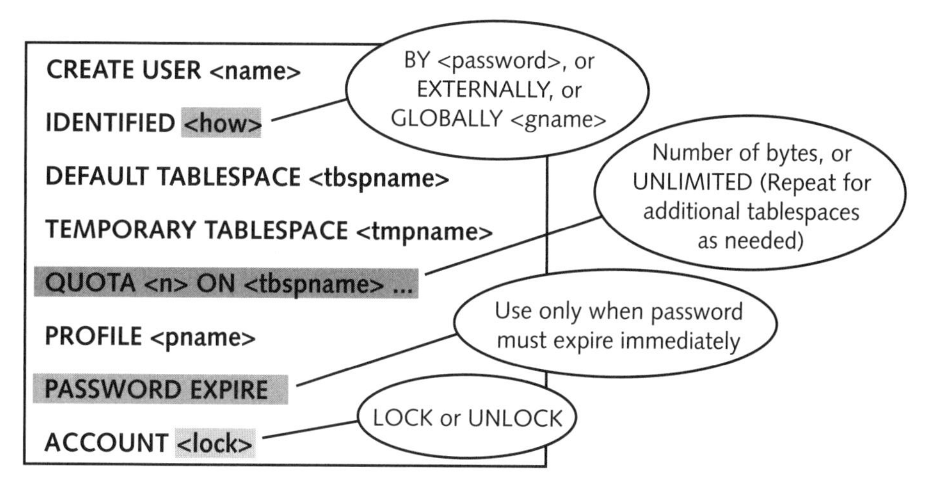

Figure 11-1 You must have the CREATE USER privilege to create a new user

The first two lines in the command syntax are actually the only required lines. The first line defines the new user's name. As you create users, keep in mind that user names are subject to the same naming conventions as tables and other database objects: a user name must be no more than 30 characters long, made up of alphanumeric characters (plus the $, #, _ symbols), and cannot start with a number. In addition, to avoid confusion, it is recommended that a user name not be identical to a keyword, such as TABLE or INDEX.

The second line defines how the database confirms the user's identification. A user's identity can be validated in three ways:

- **IDENTIFIED BY <password>:** Oracle9*i* is responsible for authenticating the user. The user must provide a password whenever logging onto the database. This type of user is called a **local user**. The password you provide must comply with the same naming conventions as other database objects. For example, you might use the password BLUESTARFISH. Many studies have shown, however, that passwords that contain only words are easier to break or guess than passwords that contain at least one number. If you want to require more complexity in passwords, Oracle9*i* provides a utility for password management. See the section on managing passwords later in this chapter.
- **IDENTIFIED EXTERNALLY:** Oracle9*i* allows a user whose user name matches his operating system logon name combined with a prefix such as OPS$ (the default). The prefix is determined by the OS_AUTHENT_PREFIX initialization parameter. The prefix can be set to null, so that the two user names match exactly. As an example, if the OS_AUTHENT_PREFIX is NT_, and the user's operating system logon name is SBAILEY, then the Oracle user name you create for SBAILY should be named NT_SBAILEY. *Users identified externally do not have to supply their user names or passwords to log on.* They simply type a slash in place of user name and password. Oracle retrieves the user's name from the operating system and assumes that the operating system has

already validated the user. For example, to log onto SQL*Plus, the user could type **sqlplus /** on the command line.

- **IDENTIFIED GLOBALLY AS <name>:** This method provides a way to identify users across distributed database systems. The advantage of **global users** is that they do not require Oracle users that have been created within the database. Instead, the user can be associated with a common user name that is shared by other users with the same privileges.

All the remaining syntax shows you optional additions you use to refine the user's capabilities in the database. The optional clauses are described here, along with suggestions on usage.

- **DEFAULT TABLESPACE:** This is only needed if the user is allowed to create database objects, such as tables or indexes. Otherwise, no default tablespace is needed. If you do not assign a default tablespace, the new user inherits the default tablespace of the user who created it. For example, the default tablespace for SYSTEM is the SYSTEM tablespace; so all users created by SYSTEM have SYSTEM as their default tablespace. Because the SYSTEM tablespace contains database objects that are critical to the functioning of the database, you should avoid creating users with SYSTEM as their default tablespace.
- **TEMPORARY TABLESPACE:** Sometimes, Oracle9*i* needs temporary space to sort query results, execute a join, or other tasks. These tasks are performed in the user's temporary tablespace. If you omit this clause, the user's temporary segments are stored in the default temporary tablespace named either in the CREATE DATABASE command or a subsequent ALTER DATABASE command. If no default temporary tablespace is designated, the user's temporary segments are stored in the SYSTEM tablespace. Storing temporary segments in the SYSTEM tablespace is inadvisable because the SYSTEM tablespace houses so many critical tables for the operation of the database. When assigning a temporary tablespace to a user, the tablespace you specify must be a temporary tablespace.
- **QUOTA:** The default **quota** (the maximum amount of storage space a user can be allocated in the tablespace) is unlimited. Allowing unlimited storage space can sometimes lead to database errors. For example, an inexperienced programmer may insert rows multiple times when testing an application and accidentally consume all the storage space available in the tablespace. In this case, other users who try to access the tablespace cannot create new tables or insert rows that cause new extents to be allocated. Remember, even when a tablespace is set to automatically allocate new extents as needed (AUTOEXTEND ON), there is a physical limitation at the point when the tablespace cannot grow because its datafiles have reached the maximum storage capacity of the physical devices. Limiting each user's total storage space prevents uncontrolled growth of the database. Limiting the temporary storage space prevents one user from

consuming so much of the temporary tablespace that other user operations are slowed or halted. Here are a few more points about quotas:

- Define quotas in any tablespace except the undo tablespace, which is used only for undo records.
- Define quotas to as many tablespaces as you want.
- Set a tablespace quota to unlimited by using the UNLIMITED keyword.
- Set a tablespace quota to zero by using the number zero as the quota.

- **PROFILE:** All users must have a profile. If you do not name a profile when you create the user, Oracle9*i* assigns DEFAULT (a predefined profile created just after the database is created) to the user.
- **PASSWORD EXPIRE:** You can set the password to automatically expire by adding the PASSWORD EXPIRE clause to the statement. This optional clause tells Oracle9*i* to create the user with the given password and then immediately cause the password to expire. An expired password makes the database prompt the user for a new password the next time he or she logs onto the database.
- **ACCOUNT:** This clause determines whether the user can log onto the database at all, even if he or she provides the correct password or is validated by the operating system. Locking a user with the ACCOUNT LOCK clause effectively locks out this user from the database. This is not usually done when a user is first created, although it is available. The default setting is ACCOUNT UNLOCK, which allows someone to log on with the new user name and password. The ACCOUNT LOCK is more often used with the ALTER USER statement to prevent an existing user from accessing the database.

Now that you have examined the syntax, it is time to dive in and create some users.

Complete the following the steps to create new users and examine the components of the CREATE USER command.

1. To start the Enterprise Manager console in Windows, on the Taskbar, click **Start/Programs/Oracle-OraHome92/Enterprise Manager console**. In UNIX, type **oemapp console** on the command line. The Enterprise Manager console login screen appears.
2. Select the **Launch standalone** radio button and click **OK**. The console appears.
3. Start up the SQL*Plus Worksheet by clicking **Tools/Database Applications/SQL*Plus Worksheet** from the top menu in the console. A background SQL*Plus process starts, and then the SQL*Plus Worksheet window appears. If the background process appears in front, simply minimize it (click the **minus sign** in the top-right corner), so that you can see the worksheet.
4. Connect as the SYSTEM user. Click **File/Change Database Connection** on the menu. A logon window appears. Type **SYSTEM** in the Username box,

the current password, *provided by your instructor*, in the Password box, and **ORACLASS** in the Service box. Leave the connection type as "Normal". Click **OK** to continue.

5. You are creating a new user named STUDENTA, who can create tables and other objects in the database. Begin by typing the first line of the command as shown here. Then press **Enter**.

   ```
   CREATE USER STUDENTA
   ```

   Next, you assign a password, or define an alternative method of authentication. For this example, use the password TRUE#1 for the user by typing this line as you continue to code the CREATE USER statement.

   ```
   IDENTIFIED BY TRUE#1
   ```

6. To create a new user, you need to supply only the user name and password. However, in this example, you examine the remaining options of the CREATE USER statement. The first optional addition is assigning a default tablespace to the user. Assume that this user can create tables, and that you want the user to create tables in the USERS tablespace. Type this line and press **Enter** to continue.

   ```
   DEFAULT TABLESPACE USERS
   ```

7. Next, assign a default temporary tablespace. This is useful even when a user cannot create tables. Type this line, and press **Enter** to assign TEMP as the temporary tablespace for this user.

   ```
   TEMPORARY TABLESPACE TEMP
   ```

   You could stop here, and the user would have access to all available space on both the USERS and the TEMP tablespaces. For this example, however, give the user 10 M in the USERS tablespace, 5 M in the USER_AUTO tablespace, and 5 M in the TEMP tablespace. Type the following lines, and press **Enter** to continue to the next step.

   ```
   QUOTA 10M ON USERS
   QUOTA 5M ON USER_AUTO
   QUOTA 5M ON TEMP
   ```

8. The next line defines the user's profile. A later section in this chapter describes how to create profiles. For now, assign the user the default profile, which, conveniently, is named DEFAULT. Type the following line and press **Enter**.

   ```
   PROFILE DEFAULT
   ```

9. The final optional setting is the ACCOUNT clause, which can block a user from logging onto the database. For this user, type the **default setting**, the **semi-colon** (;), and then click the **Execute** icon to run the command.

   ```
   ACCOUNT UNLOCK;
   ```

   Figure 11-2 shows the complete command. SQL*Plus Worksheet replies, "User created."

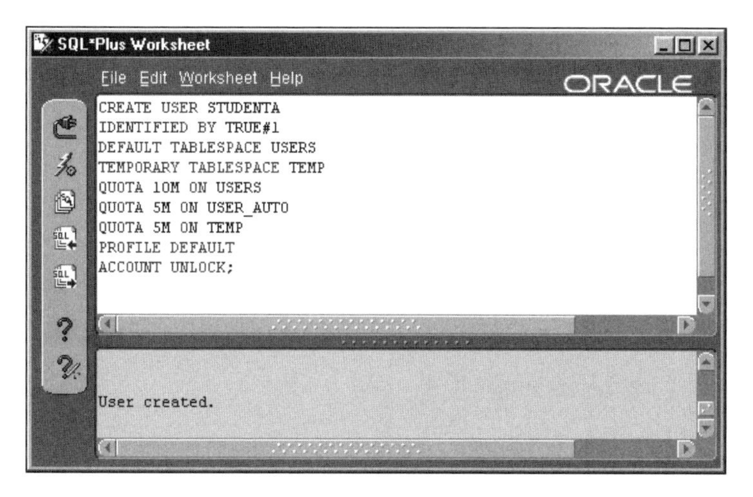

Figure 11-2 The new user STUDENTA has quotas and default tablespaces

10. Now, change the user name to STUDENTB, change the password to TRUE#2, and then execute the CREATE USER command again to create a second user named STUDENTB. The SQL*Plus Worksheet replies, "User created."

11. Type and execute one final command to give the users the system privilege needed to log onto the database.

    ```
    GRANT CREATE SESSION TO STUDENTA, STUDENTB;
    ```

    SQL*Plus Worksheet replies, "Grant succeeded."

12. Remain logged on for the next practice.

The new users you created have access to the database and quotas limiting the total storage usage, so you might think that the users can begin creating a schema; however, the users cannot create any tables until given more system privileges. Chapter 12 describes these system privileges in detail.

The default quota on all tablespaces for a user is zero unless you grant the RESOURCE role or the UNLIMITED TABLESPACE system privilege to the user.

Modifying User Settings with the ALTER USER Statement

After a user is created, you can change the initial settings using the ALTER USER statement. Figure 11-3 shows the ALTER USER syntax. As you can see, all of the settings found in the CREATE USER statement are found here as well. In addition, a new clause, DEFAULT ROLE, is available. Chapter 13 describes roles and when to modify a user's default roles.

You must have the ALTER USER system privilege to issue the ALTER USER command. There is one exception: A user can run this command to change his own password. See the section on password management for an example.

You cannot change a user's name, but instead must drop the user and create a new user with the new name.

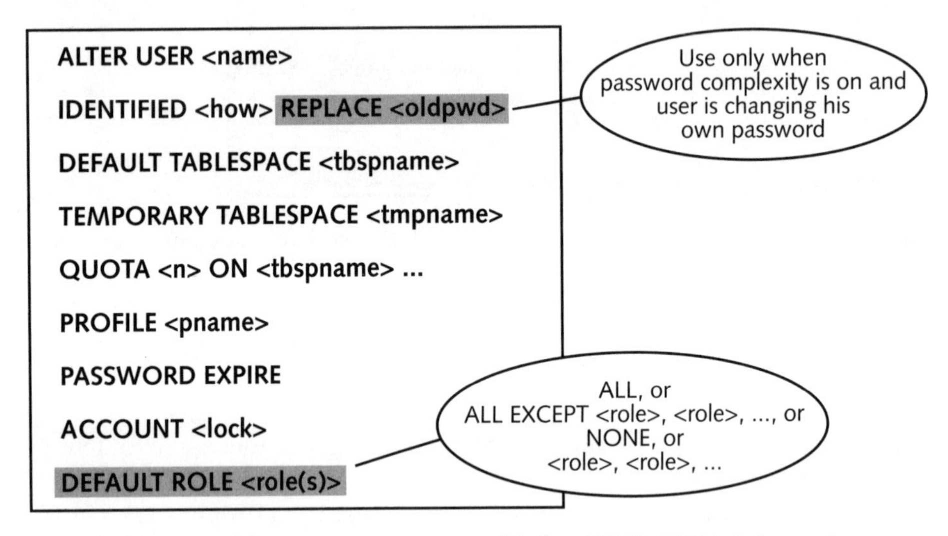

Figure 11-3 Modify most user settings with the ALTER USER statement

Follow these steps to change some of the initial settings of the STUDENTA user, which you created in the previous section.

1. Currently, the STUDENTA user has quotas on three tablespaces. To view the settings, type the following query, and run it in SQL*Plus Worksheet. The results are shown in Figure 11-4.

   ```
   SELECT USERNAME, TABLESPACE_NAME, MAX_BYTES, BYTES
   FROM DBA_TS_QUOTAS
   WHERE USERNAME = 'STUDENTA';
   ```

2. You want to allow STUDENTA to have unlimited storage on the USER_AUTO tablespace. Type and execute this command to modify the quota.

   ```
   ALTER USER STUDENTA
   QUOTA UNLIMITED ON USER_AUTO;
   ```

   SQL*Plus Worksheet replies, "User altered."

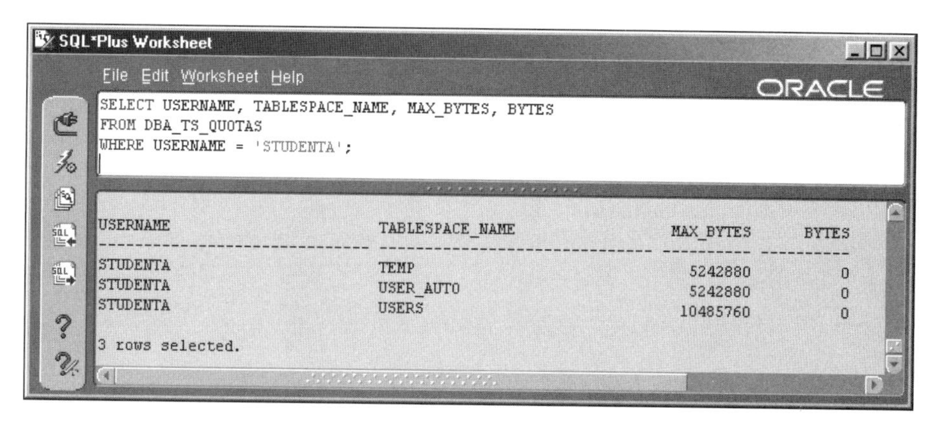

Figure 11-4 Quotas and amount of used space are listed in DBA_TS_QUOTAS

3. In addition, you have changed plans, so you want to adjust the allocated 10 M for this user on the USERS tablespace to zero. This prevents STUDENTA from creating any new tables in the USERS tablespace. Existing tables in the USERS tablespace remain intact but are prevented from allocating additional extents. Set the quota to zero by typing and running this command.

   ```
   ALTER USER STUDENTA
   QUOTA 0 ON USERS;
   ```

   The SQL*Plus Worksheet replies, "User altered."

4. To confirm the settings, run the query you ran in Step 1 again. Figure 11-5 shows the new results. As you can see, the quota on USER_AUTO is now "-1". This is the number Oracle uses to identify the quota as unlimited. The USERS tablespace no longer appears, which means the user has no quota in the USERS tablespace.
5. Remain logged on for the next practice.

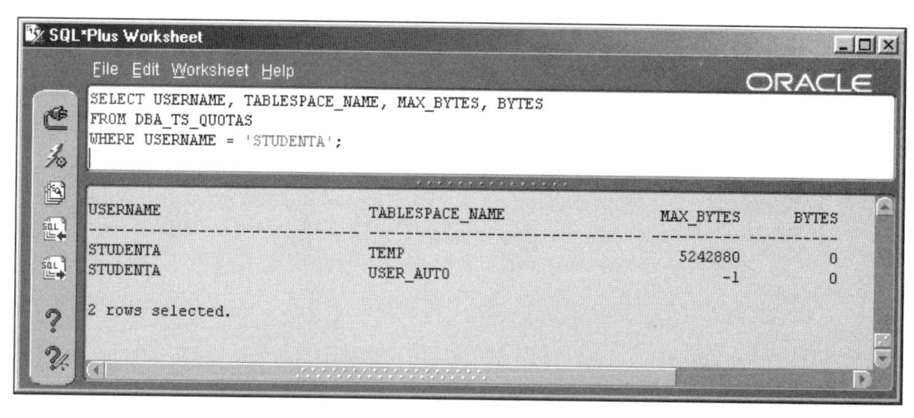

Figure 11-5 User STUDENTA has a 5 M quota on TEMP and an unlimited quota on USER_AUTO

When you change a user's quota in a tablespace, what happens to the tables that the user has already created in the tablespace? If you increase the quota, nothing happens to the tables. If you decrease the quota, the tables can still be used, but no additional extents are allocated until the storage use falls below the quota. For example, a user has three tables using five megabytes of space, and the quota is reduced to four megabytes. None of the tables can add an extent. Then, the user drops one of the tables. The remaining two tables use only three megabytes of space. Now, either table can add an extent until the total storage reaches four megabytes.

If the quota on a tablespace with existing tables is reduced to zero, these tables remain but are never allowed to allocate more space.

You have seen how to create and modify a user. Now, let's try removing a user.

Removing Users

Removing users requires the DROP USER system privilege, which the SYSTEM user has. The syntax for the command is simple:

```
DROP USER <user> CASCADE;
```

Use the CASCADE keyword if the user owns any tables or other database objects. For example, Figure 11-6 shows all the tables, indexes, and views owned by MELVIN. In addition, tables and views owned by TREVOR are shown. TREVOR has created a view that joins one of MELVIN's tables with his own. In addition, TREVOR has created a FOREIGN KEY constraint on one of his tables that references one of MELVIN's tables. The items shown in Figure 11-6 with heavy outlines will be dropped. Views belonging to TREVOR are marked invalid if they reference any of MELVIN's tables.

To drop the user named MELVIN, you would type this command. (*Do not actually execute these commands because MELVIN and SMITH are not in your database.*)

```
DROP USER MELVIN CASCADE;
```

To drop a user named SMITH, who owns no tables or other database objects, you would type this command.

```
DROP USER SMITH;
```

If a user has created other users, those users are not dropped when the creating user is dropped. The new users do not belong to the original user's schema. If a user has created tables you want to keep, do not drop the user. Instead, change the user account to LOCK status.

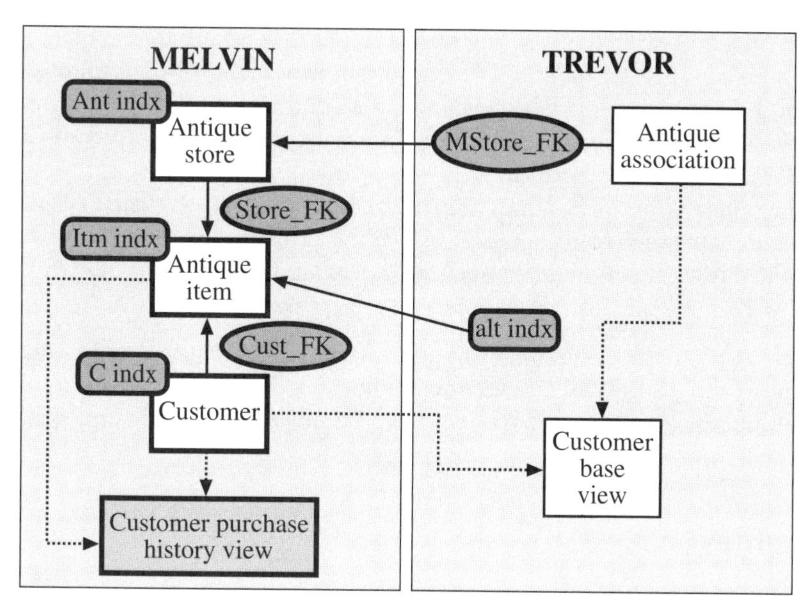

Figure 11-6 When MELVIN's user name is dropped, some of TREVOR's tables, indexes, and views are affected

In addition to determining a user's tablespace quotas, you can define several aspects of how Oracle9*i* handles the user's password (such as the number of days until it expires again) and you can set quotas on CPU time, among other resources. All this is done through creating and assigning profiles, which is the topic of the next section.

Introduction to Profiles

You saw in the previous section that you can specify a profile when you create or alter a database user. A **profile** is a collection of settings that limits the use of system resources and the database. After a profile is created, it can be assigned to any number of users. Each user can be assigned only one profile at a time. If the user has a profile and you assign a different profile, that new profile overrides the old one as soon as the user begins a new session. The user's current session is not affected by a profile change.

Even after creating and assigning profiles to users, none of the limits are enforced until the database initialization parameter RESOURCE_LIMIT is set to TRUE. By default, the parameter is set to FALSE. You can change this parameter in the **init.ora** file, then restart the database instance, or you can modify the parameter dynamically with the ALTER SYSTEM command.

The default profile, named DEFAULT, has no limits on resources or database use. This works well for smaller systems in which resources are plentiful. However, as a system grows, resources may become stretched, causing some tasks to take longer to complete. For example, your system supports an online ordering system in which response time is critical. In addition, you have background jobs that process accounting and billing running at night. Now that the online ordering is international in scale, online users are complaining that response time during night hours is very slow. Your research shows that long-running queries from online users performing generalized searches on the database have caused other online users to wait for resources. To solve the problem, you can create a profile for online users to limit the amount of data blocks read and the amount of CPU time used in one call, so that queries that are too general are terminated.

A second use of profiles involves managing passwords. A profile does not contain a user's password; however, a profile determines how often the user must change his password, how many days must pass before an old password can be reused, and other important settings.

The next section shows how to create a profile.

Creating Profiles

The syntax for creating a profile looks like this.

```
CREATE PROFILE <profile> LIMIT
<password_setting> ...
<resource_setting> <limit> ...;
```

Replace <password_setting> with the setting you want to limit, replace <resource_setting> with the name of the resource, and replace <limit> with either UNLIMITED, DEFAULT, or a number.

There are seven password settings that you can set within a profile. If you do not name one of the password settings, Oracle9*i* uses the value set for that password setting that it finds in the DEFAULT profile. If none is found, Oracle9*i* assumes the password setting is unlimited. The password settings are:

- **FAILED_LOGIN_ATTEMPTS:** The maximum number of times the user can retry the password before the account is locked
- **PASSWORD_LIFE_TIME:** The maximum number of days the password can be used without changing
- **PASSWORD_REUSE_TIME:** The minimum number of days before the same password can be used again
- **PASSWORD_REUSE_MAX:** The minimum number of times the password must change before the same password can be used again
- **PASSWORD_LOCK_TIME:** The number of days (or a fraction of a day, such as 1/24 — one hour) that the account is locked after the maximum set by FAILED_LOGIN_ATTEMPTS has been reached

- **PASSWORD_GRACE_TIME:** The number of days after the password expires in which the user is given a warning that the password is expired, and is allowed to log on with the old password
- **PASSWORD_VERIFY_FUNCTION:** The database function called to verify the complexity of the password

PASSWORD_REUSE_TIME and PASSWORD_REUSE_MAX cannot both be set at the same time. You must set one of them to UNLIMITED if you set the other to a number.

There are nine resources you can limit with a profile. If you do not name one of the resources, Oracle9*i* uses the limit set for that resource that it finds in the DEFAULT profile. If none is found, Oracle9*i* assumes this resource is unlimited.

The resources on which you can set limits include:

- **SESSIONS_PER_USER:** Set the maximum number of concurrent sessions.
- **CPU_PER_SESSION:** Set the maximum CPU time (in hundredths of a second) for a user's entire session.
- **CPU_PER_CALL:** Set the maximum CPU time (in hundredths of a second) for any single call to the database. A **call** is a single task sent to the database, such as parsing a new SQL statement or executing a query.
- **CONNECT_TIME:** Set the maximum number of minutes a user's session can last.
- **IDLE_TIME:** Set the maximum number of minutes a user's session can remain inactive. A long-running query or other task in which the user is waiting for the database does not count toward idle time.
- **LOGICAL_READS_PER_SESSION:** Set the maximum number of data blocks read from either memory or disk during a user's session.
- **LOGICAL_READS_PER_CALL:** Set the maximum number of data blocks read during a single call.
- **PRIVATE_SGA:** Set the number of bytes (in bytes, kilobytes, or megabytes) that the user's session may allocate in the shared pool of the SGA. This only applies to a shared server, in which part of the shared pool is allocated to user sessions. This setting is ignored on dedicated servers.
- **COMPOSITE_LIMIT:** Set the maximum resource cost (in service units) for a session. A service unit (defined in the "Controlling Resource Usage" section) calculates a resource cost based on a weighted sum of CPU_PER_SESSION, CONNECT_TIME, LOGICAL_READS_PER_SESSION, and PRIVATE_SGA. Oracle9*i* monitors these values for each user session. To use the COMPOSIT_LIMIT setting, you must assign a weight to one or more

11

of these four resources. By default, Oracle9*i* weights them all at zero, meaning a user never reaches a composite limit, because all the values add up to zero. You work with an example of weighting resources later in the chapter.

You can modify the DEFAULT profile in the same way you modify profiles that you create; however, you cannot drop the DEFAULT profile.

To help you experiment with profiles, begin by creating two profiles that you will use when setting both password and resource limits later in the chapter. You should still be logged onto the SQL*Plus Worksheet as the SYSTEM user. Follow these steps to create the two profiles.

1. You decide that you need two different profiles: one for programmers and one for end users. Create the first profile, named PROGRAMMER, for the programmers by typing and executing this command.

   ```
   CREATE PROFILE PROGRAMMER LIMIT
   SESSIONS_PER_USER 2;
   ```

   The SQL*Plus Worksheet replies, "Profile created."

   The resource name, SESSIONS_PER_USER, limits the number of concurrent sessions a user is allowed to maintain. You want the programmers to avoid logging on with multiple sessions, a habit that drives the database administrator crazy. (You add more limits to the profile later in this chapter.)

2. Now, create another profile for users called POWERUSER. You know you want the users to change passwords at least once every two months, so you create the profile with the password setting PASSWORD_LIFE_TIME of 60 days. Type and run the following command.

   ```
   CREATE PROFILE POWERUSER LIMIT
   PASSWORD_LIFE_TIME 60;
   ```

   The SQL*Plus Worksheet replies, "Profile created."

3. Remain logged on as you work with these profiles in the next sections.

The next section continues examining profiles, with special emphasis on the password limits available.

Managing Passwords

You saw how to assign a password to a user when creating the user. Now, you look at methods for managing a user's password. There are three different areas to examine when working with passwords:

- **Changing a password and making it expire:** These are the only password properties that you can change specifically for a user by using the ALTER USER command.

- **Enforcing password time limits, history, and other settings:** All other password properties can be changed only through profiles. You create a profile, give the profile password properties, and then assign the profile to a user.
- **Enforcing password complexity:** This uses a combination of a function and a profile. You run a predefined SQL script (or one you create yourself) to verify the complexity of a password, then adjust the PASSWORD_VERIFY_FUNCTION setting in a profile, and assign that profile to a user.

Follow these steps to examine all three of these areas. Continue in SQL*Plus Worksheet, logged on as SYSTEM.

1. Imagine that the new user created earlier, named STUDENTA, has forgotten his password. You have forgotten it as well, so you attempt to find it in the database by typing and executing this query to look at the password in the DBA_USERS table.

```
SELECT USERNAME, PASSWORD
FROM DBA_USERS
WHERE USERNAME = 'STUDENTA';
```

   Figure 11-7 shows the results of the query. As you can see, the password appears to be a string of letters and numbers, but this is not the password you assigned. The string is an encrypted form of the password that Oracle9*i* uses to store all passwords. Unless you recorded the password, you cannot recover the actual value from the database, even if you are the DBA.

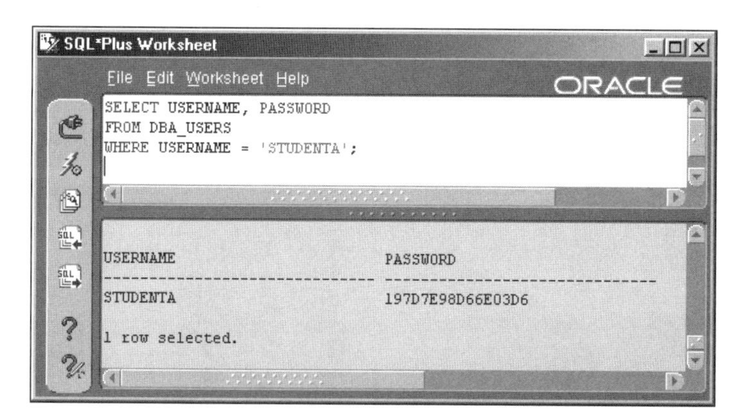

Figure 11-7 All passwords are encrypted in the database

2. Having determined that you cannot find the original password, you decide to change the password and inform the user of the new password. In addition, you make the password expire immediately so the user must choose a new

password that he finds easier to remember. Type and execute this command. *This is the first method of password management: changing the password and making it expire.*

```
ALTER USER STUDENTA
IDENTIFIED BY STUDENTA
PASSWORD EXPIRE;
```

3. Remain logged onto the SQL*Plus Worksheet, and go to a command prompt to simulate a session for STUDENTA. In Windows, on the Taskbar, click **Start/Programs/Accessories/Command Prompt**. In UNIX, go to the $ prompt.
4. Type this command at the command prompt, and press **Enter** to start up SQL*Plus:

   ```
   sqlplus studenta/studenta@oraclass
   ```

   SQL*Plus starts up and detects that the user's password has expired. SQL*Plus sends an error message and prompts for a new password.
5. Type **MYPASSWORD** as the new password, and press **Enter**. SQL*Plus prompts you to reenter the password.
6. Type **MYPASSWORD** again, and press **Enter**. SQL*Plus accepts the new password and starts up a session for STUDENTA. Figure 11-8 shows the window at this point. *This is how a user can change his own password using SQL*Plus.*
7. Exit SQL*Plus by typing **EXIT** and pressing **Enter**.
8. If you are in Windows, close the Command prompt window by clicking the **X** in the top-right corner. Return to the SQL*Plus Worksheet session you were using in Step 2.

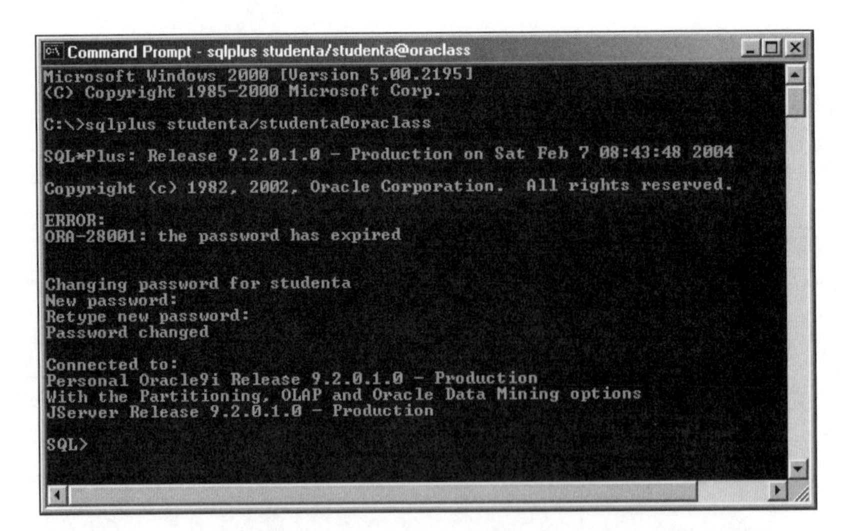

Figure 11-8 SQL*Plus prompts for a new password when the old password expires

9. Make sure that users don't reuse old passwords. This can only be done using profiles. Adjust the POWERUSER profile so that the user cannot reuse a password until he has used ten other passwords, and require the account to be closed for one day if the user attempts and fails six times in a row to log on. Type and execute this command to modify the profile.

```
ALTER PROFILE POWERUSER LIMIT
PASSWORD_REUSE_MAX 10
FAILED_LOGIN_ATTEMPTS 6
PASSWORD_LOCK_TIME 1;
```

   The SQL*Worksheet replies, "Profile altered."

10. Assign the POWERUSER profile to the STUDENTA user by typing and running this command. The rules set for passwords in the POWERUSER profile will be imposed on STUDENTA immediately. *This demonstrates the second method for password management: enforcing password time limits, history, and other settings.*

```
ALTER USER STUDENTA
PROFILE POWERUSER;
```

   The SQL*Plus Worksheet replies, "User altered."

11. Later on, you find out that STUDENTA has changed his password to "MYPASSWORD." This raises a flag in your mind, because a password like that is too common, making it very easy for an unauthorized user to guess. You ask a few other users what their passwords are, and discover that most of them use their user name or some simple word. You decide that this could be a security issue and want to require users to choose more complex passwords (thus making it harder for unauthorized users to guess correctly). Oracle9*i* provides a predefined function that you can create that specifies these rules for all passwords:

   - The password must contain at least four characters.
   - The password cannot be identical to the user name (both user name and password are *not* case sensitive).
   - The password must contain at least one alphabetic, one numeric, and one punctuation mark.
   - The password must not be a simple word, such as user, database, or password.
   - The password must differ from the previous password by at least three characters.

   To use this password management function, log on as SYS and run the script that creates the password function and places it into the DEFAULT profile, as described in the next two steps.

12. Change your connection to SYS by selecting **File/Change Database Connection** in the SQL*Plus Worksheet. The Database Connection Information window appears. Type **SYS** in the Username box, the current password for SYS, *provided by your instructor*, in the Password box, and **ORACLASS** in the Service box. Select **SYSDBA** in the Connect as box. Click **OK** to log on.

13. SQL*Plus Workshop replies, "Connected" in the lower pane. Run the Oracle-provided, password management script by clicking **Worksheet/Run Local Script**. Navigate to the **ORACLE_HOME\rdbms\admin** directory and double-click on the **utlpwdmg.sql** file. (Replace ORACLE_HOME with the full path to the main directory containing the Oracle9*i* software.) The script runs and SQL*Plus Worksheet replies, "Profile altered." Figure 11-9 shows the worksheet at this point. The script created a function named VERIFY_FUNCTION and then modified the DEFAULT profile to adjust several password settings, including the use of the new function for verifying passwords.

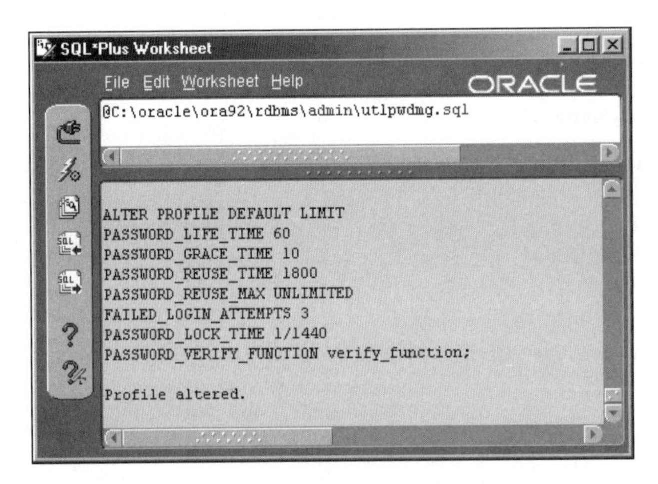

Figure 11-9 The script uses the ALTER PROFILE command

Even though the script changed the DEFAULT profile, and not the POWERUSER profile, it affects any user with the POWERUSER profile, because the setting associated with password complexity, PASSWORD_VERIFY_FUNCTION, has not been set in the POWERUSER profile. Only settings omitted from the POWERUSER profile fall back to the value in the DEFAULT profile. So, the STUDENTA user now has password complexity enforced the next time he changes his password. *You have just used the third method of managing passwords: enforcing password complexity.*

14. Reconnect as the SYSTEM user by clicking **File/Change Database Connection** and logging on as SYSTEM.

15. Test the new password verification requirements by typing and executing this command, which should produce an error, because it changes STUDENTA's password to an invalid password based on the password complexity validation rules. Remember, either the DBA or the user is able to change the user's password.

```
ALTER USER STUDENTA IDENTIFIED BY A1PLUS;
```

Figure 11-10 shows the results: An error message was returned by the VERIFY_FUNCTION function.

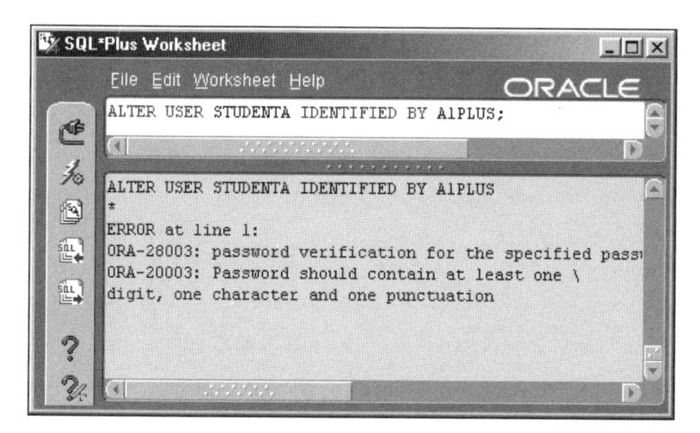

Figure 11-10 This password does not contain a symbol

16. Change the password to **A1PLU$** and try again. This time, the change succeeds. The SQL*Plus Worksheet replies, "User altered."
17. To see the password-related settings for all profiles, type this query and run it.

```
SELECT * FROM DBA_PROFILES
WHERE RESOURCE_TYPE='PASSWORD'
ORDER BY 1, 2;
```

Figure 11-11 shows the results. Notice that all parameters are listed for all the profiles. Those that are not specifically set have DEFAULT as their settings, meaning that the value in the DEFAULT profile is in force.

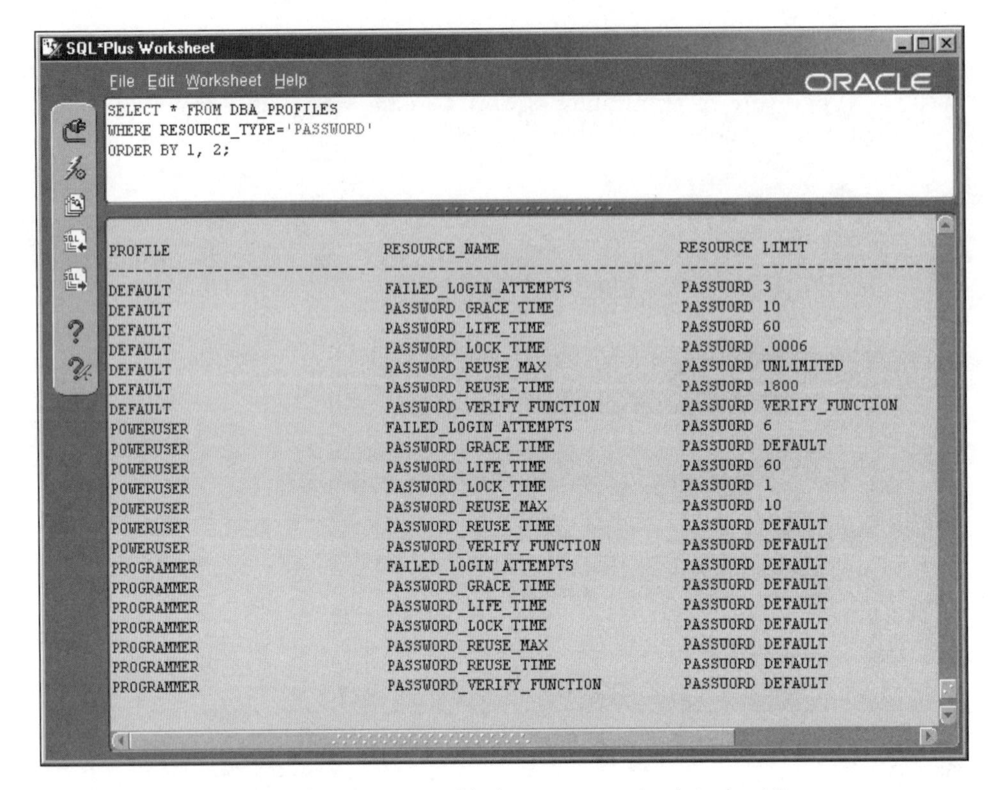

Figure 11-11 The POWERUSER profile has password-related settings

18. Remain logged on for the next practice.

Oracle documentation states that the **utlpwdmg.sql** file (used in Step 13) is provided more as an example than a real-world password validation tool. Use the source code as a model for writing your own password function.

Now that you have tried out the three ways to manage passwords, let's look at the resource limits that a profile can impose on a user session.

Controlling Resource Usage

You have already used the CREATE PROFILE and ALTER PROFILE commands. This section examines the profile settings that specifically control the use of computer resources. The syntax for ALTER PROFILE, with the resource clauses listed, looks like this.

```
ALTER PROFILE <profile> LIMIT
<password_setting> ...
SESSIONS_PER_USER <concurrent sessions>
CPU_PER_SESSION <hundredths of seconds>
CPU_PER_CALL <hundredths of seconds>
```

```
CONNECT_TIME <minutes>
IDLE_TIME <minutes>
LOGICAL_READS_PER_SESSION <data blocks>
LOGICAL_READS_PER_CALL <data blocks>
PRIVATE_SGA <bytes>
COMPOSITE_LIMIT <service units>
```

Follow these steps to try out resource settings for the PROGRAMMER profile that you created earlier in this chapter. You should still be logged onto SQL*Plus Worksheet as the SYSTEM user.

1. When you plan to adjust resource limits in any Oracle9*i* database, you must first change the initialization parameter named RESOURCE_LIMITS. The default setting of RESOURCE_LIMITS is FALSE meaning that resource limits set in profiles are ignored. Changing the parameter to TRUE means that resource limits set in profiles are enforced. Type and execute this command.

   ```
   ALTER SYSTEM SET RESOURCE_LIMIT=TRUE;
   ```

   The SQL*Plus Worksheet replies, "System altered."

2. All the resource limits in the DEFAULT profile are set to UNLIMITED. Let's say you want to limit the users assigned to the PROGRAMMER profile to 15 minutes of idle time before the system ends their sessions. In addition, you want to limit the CPU_PER_CALL to 1 second, because your current system is running at 95 percent of its CPU usage. Type the command, and execute it to set these limits. Remember, the CPU_PER_CALL setting requires hundredths of a second, so 1 second is set by stating 100 hundredths of a second.

   ```
   ALTER PROFILE PROGRAMMER LIMIT
   IDLE_TIME 15
   CPU_PER_CALL 100;
   ```

   The SQL*Plus Worksheet replies, "Profile altered."

3. Next, you want to set a composite limit for the users. Composite limits require two steps. First, you set a relative weight for the resources you want to include when Oracle9*i* calculates service units. Second, you set the maximum service units in the profile. Assume for now that you know that you want to weight the CPU_PER_SESSION at 1000 and the PRIVATE_SGA at 1. This means that 10 seconds of CPU (1000 hundredths of a second) has the same relative weight as one byte of storage in the private SGA memory. As you can tell, CPU is not the most critical resource for this particular example. To set these resource weights, type and run the ALTER RESOURCE statement as follows.

   ```
   ALTER RESOURCE COST
   CPU_PER_SESSION 1000
   PRIVATE_SGA 1;
   ```

   The SQL*Plus Worksheet replies, "Resource cost altered."

4. Add a composite limit to the PROGRAMMER profile by typing this command and executing it. You will find an explanation of how the composite limit is calculated for a user's session at the end of these steps.

```
ALTER PROFILE PROGRAMMER LIMIT
COMPOSITE_LIMIT 50000;
```

The SQL*Plus Worksheet replies, "Profile altered."

5. Assign the PROGRAMMER profile to the STUDENTB user by typing and executing this command.

```
ALTER USER STUDENTB
PROFILE PROGRAMMER;
```

The SQL*Plus Worksheet replies, "User altered."

6. To view the resource settings for the system, type and run this query. Figure 11-12 shows the results.

```
SELECT * FROM RESOURCE_COST;
```

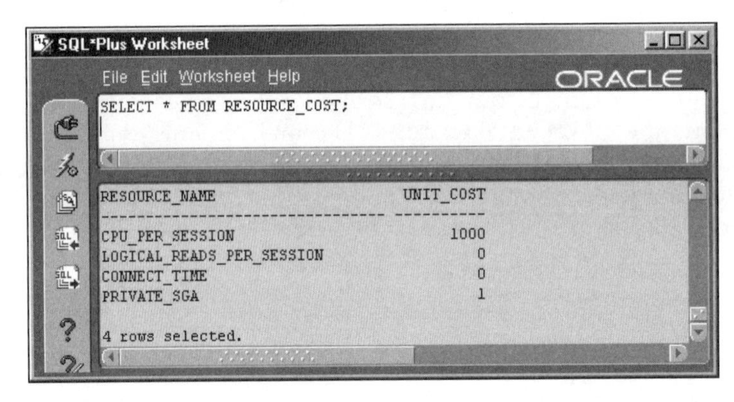

Figure 11-12 Resources can be reset to zero using the ALTER RESOURCE COST statement

7. Remain logged on for the next practice.

To decide what type of weights to use, you may want to monitor your database's activity for a while. See Chapter 12 for more information on the AUDIT SESSION command.

In our example, the composite limit monitors two resources, because only two of the four resources have been set to a nonzero cost. Let's imagine that the STUDENTB user is logged onto the database. STUDENTB has 3200 bytes of private SGA allocated. STUDENTB runs a complex query that requires 35/100 of a second to run. At this point, the composite cost of STUDENTB's session is calculated like this:

$(3200 \times 1) + (35 \times 1000) = 37{,}000$ service units

The user's session is below the maximum of 50,000 service units. A **service unit** is simply the name for the calculated value of the weighted sum of the resources consumed by the user session.

The STUDENTB user runs a series of updates, inserts, deletes, and queries that consume another 5,000 bytes of private SGA and increase the total CPU usage to one half second. Now the user session's total service units are:

$(8200 \times 1) + (50 \times 1000) = 58200$ service units

The session has exceeded the service unit cap of 50,000, so the session's transaction is rolled back and the user's session is ended with an error message.

Read about how to set realistic resource limits in the Oracle9*i*, Release 2 document named *Concepts*.

Dropping a Profile

The syntax of DROP PROFILE is similar to the syntax for dropping a user in that it includes a CASCADE parameter.

```
DROP PROFILE <profile> CASCADE;
```

You must add CASCADE if any users have been assigned the profile being dropped. Oracle9*i* automatically resets these users to the DEFAULT profile.

For example, if three users have been assigned to the ACCT_MGR profile, drop the profile like this:

```
DROP PROFILE ACCT_MGR CASCADE;
```

The final section shows how a user can view password and resource limits in his profile.

OBTAINING PROFILE, PASSWORD, AND RESOURCE DATA

You have already seen the following data dictionary views while going through the chapter:

- **DBA_USERS:** View user profile, expiration date of password, and account status.
- **DBA_TS_QUOTAS:** View the storage quotas of each user.
- **RESOURCE_COST:** View the weight setting for each resource used in calculating COMPOSITE_COST.
- **DBA_PROFILES:** View the settings for each profile.

As a user, your activity in the database is stopped if you reach a limit on a resource set in your profile. Follow these steps to run queries on the data dictionary views to learn about your own profile settings.

1. Begin by changing your connection in SQL*Plus Worksheet, so that you are logged on as STUDENTA, one of the new users you created in this chapter. Click **File/Change Database Connection**. Type **STUDENTA** in the Username box, **A1PLU$** in the Password box, and **ORACLASS** in the Service box. Click **OK** to log on. The SQL*Plus Worksheet replies, "Connected."
2. Find out what your current password settings are by running this query.

```
SELECT * FROM USER_PASSWORD_LIMITS;
```

3. Figure 11-13 shows the results. The figure shows these characteristics of the password management for the STUDENTA user.
   - The password will expire in 60 days.
   - The user may continue to use the old password for 10 days after expiration.
   - The password cannot be reused until 10 other passwords have been used. (The PASSWORD_REUSE_MAX overrides the PASSWORD_REUSE_TIME, because the latter is defined in the DEFAULT profile, not the POWERUSER profile.)
   - The user is locked out for one day if he fails to log on six times in a row.
   - The password must be validated by the VERIFY_FUNCTION function.

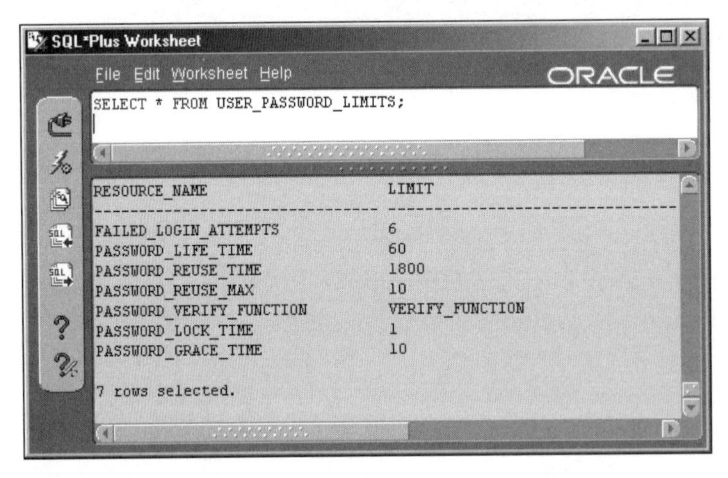

Figure 11-13 Users can view their own password limits with this view

4. Connect to STUDENTB now by clicking **File/Change Database Connection** in the menu. Type **STUDENTB** in the Username box, **TRUE#2** in the Password box, and **ORACLASS** in the Service box. Click **OK** to log on. The SQL*Plus Worksheet replies, "Connected."

5. Now query another view to see STUDENTB's resource limits.

```
SELECT * FROM USER_RESOURCE_LIMITS;
```

Figure 11-14 shows the results. From this, you know that STUDENTB has these resource limitations:

- A composite limit of 50,000 service units
- A maximum of 100 hundredths of a second (that is, one second) of CPU time per call
- No more than 15 minutes of idle time
- A limit of two sessions at a time
- Unlimited resources otherwise

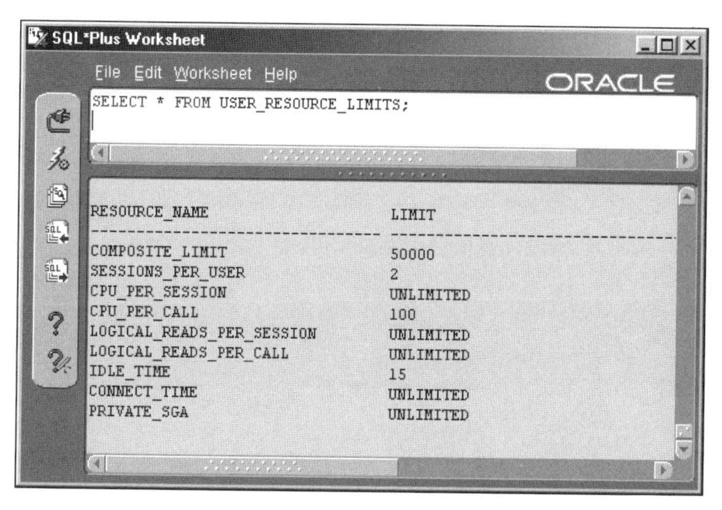

Figure 11-14 Resource limits exposed in the USER_RESOURCE_LIMITS view

6. Log off SQL*Plus Worksheet by clicking the **X** in the top-right corner.
7. You now see the Enterprise Manager console. Double-click the **ORACLASS** database, and then double-click the **Security** icon. This shows the Security Management main page, as shown in Figure 11-15.
8. Click the **Users** folder. A list of users appears in the right pane along with high-level information about the users, as shown in Figure 11-16. Scroll down until you find STUDENTA, the line that is highlighted in Figure 11-16. The columns are adjusted so you can view the profile and other details easily.

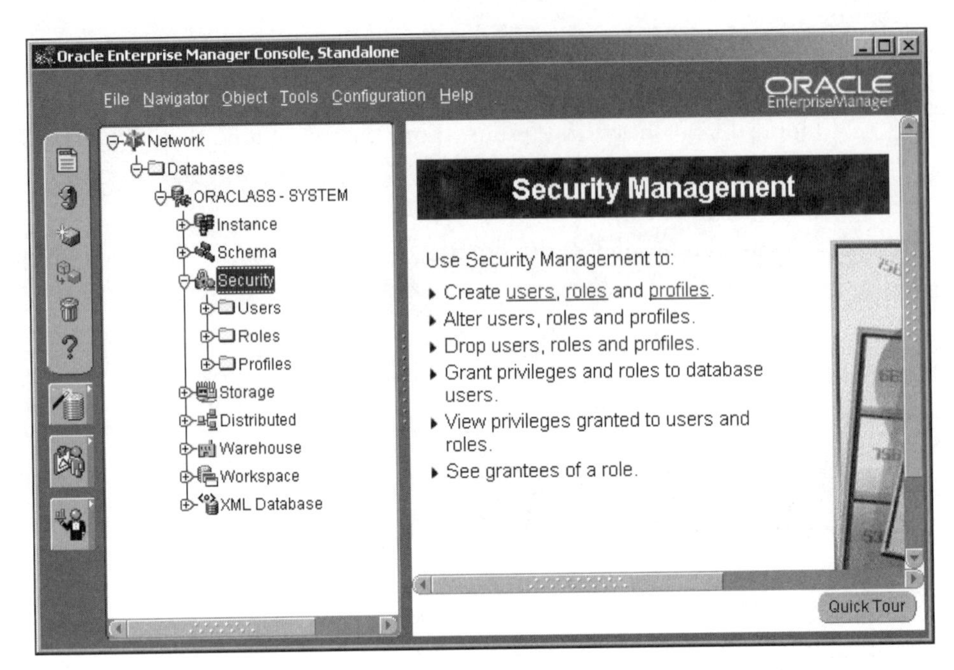

Figure 11-15 The Security Manager handles users, roles, and profiles

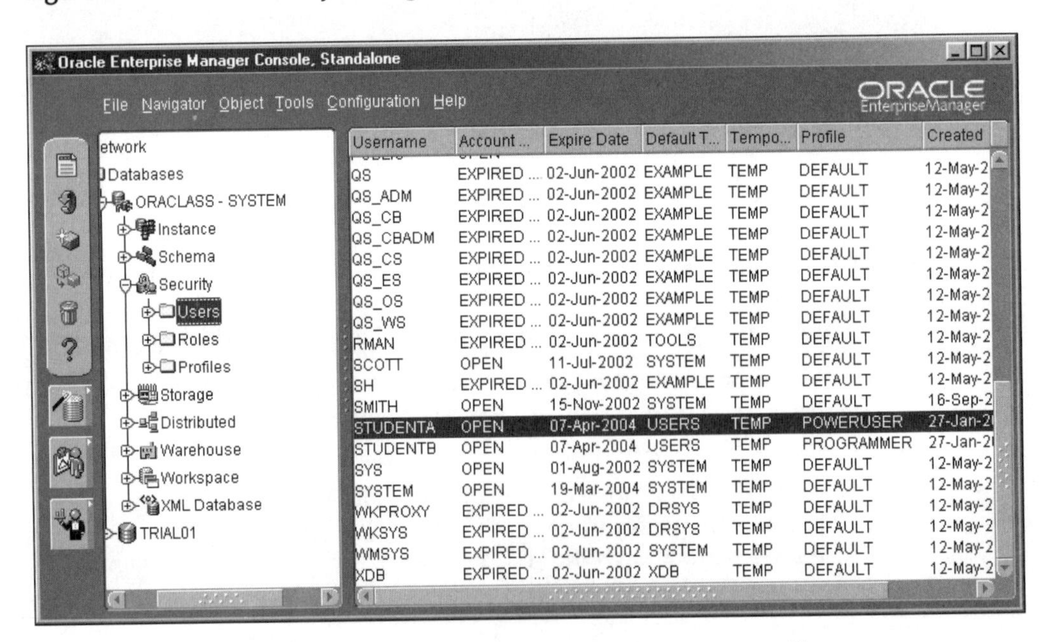

Figure 11-16 STUDENTA has been assigned the POWERUSER profile

9. Double-click **STUDENTA** in the right pane. This brings up the property sheet for the user, as shown in Figure 11-17.

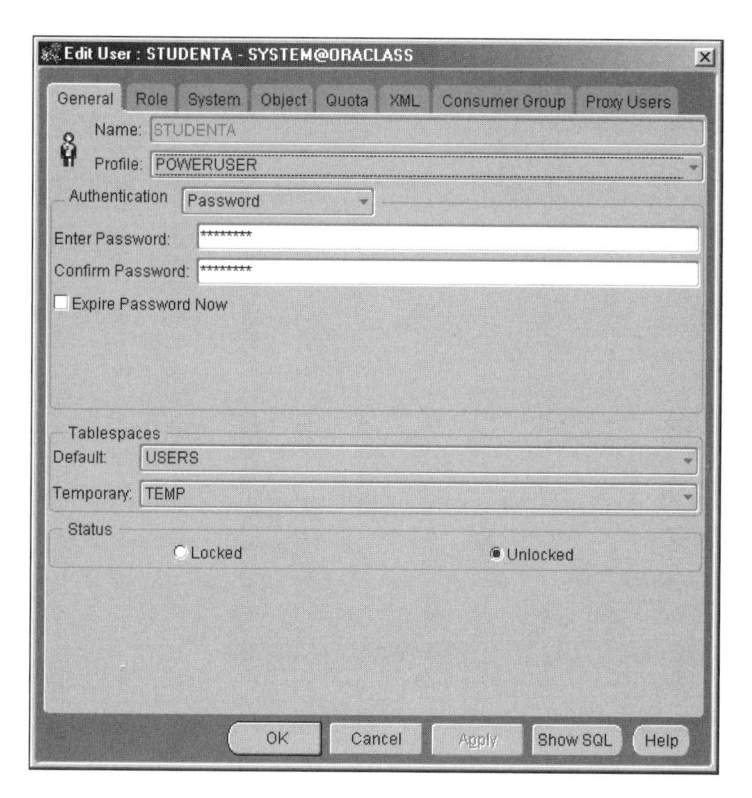

Figure 11-17 The user's property sheet displays many traits of the user

You can change the user's password, default tablespace, account status, and profile in this window without writing any code.

10. Click the **Quota** tab. The quotas for each tablespace are listed here, as you see in Figure 11-18. Your list of tablespaces may be slightly different from those shown in the figure. You can even modify the quota by selecting a tablespace row and using the radio buttons and Value box at the bottom of the window.

11. Close the property sheet by clicking the **X** in the top-right corner. This returns you to the main console window.

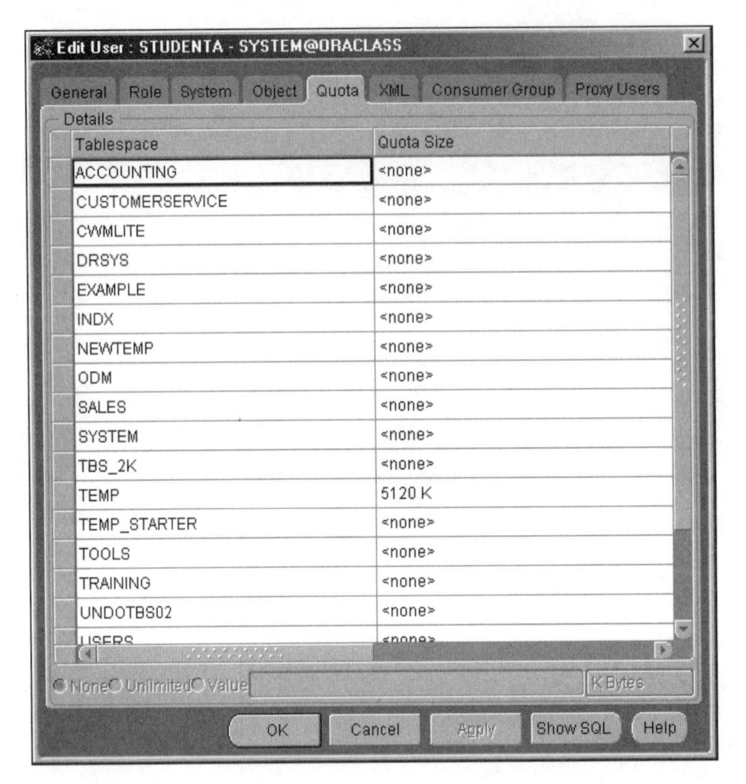

Figure 11-18 Quotas are unlimited, none, or a specific number

12. Double-click the **Profiles** folder in the left pane. A list of profiles appears in the right pane, as you see in Figure 11-19.
13. Double-click the profile named **POWERUSER** in the *left* pane. Figure 11-20 shows the property sheet that appears. You can click on any of these resource limits and choose from some suggested values in the drop-down lists, or type a value in the box.

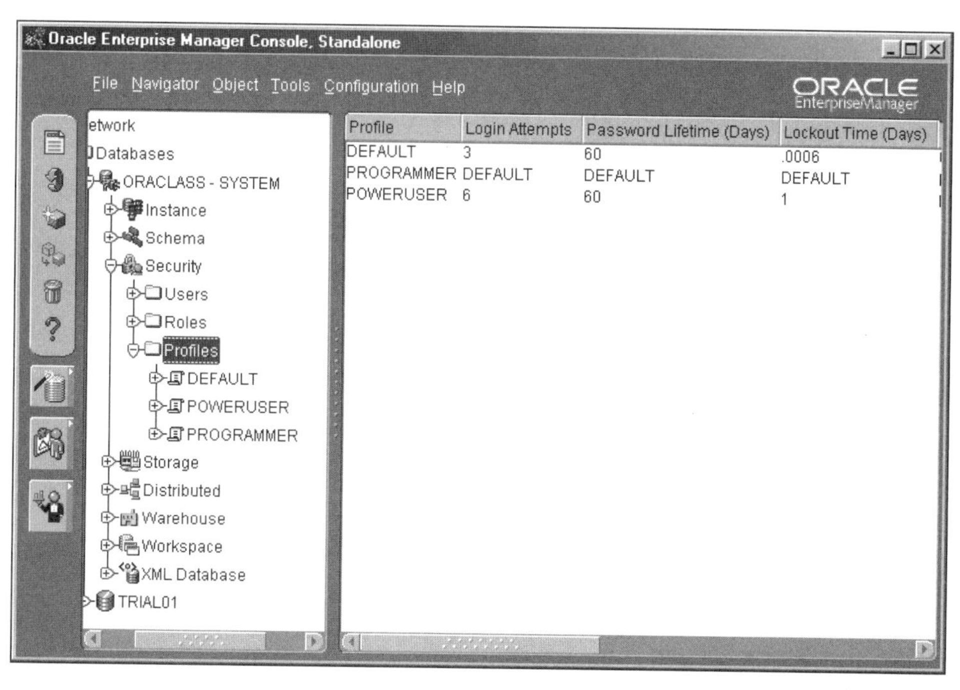

Figure 11-19 Password limits not set in a profile are marked DEFAULT

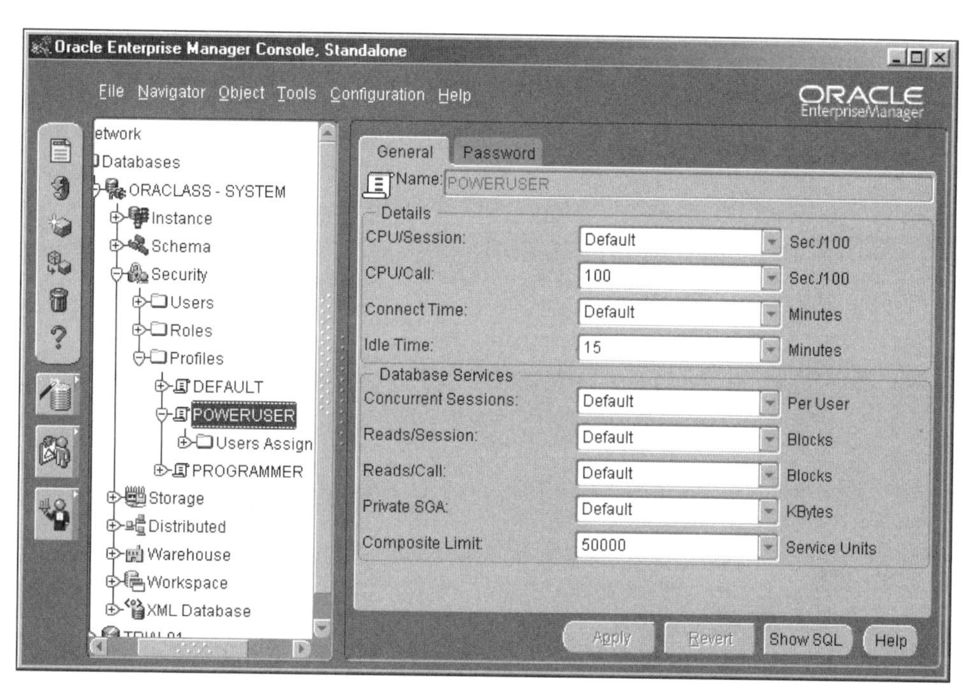

Figure 11-20 Any of the resource limits can be changed quickly

14. Click the **Password** tab. The password limits are now displayed, as shown in Figure 11-21. As you can see, you are provided with a simple click-and-type format for adjusting these settings. At times, you may find this an easier way to modify users and profiles.

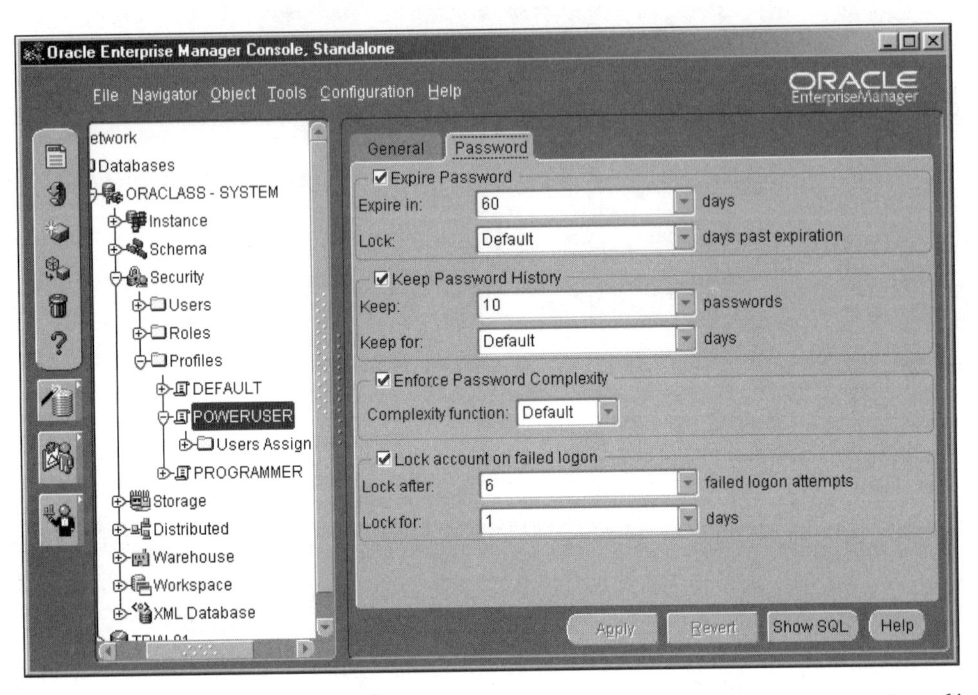

Figure 11-21 Anything marked Default reverts to the setting in the DEFAULT profile

15. Close the console by clicking the **X** in the top-right corner.

As you can see from performing these practices, the console gives you an excellent way to look at the settings for users and profiles without writing queries.

Chapter Summary

- Users are created to either own a schema or access another user's schema.
- The CREATE USER statement requires a user name and a method of identifying the user, such as a password.
- Users identified externally or globally are validated outside the database.
- Tablespace quotas limit a user's storage space.
- A new user cannot connect to the database if the account is locked or the CREATE SESSION privilege was not granted.

- Use ALTER USER to change a password, quota, or default tablespace of a user.
- DROP USER <username> CASCADE removes the user and everything he has created.
- Profiles store a collection of password and resource limits.
- The RESOURCE_LIMIT parameter must be set to TRUE before limits are enforced.
- You can modify the DEFAULT profile, but you cannot drop it.
- Passwords can be changed by the DBA and by the user.
- Profile password limits include how long a password can stay the same and when it can be reused.
- The **utlpwdmg.sql** script creates a function for password validation and alters the DEFAULT profile to use it.
- Use the profile as a method of limiting CPU usage, connect time, and more.
- The COMPOSITE_COST limit uses a weighted sum of several resources.
- DROP PROFILE <profile> CASCADE drops the profile and resets users with that profile back to the DEFAULT profile.
- The DBA_PROFILES data dictionary view displays resource limits for all profiles.
- The USER_RESOURCE_LIMITS data dictionary view displays a user's resource limits (based on the user's profile and the DEFAULT profile).

Review Questions

1. What two reasons justify creating new users?
2. The ________________ and ________________ methods of identifying a user do not require a password.
3. The OS_AUTHENT_PREFIX is null. You log onto your computer as JOEBANKS and then issue this command:

```
sqlplus /
```

   What happens next?

   a. The database prompts you for a password.

   b. The database connects you as OPS$JOEBANKS.

   c. The database cannot connect you because no prefix is specified.

   d. The JOEBANKS user must be identified EXTERNALLY.

   e. None of the above.

4. The default tablespace for SARAH is USERS, where she has a quota of 10 M. She has no other quotas. SARAH creates a table in the USER_AUTO tablespace. What system privilege does SARAH have?
   a. CREATE TABLE
   b. CREATE USER
   c. UNLIMITED TABLESPACE
   d. UNLIMITED CREATE
5. The ______________ data dictionary view lists tablespace storage limits for users.
6. What happens to existing tables when the table owner's quota is set to zero?
   a. The table remains but cannot be used.
   b. The table is dropped.
   c. The table remains but cannot increase in size.
   d. The table is marked invalid.
7. The ______________ data dictionary view lists tablespace storage limits for users.
8. In Oracle9*i*, Release 2, the default password for SYSTEM is:
   a. Defined when the database is created
   b. MANAGER
   c. CHANGE_ON_INSTALL
   d. Depends on the operating system
9. You are the DBA and you run the following SQL commands.

   ```
   CREATE USER MYEE IDENTIFIED BY MYEE
   DEFAULT TABLESPACE USERS
   PROFILE MANAGER_PROFILE
   PASSWORD EXPIRE;
   GRANT CREATE SESSION TO MYEE;
   CONNECT MYEE/MYEE;
   ```

   What happens next?
   a. You connect as the MYEE user.
   b. You are prompted to change the password for MYEE.
   c. You cannot log on because the password is expired.
   d. You cannot log on because MYEE does not have the RESOURCE role.
10. Examine the following SQL statement.

   ```
   1 CREATE USER CARL
   2 IDENTIFIED EXTERNALLY
   3 DEFAULT TABLESPACE USERS
   4 QUOTA 10M ON USERS
   5 QUOTA 0 ON SYSTEM
   6 ACCOUNT LOCK;
   ```

Which line could be removed without changing the outcome?

a. Line 6

b. Line 5

c. Line 3

d. Line 4

e. Line 2

11. Examine the following SQL statement.

```
1 ALTER USER GEORGIA
2 IDENTIFIED BY 1234STOP
3 DEFAULT TABLESPACE USER_AUTO
4 DEFAULT TEMPORARY TABLESPACE TEMP
5 QUOTA 100M ON USER_AUTO
6 ACCOUNT UNLOCK;
```

Which line has an error?

a. Line 6

b. Line 4

c. Line 3

d. Line 1

e. Line 2

12. You have a user named ALEX who owns several tables that are used by the Sales department. The user has left the company. You want to prevent the user from logging onto the database while keeping the tables intact. Which command(s) best handle this task?

a. ALTER USER ALEX ACCOUNT LOCK;

b. ALTER USER ALEX DROP ACCOUNT;

c. DROP USER ALEX CASCADE;

d. ALTER USER ALEX QUOTA 0 ON USERS; REVOKE CREATE SESSION FROM ALEX;

EXAM REVIEW QUESTIONS—DATABASE FUNDAMENTALS I (#1Z0-031)

1. You run this query with the results shown.

```
SELECT TABLESPACE_NAME, MAX_BYTES
FROM DBA_TS_QUOTAS
WHERE USERNAME = 'USER101';
TABLESPACE_NAME                   MAX_BYTES      BYTES
------------------------------ ---------- ----------
USER                                5242880     395698
USER_AUTO                                -1       5482
```

Which statements are true? Choose two.

a. USER101 had a quota on the USER_AUTO tablespace that was removed.

b. USER101 has an unlimited quota on the USER_AUTO tablespace.

c. USER101 has quotas on other tablespaces not shown.

d. USER101 can use either USERS or USER_AUTO to create a new index.

e. USER101 cannot use any more space in the USERS tablespace.

2. You issue the following SQL statements.

```
ALTER SYSTEM SET OS_AUTHENT_PREFIX = 'USA$';
CREATE USER STEVENS IDENTIFIED EXTERNALLY;
ALTER USER STEVENS QUOTA 10M ON USERS;
GRANT CONNECT TO STEVENS;
```

Which statements below are true regarding the previous commands? Choose two.

a. STEVENS is authenticated by Oracle Net.

b. STEVENS cannot log on because he is missing a privilege.

c. STEVENS is authenticated by the operating system.

d. STEVENS logs on as USA$STEVENS.

e. STEVENS can create tables in the USERS tablespace.

3. The table JOE.OLDCAR has 500 rows and uses 50 M of storage space. You issue the following SQL statements. Assume all work is done in the USERS tablespace.

```
ALTER USER FRANCIS QUOTA 100M ON USERS;
CREATE TABLE FRANCIS.NEWCAR AS SELECT * FROM JOE.OLDCAR;
ALTER USER FRANCIS QUOTA 20M ON USERS;
TRUNCATE FRANCIS.NEWCAR DROP STORAGE;
INSERT INTO FRANCIS.NEWCAR
SELECT * FROM JOE.OLDCAR WHERE ROWNUM < 10;
```

Which statements are true? Choose two.

a. All statements succeed.

b. All but the last statement succeeds.

c. FRANCIS.NEWCAR table has no rows.

d. FRANCIS.NEWCAR table has nine rows.

e. The user FRANCIS has exceeded her quota.

4. The ALTER USER command **cannot** be used to accomplish which of these tasks?

a. Change a user's password.

b. Add quotas on two tablespaces to a user with one command.

c. Revoke system privileges from a user.

d. Change a user from a local user to a global user.

5. You can create a profile to limit which of these resources? Choose three.
   a. Complexity of passwords
   b. Days until password expires
   c. Hours after failed logon until a locked account unlocks
   d. Size of the SGA
   e. Number of users per session
6. You issued the following commands.

```
ALTER PROFILE DEFAULT LIMIT
PASSWORD_LIFE_TIME 90
CPU_PER_SESSION 100;
ALTER PROFILE ACCT_PR LIMIT
SESSIONS_PER_USER DEFAULT
PASSWORD_LIFE_TIME 10
CPU_PER_CALL 10;
ALTER USER KATE PROFILE ACCT_PR;
```

   Which statements are true? Choose two.
   a. KATE's password expires in ten days.
   b. KATE can use up to one hundred seconds of CPU time per session.
   c. KATE can run an unlimited number of concurrent sessions.
   d. KATE can use up to ten seconds of CPU time per call.
   e. KATE cannot change her password for ten days.
7. Observe this query and the results.

```
SELECT * FROM RESOURCE_COST;
RESOURCE_NAME                     UNIT_COST
-------------------------------- ----------
CPU_PER_SESSION                         100
LOGICAL_READS_PER_SESSION                10
CONNECT_TIME                              0
PRIVATE_SGA                               0

4 rows selected.
ALTER PROFILE DEFAULT LIMIT
COMPOSITE_LIMIT 10000;
```

   Here are the statistics for three user sessions:

   JOE: CPU=10, Connect time=50, Logical reads=1000, Private SGA=2500

   RICH: CPU=15, Connect time=100, Logical reads=100, Private SGA=5000

   BETTY: CPU=100, Connect time=200, Logical reads=40, Private SGA=2500

Which statements are true? Choose two.

a. The RESOURCE_LIMIT parameter must be set to TRUE to enforce composite limits.

b. The user sessions for JOE and RICH have exceeded the composite limit.

c. The user session for RICH has exceeded the composite limit.

d. The user sessions for JOE and BETTY have exceeded the composite limit.

e. The user sessions for all three users have exceeded the composite limit.

8. You have created a profile called ACCOUNTING and assigned the profile to all ten accountants. Later, you create another profile called ACCT_MGR and assign this profile to five of the accountants. You then execute this command:

```
DROP PROFILE ACCOUNTING CASCADE;
```

What is the result of this command?

a. The five accountants with the ACCOUNTING profile switch to the DEFAULT profile.

b. The five accountants with the ACCOUNTING profile switch to the ACCT_MGR profile.

c. The statement fails.

d. The five accountants with the ACCOUNTING profile are dropped.

9. You want to enforce complexity rules on users' passwords. Which script do you run?

a. verify_function.sql

b. utlpwdmg.sql

c. utlpwmgt.sql

d. utlprofile.sql

10. You just fired a programmer who was using the database for his private consulting work. He has created numerous tables that you want to remove from the database. What is the best way to do this?

a. Lock the user account, write a query that generates a set of 'DROP TABLE' statements, and run the script.

b. Lock the user account, drop each table, and drop the user.

c. Drop the user with the CASCADE parameter.

d. Drop the tablespace containing all the user's tables.

HANDS-ON ASSIGNMENTS

Before working on the hands-on assignments, log onto SQL*Plus Worksheet as SYSTEM, and run the script named **setup.sql** found in the **Data\Chapter11** directory on your student disk.

1. You have three new consultants arriving to work for your company as database programmers. You want them to be able to create tables in the USER_AUTO tablespace that use up to a maximum of 10 M of space each. In addition, you want them to change their passwords every week, and you want the system to log them off the database if their sessions are idle for more than ten minutes. The three user names to create are CONS01, CONS02, and CONS03. The profile to create for the consultants is named TEMPCONSULT. Write and execute the commands needed to accomplish these goals. Save the SQL script in a file named **ho1101.sql** in the **Solutions\Chapter11** directory on your student disk.
2. The three consultants (from Assignment 1) complain that they cannot work efficiently when the system keeps logging them off. They ask you to let them have 30 minutes before timeout. You agree to take care of it. You decide that you want the consultants to stop using the USER_AUTO tablespace and start using the USERS tablespace instead. You allow them a maximum of 10 M each on the USERS tablespace. However, the tables they have already created can remain on USER_AUTO. Write and execute the script to handle this, and save it in a file named **ho1102.sql** in the **Solutions\Chapter11** directory.
3. The vice president (user name MSTEWARD) changed her password when the password expired. Now, three weeks later, she has forgotten the new password. She tells you she wants her password to be "11/12/56" (her birthday). You know that the current password validation routine will reject this password. How can you help her? Write a script that changes her password to some version of her birth date. In addition, you want to give her (and other top executives) a grace period of 45 days in which they can still enter the old password, because these people use their accounts very infrequently. You also extend their password expiration time to six months. Write and execute the script, and save it in a file named **ho1103.sql** in the **Solutions\Chapter11** directory. Hint: Create a profile named TOPEXECS.
4. You decide that the DEFAULT profile does not need the enforcement of password complexity. Instead, you want the complexity enforcement placed on the TEMPCONSULT profile and the TOPEXECS profile you created in Assignments 1 and 3. Create a profile named TOPEXECS to handle the password requirements. Write and execute a script to make this change, and save it in a file named **ho1104.sql** in the **Solutions\Chapter11** directory on your student disk.

5. The three consultants who arrived in the office earlier (refer to Assignment 1) have left. You review the database and determine that CONS01 and CONS02 have created tables that can be safely removed. The third consultant, CONS03, however, has created some useful documentation tables that you are saving. Write and execute the script to remove all three users, so that the tables of CONS03 are preserved while the remaining tables are dropped. Save the script in a file named **ho1105.sql** in the **Solutions\Chapter11** directory on your student disk.
6. You want to enable users who log directly onto the database server to access the database without supplying an additional password. This type of user should have limits on CPU time (1.75 seconds per session), logon time (no more than 30 minutes), and storage (no more than 15 M of space on the USERS tablespace). Write and execute a script that creates a profile named OSUSERS, and create a user named JPIERRE that fits the description. Save the script in a file named **ho1106.sql** in the **Solutions\Chapter11** directory on your student disk.
7. Write and execute a query that lists users whose passwords have expired or whose accounts are locked. Sort the users by create date (most recent date to oldest date). Save the script in a file named **ho1107.sql** in the **Solutions\Chapter11** directory on your student disk.
8. Write a query to display the quota of all users on all tablespaces. Write another query that calculates how much space would remain on each tablespace if all the users used up all their quotas. Eliminate users with unlimited quotas for this query. Assume that the tablespace will not expand any further than its current allocation of space. Execute both queries. Save the script in a file named **ho1108.sql** in the **Solutions\Chapter11** directory on your student disk.
9. Query the database to list each profile and the users assigned to that profile. Save the script in a file named **ho1109.sql** in the **Solutions\Chapter11** directory on your student disk.
10. Write a query that compares each profile to the DEFAULT profile. The results should list one column for the profile name, another column listing each resource limit name, a third column showing the value set in the profile, and a fourth column that states whether that value is the same as the value for the same resource limit in the DEFAULT profile. Save the script in a file named **ho1110.sql** in the **Solutions\Chapter11** directory on your student disk.

Case Project

The Global Globe has a table of employees. Each employee needs a user name and password for the database. The employee's user name is the first name plus the first three letters of the last name. There are three categories of employees who need access to the database: managers, editors, and writers. Each of these needs resource limits, so you must set up three profiles.

The MANAGERS profile has these traits: password expires after 30 days; maximum of 10 minutes idle time; reuse of password after 300 passwords; default resources otherwise.

The WRITERS profile has these traits: password expires after 45 days; maximum of 15 minutes idle time; maximum session logical reads: 1000; maximum private SGA: 256 K; default resources otherwise.

The EDITORS profile has these traits: password expires after 60 days; maximum of 45 minutes idle time; maximum time for one session: 8 hours; default resources otherwise.

Create the three profiles. Save your work in a file named **case1101.sql** in the **Solutions\Chapter11** directory on your student disk.

You can identify the category of each employee by his or her job title.

Write a query that generates the CREATE USER commands for all the managers, writers (including freelance writers), and editors. The CREATE USER command must include the appropriate profile. All the users have a password of TEMPPASS, and the password expires when the user is created.

Write another query that generates the GRANT CREATE SESSION command for all users. Run the queries, and then edit the script (if needed); run it to create all the users, and grant them the CREATE SESSION role. Save the queries in a file named **case1102.sql**, and save the script you run in a file named **case1103.sql** in the **Solutions\Chapter11** directory on your student disk.

CHAPTER

12

System and Object Privileges

In this chapter, you will:

- Identify and manage system and object privileges
- Grant and revoke privileges to users
- Understand auditing capabilities and practice using auditing commands

The previous chapter gave you a hint of how to allow users access to the database and its contents. In this chapter, you examine the types of privileges available and how to combine privileges into convenient groups. **Privileges** are tasks you are authorized to carry out in the database.

OVERVIEW OF PRIVILEGES

After a user has been created with a name and password, the user must be assigned the ability to log onto the database. Once logged on, however, the user cannot perform any other tasks unless given the privilege to do so. For example, a user must be given the privilege to create a table. It is possible to give a privilege to all users (a blanket privilege, you might say), and this can be convenient for tasks most users must do. For example, the privileges of querying most data dictionary views that begin with ALL_ have been given to all users by Oracle by default. Most privileges are given to specific users or roles. A **role** is a named group of privileges that can be assigned to a user as a set rather than individually. Find out more about roles in Chapter 13. Here is an example of how to use privileges in the database: Imagine that you are the DBA, and you have installed ten tables to support the Accounting Department's new application. You assign privileges for querying and modifying these tables to the employees in the Accounting Department.

Assigning privileges only as they are needed is a form of security for the database. A database that allows all users to query or even modify all database tables is the exception rather than the rule for Oracle databases.

There are two types of privileges that you can assign in the Oracle9*i* database:

- **System privileges:** A **system privilege** gives a user the ability to manage some part of the database system. For example, the CREATE INDEX privilege allows a user to create an index, and the ALTER TABLESPACE privilege allows a user to modify an existing tablespace using the ALTER TABLESPACE command. The DBA usually assigns system privileges.
- **Object privileges:** An **object privilege** gives a user the ability to perform certain tasks on specific tables or other objects that are owned by a schema. For example, the INSERT ON CLASSMATE.EMPLOYEE privilege allows a user to insert rows into the EMPLOYEE table in the CLASSMATE schema. The schema that owns the object usually assigns object privileges.

The next sections describe each type of privilege so that you can easily differentiate between system and object privileges.

Identifying System Privileges

The position of database administrator comes with many duties and responsibilities. These include the tasks of creating the database, creating and maintaining tablespaces, monitoring resources, and backing up the database. Each of these tasks has its own system privilege, so that the DBA can perform the tasks or delegate them to an assistant. The predefined user, SYSTEM, has already been given the privileges needed for DBA activities. When you log onto the database as the SYSTEM user, you are able to perform DBA tasks. In addition, you are allowed to assign these privileges to other users.

There are over 100 system privileges in the Oracle9*i* database. Fortunately, all of them are named very clearly, so it is easy to understand their use. Here are some of the most common and familiar system privileges.

- **SYSDBA:** You have already used this privilege when you logged onto the SQL*Plus Worksheet to create a database. This privilege allows the user to start up and shut down the database and to create an **spfile** (system initialization parameter file). The SYSTEM and SYS predefined users have this privilege.
- **SYSOPER:** This privilege is the same as SYSDBA, except it does not include the ability to create a database.
- **CREATE SESSION:** You assigned this privilege to a user in the previous chapter. This allows a user to log onto the database.
- **CREATE TABLE, CREATE INDEX, and CREATE VIEW:** These privileges allow users to create their own tables, indexes, and views.
- **CREATE USER:** The DBA must have this privilege to create new users. You logged on as SYSTEM to create users in the previous chapter.
- **CREATE ANY TABLE:** This privilege is usually reserved for the DBA, allowing him or her to create a table in any schema. The CLASSMATE user has this privilege, which enabled you to create tables in previous chapters.
- **DROP ANY TABLE:** This privilege allows the user to drop any schema's tables, except those needed by the database system itself (the data dictionary tables).
- **SELECT ANY TABLE:** Again, this privilege is usually reserved for the DBA, because it allows the user to query any table on the database.
- **GRANT ANY PRIVILEGE, GRANT ANY OBJECT PRIVILEGE:** These allow the user (preferably the DBA) to assign any system privilege or any object privilege to other users. Later in the chapter, you see how to allow users to grant specific privileges.
- **BACKUP ANY TABLE:** This privilege allows the user to use the Export utility to export any table in the database. **Exporting** is a form of backup that can be used to back up specific tables or schemas, or to back up the entire database. An alternate way to back up the database is to copy the operating system files (the datafiles) to tape or CD. You do not need any Oracle-specific privileges to perform an operating system backup of Oracle files.

12

You can see a complete list of all the system privileges in the online documentation that Oracle provides at the Oracle Technology Network site: *otn.oracle.com*. In the main window for Oracle9*i* Database Documentation, Release 2, select the link named "SQL and PL/SQL syntax and examples." Then look up the GRANT command in the alphabetical index that is displayed. You will find tables listing all system privileges and object privileges there.

You can also view and assign both system and object privileges by using the Security Manager in the Enterprise Management console. You will have some practice viewing system and object privileges in the console later in the chapter.

Using Object Privileges

Object privileges are more pinpointed than system privileges. That is, an object privilege has a much narrower focus. For example, the system privilege SELECT ANY TABLE gives the user the ability to query any table in any schema. On the other hand, the object privilege SELECT ON CUSTOMER gives the user only the ability to query the CUSTOMER table.

The user who owns a table or view is allowed by default to select, insert, update, and delete data in the table. Most of the time, users who actually log on and use the tables have their own user names and, therefore, must be assigned an object privilege for each of these tasks. The DBA user does not usually have to be assigned object privileges, because the DBA user (such as SYSTEM) already has system privileges that allow him to perform these tasks.

Object privileges always pertain to a table, function, procedure, or other object. There are several different object privileges, and some are available only for tables and views, whereas others are only available for functions, procedures, packages, or user-defined types. **Functions**, **procedures**, and **packages** are PL/SQL programs that reside in the database and can be called from SQL commands, such as SELECT and INSERT. **User-defined types** help define object columns or object tables. These types were discussed in Chapter 7 and include object types, arrays, and nested tables. Table 12-1 shows the types of object privileges available for each type of object. This is not a complete list, but it covers both the objects that you study in this book and those usually covered by introductory Oracle9*i* SQL classes or texts.

Table 12-1 Object privileges by object type

Object privilege	Table	View	Sequence	Function, procedure, package	User-defined type
ALTER	Yes		Yes		
DELETE	Yes	Yes			
EXECUTE				Yes	Yes
DEBUG	Yes	Yes			
FLASHBACK	Yes	Yes			
INDEX	Yes				
INSERT	Yes	Yes			
REFERENCES	Yes	Yes			
SELECT	Yes	Yes	Yes		
UPDATE	Yes	Yes			

Some of these privileges have obvious meanings, such as SELECT, INSERT, and UPDATE. The following list defines privileges whose meanings are less obvious.

- **EXECUTE:** Call the function, procedure, or package while running an SQL query or other command.
- **DEBUG:** Run a debugging program that looks at triggers and SQL commands using a table or view.
- **FLASHBACK:** Run a flashback query on the table or view.
- **INDEX:** Create an index on the table.
- **REFERENCES:** Create FOREIGN KEY constraints that reference the table.

How do you assign a user an object privilege or a system privilege? The next section shows you what to do.

MANAGING SYSTEM AND OBJECT PRIVILEGES

When you **grant** a privilege, you assign a privilege to a user or a role, whether it is a system privilege or an object privilege. When you revoke a privilege, you take away the privilege. Granting privileges to roles is covered in Chapter 13. In this chapter, you practice granting and revoking privileges to users.

Granting and Revoking System Privileges

The basic syntax of the GRANT command for system privileges is:

```
GRANT <systempriv>, <systempriv>,...|ALL PRIVILEGES
TO <user>,<user>...|PUBLIC
WITH ADMIN OPTION;
```

Here are some pointers about this command:

- List as many system privileges as you want, separating each with a comma. You can also substitute the phrase ALL PRIVILEGES for a list of privileges. Use ALL PRIVILEGES with caution, because it grants the user all of the 100 plus system privileges except SELECT ANY DICTIONARY.
- Add the WITH ADMIN OPTION only when you want the user to be able to grant the same system privilege to other users. For example, this is appropriate when you hire a new assistant DBA and allow him to create new users.
- List all the users to whom you want to grant the same system privileges. Alternatively, use PUBLIC instead of a specific user name to grant the privilege to all users, including users created in the future.

Revoking a system privilege is simple:

```
REVOKE <systempriv>, <systempriv>,...|ALL PRIVILEGES
FROM <user>, <user>,...|PUBLIC;
```

Complete the following steps to practice granting and revoking privileges.

1. Start up the Enterprise Manager console. In Windows, click **Start** on the Taskbar, and then click **Programs/Oracle - OraHome92/Enterprise Manager Console**. In UNIX, type **oemapp console** on the command line. The Enterprise Manager console login screen appears.
2. Select the **Launch standalone** radio button, and click **OK**. The console appears. Before granting and revoking privileges, take a look at the privileges of the STUDENTA user, who was created in the previous chapter.
3. Double-click the **Databases** folder, and then double-click the **ORACLASS** database icon. The available tools are listed below the database.
4. Double-click the **Security** icon, and then double-click the **Users** folder. This displays the current users in the ORACLASS database, as shown in Figure 12-1.

The list of users on your screen may be slightly different from the list of users shown in the figure.

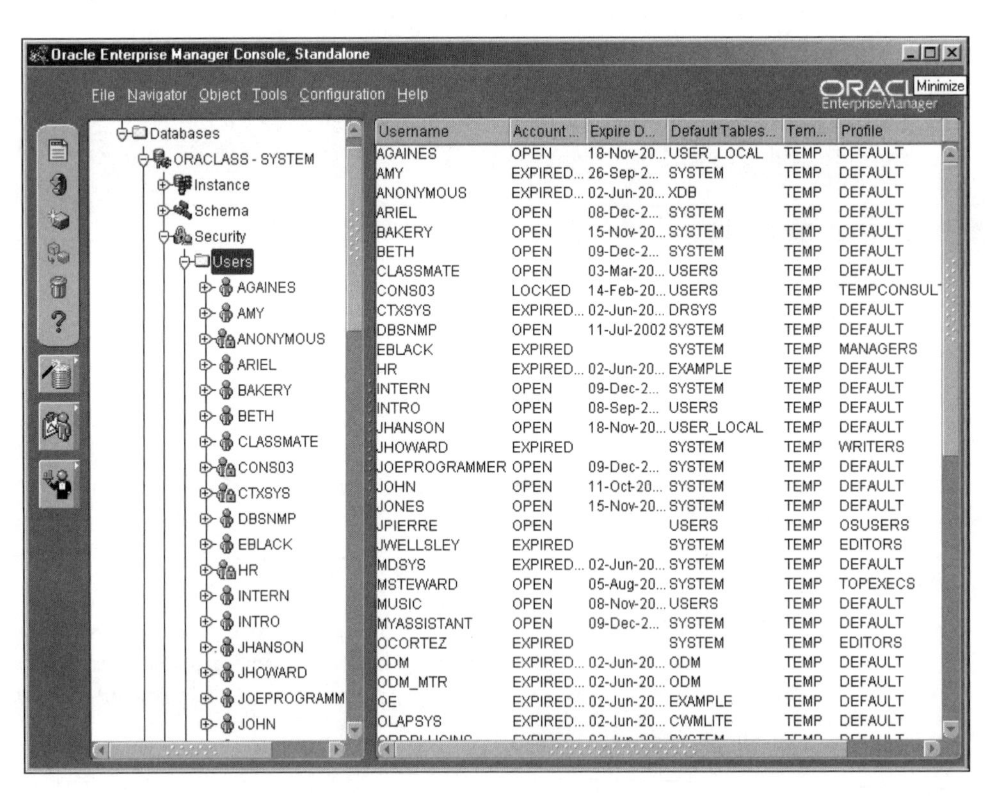

Figure 12-1 The Security Manager in the Enterprise Manager console handles users

5. Scroll down the left side to find and double-click on **STUDENTA**. The property sheet for STUDENTA appears on the right side of the console.

6. Click the **System** tab in the property sheet. This part of the property sheet displays the system privileges currently granted to the user. As you can see in Figure 12-2, CREATE SESSION is the only system privilege granted to STUDENTA. You could add more privileges right here; however, your goal is to become familiar with the GRANT command, so you will be adding privileges in SQL*Plus Worksheet instead.

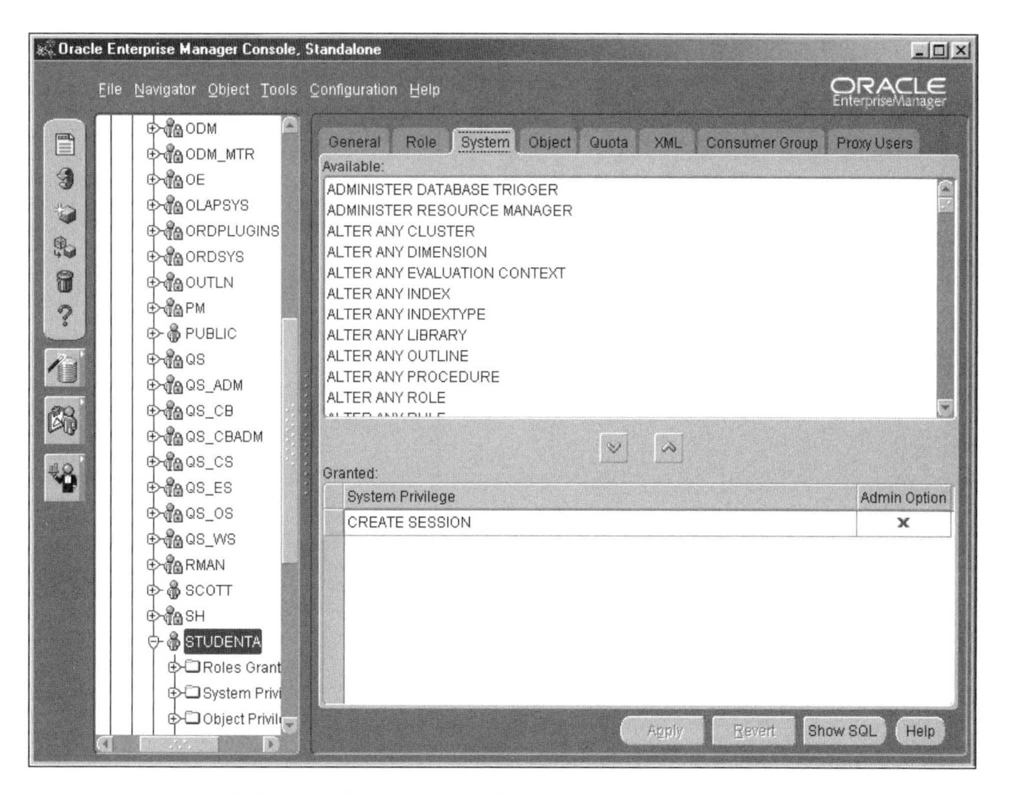

Figure 12-2 Available privileges are listed in the top panel, and granted privileges are listed in the lower panel

7. Start up the SQL*Plus Worksheet by clicking **Tools/Database Applications/ SQL*Plus Worksheet** from the top menu in the console. A background SQL*Plus process starts, and then the SQL*Plus Worksheet window appears. If the background process appears in front, simply minimize it (click the **minus sign** in the top-right corner) so that you can see the worksheet.

8. Notice that you are connected as SYSTEM. This is appropriate for granting system privileges, because the SYSTEM user has the privilege, GRANT ANY PRIVILEGE WITH ADMIN OPTION, meaning that SYSTEM can grant any system privilege to any other user. Display SYSTEM's privileges by executing this query.

12

```
SELECT * FROM SESSION_PRIVS
ORDER BY 1;
```

Figure 12-3 partially shows the results. Use this query for any user to display the currently enabled privileges.

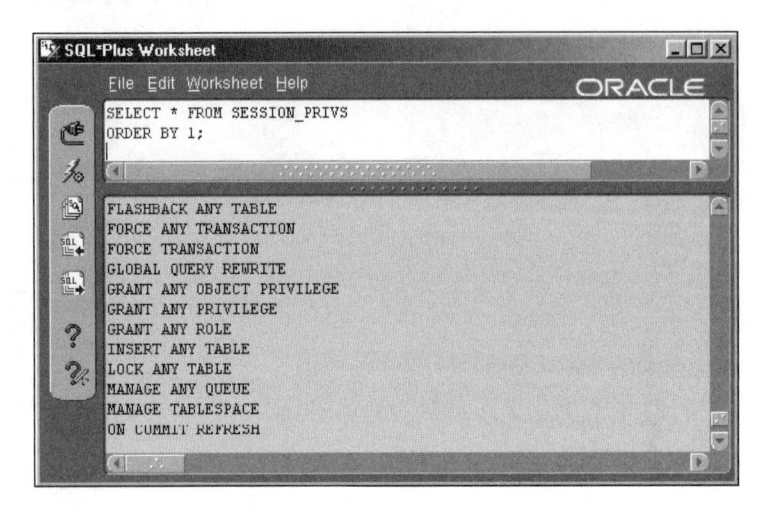

Figure 12-3 SYSTEM has nearly every system privilege available

9. Imagine that STUDENTA is going to create some tables, views, indexes, object tables, and public synonyms. This requires three system privileges. The CREATE TABLE privilege allows a user to create tables, object tables, and indexes. The CREATE VIEW privilege allows the user to create views. The CREATE PUBLIC SYNONYM privilege allows the user to create public synonyms. **Public synonyms** are like table aliases, which any user can use in place of the table's name. (A public synonym does not give users the privilege to actually select from the table: That privilege must still be granted.) Now, grant the system privileges to STUDENTA by executing these commands.

   ```
   GRANT CREATE TABLE, CREATE VIEW, CREATE PUBLIC SYNONYM
   TO STUDENTA;
   ```

10. The SQL*Plus Worksheet replies, "Grant succeeded." After granting this privilege, you find out that STUDENTA has an assistant (STUDENTB) who needs appropriate privileges.

Instead of granting the three system privileges, which you gave to STUDENTA, you decide to allow STUDENTA to grant these privileges (or only one or two of them) to STUDENTB. There is no command to alter a grant that has been given, so you must reissue the grant. You do not have to revoke the grant before reissuing it. Give STUDENTA the authority to grant the privileges by typing this command.

```
GRANT CREATE TABLE, CREATE VIEW, CREATE PUBLIC SYNONYM
TO STUDENTA WITH ADMIN OPTION;
```

The SQL*Plus Worksheet replies, "User created. Grant succeeded."

11. STUDENTA is now allowed to grant these three system privileges to other users. Connect to STUDENTA to test this out. The CONNECT command is an alternative to using the **File/Change Database Connection** menu selection. Type and execute this command.

    ```
    CONNECT STUDENTA/A1PLU$@ORACLASS
    ```

12. The SQL*Plus Worksheet replies, "Connected." Now, as STUDENTA, you can look at your own privileges and find out which ones you have with the ADMIN option by typing and executing the following command. Figure 12-4 shows the results. As you can see, STUDENTA has the ADMIN option for all but the CREATE SESSION privilege.

    ```
    SELECT * FROM USER_SYS_PRIVS;
    ```

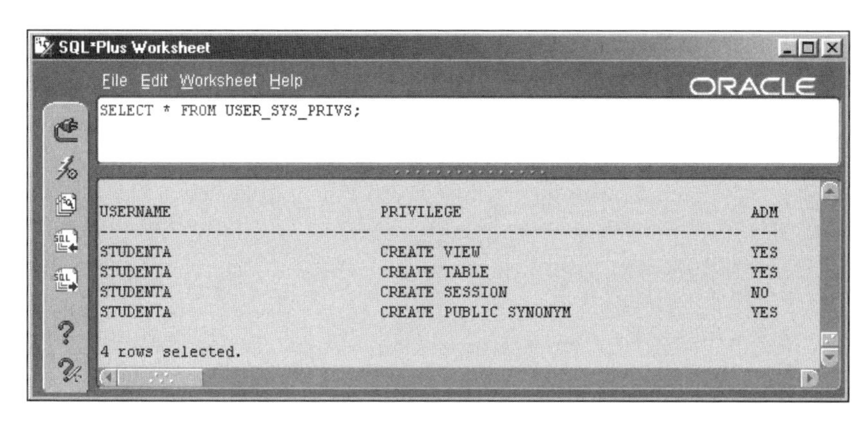

Figure 12-4 USER_SYS_PRIVS shows privileges belonging to the current user

13. The new user, STUDENTB, will be helping STUDENTA by creating tables and indexes. Grant STUDENTB the appropriate privilege by typing and executing this command.

    ```
    GRANT CREATE TABLE TO STUDENTB;
    ```

14. The SQL*Plus Worksheet replies, "Grant succeeded." Now, connect to SYSTEM again by typing and executing this command, replacing <password> with the current password for the SYSTEM user.

    ```
    CONNECT SYSTEM/<password>@ORACLASS;
    ```

The SQL*Plus Worksheet replies, "Connected."

15. Check the system privileges for the two users by typing and executing this query. Figure 12-5 shows the results.

    ```
    SELECT * FROM DBA_SYS_PRIVS
    WHERE GRANTEE LIKE 'STUDENT%';
    ```

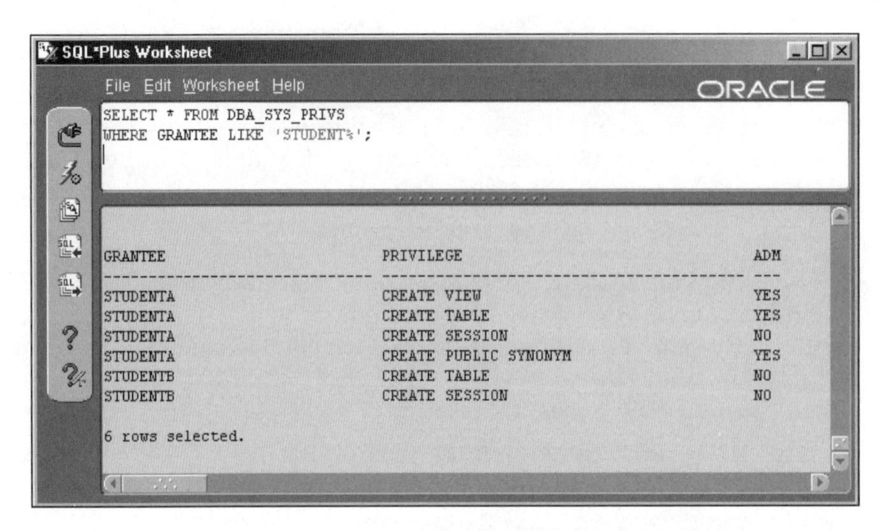

Figure 12-5 The DBA can look at the system privileges given to any user

16. Now, imagine that the project that STUDENTA and STUDENTB worked on is completed. You need to revoke the system privileges that were assigned to STUDENTA during the project. All that will be left is STUDENTA's privilege to connect to the database. Type and execute this command.

```
REVOKE CREATE TABLE, CREATE VIEW, CREATE PUBLIC SYNONYM
FROM STUDENTA;
```

17. The SQL*Plus Worksheet replies, "Revoke succeeded." The privileges have been revoked from STUDENTA, but what about STUDENTB? System privileges are not revoked when the user who issued the privilege loses his privilege. So, type and execute this command to revoke the privileges granted by STUDENTA to STUDENTB.

```
REVOKE CREATE TABLE FROM STUDENTB;
```

The SQL*Plus Worksheet replies, "Revoke succeeded."

18. Remain logged on for the next practice.

There is no distinct privilege for creating indexes on a table in your own schema. The CREATE ANY INDEX privilege allows you to create indexes on other schema's tables.

System privileges are needed to create tables in the database. The owner (schema) of a table can automatically select, insert, update, and delete data in his own tables. However, no other users (except the DBA) are allowed to even see the table's name without permission from the owner or the DBA. This is where object privileges come into play.

Granting and Revoking Object Privileges

The CLASSMATE schema has several tables that were created during previous chapters. So far, you have not worked directly with the tables, because this book focuses primarily on the DBA tasks involved in managing an Oracle9*i* database. This section discusses tasks usually handled by the schema user who owns the tables. However, DBAs can also perform these tasks, as they commonly do in some companies.

The syntax for granting object privileges looks like this:

```
GRANT <objectpriv>, <objectpriv>,...|ALL
(<colname>,...) ON <schema>.<object>
TO <user>,...|PUBLIC
WITH GRANT OPTION
WITH HIERARCHY OPTION;
```

As you can see, the preceding syntax is similar to the syntax for granting system privileges. Here are some notes on the syntax:

- **Column list:** The column list (in the second line of the preceding syntax) is used only when you want to grant a privilege for specific columns in the table or view. Although this is not usually used, you could employ it as a security feature to restrict users from updating sensitive fields that they are allowed to query but not to update. The column list can only be used to grant UPDATE, REFERENCES, and DELETE privileges.
- **PUBLIC:** You can list object privileges for one object, and you can also list users who receive those privileges. PUBLIC is substituted for user names when you want to grant the privilege or privileges to all users.
- **WITH GRANT OPTION:** This is similar to the WITH ADMIN OPTION. Use this clause when you want the user to be able to issue grants to other users.
- **WITH HIERARCHY OPTION:** This clause is a special feature, which is used for objects that have subobjects. A **subobject** is an object based on another object, for example an object type that is based on another object type. Although this book does not cover this option, the option is included for completeness of syntax. This clause instructs Oracle9*i* to grant the object privilege to the user on the object and on all its subobjects.

12

Follow these steps to practice granting and revoking object privileges.

1. In your SQL*Plus Worksheet session, connect to the CLASSMATE user by typing and executing this command.

   ```
   CONNECT CLASSMATE/CLASSPASS@ORACLASS
   ```

   The SQL*Plus Worksheet replies, "Connected."

2. To see what kind of objects CLASSMATE owns, type and execute this query. Figure 12-6 partially shows the results.

```
COLUMN OBJECT_NAME FORMAT A35
SELECT OBJECT_TYPE, OBJECT_NAME
FROM USER_OBJECTS
ORDER BY 1;
```

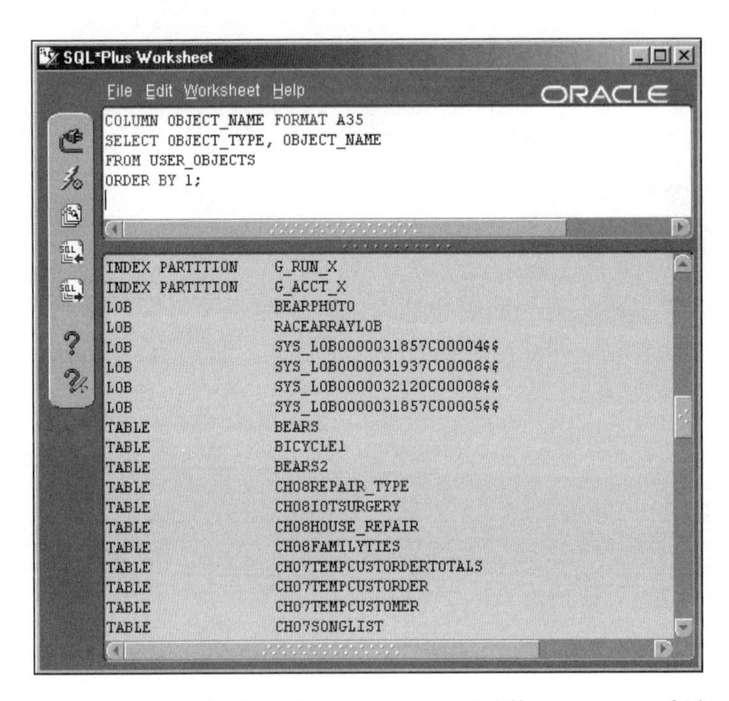

Figure 12-6 CLASSMATE owns several different types of objects

Refer to Table 12-1, which shows the types of object privileges available for different types of objects. (Table 12-1 was presented in the section titled "Using Object Privileges.") Some of the objects owned by CLASSMATE, such as individual table partitions and LOB segments, cannot be used with object privileges. The others, such as tables and views, can be used with object privileges.

3. Imagine that an application is about to be launched for the Sales Department of the Global Globe News Company. The application uses the WANT_AD_COMPLETE_VIEW view and the NEWS_ARTICLE table. For this imaginary example, there are only two users who need access to the tables when using the application: STUDENTA and STUDENTB. The application queries both the view and the table. Type and execute these commands to grant the appropriate object privileges.

```
GRANT SELECT ON WANT_AD_COMPLETE_VIEW
TO STUDENTA, STUDENTB;
GRANT SELECT ON NEWS_ARTICLE
TO STUDENTA, STUDENTB;
```

The SQL*Plus Worksheet replies, "Grant succeeded."

4. The application also allows the users to create, modify, and remove records in the NEWS_ARTICLE table. Type and execute this command to grant the privileges needed.

```
GRANT INSERT, UPDATE, DELETE ON NEWS_ARTICLE
TO STUDENTA, STUDENTB;
```

The SQL*Plus Worksheet replies, "Grant succeeded."

5. Type and execute this query to look over the grants you have created.

```
COLUMN PRIVILEGE FORMAT A10
COLUMN GRANTABLE FORMAT A10
SELECT GRANTEE, TABLE_NAME, PRIVILEGE, GRANTABLE
FROM USER_TAB_PRIVS_MADE
ORDER BY 1,2,3;
```

6. Figure 12-7 shows the results. A **grantee** is the user receiving a privilege. Each privilege for each user has a row in this table. Privileges for all types of objects are listed in this view, not just privileges for tables as its name implies. There is a similar table, named USER_TAB_PRIVS_RECD, that lists all the object privileges a user has received, rather than those the user has granted.

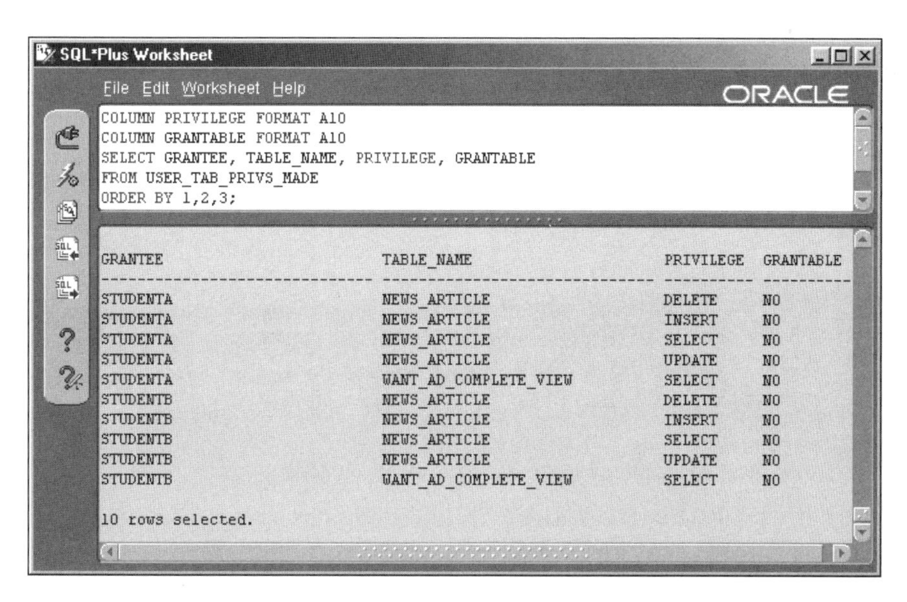

Figure 12-7 This query shows object privileges granted by the current user

7. You decide that the STUDENTA user should be allowed to grant all his privileges on both tables to other users. Therefore, type and execute these commands to provide STUDENTA with the changed privileges.

```
GRANT SELECT, INSERT, UPDATE, DELETE ON NEWS_ARTICLE
TO STUDENTA WITH GRANT OPTION;
GRANT SELECT ON WANT_AD_COMPLETE_VIEW TO STUDENTA
WITH GRANT OPTION;
```

   The SQL*Plus Worksheet replies, "Grant succeeded."

8. Now you decide that STUDENTB should not have the ability to delete rows from the NEWS_ARTICLE table, so you modify his privileges by revoking that privilege. Type and execute this command.

```
REVOKE DELETE ON NEWS_ARTICLE FROM STUDENTB;
```

   The SQL*Plus Worksheet replies, "Revoke succeeded."

9. Re-execute the query in Step 5 to confirm that STUDENTA has the GRANT option and STUDENTB has no DELETE privileges. To find the query, click the **Previous Command** icon on the left side of the SQL*Plus Worksheet. The previous command appears in the worksheet. After the correct command appears, click the **Execute** icon to execute it again. Figure 12-8 shows the results.

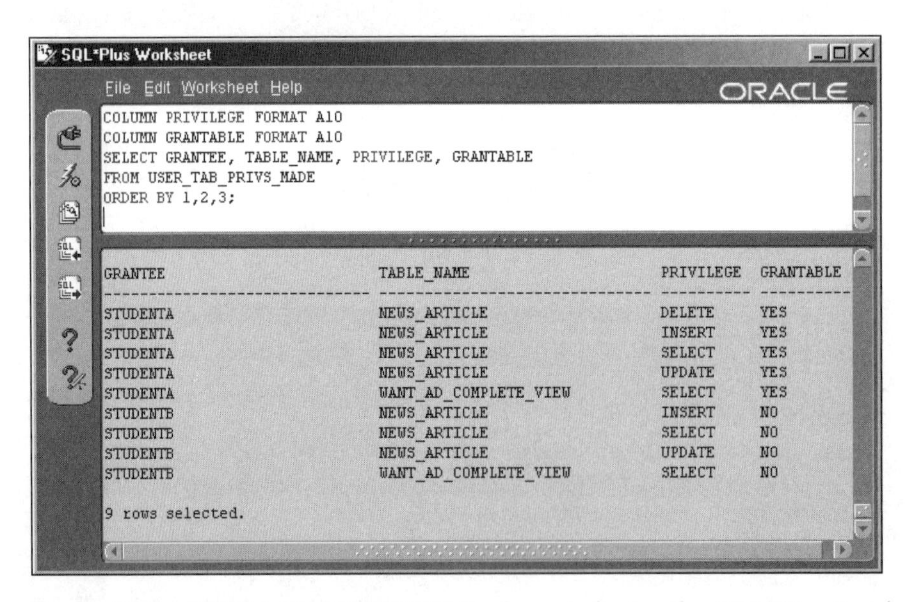

Figure 12-8 "YES" in the GRANTABLE column indicates the GRANT option has been given

10. Switch now to the STUDENTA user by typing and executing this command.

```
CONNECT STUDENTA/A1PLU$@ORACLASS
```

11. SQL*Plus Worksheet replies, "Connected." Take a look at the privileges that STUDENTA has by typing and executing this query. The results show the privileges granted to STUDENTA.

```
SELECT GRANTOR, TABLE_NAME, PRIVILEGE, GRANTABLE
FROM USER_TAB_PRIVS_RECD
ORDER BY 1,2,3;
```

12. Try querying the NEWS_ARTICLE table by typing and executing this command. Two rows are returned. Remember that you must prefix the table with the schema when you don't own the table you are querying.

```
SELECT TITLE, RUN_DATE
FROM CLASSMATE.NEWS_ARTICLE;
```

13. Try querying the EMPLOYEE table, which you don't have authority to view. Type and execute this query. Figure 12-9 shows the resulting error message.

```
SELECT EMPLOYEE_ID, FIRST_NAME, LAST_NAME
FROM CLASSMATE.EMPLOYEE;
```

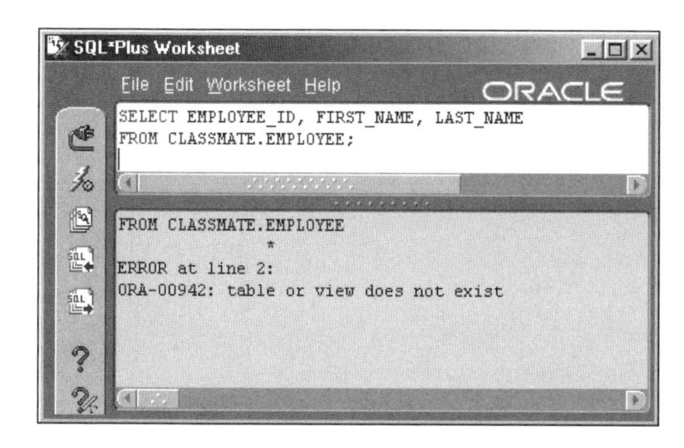

Figure 12-9 The error message is standard for object privilege violations

14. You decide that STUDENTB really does need the DELETE privilege on the NEWS_ARTICLE table. Give STUDENTB the privilege by typing and executing this command.

```
GRANT DELETE ON CLASSMATE.NEWS_ARTICLE
TO STUDENTB;
```

15. SQL*Plus Worksheet replies, "Grant succeeded." Confirm the existence of the privilege by typing and executing this command.

```
SELECT GRANTEE, TABLE_NAME, PRIVILEGE, GRANTABLE
FROM ALL_TAB_PRIVS_MADE;
```

12

16. The object privilege has been granted; however, the owner of the table (CLASSMATE) does not agree, and decides to remove STUDENTA's ability to grant privileges on the NEWS_ARTICLE table. Connect to CLASSMATE by typing and executing this command.

```
CONNECT CLASSMATE/CLASSPASS@ORACLASS
```

    The SQL*Plus Worksheet replies, "Connected."

17. You cannot change an existing GRANT privilege; you must revoke the old one and grant the new one. Now, revoke the privileges from STUDENTA by typing and executing this command.

```
REVOKE SELECT, INSERT, UPDATE, DELETE ON NEWS_ARTICLE
FROM STUDENTA;
```

    The SQL*Plus Worksheet replies, "Revoke succeeded."

18. Check the results by querying the USER_TAB_PRIVS_MADE view again. Figure 12-10 shows the results.

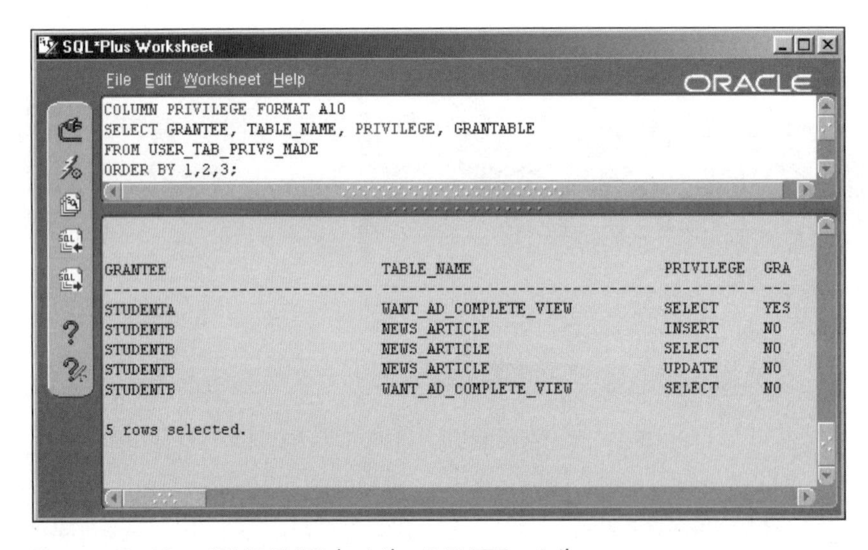

Figure 12-10 STUDENTB lost the DELETE privilege

    Notice that STUDENTA no longer has any privileges on the NEWS_ARTICLE table, and STUDENTB no longer has the DELETE privilege that was granted to him by STUDENTA. When STUDENTA lost the privilege on the table, any object privileges created by STUDENTA were also lost. This is an important difference between system privileges and object privileges.

19. Remain logged on for the next practice.

The act of revoking an object privilege cascades downwards, whereas revoking a system privilege does not. The cascading of a revoked privilege occurs when you revoke a privilege from a user who has previously granted this privilege to other users. The privilege is revoked for the other users automatically.

You have examined how to grant and revoke system and object privileges. You have also queried several data dictionary views. Table 12-2 summarizes the data dictionary views and dynamic performance views related to system and object privileges. You have tried out most of these views during the practices in the chapter.

Table 12-2 Data dictionary views and dynamic performance views for privileges

Name	Description
DBA_SYS_PRIVS	All system privileges granted
DBA_TAB_PRIVS	All object privileges granted
USER_TAB_PRIVS_MADE	All object privileges granted by the user
USER_TAB_PRIVS_RECD	All object privileges when the user is the grantee
DBA_COL_PRIVS	All object privileges on column lists
SESSION_PRIVS	User's privileges currently enabled (see next chapter for more on enabling and disabling privileges through roles)

You have seen how Oracle9*i* security works in giving out privileges only as they are needed. Very few privileges are given to PUBLIC (all users). You, as the DBA, control access to the database by restricting a user's authority to perform tasks, limiting users to only the tasks you assign. You also restrict a user's ability to query or modify data to only those objects required by the job. What if you have a large company in which many users have access to many tables? If a table's data appears to be corrupted by a malicious user, how do you identify the culprit? You can narrow it down by finding out which users are allowed to modify that table, but how can you pinpoint each user's activity? The next section shows how to use auditing to handle this situation.

DESCRIPTION OF AUDITING CAPABILITIES

Monitoring activity on the database is called **auditing**. Auditing is most frequently used to determine who is making unauthorized updates or deletions to sensitive data. There are three types of auditing that Oracle9*i* can run automatically.

- **Statement auditing:** Audits types of SQL commands. You must have the AUDIT SYSTEM privilege to use this type of auditing. For example, an accountant with access to the CUSTOMER_ORDER table could use this access to adjust the balance due to zero for friends who have made orders. You can set up an audit that monitors all UPDATE commands performed by this accountant. The command looks like this (do not actually run this command):

```
AUDIT UPDATE TABLE BY JACK;
```

- **Privilege auditing:** Audits use of particular privileges. You must have the AUDIT SYSTEM privilege to use this type of auditing. As another example, imagine that your company works from 9:00 to 5:00, and you suspect that one of the programmers is moonlighting and using the database for personal projects. You cannot find any evidence left in the database (the programmer knows enough to drop any tables created for personal use). However, you can set up auditing to track anyone who creates tables and record what time this activity is happening. The command looks like this (do not actually run this command):

  ```
  AUDIT CREATE TABLE;
  ```

Object auditing: Audits activity on a certain object. You can use the AUDIT command to set up object auditing for any object you own. Otherwise, you must have the AUDIT ANY privilege to audit objects. For example, you have a table of employee pay rates and work history called EE_PRIVATE. Unauthorized access to the table would reveal confidential information. You set up auditing to monitor all queries of the table. The command looks like this (do not actually run this command):

```
AUDIT SELECT ON EE_PRIVATE;
```

Use auditing with caution, because it generates a high volume of records. It is best to start auditing when you have narrowed the area you want to monitor to a few tables, users, or privileges. This keeps the SYS.AUD$ table from growing too large too quickly.

Auditing commands have no effect until you set the AUDIT_TRAIL initialization parameter. You can set it to the values in the following list by modifying the **init.ora** file or the **spfile**. In either case, you must restart the database for the setting to take effect. Here are the valid settings for AUDIT_TRAIL.

- **TRUE** or **DB:** Starts auditing and places the audit trail records into the SYS.AUD$ table.
- **FALSE** or **NONE:** Turns off auditing. The default is NONE.
- **OS:** Starts auditing and places the audit trail records into an operating system file in the directory named in the AUDIT_FILE_DEST initialization parameter. The default setting for AUDIT_FILE_DEST is ORACLE_HOME\rdbms\audit. Recall that ORACLE_HOME stands for the full path on your computer where the Oracle Software is installed, such as **C:\oracle\ora92**.

Another initialization parameter, AUDIT_SYS_OPERATIONS, can be set to TRUE or FALSE to turn on or off monitoring of any activity of the SYS user and of other users logged on with SYSDBA privileges.

The syntax of the AUDIT command for object auditing is:

```
AUDIT <objpriv>,<objpriv>,...|ALL
ON <schema>.<object>|DEFAULT|NOT EXISTS
```

```
BY SESSION|BY ACCESS
WHENEVER SUCCESSFUL|WHENEVER NOT SUCCESSFUL;
```

To set this auditing for the automatic turn on of any new object that is created, substitute DEFAULT for an object name. This can save time later on, if you decide to audit certain activities on every object. With DEFAULT, the auditing is automatically performed as new objects are created.

Substitute NOT EXISTS for an object name, and Oracle9*i* creates an audit trail record for attempted actions that fail with the "object does not exist" error. This can be useful in determining who might be trying to access sensitive tables, because the "object does not exist" error is generated when you attempt to access an existing object on which you have no privileges.

The syntax for the AUDIT command for auditing SQL statements is:

```
AUDIT <priv>,<priv>,...|ALL PRIVILEGES|CONNECT|RESOURCE|DBA
BY <username>
BY SESSION|BY ACCESS
WHENEVER SUCCESSFUL|WHENEVER NOT SUCCESSFUL;
```

When you specify CONNECT, RESOURCE, or DBA instead of a list of privileges, Oracle9*i* audits all privileges granted to that role. See Chapter 13 for more information about these three roles.

The syntax for auditing SQL statements is:

```
AUDIT <sql>,<sql>...|ALL
BY <username>
BY SESSION|BY ACCESS
WHENEVER SUCCESSFUL|WHENEVER NOT SUCCESSFUL;
```

12

All three types of AUDIT commands are similar. You can narrow the focus of any AUDIT command by specifying any of these optional clauses:

- **BY SESSION:** This tells Oracle9*i* to write one record to the audit trail for each session for the same SQL or privilege on the same object. This saves space in the audit trail.
- **BY ACCESS:** This tells Oracle9*i* to write one record to the audit trail for every occurrence of the audited event. This is the default. You can specify either BY ACCESS or BY SESSION, but not both.
- **WHENEVER SUCCESSFUL:** This tells Oracle9*i* to write a record to the audit trail only when the operation is successful.
- **WHENEVER NOT SUCCESSFUL:** This tells Oracle9*i* to write a record to the audit trail only when the operation is not successful. If you don't specify this or the previous clause, Oracle9*i* writes a record for the operation it is auditing regardless of whether it succeeds.

Follow these steps to practice creating some auditing commands.

1. Begin by connecting to the database as the SYSTEM user, because SYSTEM has the privileges needed for auditing. Type and execute this command, replacing <password> with the actual password.

```
CONNECT SYSTEM/<password>@ORACLASS
```

2. SQL*Plus Worksheet replies, "Connected." The first thing to check is the setting of the AUDIT_TRAIL initialization parameter. Type and execute this SQL*Plus command. Figure 12-11 shows the results.

```
SHOW PARAMETERS AUDIT
```

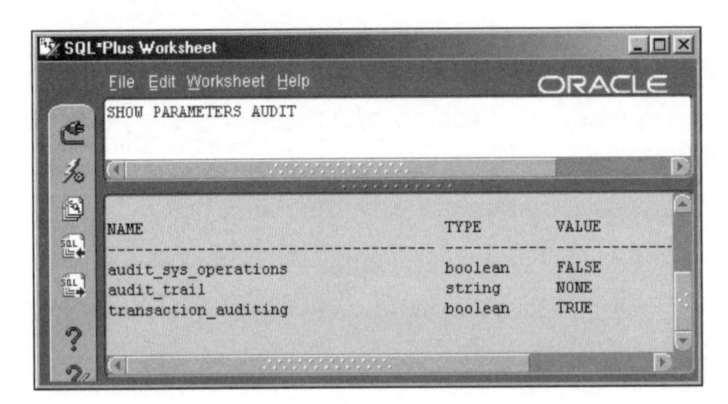

Figure 12-11 The TRANSACTION_AUDITING parameter is not involved in auditing

3. Auditing commands are accepted but have no effect unless the AUDIT_TRAIL parameter is set to a value other than "NONE." AUDIT_TRAIL is a static parameter, so you cannot change it in the current session. However, you can modify its value in the **spfile** and then restart the database to activate the parameter. Modify the **spfile** by typing and executing the following command. You are setting the parameter to "DB" to turn on auditing and to have Oracle9*i* place audit trail records in the database.

```
ALTER SYSTEM SET AUDIT_TRAIL = 'DB' SCOPE=SPFILE;
```

The SQL*Plus Worksheet replies, "System altered."

4. Now, you must shut down and restart the database to put the parameter into effect. You must be logged on with SYSDBA privileges to shut down and start up the database, so connect to SYS by typing and executing the following command. Replace <password> with the current password of SYS, provided by your instructor.

```
CONNECT SYS/<password>@ORACLASS AS SYSDBA
```

The SQL*Plus Worksheet replies, "Connected."

5. Shut down and start up the database by typing and executing these commands. The **spfile** is automatically used for initialization parameters. This takes a minute or so. The last command displays the parameters again to confirm the change in the AUDIT_TRAIL parameter. Figure 12-12 shows the results.

```
SHUTDOWN IMMEDIATE
STARTUP
SHOW PARAMETERS AUDIT
```

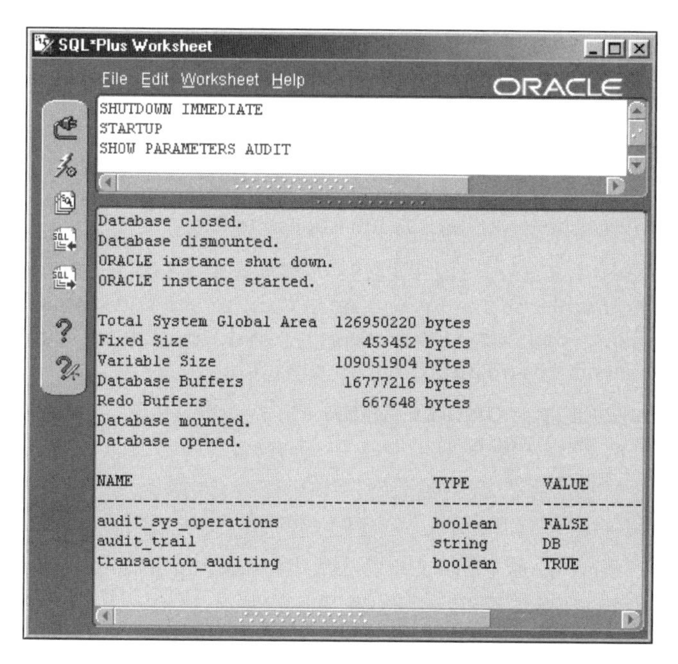

Figure 12-12 Restarting the database causes the new values in the **spfile** to take effect

6. There is no need to be logged on as SYSDBA at this point, and Oracle recommends using SYS and SYSDBA privileges only when absolutely necessary. Therefore, return to the SYSTEM user by typing and executing this command, replacing <password> with the actual password.

```
CONNECT SYSTEM/<password>@ORACLASS
```

The SQL*Plus Worksheet replies, "Connected."

7. It is time to set up some auditing. Imagine that you are concerned about the STUDENTB user and want to monitor everything he does. You want to see a record for every action taken. Type and execute this command to audit all activity of STUDENTB.

```
AUDIT ALL BY STUDENTB;
```

The previous command is a statement auditing command. If it were a privilege auditing command, it would use "ALL PRIVILEGES" instead of "ALL." It does not name an object or display "ON DEFAULT," so it is not an object auditing command. (Object auditing commands always contain the word "ON" followed by either "DEFAULT" or an object name.)

The SQL*Plus Worksheet replies, "Audit succeeded."

8. In addition, you want to monitor all updates to the EMPLOYEE table that are made by anyone. You only need to see one record for a user's session, even if she updates the table ten times in the session. This requires an object auditing command, so type and execute the following command.

```
AUDIT SELECT, UPDATE ON CLASSMATE.EMPLOYEE
BY SESSION;
```

The SQL*Plus Worksheet replies, "Audit succeeded."

9. And finally, you want to audit the CLASSMATE and STUDENTA users when they create a table or a view. You only want to see auditing records when the commands succeed. This requires a privilege auditing command. Type and execute this command to set up the audit.

```
AUDIT CREATE TABLE, CREATE VIEW
BY CLASSMATE, STUDENTA
WHENEVER SUCCESSFUL;
```

The SQL*Plus Worksheet replies, "Audit succeeded."

10. You can use data dictionary views to display the auditing settings you have created. There is a view for each type of audit statement: DBA_OBJ_AUDIT_OPTS for object auditing, DBA_PRIV_AUDIT_OPTS for privilege auditing, and DBA_STMT_AUDIT_OPTS for statement auditing. There is also a special view for object auditing in which you set the auditing as the default for all objects: ALL_DEF_AUDIT_OPTS. Query the view for privilege auditing to confirm that you have created an audit trail for CREATE VIEW and CREATE TABLE for the CLASSMATE and STUDENTB users. Type and execute this query. Figure 12-13 shows the results.

```
COLUMN PRIVILEGE FORMAT A12
SELECT USER_NAME, PRIVILEGE, SUCCESS, FAILURE
FROM DBA_PRIV_AUDIT_OPTS
ORDER BY 1, 2;
```

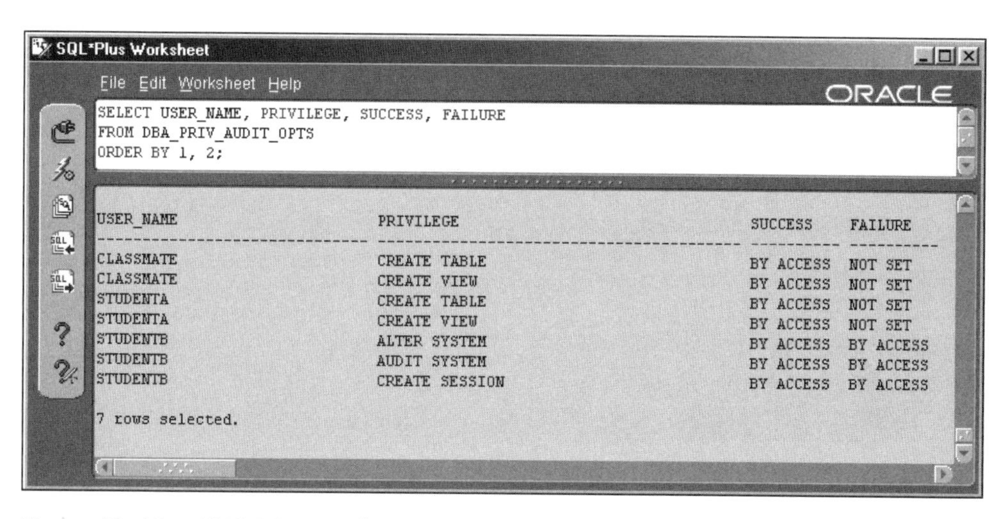

Figure 12-13 "NOT SET" in the FAILURE column indicates no auditing for failed commands

11. To simulate activity that would be audited, you are going to connect to the CLASSMATE user and run a script. Begin by typing and executing this command to switch to the CLASSMATE user.

    ```
    CONNECT CLASSMATE/CLASSPASS@ORACLASS
    ```

    The SQL*Plus Worksheet replies, "Connected."

12. Now, run the script that performs some updates on the EMPLOYEE table, creates a table and a view, and then drops the table and view. One CREATE VIEW statement fails, and one works. One DROP TABLE fails, and one works. Click **Worksheet/Run Local Script**. A window appears in which you can select a file from your computer.

13. Navigate to the **Data\Chapter12** directory, and select the **audit.sql** file. Click **Open** to retrieve the file, and run it in the worksheet.

14. Connect to SYSTEM again by typing and executing this command, replacing <password> with the actual password.

    ```
    CONNECT SYSTEM/<password>@ORACLASS
    ```

    The SQL*Plus Worksheet replies, "Connected."

15. The audit trail records are written to the SYS.AUD$ table. You can query this table directly, or you can use the data dictionary views to examine audit trail records. Query the object auditing records by typing and executing this query. Figure 12-14 shows the results.

    ```
    SELECT USERNAME, ACTION_NAME,
    SES_ACTIONS FROM DBA_AUDIT_OBJECT
    WHERE OBJ_NAME = 'EMPLOYEE';
    ```

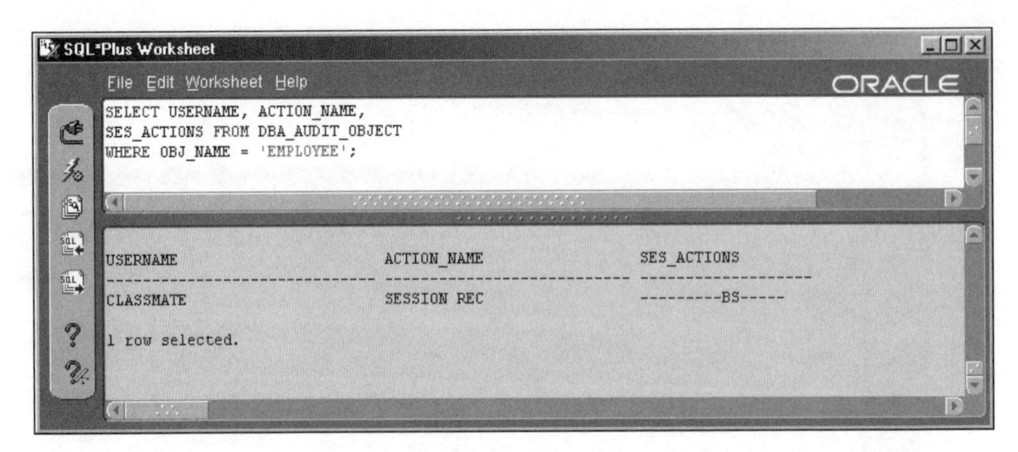

Figure 12-14 Activity on the EMPLOYEE table is consolidated into one record

The SES_ACTIONS column has a set of codes. Each of the 16 positions represents one type of activity that can occur on the table. Your results may show a different combination of letters than the "B" and "S" shown in the Figure, depending on other activities in the database. The tenth and eleventh positions are for SELECT and UPDATE. A dash (-) means no activity, a letter "S" means the action succeeded, a letter "F" means the action failed, and a letter "B" means that it failed and succeeded during this session. So, the CLASSMATE session included successful and failed SELECT actions and successful UPDATE actions. An update command has an implicit SELECT that the audit trail records. One of the update commands failed because it attempted to update a nonexistent record. In this case, the implicit SELECT failed.

16. Remain logged on for the next practice.

You used the DBA_AUDIT_OBJECT data dictionary view in the last practice. You can query several other data dictionary views for audit trail results. The complete list includes:

- **DBA_AUDIT_EXISTS:** Audit trail records generated by object auditing of non-existent objects
- **DBA_AUDIT_OBJECT:** Audit trail records generated by object auditing
- **DBA_AUDIT_SESSION:** Audit trail records generated by session auditing
- **DBA_AUDIT_STATEMENT:** Audit trail records generated by statement auditing
- **DBA_AUDIT_TRAIL:** All audit trail records

All of these have a corresponding USER_counterpart, except DBA_AUDIT_EXISTS.

You may want to turn off auditing or change what you are auditing. This is done with the NOAUDIT command. Its structure is exactly like the AUDIT command; it turns off the auditing it names. You can use it to turn off selective portions of the auditing you have set up. Follow these steps to explore the NOAUDIT command.

1. Recall that you have set up auditing of all statements made by STUDENTB. You can adjust this so that no audit records are generated by the SELECT command. Type and execute this command.

   ```
   NOAUDIT SELECT TABLE BY STUDENTB;
   ```

   The SQL*Plus Worksheet replies, "Noaudit succeeded."

   The other statements will still be audited.

2. To turn off the auditing you created for the EMPLOYEE table, type the same command you used to create the auditing, except use NOAUDIT instead of AUDIT. Another difference between AUDIT and NOAUDIT is that you do not specify BY SESSION or BY ACCESS in the NOAUDIT command. Type and execute this command.

   ```
   NOAUDIT SELECT, UPDATE ON CLASSMATE.EMPLOYEE;
   ```

   The SQL*Plus Worksheet replies, "Noaudit succeeded."

3. Log off by clicking the **X** in the top-right corner of the SQL*Plus Worksheet window.

4. Close the console by clicking the **X** in the top-right corner of the console window. You have seen how granting privileges and auditing database activity can be used to enhance security on the Oracle9*i* database. Building on the base established in this chapter, the next chapter covers another security technique: using roles to manage privileges.

Chapter Summary

- System privileges allow a user to manage some part of the database system.
- Object privileges allow a user to work with an object.
- SYSDBA and SYSOPER are system privileges that allow a user to start up and shut down the database, as well as other high-level tasks.
- The CREATE SESSION system privilege is needed to log onto the database.
- Typical object privileges for a table include SELECT, INSERT, UPDATE, and DELETE.
- The GRANT and REVOKE commands are used for both system and object privileges.
- Use WITH ADMIN OPTION when granting system privileges to allow the user to grant that privilege to others.

- A grant made to PUBLIC gives all users the privilege.
- Revoked system privileges do not cascade to other users.
- Use WITH GRANT OPTION when granting object privileges to allow the user to grant that privilege to others.
- Revoked object privileges cascade to other users.
- Object privileges can be granted on columns.
- The owner of a table can grant object privileges on that table.
- The grantor grants the privilege and the grantee receives the privilege.
- Querying an object without privileges to query causes an error stating that the object does not exist.
- Statement auditing is the monitoring of activity on a particular type of statement, such as SELECT.
- Privilege auditing audits any command that is authorized by the privilege, such as CREATE TABLE.
- Object auditing generates audit trail records as soon as the object is used, such as with SELECT or DELETE statements.
- The SYS.AUD$ table holds auditing records unless the AUDIT_TRAIL initialization parameter is set to "OS."
- AUDIT_SYS_OPERATIONS is an initialization parameter that, when set to "TRUE," causes Oracle9*i* to audit all activity by SYS or users with SYSDBA privileges.
- BY ACCESS or BY SESSION tell Oracle9*i* whether to write a record for each occurrence of an audited event or a summary record for the session.
- The following clauses limit the writing of audit trail records: WHENEVER SUCCESSFUL and WHENEVER NOT SUCCESSFUL.
- AUDIT_TRAIL is a static parameter, so you must restart the database after changing it.
- A group of data dictionary views shows audit trail records for each type of auditing.
- Use the NOAUDIT command to stop specific auditing activities.

Review Questions

1. Which of these is not a system privilege?
   a. SELECT TABLE
   b. ALTER TABLESPACE
   c. CREATE INDEX
   d. SYSDBA

2. Object privileges must be granted by the DBA. True or False?
3. To create a table in another schema, a user must have the ________________ system privilege.
4. Which of the following object privileges **cannot** be granted on a view?
   a. SELECT
   b. UPDATE
   c. FLASHBACK
   d. EXECUTE
5. Explain why you would grant a system privilege to PUBLIC.
6. If user SMITH has the CREATE TABLE privilege and you want to revoke it, which statement would you use?
   a. REVOKE CREATE TABLE FROM USER SMITH;
   b. REVOKE CREATE TABLE FROM SMITH;
   c. REVOKE ANY PRIVILEGE FROM SMITH;
   d. REVOKE CREATE ANY TABLE FROM SMITH;
7. Which of these is a valid statement to grant object privileges?
   a. GRANT SELECT ON CUSTOMER TO SMITH;
   b. GRANT SELECT ANY TABLE TO SMITH;
   c. GRANT ALL PRIVILEGES TO SMITH;
   d. GRANT ALL ON CUSTOMER;
8. You have a static table that is commonly queried by most users and applications. What would be an appropriate action to handle privileges for the table?
   a. Grant all object privileges on the table to PUBLIC.
   b. Grant the SELECT object privilege to PUBLIC.
   c. Create a public synonym for the table.
   d. Grant the SELECT object to all users.
9. The ________________ clause of the GRANT command changes the audit trail so that it writes one record per session.
10. Which of the following data dictionary views should a user query to view any object privileges he has granted to other users?
    a. DBA_SYS_PRIVS
    b. USER_TAB_PRIVS_MADE
    c. USER_TAB_PRIVS_RECD
    d. USER_OBJ_PRIVS_MADE

Exam Review Questions—Database Fundamentals I (#1Z0-031)

1. Examine the following SQL statement.

```
1 GRANT CREATE USER, BECOME ANY USER,
2 CREATE TABLE, CREATE SESSION
3 TO NEWDBA, ADMINDBA
4 WITH GRANT OPTION;
```

Which line causes the statement to fail?

a. Line 1

b. Line 2

c. Line 3

d. Line 4

e. None of the lines cause an error

2. You are the DBA at a large company. Your assistant DBA is taking over for you while you are on vacation. You want your assistant to be able to create the schema needed for a new application system and grant the appropriate privileges to the end users. Which privileges will your assistant need? Choose three.

a. CREATE SCHEMA

b. CREATE USER

c. CREATE ANY TABLE

d. GRANT ANY OBJECT PRIVILEGE

e. GRANT ANY PRIVILEGE

3. Examine the following SQL script.

```
SQL> GRANT SELECT, INSERT, UPDATE
2     ON  CLASSMATE.CUSTOMER
3     TO JIM
4     WITH GRANT OPTION;
Grant succeeded.
SQL> CONNECT JIM/jimmy@ORACLASS
Connected.
SQL> GRANT SELECT ON CLASSMATE.CUSTOMER
2  TO SMITH;
SQL> CONNECT SYSTEM/MANAGER@ORACLASS
Connected.
SQL> REVOKE SELECT ON CLASSMATE.CUSTOMER
2  FROM JIM;
Revoke succeeded.
```

What will happen if SMITH queries the CUSTOMER table?

a. The query fails with the error: "Insufficient privileges."

b. The query fails with the error: "Table does not exist."

c. The query succeeds.

d. SMITH is logged off immediately.

4. Which of the following queries displays the system privileges given out by the ORADBA user?

   a. SELECT * FROM DBA_SYS_PRIVS WHERE GRANTEE = 'ORADBA';

   b. SELECT * FROM DBA_TAB_PRIVS WHERE GRANTEE = 'ORADBA';

   c. SELECT * FROM DBA_SYS_PRIVS WHERE GRANTOR = 'ORADBA';

   d. SELECT * FROM SESSION_PRIVS WHERE GRANTOR = 'ORADBA';

5. You logged on as SYSTEM and granted CREATE TABLE WITH ADMIN OPTION to Susan. Susan then granted CREATE TABLE WITH ADMIN OPTION to Henry and Albert. You then revoked the CREATE TABLE privilege from Susan. Who has the CREATE TABLE privilege now?

   a. Henry and Albert

   b. SYSTEM, Henry, and Albert

   c. Only SYSTEM

   d. SYSTEM and Susan

6. You grant the SYSDBA privilege to your new DBA coworker. What tasks can the new DBA perform? Choose two.

   a. Query any table

   b. Create new users

   c. Shut down the database

   d. Create an **spfile**

   e. Grant SYSDBA to another user

7. You plan to start auditing the database to investigate some unusual changes to the data. You change the AUDIT_TRAIL parameter with this command:

```
ALTER SYSTEM SET AUDIT_TRAIL = 'DB' SCOPE=SPFILE;
```

   What is your next step to implement auditing?

   a. Execute the AUDIT command

   b. Shut down and restart the database

   c. Set the AUDIT_FILE_DEST parameter

   d. Back up the database

8. You are suspicious that the user named JOEY has been snooping and attempting to query sensitive tables that he is unauthorized to view. You want to monitor this type of activity. Which of these AUDIT commands is appropriate?
   a. AUDIT SELECT ON NOT EXISTS BY JOEY;
   b. AUDIT SELECT BY JOEY WHENEVER NOT SUCCESSFUL;
   c. AUDIT SELECT TABLE BY JOEY WHENEVER SUCCESSFUL;
   d. AUDIT SELECT ON JOEY WHENEVER NOT SUCCESSFUL;
9. Examine the following query.

```
SELECT USERNAME, ACTION_NAME,
SES_ACTIONS FROM DBA_AUDIT_STATEMENT
ORDER BY 1;
```

   Which of the following AUDIT commands will generate records seen by the query?
   a. AUDIT INSERT ON CUSTOMER WHENEVER SUCCESSFUL;
   b. AUDIT SELECT TABLE BY STUDENTA WHENEVER NOT SUCCESSFUL;
   c. AUDIT SELECT ANY TABLE BY ACCESS;
   d. AUDIT UPDATE ON CUSTOMER BY SESSION;
10. Examine the following command.

```
GRANT SELECT, INSERT, UPDATE, DELETE
ON CLASSMATE.CUSTOMER TO JOEY, SAMUEL;
GRANT ALL ON CLASSMATE.CUSTOMER TO MARTIN;
REVOKE SELECT, DELETE
ON CLASSMATE.CUSTOMER FROM SAMUEL, MARTIN;
GRANT SELECT ON CLASSMATE.CUSTOMER TO PUBLIC;
```

   Which of the users (JOEY, MARTIN, and SAMUEL) can successfully execute this query?

```
DELETE FROM CLASSMATE.CUSTOMER
WHERE FIRST_NAME = 'Mark';
```

   a. JOEY and MARTIN
   b. SAMUEL and JOEY
   c. JOEY
   d. JOEY, MARTIN, AND SAMUEL

HANDS-ON ASSIGNMENTS

1. Create a user named VICTOR who can create tables, views, indexes, and can grant object privileges to other users on any object that he creates.

   Save your work in a script named **ho1201.sql** in the **Solutions\Chapter12** directory on your student disk.

2. You want to share some of the DBA tasks with your coworker, Amy Schultz. Create a user named ASCHULTZ, and assign her the appropriate privileges to do the following tasks:

   - Back up the database
   - Shut down and start up the database
   - Query all schema tables and views
   - Audit SQL statements

   (*Hint*: You must be logged on as SYSDBA to grant the SYSDBA privilege.) Save your work in a script named **ho1202.sql** in the **Solutions\Chapter12** directory on your student disk.

3. You decide that all application developers except one have made too many mistakes when moving their Oracle9*i* schemas into the production database. They are constantly forgetting to create a table, or to grant DELETE privileges to a user. You assign the task of installing every developer's schemas to the one developer who seems to be able to handle the job. The developer's user name is K_ITO. Create the user K_ITO, and then grant the user all the appropriate privileges to handle her new responsibilities. The schemas have tables, indexes, and sequences. The developers usually grant SELECT, INSERT, UPDATE, and DELETE to end users.

   Save your work in a script named **ho1203.sql** in the **Solutions\Chapter12** directory on your student disk.

4. Your job as DBA in Local Locale Company requires you to help the staff of application programmers who work with the database. One common complaint from the programmers is: "I keep getting the error message, 'Table does not exist' when I know it is just a problem with permissions." The error does not appear for the programmer, because she works on development while logged onto the database as the table owner. The error shows up when the user logs on and runs the application. The application queries the CLASSMATE.CLASSIFIED_SECTION and CLASSMATE.CLASSIFIED_AD tables. Write an auditing statement that sets up auditing on these two tables, so that you can find out which users get the "Table does not exist" error message. Then write a query on the appropriate data dictionary view to list records in the audit trail.

   Save your work in a script named **ho1204.sql** in the **Solutions\Chapter12** directory on your student disk.

5. There is a new Web-database program that will be available to the registered users (there are thousands) on the Local Locale Company Web site. The program allows users to log onto the database and query the CLASSIFIED_SECTION table and to view their own records in the CUSTOMER table and update their own e-mail addresses and phone lists in the CUSTOMER table. Assume that the program will display the correct record from the CUSTOMER table for the users to view and update. Write the appropriate GRANT commands to handle these requirements.

Save your work in a script named **ho1205.sql** in the **Solutions\Chapter12** directory on your student disk.

6. Write a script that spools out a REVOKE command for every object privilege in the system. (*Hint*: The SQL*Plus command "spool <path\filename>" writes the results of a query to a file.)

   Save your work in a script named **ho1206.sql** in the **Solutions\Chapter12** directory on your student disk.

7. Write a query that lists all the current privilege auditing. Use the query to generate a NOAUDIT command for each auditing record, and spool the results to a file. Run the resulting NOAUDIT commands. Make corrections to your query if needed.

   Save your work in a script named **ho1207.sql** in the **Solutions\Chapter12** directory on your student disk.

8. Give every employee in the Local Locale Company the privileges needed to query and update the NEWS_ARTICLE table and the CLASSIFIED_AD table. Write only two commands to accomplish this task.

   Save your work in a script named **ho1208.sql** in the **Solutions\Chapter12** directory on your student disk.

9. Look at Table 12-1. Create a list of these CLASSMATE schema objects and the privileges you can grant to them: CUSTOMER, EDITOR_INFO, CLIENT_VIEW, and CUSTOMER_ADDRESS.

   Save your work in a file named **ho1209.txt** in the **Solutions\Chapter12** directory on your student disk.

10. There is an employee who has given notice and is going to another job in two weeks. You are worried that he might try to damage some of the database data. He has access to several different user names in the database, so you want to audit all those user names.

    Write an audit command to audit these three user names: STUDENTA, STUDENTB, and CLASSMATE. The audit should write an audit trail record every time one of these users attempts an update or delete on any table.

    Write a query on the audit trail records that will show not only the user name, but also the operating system name and the terminal from which the commands originated. (*Hint*: Find all the columns in a table or view by using the DESCRIBE command in SQL*Plus, which can be shortened to DESC. For example, DESC DBA_AUDIT_SESSION lists all the columns in the DBA_AUDIT_SESSION table.)

    Save your work in a file named **ho1210.sql** in the **Solutions\Chapter12** directory on your student disk.

CASE PROJECT

The Global Globe Company has editors who will be using a new application to enter their classified ads. The editors will need to be able to perform these tasks:

- Create new ads (recorded in the CLASSIFIED_AD table)
- Modify the ads that they have created
- Look up the description of want ad sections in the CLASSIFIED_SECTION table
- Update customer information, but not create or remove customer records (in the CUSTOMER table)

Head editors can do all these tasks and, in addition, are allowed to delete ads from the CLASSIFIED_AD table.

Write a script to grant all the privileges to the appropriate users. To help you find the user names, recall from Chapter 11 that all the editors have "EDITOR" in their job title and that all user names are the first initial followed by the last name of the employee. The head editor's job title is "CHIEF EDITOR." Save your work in a script named **case1201.sql** in the **Solutions\Chapter12** directory on your student disk.

CHAPTER

13

DATABASE ROLES

In this chapter, you will:

- Discover when and why to use roles
- Learn how to create, modify, and remove roles
- Learn how to assign roles
- Examine data dictionary views of roles
- Assign roles and privileges using the Enterprise Management console

In the previous chapters, you have learned how to create users and assign system and object privileges to those users. In this chapter, you learn how to make groups of privileges, called roles. You'll find out how roles can simplify security and what predefined roles Oracle9*i* provides.

INTRODUCTION TO ROLES

A **role** is a collection of privileges that is named and assigned to users or even to another role. A role can help you simplify database maintenance by giving you an easy way to assign a set of privileges to new users.

How to Use Roles

The primary use of roles is to help simplify security. For example, in a typical database system, you have a schema with 25 tables and a group of 50 users who need access to the schema. Figure 13-1 shows a miniversion of a database schema and its users, with the privileges they each require. As you can see, even scaled down to five tables and six users, you must grant 59 privileges. Even if you group them together in sets of users with the same sets of grants, you have at least 20 combinations to deal with.

Privileges	Tables				
	CUSTOMER	PAYMENT	ORDER	STORE	INVENTORY
SELECT	Joe Amy Henry Sarah John	Joe Amy Henry Sarah John	Joe Amy Henry Sarah John	Joe Amy Henry Sarah John Martin	Joe Amy Henry Sarah John
INSERT	Joe Amy	Joe Amy	Henry Sarah John	Martin	Henry Sarah John
UPDATE	Joe Amy	Joe Amy	Henry Sarah John	Martin	Henry Sarah John
DELETE	Joe Amy	Joe Amy	Henry Sarah John	Martin	Henry Sarah John

Figure 13-1 A matrix of privileges, tables, and users needing those privileges

If you look closely at the table in Figure 13-1, you can see a pattern of users who share the same set of grants. Figure 13-2 highlights the patterns, or groups of users who share the same set of privileges. This is where roles come into play. You can create a role for each group of users, grant the privileges to the role, and then grant the role to each user in each group.

Privileges	Tables				
	CUSTOMER	PAYMENT	ORDER	STORE	INVENTORY
SELECT	Joe Amy Henry Sarah John	Joe Amy Henry Sarah John	Joe Amy Henry Sarah John	Joe Amy Henry Sarah John Martin	Joe Amy Henry Sarah John
INSERT	Joe Amy	Joe Amy	Henry Sarah John	Martin	Henry Sarah John
UPDATE	Joe Amy	Joe Amy	Henry Sarah John	Martin	Henry Sarah John
DELETE	Joe Amy	Joe Amy	Henry Sarah John	Martin	Henry Sarah John

Figure 13-2 The six users fall into three groups

At first, creating roles may seem like an extra step. After you begin to add new users, however, you save time, because you must grant roles only to new users. In addition, you can revise the privileges in a role, and the change is automatically reflected for every user who has the role.

13

Using Predefined Roles

Recall from Chapter 12 that there are over one hundred system privileges, and as you create more tables, the number of possible object privileges grows as well. Even so, as you become familiar with these privileges, you quickly realize that many jobs require a typical set of privileges. The job of DBA, for example, requires nearly all of the system privileges. The job of applications developer requires the subset of the DBA privileges that enables the developer to create tables, procedures, and other database objects.

Oracle9*i* provides predefined roles for common job titles to speed up your ability to get your users up and running with the privileges they require. Table 13-1 shows some of the commonly used predefined roles and what they are intended to handle.

Table 13-1 Predefined roles

Role name	Description
CONNECT	Logs onto the database and performs limited activities within the user's own schema, such as creating tables, views, synonyms, and database links.
DBA	Manages the database, including these tasks: creates users, profiles, and roles, and grants privileges; manages storage and security; starts up and shuts down the database.
DELETE_CATALOG_ROLE	Gives the user the ability to delete from tables owned by SYS. This role was added because the system privilege DELETE ANY TABLE specifically excludes deleting from tables owned by SYS.
EXECUTE_CATALOG_ROLE	Enables the user to execute any package supplied by Oracle that is owned by SYS. Most supplied packages are owned by SYS, and those most commonly used already allow users to execute them. If additional packages are needed, grant the user this role.
EXP_FULL_DATABASE	Exports the database using the EXPORT utility.
IMP_FULL_DATABASE	Imports the database using the IMPORT utility.
RESOURCE	Provides more extensive abilities to create objects, such as procedures, triggers, and object types, for users who need to create their own objects. If you assign the CONNECT role to a user, you should grant a RESOURCE as well.
SELECT_CATALOG_ROLE	Allows the user to query any data dictionary view or table owned by SYS. This can give a user more access to certain data dictionary views, although usually a user can already access those he needs, because the most common data dictionary views are viewable by all users.

When you install and create the database, there are a dozen or so additional roles created that are for internal use by database tools and utilities. For example, there is a role called AQ_ADMINISTRATOR_ROLE, containing privileges specifically used for managing Oracle advanced queues, which is an Oracle feature for communicating between databases.

Oracle recommends that you create your own roles instead of using the CONNECT, RESOURCE, and DBA roles, because these three roles will be dropped in a future release.

Complete the following steps to query the data dictionary views and view the privileges that some of these roles include.

1. Start up the SQL*Plus Worksheet. In Windows, click **Start/Programs/ Oracle - OraHome92/Application Development/SQLPlus Worksheet**. In UNIX, type **oemapp worksheet** on the command line. The standard Oracle Enterprise Manager Login screen appears.
2. Select the **Connect directly to a database** radio button, and type **SYSTEM** in the Username box. In the Password box, type the current password for SYSTEM that is *provided by your instructor*. Type **ORACLASS** in the Service box, and leave the default selection of "Normal" in the Connect as box. Click **OK**. A background SQL*Plus process starts, and then the SQL*Plus Worksheet window appears. If the background process appears in front, simply minimize it (click the **minus sign** in the top-right corner), so that you can see the worksheet.
3. Type and execute this query to list the system privileges granted to the IMP_FULL_DATABASE role. Figure 13-3 partially shows the results.

```
SET LINESIZE 100
SELECT * FROM ROLE_SYS_PRIVS
WHERE ROLE = 'IMP_FULL_DATABASE'
ORDER BY 1, 2;
```

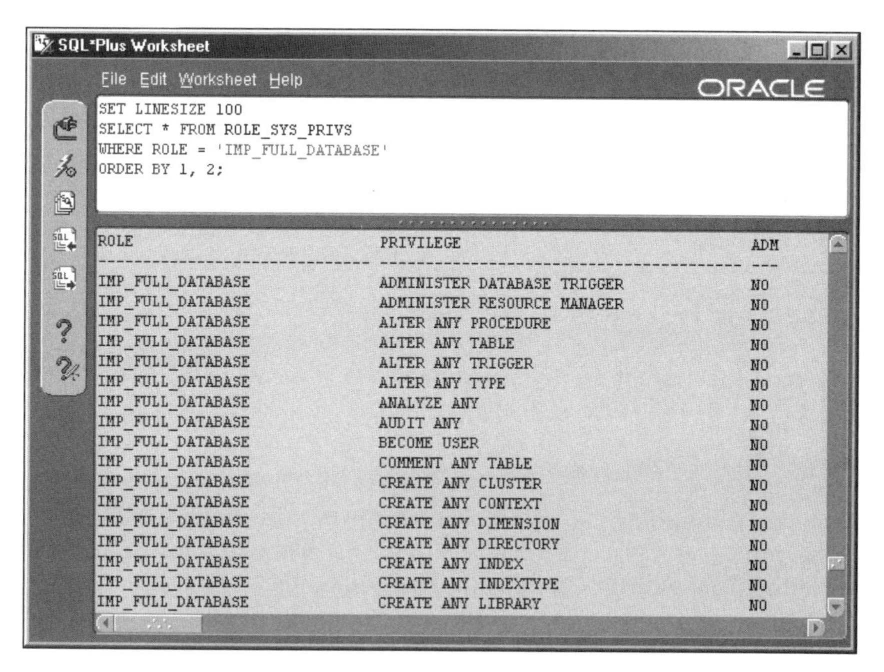

Figure 13-3 Roles are often granted system privileges

This role has over 60 privileges that are needed to perform a full database import. An **import** is a process that uses a previously created snapshot of the database and re-creates all or part of it in the current database. An import can create users, tables, views, and triggers, add rows to existing tables, and much more.

4. Remain logged on for the next practice.

   You have seen Oracle's predefined roles, which give you an idea of how roles can be used. The next section shows you how to create your own roles.

CREATING AND MODIFYING ROLES

Roles are used to consolidate a group of system or object privileges so that you, the DBA, can assign the role to a user, rather than assigning all the underlying system and object privileges. You can also assign the role to another role.

The first step in this process is creating the role itself. The syntax is:

```
CREATE ROLE <name>
NOT IDENTIFIED|IDENTIFIED BY <password>
```

The NOT IDENTIFIED clause means that no additional authorization is required. This is the default, so omitting the clause is the same as including NOT IDENTIFIED. The alternative option is the IDENTIFIED BY <password> clause, which means that the user must provide the correct password to be able to use the privileges within that role.

The next step in working with roles is to assign privileges to the role. This is done with the GRANT command in exactly the same way as granting privileges to a user. Chapter 12 dealt with granting privileges to users. In this chapter, you'll see the same command, except that the user name is replaced by a role name. The syntax is:

```
GRANT <privilege> TO <role>;
```

You cannot grant a privilege and add WITH ADMIN OPTION or WITH GRANT OPTION when granting to a role.

The final step is assigning the role to a user. Again, this uses the GRANT command, except this time, you replace the privilege with the name of a role. You can also grant a role to another role with this command. The syntax is:

```
GRANT <role> TO <user>|<role>
WITH ADMIN OPTION;
```

Include the WITH ADMIN OPTION only when you want the user to be able to grant the role to other users. If you grant a role to a second role with the WITH ADMIN OPTION, any user who is granted the second role is allowed to grant the first role to others.

The only part of a role you can change is whether it uses a password. The syntax of the ALTER ROLE command is:

```
ALTER ROLE <name>
NOT IDENTIFIED|IDENTIFIED BY <password>
```

For example, to change the UPDATEALL role so that it requires a password, you use the following code:

```
ALTER ROLE UPDATEALL
IDENTIFIED BY U67DATR
```

When a role switches to requiring a password, users currently logged on who are granted the role are unaffected until they log off and back on again.

Creating and Assigning Privileges to a Role

You have decided to create a role that allows a user to query the Global Globe tables that are needed to use an application that helps employees build their own queries. Follow along with these steps to create a new role and assign privileges to the role.

1. Create the role, named SELALL by typing and executing this command.

   ```
   CREATE ROLE SELALL;
   ```

2. SQL*Plus Worksheet replies, "Role created." Now, grant the SELECT object privilege to the role for the Global Globe tables by typing and executing these GRANT commands.

   ```
   GRANT SELECT ON CLASSMATE.CLASSIFIED_AD TO SELALL;
   GRANT SELECT ON CLASSMATE.CLASSIFIED_SECTION TO SELALL;
   GRANT SELECT ON CLASSMATE.CUSTOMER TO SELALL;
   GRANT SELECT ON CLASSMATE.CUSTOMER_ADDRESS TO SELALL;
   GRANT SELECT ON CLASSMATE.NEWS_ARTICLE TO SELALL;
   GRANT SELECT ON CLASSMATE.EMPLOYEE TO SELALL;
   ```

3. SQL*Plus Worksheet replies, "Grant succeeded" after it executes each command.
4. Remain logged on for the next practice.

Other users are allowed to grant privileges to a role created by the DBA. The user must, of course, have the authority to grant the privilege to others. For example, the CLASSMATE user could grant UPDATE ON EMPLOYEE to the SELALL role. If it makes sense to do so, users can grant privileges to the pre-defined roles as well.

Assigning Roles to Users and to Other Roles

Now that you have a role, you can assign (grant) the role to any user you want. In addition, a role can be granted to another role. The grantee role inherits all the privileges of the grantor role.

Continuing with the Global Globe example, follow these steps to grant SELALL to users and to another role.

1. For this exercise, you can use the two users created in the previous chapter, STUDENTA and STUDENTB. Type and execute this command to grant each of them the SELALL role.

   ```
   GRANT SELALL TO STUDENTA, STUDENTB;
   ```

2. SQL*Plus Worksheet replies, "Grant succeeded."
3. Connect to STUDENTA, and check to see what privileges he has. Type and execute these commands. Figure 13-4 shows the results.

   ```
   CONNECT STUDENTA/A1PLU$@ORACLASS
   SELECT TABLE_NAME, PRIVILEGE FROM USER_TAB_PRIVS;
   ```

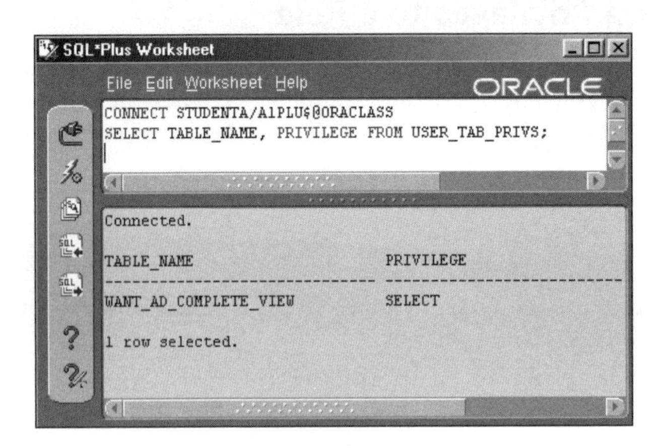

Figure 13-4 STUDENTA appears to have SELECT privileges on only one table

4. What went wrong? Nothing, really. However, when you grant a role, the user inherits the privileges granted to the role even though the privileges don't appear in the USER_TAB_PRIVS view. The following query shows that STUDENTA has been granted the SELALL role. Type and execute the query.

   ```
   SELECT * FROM USER_ROLE_PRIVS;
   ```

5. Here is another query, in which the object privileges of the SELALL role and the object privileges of the STUDENTA user are combined to show a complete list of STUDENTA's privileges. Type and execute this query. Figure 13-5 shows the results.

   ```
   COLUMN OWNER FORMAT A10
   COLUMN PRIVILEGE FORMAT A10
   SELECT OWNER, TABLE_NAME, PRIVILEGE FROM USER_TAB_PRIVS
   ```

```
UNION
SELECT OWNER, TABLE_NAME, PRIVILEGE FROM ROLE_TAB_PRIVS
WHERE ROLE IN (SELECT ROLE FROM USER_ROLE_PRIVS)
ORDER BY 1, 2;
```

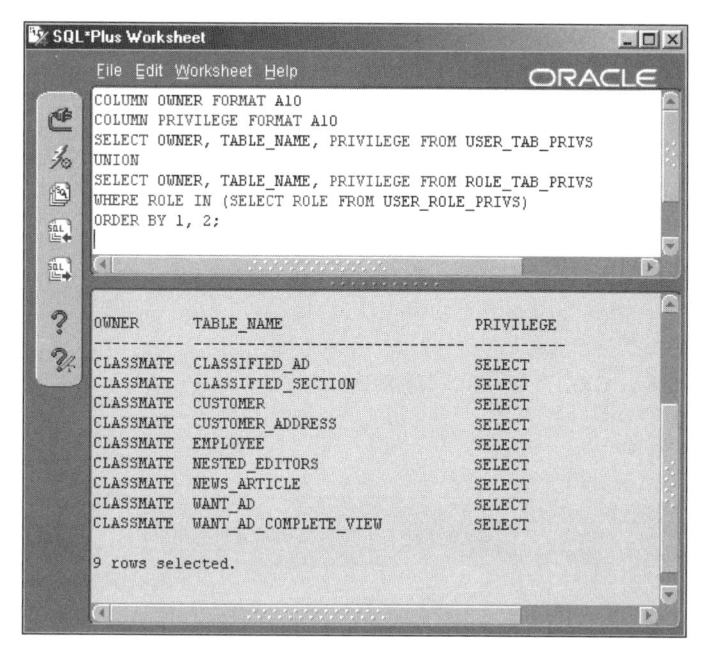

Figure 13-5 Role-owned privileges combine with user-owned privileges in this query

6. The previous query works well, unless a role has been granted to another role, in which case, you'd need a more complex query to derive the privileges of the underlying role. Reconnect to the SYSTEM user to create another role and grant one role to another. Type and execute this command, replacing <password> with the current password for SYSTEM.

   ```
   CONNECT SYSTEM/<password>@ORACLASS
   ```

   The SQL*Plus Worksheet replies, "Connected."

7. You want another role for users who are allowed to update the EMPLOYEE and CUSTOMER tables. These users should also be able to query all the tables listed in the SELALL role. The new role, called UPDPEOPLE, is more sensitive, so you add a password requirement to the role. Type and execute this command.

   ```
   CREATE ROLE UPDPEOPLE IDENTIFIED BY M6XABC;
   ```

8. SQL*Plus replies, "Role created." Add the two privileges for the EMPLOYEE and CUSTOMER tables by typing and executing these commands.

```
GRANT UPDATE ON CLASSMATE.CUSTOMER TO UPDPEOPLE;
GRANT UPDATE ON CLASSMATE.EMPLOYEE TO UPDPEOPLE;
```

9. SQL*Plus replies, "Grant succeeded" after it executes each statement. To assign the privileges of the SELALL role to the UPDPEOPLE role, type and execute this command.

```
GRANT SELALL TO UPDPEOPLE;
```

   The SQL*Plus Worksheet replies, "Grant Succeeded."

10. You can see which roles have been granted to another role by querying the ROLE_ROLE_PRIVS view. Type and execute this query. Figure 13-6 shows the results, in which you can see that the SELALL role was granted to the UPDPEOPLE role.

```
SELECT ROLE, GRANTED_ROLE, ADMIN_OPTION
FROM ROLE_ROLE_PRIVS
ORDER BY 1,2;
```

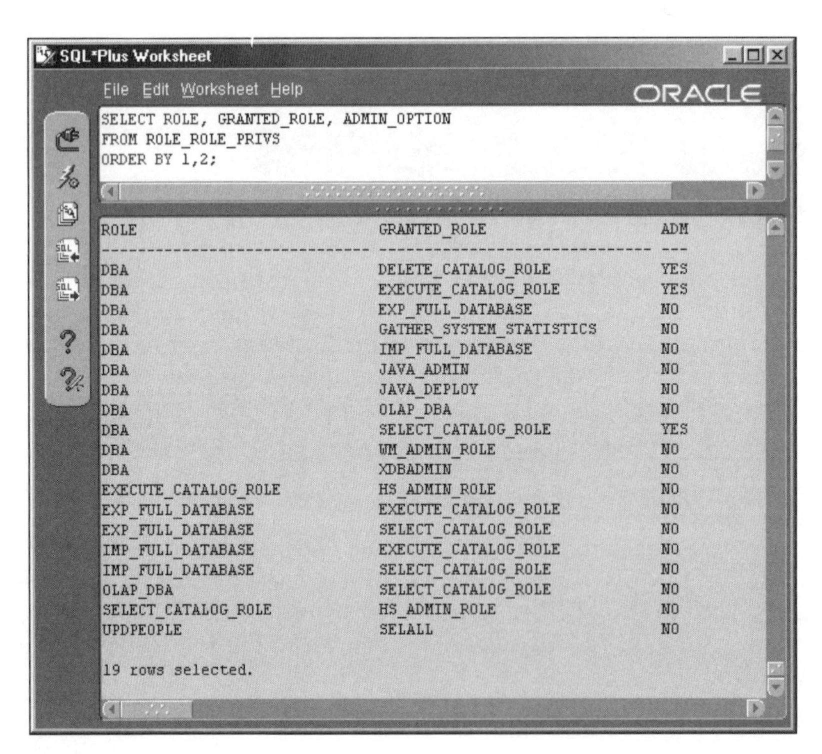

Figure 13-6 Roles that contain granted roles are shown in the ROLE_ROLE_PRIVS view

11. You can grant the WITH ADMIN OPTION to either a role or a user. This allows the user (or a user with the role granted the WITH ADMIN OPTION) to grant that role to other users or roles. Change the UPDPEOPLE role so that it can administer the SELALL role by typing and executing this command:

    ```
    GRANT SELALL TO UPDPEOPLE WITH ADMIN OPTION;
    ```

    The SQL*Plus Worksheet replies, "Grant succeeded."

12. Now, a user with the UPDPEOPLE role can assign the SELALL role to other users. Try this out by granting the UPDPEOPLE role to Amanda Gaines, the chief editor whose user name is AGAINES. Type and execute this command:

    ```
    GRANT UPDPEOPLE TO AGAINES;
    ```

    The SQL*Plus Worksheet replies, "Grant succeeded."

13. Remain logged on for the next practice.

Roles are dynamic. After a user has a role, you can add more privileges to the role, and the user receives the privilege immediately.

The next session shows you how to enable (bring into effect) a role with a password, and how to control which roles are enabled.

Limiting Availability and Removing Roles

Sometimes, a role does not need to be in force (**enabled**) all the time. You can control when a role becomes enabled for a user in these ways:

13

- **Default roles:** The role's creator or the DBA can adjust the default roles for a user using the ALTER USER command. **Default roles** are roles that are automatically enabled when the user logs onto the database. Even roles that require a password are enabled (without the user specifying the role's password) when the user logs on.
- **Enable roles:** The user with a role can enable or disable his role with the SET ROLE command.
- **Drop roles:** The DBA can drop the role from the database entirely and thereby cancel the role for all users who had it.

The syntax for changing a user's default roles is:

```
ALTER USER <username> DEFAULT ROLE
<role>,...|ALL|ALL EXCEPT <role>,...|NONE
```

The DBA can issue this command to adjust the default roles for a user. When it is granted to a user, the role is automatically in the list of default roles. The only way to remove the role from the user's default roles is by issuing the ALTER USER command. To remove all the roles at once, use the NONE clause.

The syntax for the SET ROLE command is:

```
SET ROLE
<role> IDENTIFIED BY <password>,...|ALL|ALL EXCEPT|NONE|
```

The *user* can issue this command to adjust his enabled roles. To enable roles with passwords, include the IDENTIFIED BY <password> clause; otherwise, simply list the role. Any role not listed is disabled. Enable all roles by using ALL, and disable all roles by using NONE.

The roles remain enabled or disabled until the user issues another SET ROLE command, or until the user logs off. When the user logs on again, his roles are reset to the default roles dictated by the DBA.

Dropping a role revokes it from all users and roles, except those who are currently logged on and have the role enabled. The role is revoked for those users when their sessions end. The syntax of DROP ROLE is simply:

```
DROP ROLE <role>
```

Follow along to work with enabling, disabling, and dropping roles. Remember that in the previous practice, the UPDPEOPLE role was granted to AGAINES.

1. Add the CONNECT role and take the UPDPEOPLE role out of the default list for AGAINES by typing these commands:

   ```
   GRANT CONNECT TO AGAINES;
   ALTER USER AGAINES DEFAULT ROLE ALL EXCEPT UPDPEOPLE;
   ```

   The SQL*Plus Worksheet replies, "User altered."

2. Type and execute the following query to list the roles and to see whether the roles are default or not. Figure 13-7 shows the results. As you can see, the CONNECT role is a default role, and the UPDPEOPLE role is not a default role.

   ```
   SELECT * FROM DBA_ROLE_PRIVS
   WHERE GRANTEE = 'AGAINES';
   ```

3. Connect to AGAINES by typing and executing this command:

   ```
   CONNECT AGAINES/AGAINES@ORACLASS;
   ```

4. The only role currently enabled for AGAINES is the CONNECT role. To verify this, type and execute this query to see the currently enabled roles:

   ```
   SELECT * FROM SESSION_ROLES;
   ```

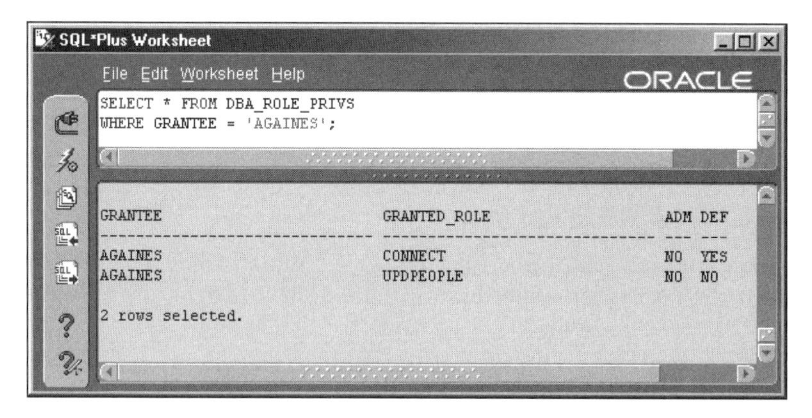

Figure 13-7 View roles assigned to users and the default setting of each role

5. When you enable it, the UPDPEOPLE role requires a password. Type and execute this command to enable the role by providing the correct password:

```
SET ROLE CONNECT, UPDPEOPLE IDENTIFIED BY M6XABC;
```

6. SQL*Plus Worksheet replies, "Role set." The CONNECT role was also listed to keep it enabled. Otherwise, it would have been disabled. Rerun this query to verify the enabled roles. Figure 13-8 shows the results.

```
SELECT * FROM SESSION_ROLES;
```

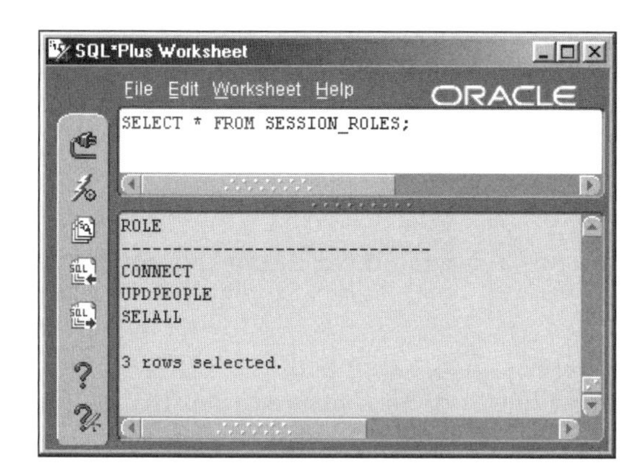

Figure 13-8 The current roles for the user include roles granted to other roles

The SELALL role appears, because it was granted to the UPDPEOPLE role.

7. While the UPDPEOPLE role is enabled, this user has authority to grant SELALL to another user. Try this by typing and executing this command:

```
GRANT SELALL TO JHANSON;
```

The SQL*Plus Worksheet replies, "Grant succeeded."

8. Let's say that AGAINES is finished working on the EMPLOYEE table and wants to disable that role in her session. The SET ROLE command can be used to either enable or disable roles by including or excluding a role in the command. Type and execute this command to disable only the UPDPEOPLE role.

```
SET ROLE ALL EXCEPT UPDPEOPLE;
```

The SQL*Plus Worksheet replies, "Role set."

9. Requery the SESSION_ROLE view to see which roles remain. Figure 13-9 displays the results, showing that only the CONNECT role is enabled now.

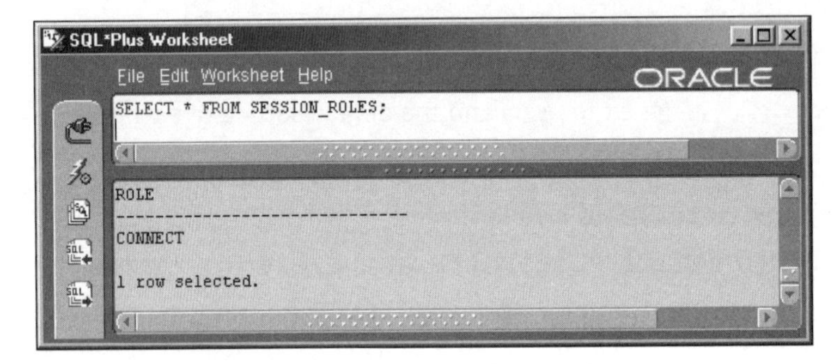

Figure 13-9 The SELALL role is disabled along with the UPDPEOPLE role

10. A role can be removed from the database by running the DROP ROLE command. This must be done by the DBA, so type and execute this command to connect as the SYSTEM user. Replace the <password> variable with the current SYSTEM password.

```
CONNECT SYSTEM/<password>@ORACLASS
```

The SQL*Plus Worksheet replies, "Connected."

11. Drop the UPDPEOPLE role by typing and executing this command:

```
DROP ROLE UPDPEOPLE;
```

The SQL*Plus Worksheet replies, "Role dropped." The UPDPEOPLE role is thereby removed from the privileges of all users and roles that had been granted it. If a user has a session with the UPDPEOPLE role enabled when the DROP ROLE command is run, that role remains enabled for the user until his session ends.

12. Close your SQL*Plus Worksheet session by clicking the **X** in the top-right corner of the window.

Roles are similar to system privileges in that a role granted by a user is not revoked if the granting user loses the role.

Data Dictionary Information about Roles

You have used many of the data dictionary views while practicing the display of information about roles. Table 13-2 shows all the views, including those that you have already queried.

Table 13-2 Data dictionary views about roles

Name	Description
ALL_TAB_PRIVS_MADE	All object privileges granted and by whom
DBA_ROLE_PRIVS	All roles and grantees including users and roles
DBA_ROLES	All the roles in the database
DBA_SYS_PRIVS	All system privileges granted to users or roles
DBA_TAB_PRIVS	All object privileges granted to users or roles
ROLE_ROLE_PRIVS	Roles granted to other roles that the current user can enable
ROLE_SYS_PRIVS	System privileges granted to roles that the current user can enable
ROLE_TAB_PRIVS	Object privileges granted to roles that the current user can enable
SESSION_ROLES	Roles currently enabled in your session

Remember that the DBA_ prefixed views have corresponding USER_ and ALL_ prefixed views.

Roles in the Enterprise Manager Console

The Enterprise Manager console's Security Manager can be used to create and manage roles very easily. Take a quick look at this by following these steps.

1. Start up the Enterprise Manager console. In Windows, click **Start/ Programs/Oracle - OraHome92/Enterprise Manager Console**. In UNIX, type **oemapp console** on the command line. The Enterprise Manager console login screen appears.
2. Select the **Launch standalone** radio button, and click **OK**. The console appears.
3. In the main window of the Enterprise Manager console, navigate starting at the **Databases** icon to **ORACLASS/Security/Roles**. Your screen should look similar to Figure 13-10.

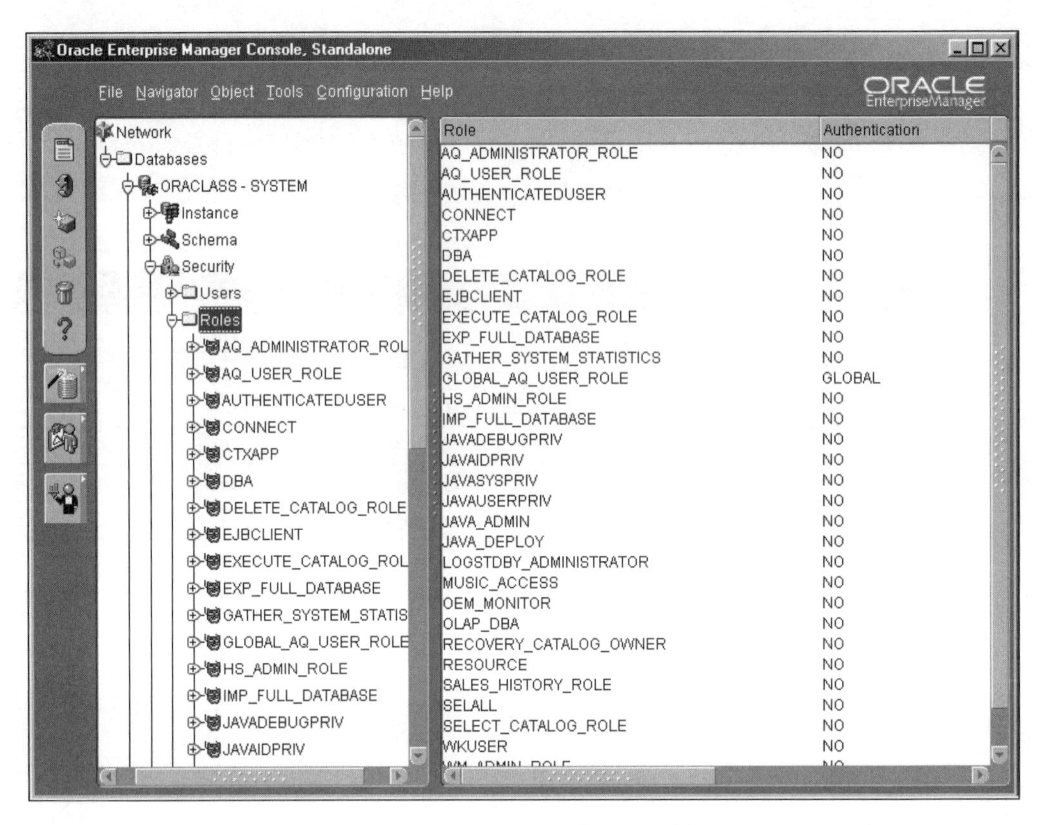

Figure 13-10 The Security Manager has an excellent tool for managing roles

4. Click the **CONNECT** role on the left side of the screen. This brings up the property sheet for the CONNECT role.
5. Click the **System** tab, which displays all the system privileges assigned to the CONNECT role. Figure 13-11 shows the screen. You can add or remove privileges using this screen by highlighting a privilege and using the arrow buttons in the center to move the privilege up (revoking the privilege from the role) or down (granting the privilege to the role).

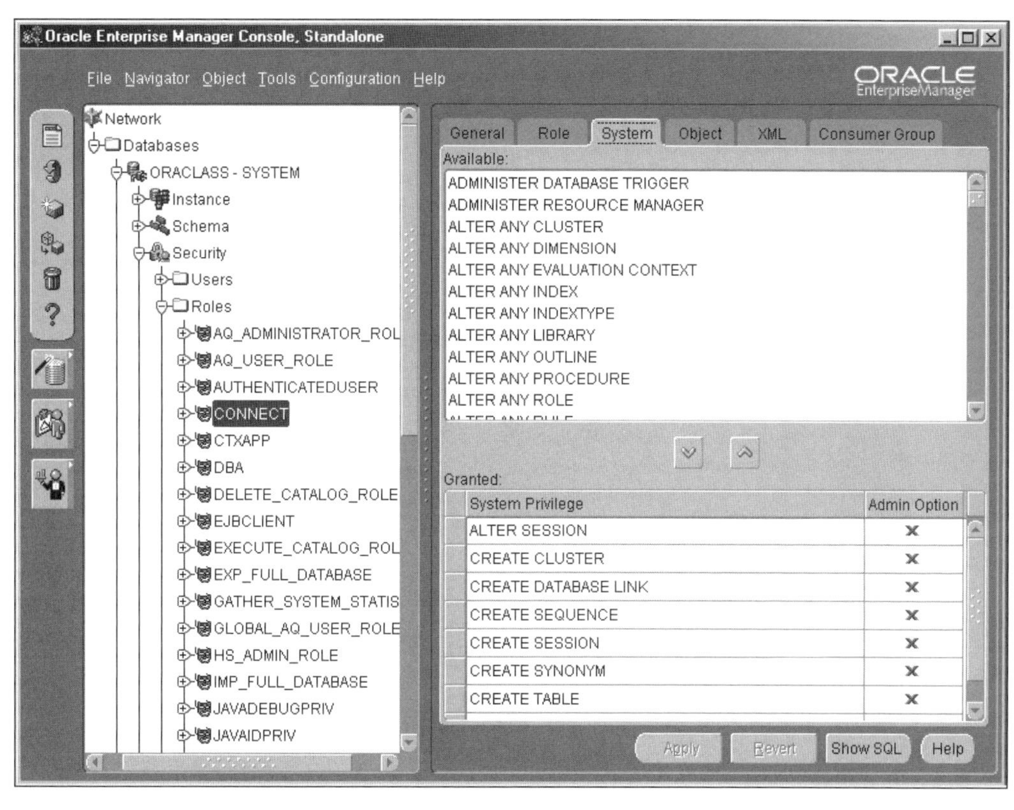

Figure 13-11 Available privileges are on the top, and granted privileges are on the bottom

6. Scroll down on the left and click the **SELALL** role. Then click the **Object** tab to display the object privileges assigned to this role, as shown in Figure 13-12. Notice the navigation tree in the top-right section of the screen. You can add new object privileges by navigating through this tree in the same way you use the Schema Manager.

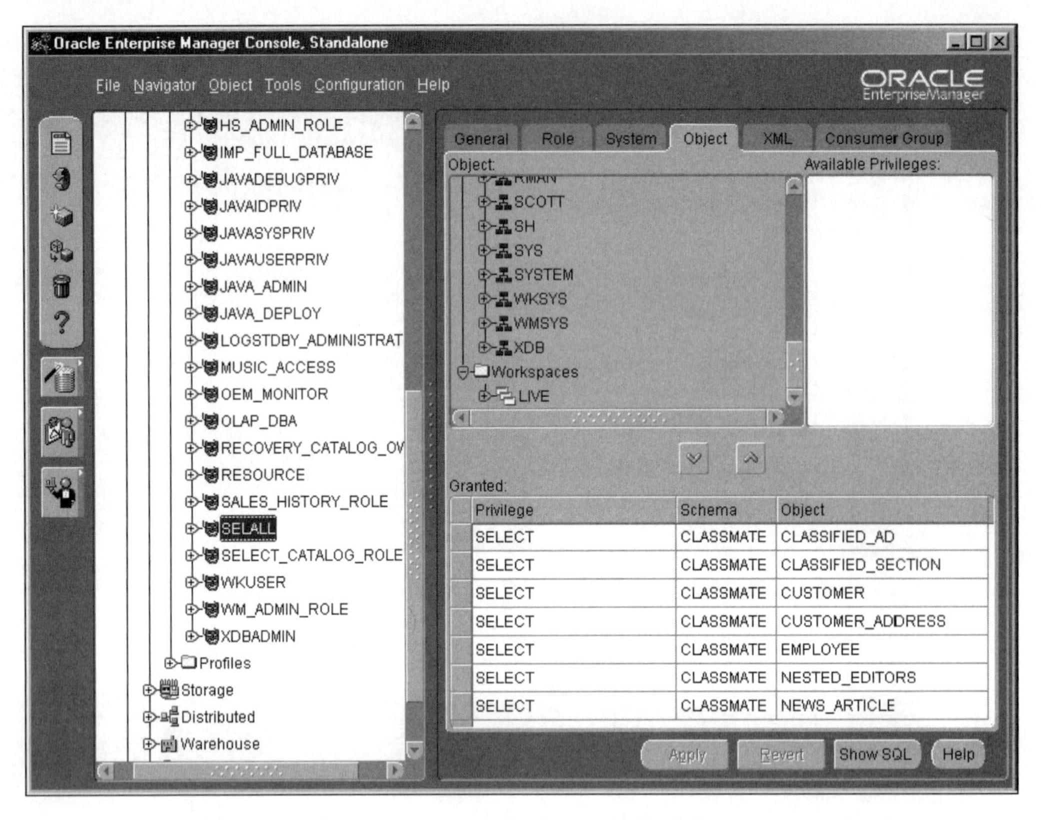

Figure 13-12 Object privileges appear in the lower half of the property sheet

7. Scroll up the navigation tree in the upper-right corner of the screen until you find the CLASSMATE schema. Double-click on **CLASSMATE**.
8. The list of available object types appears below the CLASSMATE schema. Double-click the **Tables** folder.
9. A list of tables owned by CLASSMATE is displayed below the Tables folder. Scroll down and select **WANT_AD**. A list of available privileges for the WANT_AD table appears in the box labeled "Available Privileges." This list changes according to the type of object you select. Figure 13-13 shows the screen at this point.

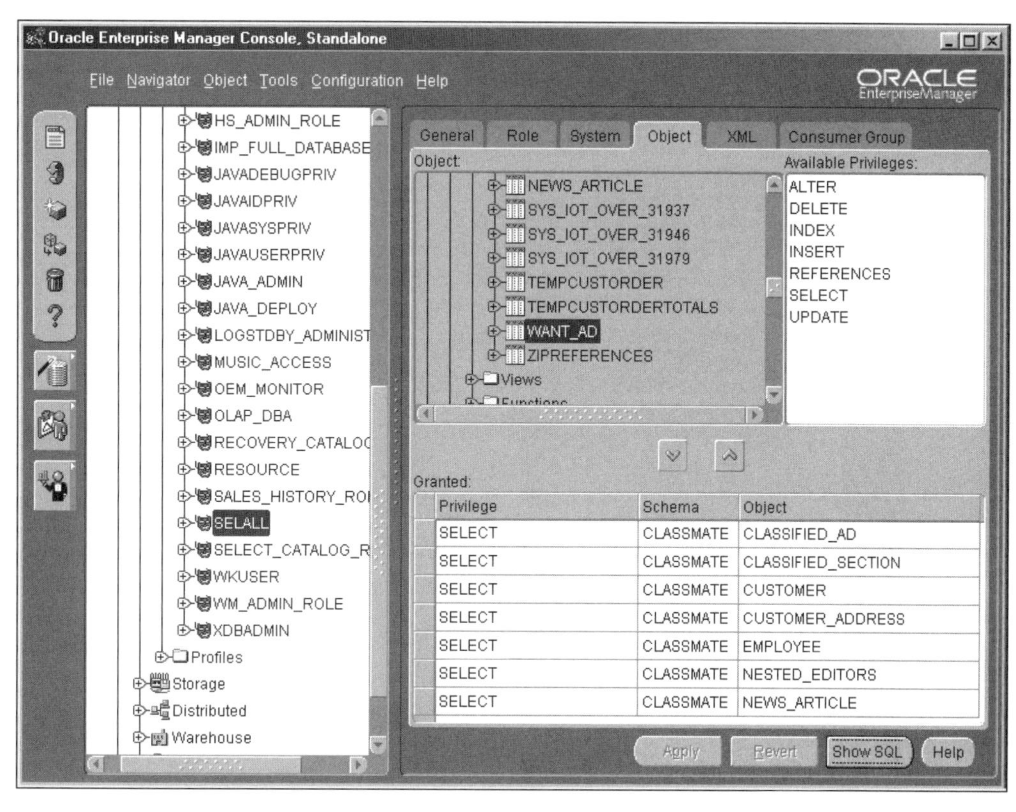

Figure 13-13 You have a choice of seven object privileges for the WANT_AD table

10. Select the **SELECT** privilege, and click the **down arrow** button that is displayed in the middle of the property screen.
11. To see the change in the assigned privileges, scroll down to the bottom of the list, as shown in Figure 13-14. You see that the bottom line shows the new privilege for the WANT_AD table. This line has a plus sign in the far left column, meaning that this privilege has not yet been added.

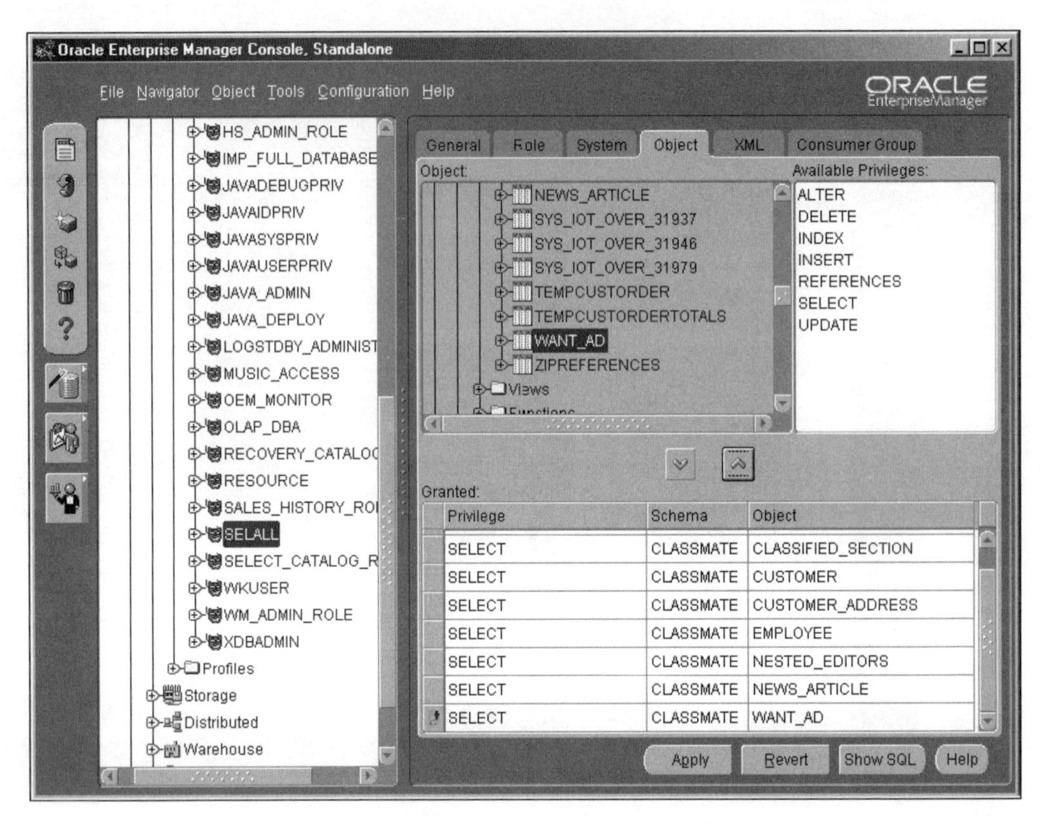

Figure 13-14 A privilege that has been selected but not yet been granted has a plus sign

12. Click the **Apply** button to grant the new privilege to the SELALL role. The plus sign disappears, and the screen reverts to its original display mode (similar to Figure 13-12). The SELECT on WANT_AD privilege appears in the list of granted object privileges.

13. You can also see which users have been assigned a role in the Security Manager. To find the list, right-click on the **SELALL** role on the left side of the console. A popup menu appears.

14. Select **Show Grantees** from the popup menu. A window displays the users who were granted this role, as shown in Figure 13-15. Notice that SYSTEM is shown with ADMIN OPTION for the role. This is because SYSTEM created the role. You could grant SELALL to more users (or roles) using this screen.

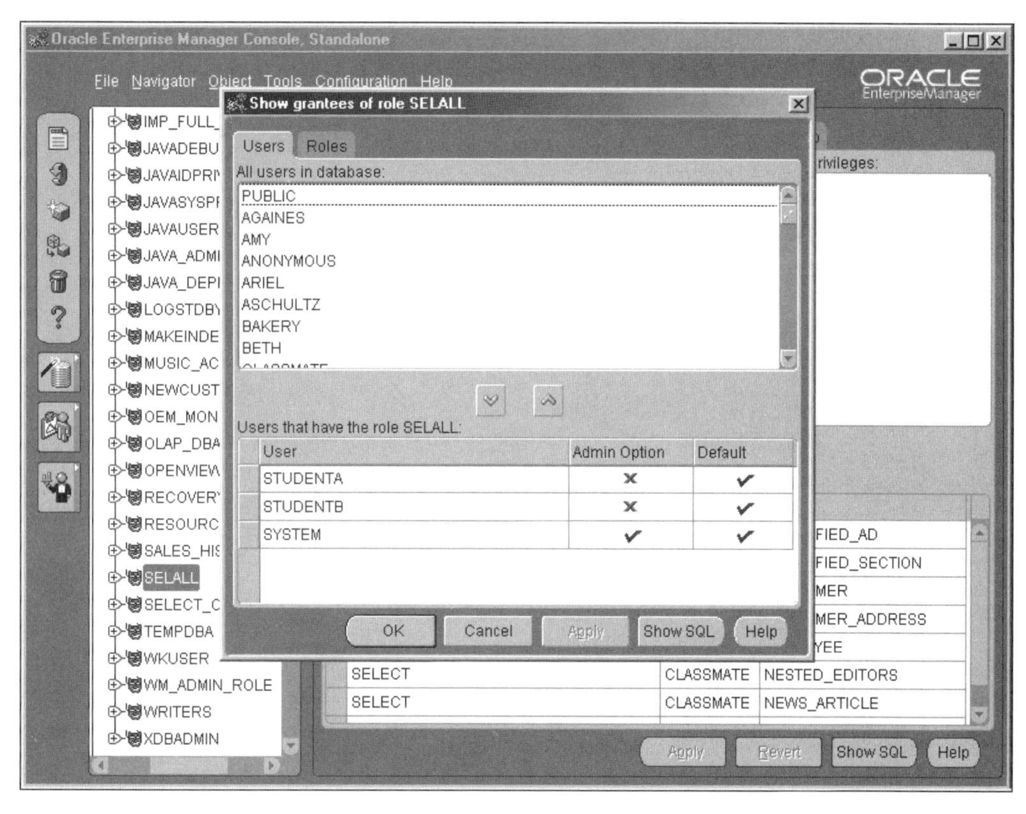

Figure 13-15 Users granted the SELALL role are listed in the lower half of the screen

15. Click the **Cancel** button to close the additional screen.
16. Scroll up on the left side of the console, and double-click the **Users** folder. A list of all the users in the database appears below the folder.
17. Click the **AGAINES** user. A property window appears on the right showing the attributes of the user.
18. Click the **Role** tab. Figure 13-16 shows the screen. As you can see, the AGAINES user has the CONNECT role.

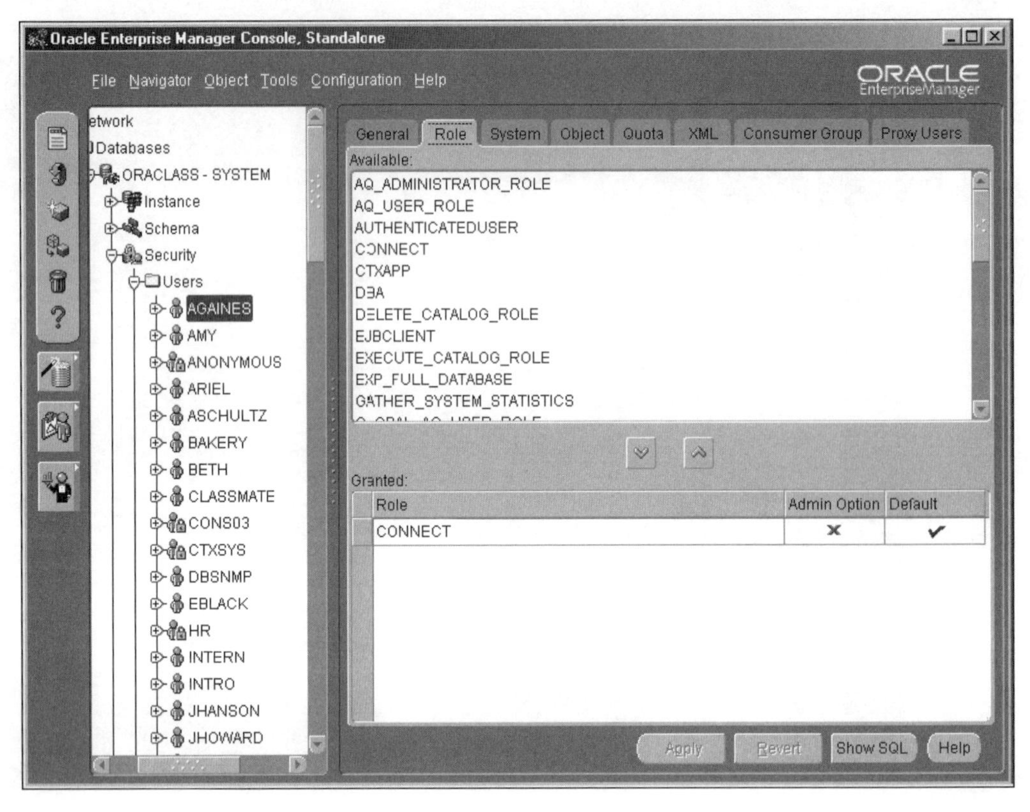

Figure 13-16 AGAINES has only one role assigned

19. Scroll until you find the SELALL role in the Available roles list. Then select the **SELALL** role, and click the **down arrow**.

20. The SELALL role is added to the list of granted roles. Click the **Apply** button to complete the process, and grant the role to AGAINES. Figure 13-17 shows the results.

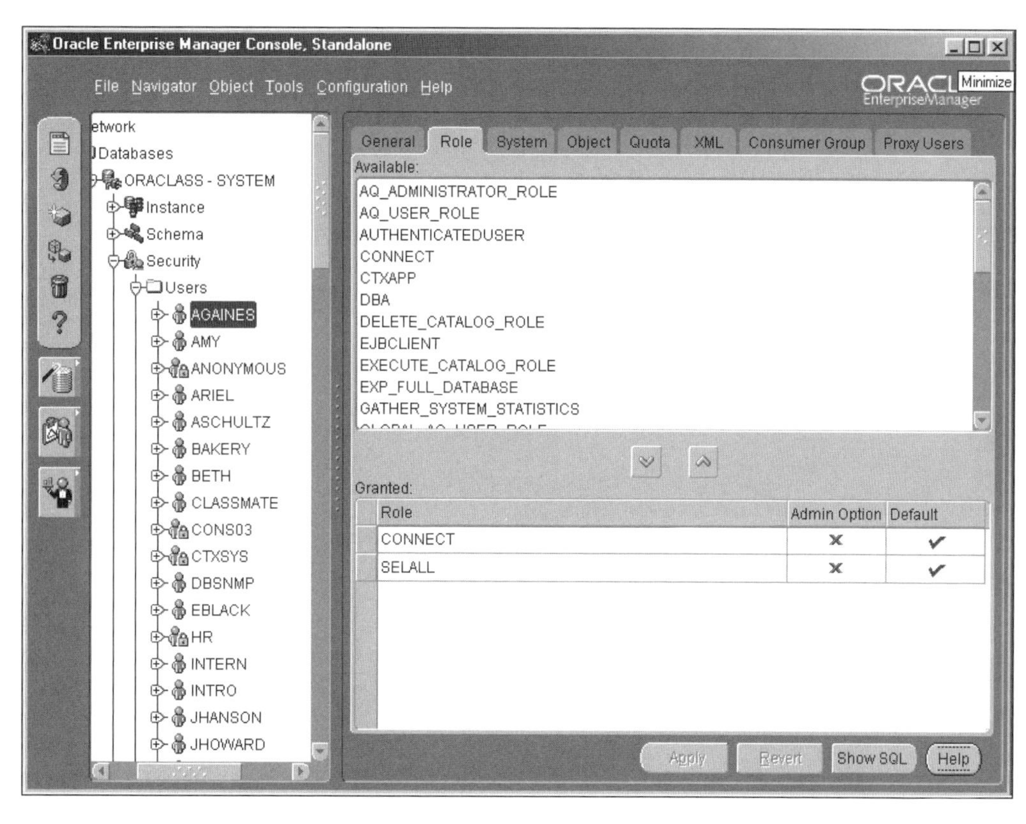

Figure 13-17 AGAINES now has two roles

21. Close the console by clicking the **X** in the top-right corner of the window.

You have seen how easy it is to grant privileges using the console. A real bonus of the console is the simple way it lists all the available system and object privileges for you. When they are so conveniently listed, you don't have to remember the exact names of the system or object privileges. The console can be a timesaving tool when you are working with roles.

Chapter Summary

- Roles simplify security administration.
- Roles can be granted other roles, system privileges, and object privileges.
- Predefined roles help speed up administration by providing basic groupings of roles.
- Roles with passwords add security to the roles.
- You can grant system privileges and object privileges to a role, but you cannot use the WITH ADMIN OPTION or WITH GRANT OPTION clauses.

- You can grant a role to a role and optionally include the WITH ADMIN OPTION clause.
- Create a role with the CREATE ROLE command.
- Change a role with the ALTER ROLE command.
- Grant privileges to a role with the GRANT command.
- USER_TAB_PRIVS does not list privileges granted to the user's roles.
- DBA_TAB_PRIVS includes privileges granted to roles.
- Use the ROLE_ROLE_PRIVS view to find roles granted to other roles.
- After a user has been granted a role, subsequent grants to the role are effective immediately for the user.
- Default roles are roles enabled when you log on.
- Use ALTER USER to change the default roles for a user.
- Use SET ROLE to enable or disable roles in your current session.
- Use DROP ROLE to drop a role.
- Dropped roles are automatically revoked from users and other roles.
- DBA_TAB_PRIVS lists all object privileges granted to users and roles.
- The console displays roles and privileges within the Security Manager.

Review Questions

1. List the three categories of privileges that a role may be granted.
2. Which of the following predefined roles enables a user to create other users?
   a. RESOURCE
   b. EXP_FULL_DATABASE
   c. CONNECT
   d. DBA
3. To make a new role, use the ________________ command.
4. The ________________ clause of the CREATE ROLE command is optional and the default.
5. To give a user permission to enable a role, use the GRANT command. True or False?
6. Assuming that BOATROLE and SKIPPERROLE are both roles, which of the following commands is invalid?
   a. GRANT SELECT ON SHOWBOAT TO BOATROLE;
   b. GRANT CREATE TABLE TO BOATROLE;

c. GRANT BOATROLE TO SKIPPERROLE WITH ADMIN OPTION;

d. GRANT SKIPPERROLE TO BOATROLE IDENTIFIED BY CAPTAIN;

7. You want to rename a role. How do you do it?

   a. Use the ALTER ROLE command.

   b. Drop and re-create the role.

   c. Revoke and re-create the role.

   d. Use the CREATE ROLE command.

8. You want to enable all roles in your session. What command do you use?

   a. SET ROLE ALL;

   b. SET ROLE ENABLE ALL;

   c. ALTER SESSION ENABLE ALL ROLES;

   d. ALTER USER DEFAULT ROLE ALL;

9. What Data Dictionary role should you query to determine which roles are available to you, regardless of whether they are enabled?

10. You have granted the role ACCTMGR to many users. Which command revokes the role from all the users simultaneously and removes the role from the database?

EXAM REVIEW QUESTIONS—DATABASE FUNDAMENTALS I (#1Z0-031)

1. Which of the following roles are predefined? Choose two.

   a. IMP_FULL_DATABASE

   b. POWER_USER

   c. CONNECT

   d. SYSOPER

   e. CREATE_SESSION

2. You want to find out which object privileges have been granted to the ADMINACT role. Which query is the most useful?

   a. SELECT * FROM ROLE_SYS_PRIVS
      WHERE ROLE = 'ADMINACT';

   b. SELECT * FROM SESSION_ROLES
      WHERE ROLE = 'ADMINACT';

   c. SELECT * FROM DBA_TAB_PRIVS
      WHERE GRANTEE = 'ADMINACT';

   d. SELECT * FROM ROLE_OBJ_PRIVS
      WHERE ROLE = 'ADMINACT';

3. You plan to create a role named SALESREAD that does not require a password and gives read-only privileges to users. Which of the following commands creates the role?
   a. CREATE ROLE SALESREAD IDENTIFIED BY READ ONLY;
   b. CREATE ROLE SALESREAD NOT IDENTIFIED;
   c. CREATE READ ONLY ROLE SALESREAD;
   d. CREATE ROLE SALESREAD IDENTIFIED BY S123ALES;

   Look at the following SQL script output.

```
GRANT SUPERUSER TO JONES WITH ADMIN OPTION;
Grant succeeded.
CONNECT JONES/XXX
Connected.
GRANT SUPERUSER TO HAROLD;
Grant succeeded.
CONNECT SYSTEM/XXX
Connected.
REVOKE SUPERUSER FROM JONES;
```

4. Assuming that HAROLD is logged on when the last command is executed, what happens to HAROLD?
   a. HAROLD loses the SUPERUSER role immediately.
   b. HAROLD loses the SUPERUSER role when his session ends.
   c. HAROLD keeps the SUPERUSER role.
   d. HAROLD loses the SUPERUSER if it is enabled.
5. Which commands are valid, assuming that ROLEA and ROLEB are roles and USER1 is a user? Choose two.
   a. GRANT SELECT ON CUSTOMERS TO ROLEA WITH GRANT OPTION;
   b. GRANT INSERT ON CUSTOMERS TO ROLEB, ROLEC, USER1;
   c. GRANT ROLEA TO ROLEB WITH ADMIN OPTION;
   d. GRANT CREATE VIEW TO ROLEA WITH ADMIN OPTION;
   e. GRANT CREATE VIEW TO USER1 WITH GRANT OPTION;
6. The TOPDOG role currently requires a password. The user JAMESK is logged on and has the TOPDOG role enabled. You execute the following command.

```
ALTER ROLE TOPDOG NOT IDENTIFIED;
```

   What happens in JAMESK's session?
   a. The TOPDOG role becomes disabled.
   b. The TOPDOG role becomes his default role.

c. The TOPDOG role stays enabled.

d. The TOPDOG role asks for a password.

e. CREATE_SESSION

7. You execute the following commands.

```
GRANT DEALERROLE TO JOHNQ;
GRANT COLLECTORROLE TO JOHNQ;
ALTER USER JOHNQ DEFAULT ROLES NONE;
GRANT MANAGERROLE TO JOHNQ;
```

Which of the three roles are default roles for JOHNQ?

a. DEALERROLE and COLLECTORROLE

b. None of the roles

c. MANAGERROLE only

d. All three of the roles

e. CREATE_SESSION

8. You are logged on as the user SMITH, and two roles you have are currently disabled. They were created with these commands:

```
CREATE ROLE MGR IDENTIFIED BY MGR123;
CREATE ROLE SALES IDENTIFIED BY SALES789;
```

Which of the following commands successfully enables both roles?

a. SET ROLE MGR, SALES IDENTIFIED BY MGR123, SALES789;

b. SET DEFAULT ROLE MGR, SALES;

c. GRANT MGR IDENTIFIED BY MGR123; GRANT SALES IDENTIFIED BY SALES789 TO SMITH;

d. SET ROLE MGR IDENTIFIED BY MGR123, SALES IDENTIFED BY SALES789;

9. The user MARTHA logs onto the database. You, as the DBA, create a role called ADM and grant the CREATE TABLE privilege to the role. You grant the role to MARTHA. Then, you grant the CREATE VIEW privilege to the TOPDOG role. What happens to MARTHA?

a. MARTHA can now create tables and views.

b. MARTHA cannot create tables or views until she starts a new session.

c. MARTHA can create tables but not views.

d. MARTHA cannot create tables or views even when she starts a new session.

10. What are the advantages of granting privileges to roles rather than to users? Choose three.
    a. Better security on the database
    b. Easier to grant additional privileges to users
    c. Simplifies security administration
    d. Provides additional security through extra passwords
    e. Easier for users

HANDS-ON ASSIGNMENTS

1. Look at Figure 13-18. Your job is to write the code that implements the matrix using three roles: CHIEF_EDITOR, WRITERS, and EDITORS. Create the roles, and grant the appropriate privileges to the roles. (*Hint*: AGAINES is the Chief Editor, OCORTEZ is a writer, and JHANSON is an editor.) Save your work in the **Solutions\Chapter13** directory in a file named **ho1301.sql**.

Privileges	Tables				
	WANT_AD	CLASSIFIED_AD	CLASSIFIED_SECTION	EMPLOYEE	CUSTOMER
SELECT	JWELLSLEY OCORTEZ	JWELLSLEY OCORTEZ	JWELLSLEY OCORTEZ	JWELLSLEY OCORTEZ	JWELLSLEY OCORTEZ
	JHANSON JHOWARD	JHANSON JHOWARD	JHANSON JHOWARD	JHANSON JHOWARD	JHANSON JHOWARD
				AGAINES	
INSERT	JWELLSLEY OCORTEZ	JWELLSLEY OCORTEZ	OCORTEZ JHANSON JHOWARD	AGAINES	OCORTEZ JHANSON JHOWARD
UPDATE	JWELLSLEY OCORTEZ	JWELLSLEY OCORTEZ	OCORTEZ JHANSON JHOWARD	AGAINES	OCORTEZ JHANSON JHOWARD
DELETE	JWELLSLEY OCORTEZ	JWELLSLEY OCORTEZ	OCORTEZ JHANSON JHOWARD	AGAINES	OCORTEZ JHANSON JHOWARD

Figure 13-18 These users and tables are already in the ORACLASS database

2. Building on Hands-on Assignment # 1, grant the appropriate roles to the users shown in Figure 13-18. Save your work in the **Solutions\Chapter13** directory in a file named **ho1302.sql**.
3. The boss has discovered that even though the NEWCUST role was created to require a password (NC2002), users are not being required to provide the password. Something must change in the database so that all users must provide the password before being able to enable the role. Create the NEWCUST role, and assign it to the five users you used in Hands-on Assignment # 2. Then add commands to ensure that the users must provide the password. Save your work in the **Solutions\Chapter13** directory in a file named **ho1303.sql**.
4. You are going on vacation for two weeks, leaving your assistant in charge. Create a role called TEMPDBA. Write a set of commands that grants all the roles that SYSTEM has, including the ADMIN OPTION when SYSTEM has that option. Save your work in the **Solutions\Chapter13** directory in a file named **ho1304.sql**.
5. Change the TEMPDBA role created in Hands-on Assignment # 4 so that it requires a password (JJ22BIRD) to be enabled. Assign the role to the STUDENTA user, but do not make it one of the user's default roles. Save your work in the **Solutions\Chapter13** directory in a file named **ho1305.sql**.
6. Log on as STUDENTA, and enable the TEMPDBA role. Write a query that displays all the roles that are granted to you through the TEMPDBA role. Save your work in the **Solutions\Chapter13** directory in a file named **ho1306.sql**.
7. You return from vacation and want to keep the TEMPDBA role around for next year, but you want to change the password to SAVE4LATER and revoke the role from STUDENTA. Write a script to carry out these two tasks. Save your work in the **Solutions\Chapter13** directory in a file named **ho1307.sql**.
8. Write a query on one or more data dictionary views that lists each user, the roles each user has been granted, and the object privileges each of those roles has been granted. Narrow your query to list only objects owned by CLASSMATE. Save your work in the **Solutions\Chapter13** directory in a file named **ho1308.sql**.
9. You have some tables that contain data that is available to all users. The tables are named NEWS_ARTICLE, WANT_AD, and CLASSIFIED_AD. Create a role called OPENVIEW, and grant SELECT on each of these tables to the role. Then grant the role to all current users and future users. Save your work in the **Solutions\Chapter13** directory in a file named **ho1309.sql**.
10. Create a role called MAKEINDEX, and grant it the privileges needed to create indexes or views that query these CLASSMATE tables: CUSTOMER, CUSTOMER_ADDRESS, and WANT_AD. Grant the role to STUDENTB. Also grant the CONNECT role to STUDENTB, which enables the user to create various objects in the database. Connect to STUDENTB (password STUDENTB), and create a simple view on the CUSTOMER table. Save your work in the **Solutions\Chapter13** directory in a file named **ho1310.sql**.

CASE PROJECT

1. The delivery people in the Global Globe Company have just been given access to an off-site computer that can reach the central database. These employees need Oracle user accounts. The delivery people are authorized to view the CUSTOMER and CUSTOMER_ADDRESS tables. They are also authorized to make changes to addresses stored in the CUSTOMER_ADDRESS table. Create the new users. The new user names will be the first initial and last name of employees in the EMPLOYEE table who have the job of "DELIVERY PERSON". You may create the users so that their user names and passwords are identical. Create a role named DELIVERY, and grant it the appropriate privileges. Include the CREATE SESSION privilege in the DELIVERY role as well as the privileges that are described here. Then grant the role to the new users. Save your work in the **Solutions\Chapter13** directory in a file named **case1301.sql**.
2. Refer to the Case Project found in Chapter 12. Instead of granting privileges to the users, you could now create roles, grant privileges to the roles, and grant the roles to the users. Write a script to handle this task. Create two roles named CH12EDITOR and CH12CHIEF. In addition, revoke all the privileges granted directly to the users. Save your work in the **Solutions\Chapter13** directory in a file named **case1302.sql**.

CHAPTER

14

GLOBALIZATION SUPPORT IN THE DATABASE

In this chapter, you will:

- Examine how globalization support is implemented in the database
- Use globalization parameters and variables
- View globalization support information in data dictionary views

Oracle has supported international languages, called **National Language Support** (NLS), in its database since the Oracle 7 release. NLS makes it possible for tables to store data in Japanese, Chinese, French, and so on. Each language has its own special characters and complexities, such as being read from the right to the left.

A need to communicate globally grew as large corporations started doing business in multiple countries. This need grew even stronger when more and more businesses began to use the Internet to communicate with customers and business partners across the globe. The trend toward multinational business shows no signs of slowing down, and in response to this trend, Oracle has expanded its capability to handle multiple languages.

The concept of supporting languages in many countries has been expanded in Oracle9*i*. NLS is now a subset of the **globalization support** feature. Whereas NLS allows you to use the database in your native language, globalization support allows you to design a database that supports users from many nationalities, converting data to each user's native language.

You, as the DBA, are most interested in the database itself, not the applications, so this chapter introduces the new globalization support features, but primarily focuses on the National Language Support (NLS) area of globalization support, showing you how to work with the NLS parameters and variables in Oracle9*i*.

INTRODUCTION TO GLOBALIZATION SUPPORT

The aim of globalization support is to create a multilingual database. This is important in today's business world, because the Internet makes it possible for just about anyone to reach customers in many countries. As the trend to internationalize business enterprises grows, you'll find more and more businesses needing a database that can handle the translation and storage of multiple languages.

Oracle9*i*, Release 2, has many new features for globalization support:

- Oracle now has support for the Euro, which is now the default currency for all countries participating in the European Monetary Union (EMU).
- The ALTER TABLE MODIFY command can now convert a CHAR column to an NCHAR column.
- New Oracle-supplied packages support conversion of BFILE data into NLS character data when loading a NCLOB column.
- Oracle's database structure now supports storing data in **unicode** (universal encoded character set), a character set that contains codes for any language and, therefore, can be used to translate more quickly.

The capability to store and display data in native languages is not the only feature of globalization support. Error messages, sort order, date format, and currency format are also affected. Internally, however, some portions of the database are not affected, as described in the next section.

Language-dependent Behavior in the Database

Imagine that you are Japanese and have set up your database with Japanese as its primary language. You can create tables that store Japanese characters correctly. Your applications display Japanese instructions and error messages.

Behind the scenes, the database stores data using bytes, as it does with English. Some languages with complex characters, like Chinese, use several bytes to represent a single character.

Even though the database stores data in any language, it requires that you store SQL and PL/SQL in English. For example, you might use French for your CUSTOMER table's data; however, if you want to create a view on the table, you must write the CREATE VIEW command in English.

The database can now support multiple languages simultaneously, making your applications user-friendly all over the world.

How Language-dependent Settings Affect Applications

Any application that reaches the database is seen by the database as a client. A client can have its own environmental settings for language that may be different from the language used by the database. With globalization support, the database can convert data to and from different languages, so that each client sees the appropriate language. Figure 14-1 illustrates a database supporting three clients in three different languages.

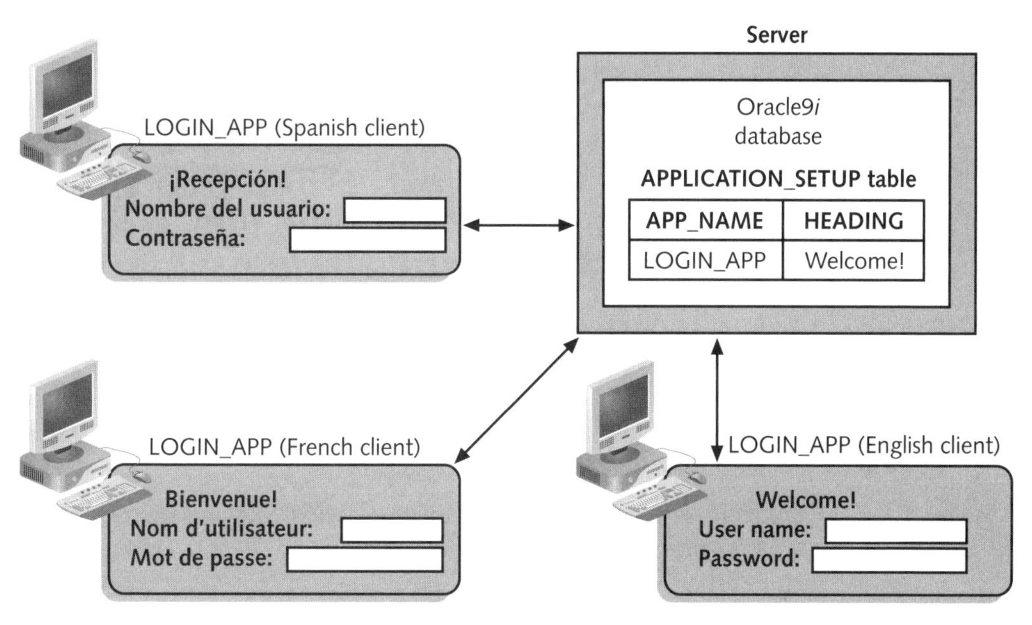

Figure 14-1 Language-independent applications can support multiple languages

Figure 14-1 shows one application named LOGIN_APP that is running on three clients, one English, one French, and one Spanish. The database's primary language is English. A table named APPLICATION_SETUP is stored in the database. The table has a column named APP_NAME, for application name, and a column named HEADING that stores the heading displayed on the application screen. When the client starts up the LOGIN_APP application, it queries the APPLICATION_SETUP table for the heading to display on the screen. Before the data ("Welcome!" in this case) is returned to each client, Oracle9*i* translates the data into the appropriate language for each client. The database knows which language to use, because each client has an environment variable called NLS_LANG that was installed with the Oracle9*i* client software. Multilanguage support of this kind is very powerful, but designing and building language-independent applications requires planning and careful testing.

This section has introduced you to the overall concept of globalization support with an application system linked to a database. Although it is important to understand these concepts, you, as the DBA, are more likely to be involved in the NLS area of globalization support, because NLS support is set up within the database using initialization parameters and variables. The remainder of the chapter gives you detailed information on how to use NLS support in an Oracle9*i* database.

USING NLS PARAMETERS AND VARIABLES

An Oracle9*i* database has a base language, such as English, that is called its **database character set**. In addition, you can choose a second language that can be used to store data in columns that are defined as one of the three NLS data types (NCHAR, NVARCHAR2, NCLOB). This second language is called the **national character set**. Both character sets are designated when you create the database by specifying the CHARACTER SET and NLS CHARACTER SET clauses in the CREATE DATABASE command. After the database is created, you cannot change these two parameters and must re-create the database if you want to modify these parameters. There are a few exceptions, in which you can use the ALTER DATABASE command to modify the character sets, but they are seldom used. For example, you can change the character set using the ALTER DATABASE command if the new character set is a subset contained entirely within the current character set. Changing to a character set that is not a subset of the original causes errors when converting data from the old character set to the new character set, a process that occurs during the ALTER DATABASE process to complete the switch from one character set to another.

Use initialization parameters to fine-tune the database's behavior regarding the NLS character set you have chosen. For example, the NLS_SORT parameter determines the default sorting order of NLS data.

Choose the character set and national character set carefully when you create your database. Consider what language you normally use and what other languages you might want to have stored in the database now and in the future. Also consider what machine language your database's server uses, because this limits the sets available to you. The machine language is the encoding standard for the numerical code inside individual bytes of data in the machine. EBCDIC and ASCII are the most common machine languages. A character set maps each character of a language (or several closely related languages) to a numeric code in machine language. All the values for CHARACTER SET and NLS CHARACTER SET are listed in the *Oracle9i Globalization Support Guide*. Table 14-1 lists a few of them to give you an idea of what they look like and the variety of choices available.

Table 14-1 Sample listing of character set names and descriptions

Character set name	Description
JA16EUC JA16EUCTILDE JA16EUCYEN	Three variations of Japanese character sets
ZHS16CGB231280	Simplified Chinese

Table 14-1 Sample listing of character set names and descriptions (continued)

Character set name	Description
ZHT16CCDC	Traditional Chinese
US7ASCII	American
D7DEC	German
F7DEC	French
AR8APTEC715	Latin/Arabic
IW8ISO8859P8	Latin/Hebrew

Although you can use most of the character sets as the value for either the CHARACTER SET or the NLS CHARACTER SET, a few are allowed only in the NLS CHARACTER SET.

Here is an example of a CREATE DATABASE command that sets the character set to American English and the NLS character set to French. (Do not actually execute this example.)

```
CREATE DATABASE MYNEWDB
DATAFILE 'D:\oracle\oradata\mynewdb\system01.dbf'
   STORAGE(INITIAL 300M NEXT 10M MAXEXTENTS UNLIMITED)
UNDO TABLESPACE UNDOTBS
      DATAFILE 'D:\oracle\oradata\mynewdb\undo01.dbf'
LOGFILE GROUP 1('D:\oracle\oradata\mynewdb\logfile1.log',
               'E:\oracle\logs\logfile2.log')
LOGFILE GROUP 2('D:\oracle\oradata\mynewdb\logfile3.log',
               'E:\oracle\logs\logfile4.log')
CHARACTER SET US7ASCII
NATIONAL CHARACTER SET F7DEC;
```


After the database is created, you can use French in the following data types:

- NCHAR for fixed-length character data
- NVARCHAR2 for variable-length character data
- NCLOB for large character data objects

The national character set UTF8 is a **universal language character set**. This means that it contains codes for every language supported by Oracle9*i*. Using this character set can simplify translations from one language to another when the database has multilingual clients.

US7ASCII is a **single-byte character set**, because each character in the language (such as the letter "A") is represented with a single byte code in the base operating system's machine language. Many other character sets are single-byte character sets, especially for languages similar to English, such as French, Latin, Spanish, Italian, and many more Western European languages.

UTF8 is a **multibyte character set**, because it uses more than one byte to represent a single character in a language. This is needed for languages with complex symbols, such as Chinese. A multibyte character set can be either **fixed-length** or **variable-length**. Even though these are the same terms used to describe data types in the context of character sets, these terms have a different meaning. A fixed-length, multibyte character set uses a fixed number of bytes for each character in the character set. Calculating the length of a data string is easier with fixed-length character sets; however, they take up more space. A variable-length, multibyte character set uses a variable number of bytes to represent one character in the character set. Variable-length character sets save space but take longer for processes such as calculating the length of a data string.

Besides the two clauses in the CREATE DATABASE command, there are many initialization parameters that affect the behavior of the database when it works with national character data. The next section describes the most commonly used parameters.

Adjusting Globalization Initialization Parameters

There are many NLS-related initialization parameters in Oracle9*i*, and all of them affect how the database handles national language data. For example, the NLS_DATE_LANGUAGE parameter tells Oracle9*i* how to format dates.

In addition to being modified in the **init.ora** file, NLS initialization parameters can be modified at two levels:

- **SQL level:** Changing the value of an initialization parameter within an SQL function or an SQL statement alone. Use special NLS clauses to accomplish this.
- **Session level:** Changing the value of an initialization parameter in the current session only. Use the ALTER SESSION command to do this.

The value of a parameter at the narrower level always overrides the value of the same parameter at the broader level. For example, the value of NLS_SORT in an SQL statement overrides the values of NLS_SORT in the user's session, and the value of NLS_DATE_LANGUAGE of the user's session overrides the default value set in the **init.ora** file. If a parameter is not set in any of these three areas, it defaults to a value defined by the NLS character set of the database.

Follow these steps to examine a variety of NLS-related parameters in the console.

1. Start up the Enterprise Manager console. In Windows, click **Start** on the Taskbar, and then click **Programs/Oracle - OraHome92/Enterprise Manager console**. In UNIX, type **oemapp console** on the command line. The Enterprise Manager console login screen appears.
2. Select the **Launch standalone** radio button and click **OK**. The console appears.
3. In the main window of the Enterprise Manager console, navigate starting at the **Databases** icon to **ORACLASS/Instance/Configuration**. Your screen should look similar to Figure 14-2.

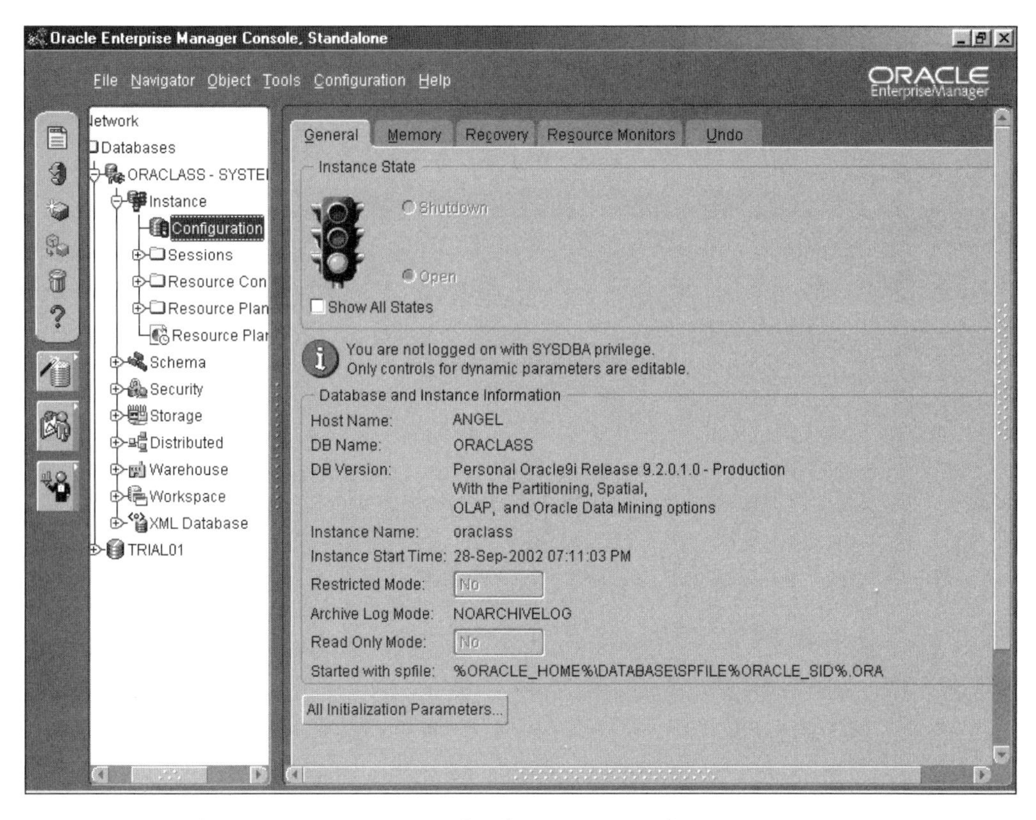

Figure 14-2 The Instance Manager's Configuration window

4. Click the **All Initialization Parameters** button. A new window appears with a tab labeled "All Parameters". The initialization parameters currently in force in the database are displayed in alphabetical order.

5. Scroll down until you find a set of parameters that begin with "nls_". These are the 17 parameters affecting NLS support in the database. Most of these are intended for use with the more expanded globalization support and are outside the scope of this book. Your focus in this chapter is on a subset of the parameters, which includes those that affect NLS support in the database.

6. Select NLS_LANGUAGE, and click the **Description** button. You may need to scroll down to see the button, and you may need to scroll again to view the description. Figure 14-3 shows the results. The NLS_LANGUAGE parameter is a key parameter for NLS support. As the description says, this parameter affects the language used for messages, day and month names, and sorting order, among other things. The current NLS_LANGUAGE parameter is AMERICAN. The default value of this parameter is derived from the NLS_LANG variable established when Oracle9*i* is installed. NLS_LANG is stored in the Windows Registry in Windows systems and as an environmental variable in UNIX systems.

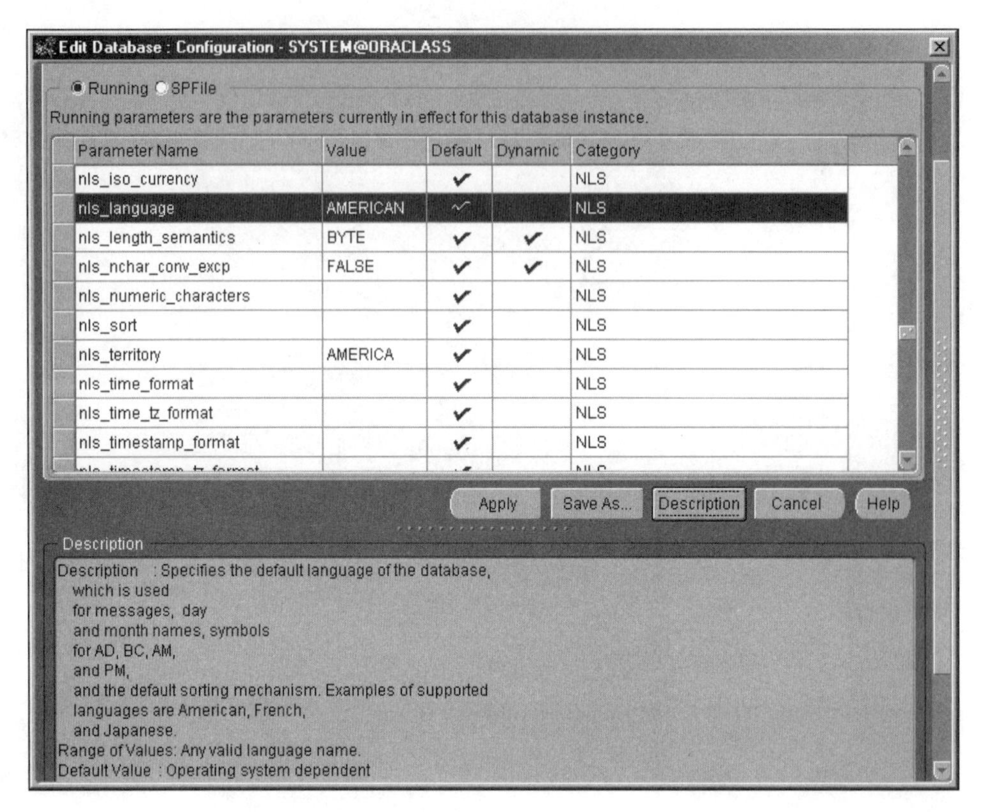

Figure 14-3 NLS_LANGUAGE defines most language-dependent database behaviors

7. The other key parameter is called NLS_TERRITORY. Select **nls_territory**, and read the description that appears. This parameter affects how dates are formatted, what currency is used by default, and the numbering format of days and weeks.
8. Click **Cancel** to close the parameter window. You return to the console's main window. Remain logged on for the next practice.

In addition to the NLS_LANGUAGE and NLS_TERRITORY parameters, there are parameters that can override specific parts of the default settings that are made by these two parameters. These are:

- **NLS_CURRENCY:** Changes the symbol used for currency (for example, "$" is the default when NLS_TERRITORY is AMERICA).
- **NLS_DATE_FORMAT:** Changes the default format of dates, as set by NLS_TERRITORY. The format defines the expected format of dates in the TO_DATE function.

- **NLS_DATE_LANGUAGE:** Changes the spelling of days, months, and abbreviations for time and date, such as AM, PM, and AD. The default for this parameter is defined by the NLS_LANGUAGE parameter.
- **NLS_ISO_CURRENCY:** Changes the international currency symbol set by the NLS_TERRITORY parameter.
- **NLS_NUMERIC_CHARACTERS:** Changes the symbols used as decimals and separators in numbers. The default value is set by the NLS_TERRITORY parameter. For example, in the AMERICA territory, a comma (,) is the separator and a period (.) is the decimal point. In the FRANCE territory, a period (.) is the separator and a comma (,) is the decimal point.
- **NLS_SORT:** Changes the method used to sort character data. The default is set by the NLS_TERRITORY parameter, so that sorting is done according to the language of the data. BINARY is the default for the NLS_TERRITORY of AMERICA; however, other values, such as JAPANESE, reflect the name of the language. In the case of the English language, the order of the alphabet is reflected by the order of the numeric values stored for the symbol for each letter, so the sort is based on the numerical values of the characters, hence the name BINARY.

The remaining NLS parameters pertain to globalization support features that are not addressed in this book.

How do you see the current values of these parameters and modify them? The next section shows you how to make changes to the parameters in your SQL statements.

Using NLS Parameters in SQL

Let's say you have a database in New York City, and you are preparing a report for a client in Norway. You want to provide your client with unmistakable information by formatting the dates and numbers in your report by the method Norwegians use without making any changes to your database or your session. The best way to do this is to use NLS parameters directly in the SQL code itself.

14

You can add NLS parameters to the TO_CHAR, TO_DATE, and TO_NUMBER functions to use language-dependent values, such as Norwegian words for the months.

You can also use the NLSSORT function in the ORDER BY clause to invoke language-dependent sorting of your query results.

Follow these steps to use NLS parameters in dates, numbers, and sorting order.

1. Start a SQL*Plus Worksheet session by clicking **Tools/Database Applications/SQL*Plus Worksheet** from the menu in the console.
2. Connect to the CLASSMATE user by clicking **File/Change Database Connection** from the menu. The Database Connection Information window appears.

3. Type **CLASSMATE** in the Username box, **CLASSPASS** in the Password box, and **ORACLASS** in the Connect as box. Then click **OK** to log on. You return to the SQL*Plus Worksheet window.

4. Type and execute the following query to view the current settings of the NLS-related initialization parameters. Figure 14-4 shows the results.

```
SELECT * FROM NLS_DATABASE_PARAMETERS;
```

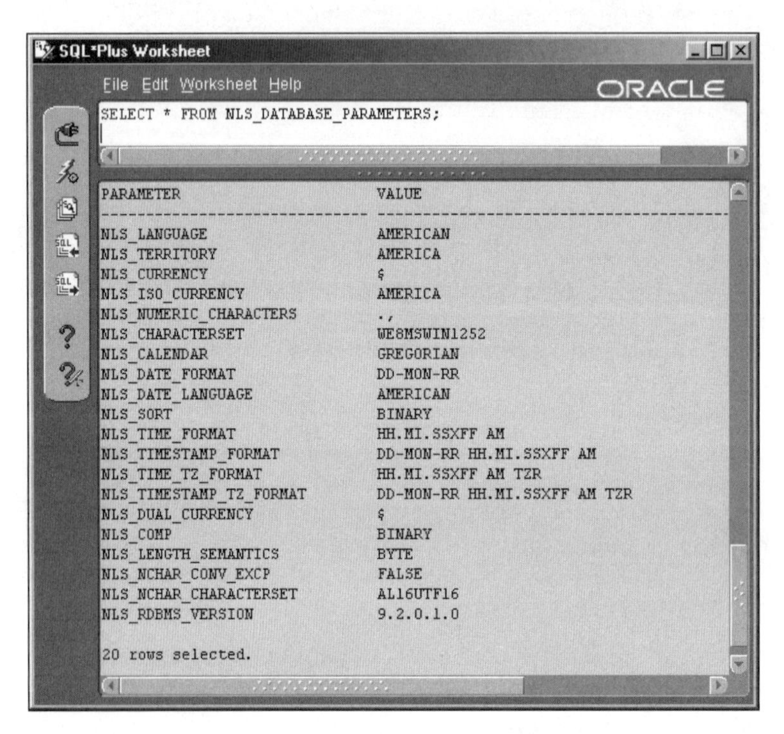

Figure 14-4 The data dictionary view shows the formatting and sorting options currently in place

Notice that there is a value listed for NLS_CHARACTERSET, which is not an initialization parameter. Recall that the NLS character set is established by the CREATE DATABASE command. The value shown in Figure 14-4 is "WE8MSWIN1252." If you look this up in the *Oracle9i Globalization Reference Guide*, you find that the description is "MS Windows Code Page 1252 8-bit West European." Western European character sets support multiple languages that share common Latin-based roots and originate in Western Europe. Languages such as Italian, French, Spanish, Dutch, and German are all supported.

5. Type and execute the following query in your SQL*Plus Worksheet session. Figure 14-5 shows the results, with the day and month spelled

out in Norwegian. Adding the NLS_DATE_LANGUAGE parameter to the TO_CHAR function causes the language of the date values to change.

```
SELECT TO_CHAR(PLACED_DATE,'Day, DD Month, YY',
       'NLS_DATE_LANGUAGE=NORWEGIAN') "Date Placed"
FROM CLASSIFIED_AD;
```

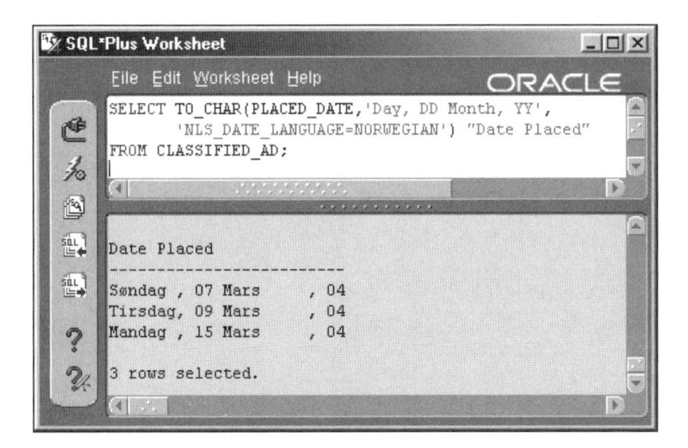

Figure 14-5 Norwegian can be used because the database's NLS character set supports it

6. You also want to change the format of numbers, so that the decimal is a comma, the group separator (the symbol between hundreds and thousands and so on) is a period, and the local currency symbol is "Kr". When using SQL to modify these values, you use the TO_CHAR function for converting numbers to a formatted character value. The syntax of the TO_CHAR function is:

```
TO_CHAR(<colname>,'<format>','<nls_parameters>')
```

You can specify up to three parameters in the TO_CHAR function to modify the default values set up by the NLS_TERRITORY parameter. You must use the appropriate placeholder in the format for each parameter. The parameters and their placeholders are:

- **NLS_CURRENCY:** Sets the symbol of the local currency. Placeholder is L.
- **NLS_NUMERIC_CHARACTERS:** Sets the symbols for the decimal point (placeholder is D) and group separator (placeholder is G).
- **NLS_ISO_CURRENCY:** Sets the symbols for the international (ISO) currency. Placeholder is C.

Add the PRICE column to the query on CLASSIFIED_AD, and use the special formatting capabilities of the TO_CHAR function. The completed query looks like this. After making the change, execute this query and remain logged on for the next practice. Figure 14-6 shows the results.

```
SELECT TO_CHAR(PLACED_DATE,'Day, DD Month, YY',
       'NLS_DATE_LANGUAGE=NORWEGIAN') "Date Placed",
TO_CHAR(PRICE,'L999G999G999D99',
   'NLS_NUMERIC_CHARACTERS='',.''
    NLS_CURRENCY=''Kr.''') PRICE
FROM CLASSIFIED_AD;
```

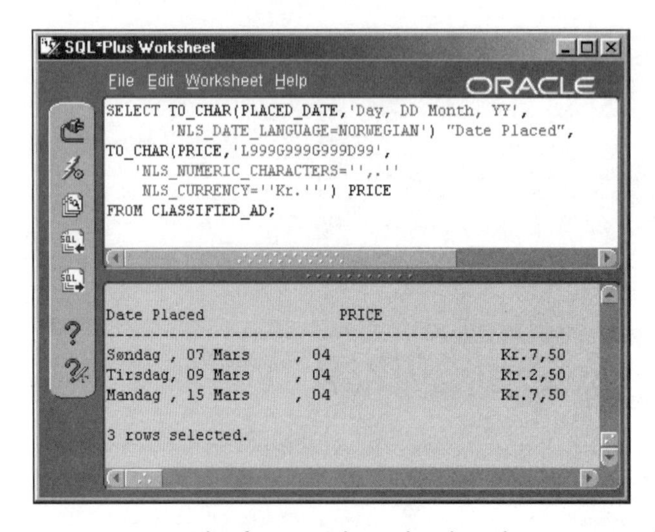

Figure 14-6 The formatted number has the currency, decimal, and group separators in the Norwegian format

Another capability within the SQL command involves the ORDER BY clause and the NLS_SORT parameter. You can specify the language-dependent sort order within the ORDER BY clause. The syntax is:

```
ORDER BY NLSSORT(<sort_column>,'NLS_SORT=<language>');
```

The ORDER BY clause sorts by using the results of the NLSSORT function that changes a column's data into an alphanumeric string that sorts according to the language-dependent properties of whatever language you specify. Complete the following steps to see how it works.

1. First, run a script that defines a table containing French letters by clicking **Worksheet/Run Local Script**. Navigate to the **Data\Chapter14** directory, and select the **setup.sql** file. Click **Open**. The script is read and executed in the worksheet.

2. Type and execute the following query to see how the letters sort with the default setting (American) for language. Figure 14-7 shows the results. Notice

that the uppercase letters sort first, followed by the lowercase letters, and finally the letters with accent symbols.

```
SELECT LETTER
FROM CH14_FRENCH
ORDER BY LETTER;
```

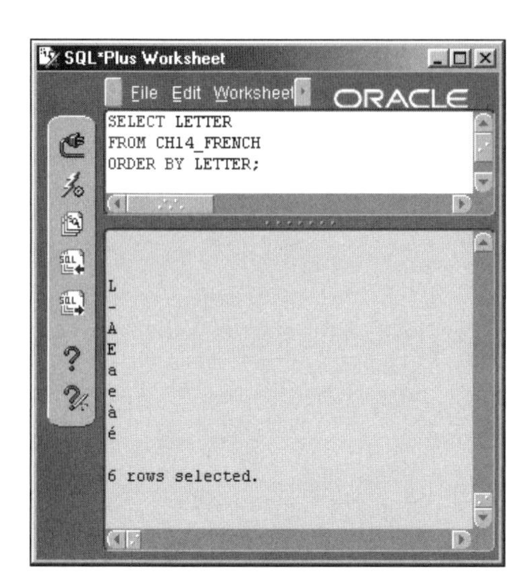

Figure 14-7 Although most French letters are the same as American letters, others are similar but include accent marks

3. Now, adjust the ORDER BY clause so that it sorts the letters according to the sorting order defined for the French language. The complete query looks like this.

```
SELECT LETTER
FROM CH14_FRENCH
ORDER BY NLSSORT(LETTER,'NLS_SORT=FRENCH');
```

4. Execute the query by clicking the **Execute** icon. Figure 14-8 shows the results. As you can see, the French sorting order includes the letters with accents in their appropriate sequence. In addition, the French sorting order sorts a lowercase "a" before an upper case "E", unlike American English, in which all the uppercase letters sort before all the lowercase letters. Remain logged on for the next practice.

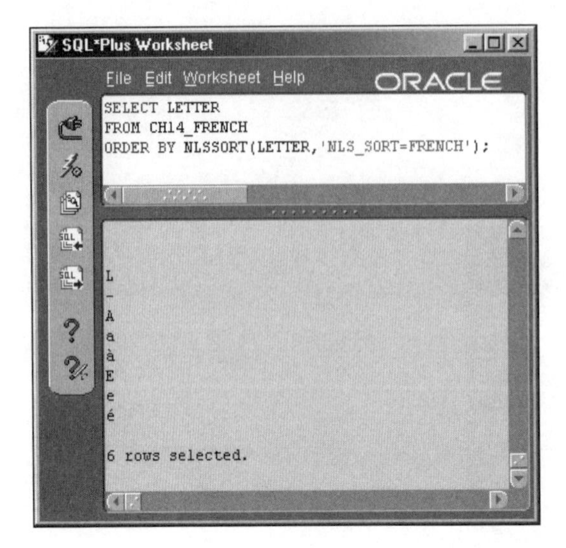

Figure 14-8 Sorting with the French language-dependent sorting order changes the order of all but the first letter in this list

You have seen how to adjust some of the NLS parameters inside your SQL commands, when you don't want to change anything else. You can also change the NLS parameters in your session so that all queries you perform use your session settings and no one else using the database is affected.

Changing NLS Parameters in Your Session

The best way to use the NLS parameters in your session is by adjusting the two primary parameters (NLS_LANGUAGE and NLS_TERRITORY), which in turn adjust the others (NLS_SORT, NLS_CURRENCY, and so on) automatically. Follow these steps to experiment with the ALTER SESSION command.

1. Adjust the NLS_LANGUAGE and NLS_TERRITORY parameters in your session by typing and executing these commands:

```
ALTER SESSION SET NLS_LANGUAGE='FRENCH';
ALTER SESSION SET NLS_TERRITORY='FRANCE';
```

2. The SQL*Plus Worksheet replies, "Session altered" twice in a row. To view the changes to all the parameters, type and execute this query. You are looking at the *session* parameters now, instead of the *database* parameters. The database parameters have not changed. Figure 14-9 shows the results.

```
SELECT * FROM NLS_SESSION_PARAMETERS;
```

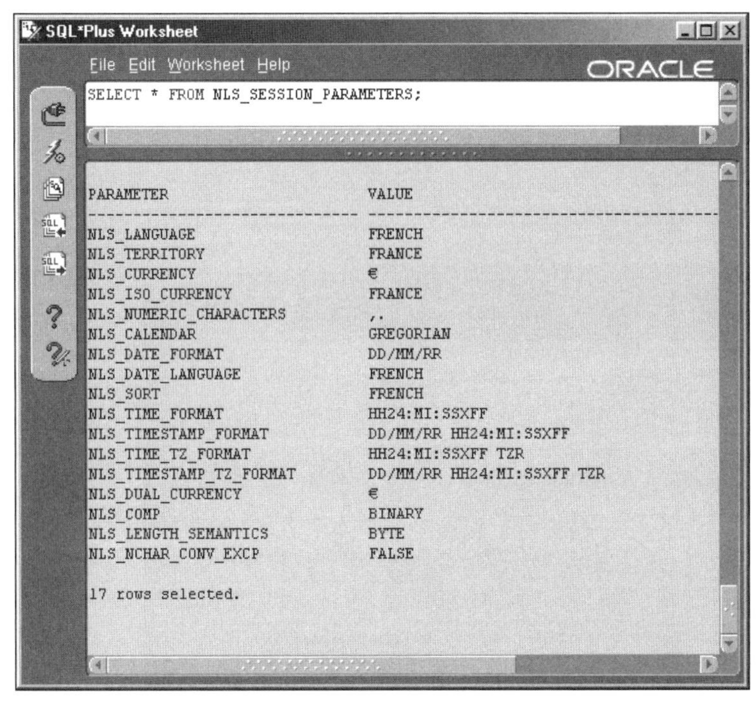

Figure 14-9 The data dictionary view shows the formatting and sorting options currently in place for your session

3. Type and execute this query to see how the spelling of days and months has changed. Figure 14-10 shows the results. As you can see, the day and the month are in French instead of English.

```
SELECT TO_CHAR(SYSDATE,'DAY, DD MONTH, YYYY') FROM DUAL;
```

14

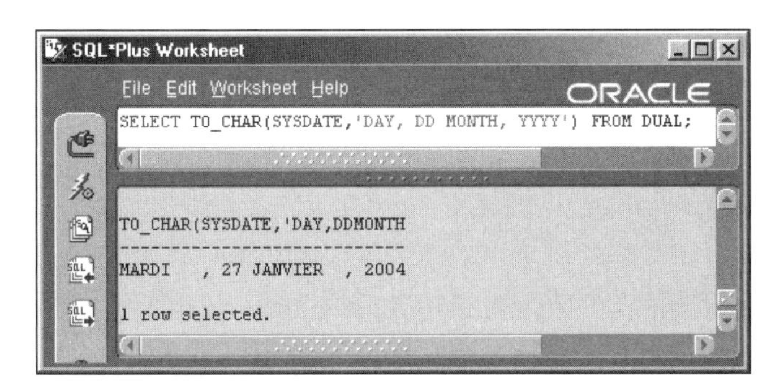

Figure 14-10 MARDI is French for TUESDAY

4. Type and execute the following query to see how the NLS parameters affect numbers, dates, and text data. Figure 14-11 shows the results.

```
COLUMN AD_TEXT FORMAT A40 WORD_WRAP
SELECT PLACED_DATE,
TO_CHAR(PRICE,'L999G999D00') PRICE,
AD_TEXT
FROM CLASSIFIED_AD;
```

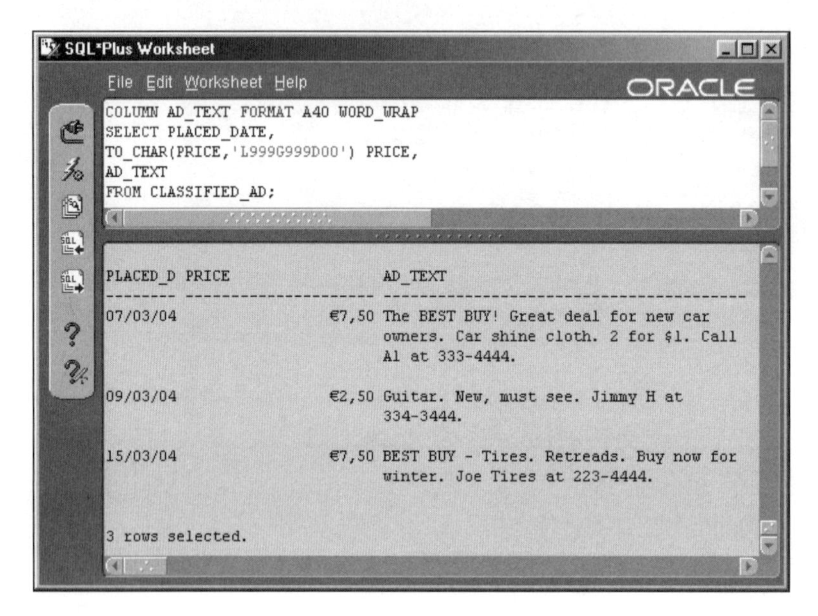

Figure 14-11 The default format of dates, currency, and decimal point symbols has changed

Notice that the words inside the AD_TEXT column are still in English. This is because the NLS parameters do not handle translating text data from one language to another. NLS parameters are not intended as interpreters of text data.

5. Return the two parameters to their original state by typing and executing these commands, and remain logged on for the next practice.

```
ALTER SESSION SET NLS_LANGUAGE='AMERICAN';
ALTER SESSION SET NLS_TERRITORY='AMERICA';
```

The SQL*Plus Worksheet replies, "Session altered" twice.

Modifying your SQL command or your session handles the unusual requests you may come across when helping your users or creating reports for management. The initialization parameters in the **init.ora** file should be set up to handle the NLS requirements for typical requests and reports. While practicing in this chapter, you have seen several data dictionary views that show you NLS parameters. The next section describes all the views that are available for viewing NLS parameters.

VIEWING NLS PARAMETERS IN DATA DICTIONARY VIEWS

There are only three data dictionary views and two dynamic performance views related to NLS parameters. Table 14-2 lists them all, along with a description.

Table 14-2 Views about NLS parameters

View name	Description
NLS_DATABASE_PARAMETERS	Lists NLS parameter settings for the database, including NLS_CHARACTER_SET.
NLS_INSTANCE_PARAMETERS	Lists NLS parameter settings for the database instance. Only shows values that have been set in the **init.ora** file. Parameters left to default are shown as null values.
NLS_SESSION_PARAMETERS	Lists NLS parameters in force for the current session. Does not include NLS_CHARACTER_SET.
V$NLS_PARAMETERS	Lists current values of NLS parameters.
V$NLS_VALID_VALUES	Lists all valid values for these NLS parameters: CHARACTERSET, LANGUAGE, SORT, TERRITORY.

Use these views in queries just as you use any other view. Follow these steps to try out one of the views.

1. You want to change the value of the NLS_SORT parameter. List all the possible settings by typing and executing the following query:

   ```
   SELECT VALUE FROM V$NLS_VALID_VALUES
   WHERE PARAMETER='SORT'
   ORDER BY 1;
   ```

   Figure 14-12 shows some of the results.

2. Exit SQL*Plus Worksheet by clicking the **X** in the top-right corner of the window.
3. Log off the console by clicking the **X** in the top-right corner of the window.

Remember, with few exceptions, you choose the default character set and the NLS character set at database creation and never change them. Other NLS parameters can be modified by changing the initialization parameters, adjusting them in your session, or adding them to an SQL statement.

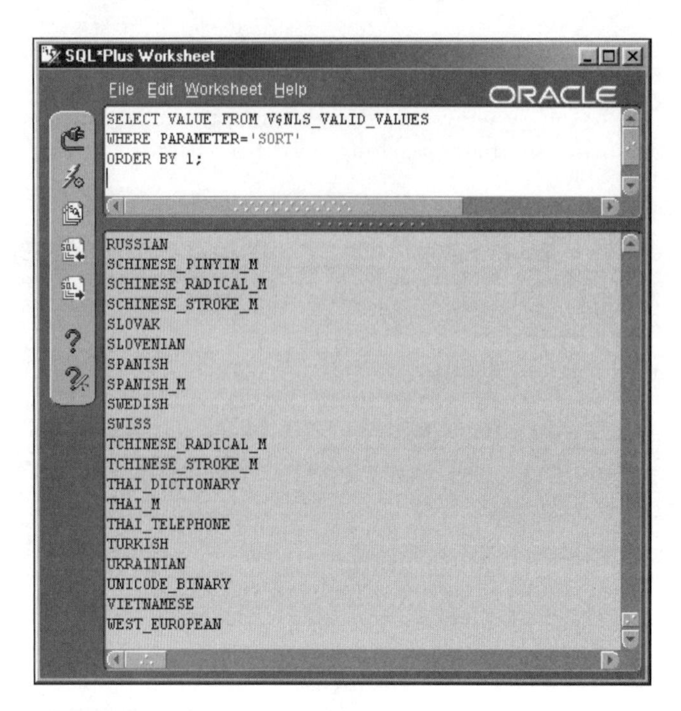

Figure 14-12 Over 80 languages are available in Oracle9*i*

CHAPTER SUMMARY

- National Language Support (NLS) allows one database to store multiple languages of data.
- Globalization support includes NLS support and more capabilities aimed at producing a multilingual database.
- Globalization support has improved with new features in Oracle9*i*.
- SQL and PL/SQL must be entered in English, even if the native language is not English.
- The database character set is the default character set used for all but NLS data types.
- NLS data types are NCHAR, NVARCHAR2, and NCLOB.
- The national character set is used for NLS data types.
- Character sets are defined when the database is created.
- A character set is a map of characters to machine language codes.
- UTF8 is a universal language character set capable of supporting all languages supported by Oracle9*i*.

- Character sets are either single-byte or multibyte.
- Multibyte character sets are either fixed-length or variable-length.
- NLS-related initialization parameters can be set in the **init.ora** file, a session, or a specific SQL statement.
- Use the Instance Manager to examine NLS parameters and their definitions.
- Add NLS parameters to the TO_CHAR, TO_DATE, and TO_NUMBER functions to use values other than defaults.
- Use language-dependent sorting through the NLSSORT function in the ORDER BY clause.
- Number formatting can be done with the TO_CHAR function and includes changing the currency, decimal point, and group separator symbols.
- American English and French sorting orders are different.
- Adjust sorting order by including NLSSORT in the ORDER BY clause.
- The NLS_SESSION_PARAMETERS view lists NLS parameters currently set for your session.
- Changing NLS_LANGUAGE does not translate text data into another language.
- V$NLS_VALID_VALUES lists all valid values for some of the NLS parameters.

Review Questions

1. National Language Support is a subset of ________________ support.
2. Which data types support NLS characters?
3. What kind of character set supports all languages supported by Oracle9*i*?
4. Which of these can be stored in Japanese in Oracle9*i*?
   a. PL/SQL stored procedure
   b. Database trigger
   c. Large documents
   d. Source code for a view
5. A single application can send error messages to one client in French and another client in Spanish. True or False?
6. List the two types of character sets established when the database is created.
7. You can change the value of the NLS_SORT parameter at the database level. True or False?
8. A single-byte character set, like US7ASCII, can be either fixed-length or variable-length. True or False?

9. A ______________ length, ______________ byte character set uses less space by allowing different numbers of bytes for different characters.

10. Which of the following parameters affects the language used for days of the week? Choose two.

    a. NLS_DATE_FORMAT

    b. NLS_TERRITORY

    c. NLS_DATE_LANGUAGE

    d. NLS_LANGUAGE

    e. NLS_CHARACTER_SET

Exam Review Questions - Database Fundamentals I (#1Z0-031)

1. Examine the following statement. Which line defines the character set used in the NCHAR data type?

```
1 CREATE DATABASE MYNEWDB
2 DATAFILE 'D:\oracle\oradata\mynewdb\system01.dbf'
3   STORAGE(INITIAL 300M NEXT 10M MAXEXTENTS UNLIMITED)
4      DATAFILE 'D:\oracle\oradata\mynewdb\undo01.dbf'
5 NATIONAL CHARACTER SET WE8MACROMAN8S
6 CHARACTER SET TH8MACTHAI;
```

   a. Line 1

   b. Line 5

   c. Line 6

   d. Line 5 and 6

2. Which of the following languages is most likely supported with a multibyte character set?

   a. Finnish

   b. Latin

   c. Chinese

   d. Spanish

3. The **init.ora** file has a value of FRENCH for the NLS_DATE_LANGUAGE. You issue the following two statements in SQL*Plus:

```
ALTER SESSION SET NLS_DATE_LANGUAGE='GERMAN';
SELECT 'Today is '||TO_CHAR(SYSDATE,'Day',
       'NLS_DATE_LANGUAGE=ITALIAN')||'.'
FROM DUAL;
```

The results of the query will be:

a. Error: the date format is invalid

b. In Italian

c. In German

d. In French

4. The **init.ora** file has a value of AMERICA for the NLS_TERRITORY parameter. You execute these statements using the data shown in Table 14-3:

```
SELECT TO_CHAR(TOTAL_FEE,'C999G990D00')
FROM CUSTOMER_ACCOUNT
WHERE ACCOUNT_ID=3;
```

Table 14-3 CUSTOMER_ACCOUNT table data

ACCOUNT_ID	CUSTOMER_NAME	TOTAL_FEE
1	John S. Silverton	5352.25
3	Mary Jo Kilpatrick	6502.4

The results of the query will be:

a. 5352.25

b. 6502.4

c. $6,502.4

d. $5,352.25

e. $6,502.40

5. The **init.ora** file has a value of AMERICA for the NLS_TERRITORY parameter. You execute these statements on July 5, 2003:

```
ALTER SESSION SET NLS_ISO_CURRENCY='f.';
ALTER SESSION SET NLS_DATE_FORMAT='DD.MM.YYYY'
SELECT SYSDATE, TO_CHAR(150.4,'L999D99') FROM DUAL;
```

What does the output look like?

a. 07.05.2004 $150.40

b. 07.05.2004 f.150.4

c. 07.05.2004 $150.4

d. 05.07.2004 f.150.40

6. Your database character set is UTF8, your NLS_LANGUAGE is 'AMERICAN', and you want to sort the FULLNAME column using the German sorting sequence. Which query is the best solution?
   a. SELECT * FROM CUSTOMER

      ORDER BY FULLNAME (GERMAN);
   b. SELECT * FROM CUSTOMER

      ORDER BY TO_CHAR(FULLNAME,'NLS_LANGUAGE='GERMAN');
   c. ALTER SESSION SET NLS_SORT='GERMAN';
   d. SELECT * FROM CUSTOMER

      ORDER BY NLSSORT(FULLNAME,'NLS_SORT=GERMAN');
7. Which views show you the current settings of the NLS parameters after you have made changes with the ALTER SESSION command? Choose two.
   a. NLS_TERRITORY
   b. V$NLS_PARAMETERS
   c. V$NLS_VALID_VALUES
   d. USER_SESSION_PARAMETERS
   e. NLS_SESSION_PARAMETERS
8. Your client needs a report, based on a query sent to Spain that displays dates in Spanish. Your NLS character set is UTF8. What do you do? Choose two.
   a. Write the query using the NLS_DATE_LANGUAGE='SPANISH', in the TO_CHAR function of the date column.
   b. Change the database's NLS_LANGUAGE parameter to 'SPANISH', and then write the query.
   c. Change your session's NLS_LANGUAGE parameter to 'SPANISH', and then write the query.
   d. Change your session's NLS_CHARACTER_SET parameter to 'SPANISH', and then write the query.
   e. Change your session's NLS_DATE_FORMAT parameter to 'SPANISH', and then write the query.
9. Which of the following NLS parameters cannot be used in an SQL statement?
   a. NLS_TERRITORY
   b. NLS_DATE_LANGUAGE
   c. NLS_SORT
   d. NLS_NUMERIC_CHARACTERS
   e. NLS_CURRENCY

10. Which of these NLS parameters can be used within the TO_CHAR function when converting a number to characters? Choose two.
    a. NLS_LANGUAGE
    b. NLS_DATE_LANGUAGE
    c. NLS_SORT
    d. NLS_NUMERIC_CHARACTERS
    e. NLS_CURRENCY

HANDS-ON ASSIGNMENTS

1. You have a new customer who wants the dates in your reports to be in the same format as this date: 2004-January-19. In addition, this customer wants the month spelled out in Danish. Write a query using the CH08HOUSE_REPAIR table that lists the name of the house, the repair done, and the start and end dates of the repair work. Format the two dates as your new customer requests by using NLS parameters in the SQL query. Save your work in a file named **ho1401.sql** in the **Solutions\Chapter14** directory on your student disk.
2. Write another query that returns the same results as Hands-on Assignment 1, except this time, don't use any NLS parameters in the SQL query. Instead, change the NLS parameters in your session. Reset your NLS parameters to their original setting after running the query. Save all the SQL commands you used in a file named **ho1402.sql** in the **Solutions\Chapter14** directory on your student disk.
3. Write a query that displays only the names and values of the NLS parameters affected when you change the NLS_TERRITORY parameter in your session to 'ITALY'. Save your work in a file named **ho1403.sql** in the **Solutions\Chapter14** directory on your student disk.
4. Write a query that displays only the name(s) and value(s) of the NLS parameters affected when you change the NLS_LANGUAGE parameter in your session to 'ITALIAN'. Save your work in a file named **ho1404.sql** in the **Solutions\Chapter14** directory on your student disk.
5. Write a query that shows the values of the two NLS-related parameters set when the database was created. Save your work in a file named **ho1405.sql** in the **Solutions\Chapter14** directory on your student disk.
6. Change the NLS_LANGUAGE of your session to 'FRENCH', and the NLS_TERRITORY to 'CANADA'. Write a command to insert one row in the CH08HOUSE_REPAIR table without using a TO_DATE function. There must be a value added to each column. Commit the inserted row. Reset your NLS parameters to their previous settings. Save your work in a file named **ho1406.sql** in the **Solutions\Chapter14** directory on your student disk.

7. Write a query that lists all the variations of CHINESE allowed for the NLS_LANGUAGE parameter. Change the NLS_LANGUAGE parameter in your session to one of these. Display the NLS parameters. What is different, compared to the default parameter settings? Reset your session NLS_LANGUAGE parameter. Save your SQL code and your answer to the question in a file named **ho1407.sql** in the **Solutions\Chapter14** directory on your student disk.

CASE PROJECT

The Global Globe News Company has begun printing newspapers in France! Your job is to create a table for classified ads named FRENCH_CLASSIFIED_AD and one for classified section names named FRENCH_CLASSIFIED_SECTION that can store data entered in French. Create the two tables with the same columns as the original tables, but use NLS data types for the AD_TEXT and SECTION_TITLE columns. Insert French translations of the classified section names in the table. Commit the changes to the database. (*Hint*: For data in the new FRENCH_CLASSIFIED_AD table, start with the INSERT commands in the **ch09_handson_setup.sql** file in the **Data\Chapter09** directory on your student disk. For data in the new FRENCH_CLASSIFIED_SECTION table, start with the INSERT commands in the **ch10_handson_setup.sql** file in the **Data\Chapter10** directory on your student disk. To translate English ad text and section names to French, use translations available at *http://babelfish.altavista.com/*. Save your work in a file named **case1401.sql** in the **Solutions\Chapter14** directory on your student disk.

Glossary

The terms in this glossary are defined in the context of the Oracle9*i* database structures rather than in the context of general computer technology. This approach should make the glossary more useful to you as a student of Oracle9*i*.

add-on — Components available in Oracle that can enhance the database system. Some of these access the database through Oracle Net, while others access the database through the Java/Web connectors.

algorithm — A set of steps or a mathematical calculation for solving a problem.

analyze (a table) — To issue a command that causes Oracle9*i* to read the table's structure and update the table's metadata with current information about the size of the table, average row length, total number of rows, and free space remaining in extents.

ARCHIVELOG mode — A database in ARCHIVELOG mode automatically archives redo log files. The opposite is NOARCHIVELOG mode.

ARCn — Archiver background process. "n" indicates that multiple archiver processes can run concurrently. Used only when database is in ARCHIVELOG mode.

automatic undo management mode — New method of managing rollback segments in which Oracle9*i* automatically handles undo records using an undo tablespace. Opposite of manual undo management mode.

BFILE locator — The pointer that directs the database to the actual location of an external LOB file.

binary tree — A type of data structure that uses a branching formation that always divides a list of data into two halves. Each half is then divided in half again, and so on. A binary tree contains a root node, branch nodes, and leaf nodes.

bitmap index — An index that does not use the b-tree method to store index key values and rowids. Instead, it uses a bitmap to store index key values and row locations.

branch — An element of a data structure. A branch is the node of a binary tree that has other nodes below it and above it. *See also* binary tree.

b-tree index — An index structure in which data is divided and subdivided based on the index key values to minimize the look-up time required when searching for a key value. The b-tree is so named, because it is the complex variation of a binary tree.

business rules — Statements defined during the design of a database system that inform both the database designer and the application programmer of how data is used to support the business.

chained row — A table row that spans multiple blocks.

change vector — Component of a redo entry. Describes a change made to a data block. Many change vectors can be contained within one redo entry.

character set — A standardized definition of the numeric bit values used to represent each character in a language. For example, in the American character set, the letter "Q" is stored as the bit value 81.

checkpoint — A moment in time when the CKPT background process signals that all used memory buffers are to be written to a disk.

chunk — Oracle9*i* writes and reads one chunk from a large object (LOB) value at a time. A chunk can be from one data block that is up to 32K in size.

clearing (a group) — Removes all the corrupted data from the online redo log group.

cluster — A collection of information that is organized for efficient retrieval, for example name, address, and phone number. Technically, a cluster is not a table, but is a group of tables stored together as if they were one table. Clusters store data of multiple tables in one segment.

cluster key — The set of columns from each table in a cluster that make up the index for the cluster.

cluster manager — A software application that coordinates all the instances and tasks that each cluster handles. The cluster manager can reside on a separate computer or on one of the cluster's nodes.

clustered rows — Rows from each table in a cluster that correspond to one unique value in the cluster key per block.

clustered server — Clustered server configurations include several computers (database nodes), each containing a database instance. Another computer (file server) houses the database files, which are shared by all the database instances located on the database nodes.

coalesce — The combining of multiple adjacent free extents into a single contiguous free extent. For example, an index is coalesced to consolidate fragmented storage space in its leaf blocks.

collection data type — A type of data that repeats and is contained within one column. A collection can be defined as an array (called varray in Oracle9*i*) or as a nested table.

column constraint — A restriction that applies to a single column and appears inline with the column.

column data — Consists of a 1- to 3-byte section of a table containing the length of the column's data followed by the actual data for the column.

common and variable header — The section of a table that contains identifying information, such as the type of block and block location.

composite index — An index that contains multiple columns that can be used in more than one index, although no two indexes on a table can contain the same combination and order of columns.

composite range-hash — A method of table partitioning that is used when the data within a table is being partitioned and then the data within each partition is partitioned (sub-partitioning).

composite range-list — A method used to partition and sub-partition a table. The data is partitioned using a range and the sub-partitioned using a list. *See also* range partitioning and list partitioning.

compound key — A field used to sort data that is composed of multiple columns.

connection — The link from the user session, through the server session to the database instance.

constraint — A rule that defines data integrity for a column or group of columns in a table.

constraint state — An attribute that tells Oracle9*i* how to use a constraint when data is added to a table. A constraint state can also tell Oracle9*i* how to use a constraint on existing data when a constraint changes from DISABLE to ENABLE.

control files — Files that contain the names and locations of all the datafiles and redo log files belonging to a database instance.

copy semantics — Internal large objects (LOBs) use copy semantics, which means that the LOB value is copied like other data when you copy the LOB into a new location. *See also* reference semantics.

cost-based optimizer — The default optimizer that uses statistics on the actual volume and distribution of table data to determine the best path to retrieve the data. Takes factors into account such as the relative costs of I/O, CPU time, and execution time. *See also* rule-based optimizer.

data block — A data block is made up of a set number of contiguous physical bytes in a physical file. A data block is the smallest size of a logical unit in the database.

Data Dictionary view — A method of looking at a database that provides system-related information by querying the database's internal management tables and presenting the data as a view.

data segment — An area used to store data for objects such as tables, object tables, materialized views, and triggers.

data type — A predefined form of data. Oracle has twelve standard data types, including numbers and dates.

database — A group of files containing data, meta data, undo records and redo log entries. Accessed through a database instance. *See also* database instance and database server.

database character set — Each database has a base language, such as English, that is called its database character set. *See also* character set.

database cluster — A database cluster combines multiple databases distributed on multiple computers but sharing a central memory area. Allows a long-running task to be spread among several computers, thereby speeding up processing time.

database instance — A running database that has its own allocation of memory (called the System Global Area), its own set of background processes, and its own set of control files. Used to read and manipulate the database. *See also* database and database server.

database object — Any database item that can be individually set and manipulated. A database object is not only an object table, but also any other structure held in the database, such as a relational table, an index, a PL/SQL procedure, a Java servlet, or even a customized attribute with methods and rules attached.

database server — The combination of a database (the files) and a database instance (the SGA and the background processes).

datafile — A physical file that is located on an operating system and contains data. It is the primary physical structure that contains all the data stored in the database.

DBA — DBA stands for both database administration and database administrator.

DBA authentication method — A method used to validate the log-in of users with SYSDBA (for SYSDBA privileges) or SYSOPER (for SYOPER privileges).

deadlock — Occurs when two transactions cannot resolve their simultaneous requirements for the same data.

dedicated server — A mode of database server that connects one user process with one server process. *See also* shared server.

default role — Any role assigned to a user that is automatically enabled when the user logs onto a database. *See also* role.

default value — A value that Oracle inserts into a column when an inserted row does not specify the value for that column.

deferred state — A type of constraint state in which the constraint is validated only when a COMMIT command is executed. *See also* immediate state.

dictionary-managed tablespace — A tablespace whose free space and storage allocation information is stored inside the data dictionary tables in the SYSTEM tablespace. *See also* locally managed tablespace.

dirty buffer — Used memory buffers in the System Global Area (SGA) are called dirty buffers.

disabled — Disabled means not in force, or turned off. For example, a disabled role gives the user no privileges. A disabled constraint does not enforce its data integrity rule. *See also* enabled.

distributed database system — A distributed database system has multiple instances that are used as if they were one instance. Data on any of the databases can be modified by any of the instances.

enabled — Enabled means in force. For example, a database role does not have to be enabled all the time. A constraint can also be enabled or disabled. *See also* disabled.

exporting — A form of backup that can be used on specific tables or schemas, or to back up the entire database. Creates a file in Oracle-proprietary format that can only be used with Oracle's importing process. *See also* importing.

extensible markup language (XML) — A programming language for the Internet that provides both the data and the display details to a web page or to an application. Oracle supports storing XML formatted data in tables.

extent — A contiguous group of data blocks that are assigned to a segment. The number of data blocks in one extent is determined by the storage parameters of the object or the tablespace.

external LOB — The data of an external large object (LOB) is stored outside the database in an operating system file. Commonly used for imagery or multimedia. The data type of an external LOB column is BFILE. *See also* internal LOB.

external table — A table that is defined in the database but whose data is stored in a file that is outside the database. Commonly used to copy external data into a table inside the database.

external user — An Oracle user whose name and password are authenticated outside the database by the operating system or a third-party service. *See also* local user and global user.

fast commit — Oracle uses a fast commit mechanism to speed up performance. When a transaction is committed, the changed data blocks are kept in the buffer cache and are not written to the datafiles until later.

fixed-length — A multibyte character set can be either fixed-length or variable-length. A fixed-length, multibyte character set uses a fixed number of bytes for each character in the character set. This simplifies retrieval but uses more space to store data. *See also* variable-length, and multibyte character set.

flushing the buffer — The process of copying the memory buffer contents to the appropriate datafile, or to the archive log file, and then deleting the memory buffer's contents.

foreign key — The name of a relationship between two tables, in which one is the parent and the other is the child. Usually enforced with a foreign key constraint defined on the child table and a primary key constraint defined on the parent table. *See also* primary key.

free space — The contiguous blocks of storage space in a tablespace that are unallocated.

freelist — A list of individual blocks within an extent that have room available for inserting new rows. To be included in the freelist, the percentage of available space in a block must be greater than the PCTFREE value for the table.

function-based index — An index that uses the value of an expression or function instead of the value of a column as the indexed column For example, the index uses TO_CHAR(SALES_DATE,'MONTH') as one of its indexed column values.

functions, procedures, and packages — PL/SQL programs that reside in the database and can be called from SQL commands, such as SELECT and INSERT. Oracle9*i* provides many pre-defined functions, procedures, and packages. You can also create your own. *See also* PL/SQL.

global non-partitioned index — A normal index that is created on a partitioned table and contains data from all the partitions in the table. *See also* global partitioned index and local partitioned index.

global partitioned index — An index that is partitioned. Can be created on a non-partitioned or partitioned table. When created on a partitioned table, the index is partitioned differently than the table. *See also* local partitioned index.

global user — An Oracle user whose name and password are authenticated outside the database by the enterprise directory service. *See also* local user and external user.

globalization support — A characteristic of Oracle9*i* that enables database administrators to design a database that supports users from many nationalities by converting data to each user's native language and by storing data in multiple languages.

grant — When a database administrator grants a privilege, he assigns the ability to access an object or use a system privilege to a user or to a role. *See also* object privilege and system privilege.

grantee — A user or role who receives (is granted) an object privilege, system privilege, or a role.

grantor — A user who executes the GRANT command, which conveys (grants) a privilege or role to a user or role.

group function — A function that acts on sets of column data in rows. For example, SUM(SALES_AMOUNT) adds the value of SALES_AMOUNT in a set, or group, of rows.

hash partitioning — A method of table partitioning, in which the database administrator specifies the partitioning key and the number of partitions and lets Oracle9*i* use a hash value (calculated on the partitioning key) to divide the data evenly among the set of partitions.

heap-organized table — A table that is not stored in index order. This is the normal method of storing tables. *See also* index-organized table.

high watermark (HWM) — The boundary between used data blocks (blocks formatted for data) and unused data blocks (blocks allocated but not yet formatted for data) in a table. Used blocks can contain no data if the rows they previously stored were deleted. You cannot deallocate blocks below the HWM.

immediate state — A type of constraint state in which the constraint is validated as soon as an insert, update, or delete statement is executed. *See also* deferred state.

importing — A process that uses a previously created snapshot of a database (or part of a database such as schema or a table) and recreates all or part of it in the current database. An import can perform many functions including adding rows to existing tables and creating users, tables, views, and triggers. *See also* exporting.

index — A database structure that is associated with a table or a cluster and speeds up data retrieval when the table or cluster is used in a query. *See also* bitmap index and b-tree index.

index segment — The logical structure that stores all the index data, except when the index is partitioned. For partitioned indexes, there is one index segment per partition.

index-organized table — A relational table with rows stored in the physical order of the primary key values, in the same way an index is stored. Can improve performance for tables that are always queried by a primary key. *See also* hash-organized table.

inheritance — The act of one object type obtaining characteristics from another object type. For example, an animal object type has attributes inherited by a mammal object type, which is a sub-type of animal.

inline constraint — An integrity rule (constraint) that, when defined, appears immediately next to the column to which it applies. *See also* out of line constraint.

integrity constraint — A rule that defines restrictions or relationship requirements on a column or a set of columns in a table. *See also* constraint, foreign key, and primary key.

internal LOB — A large object (LOB) column with data stored inside the database. The column's datatype is BLOB, CLOB, or NCLOB. *See also* external LOB.

Java Virtual Machine — An internal component of the database that enables the storing, parsing, and executing of Java applets, servlets, and stored procedures within the database.

leaf — The bottom level of a binary tree. A node with no nodes below it. *See also* binary tree, and b-tree index.

list partitioning — A method of table partitioning in which a distinct list of partitioning key values is set up, and values are defined to go into each partition. Useful when the partitioning key has a small number of distinct values, such as state abbreviations.

listener — A database service that waits for incoming requests for the database server and responds to them.

LOB — A value stored in a column that can be up to 4 gigabytes in size. LOB stands for "large object." LOBs usually contain multimedia, audio, or imagery information. *See also* LOB value, internal LOB and external LOB.

LOB locator — The pointer that directs the database to the actual location of the value is called the large object (LOB) locator or internal LOBs and the BFILE locator for external LOBs. The LOB locator resides in the table, while the LOB value resides in a separate location (either inside or outside the database.) Some LOB values are stored inside the table (no pointer required).

LOB value — The data stored in a column with one of the LOB data types. LOB values can be stored with the rest of the table data (inline) or in a separate segment (out of line). *See also* LOB pointer. LOB data types are BLOB, CLOB, NCLOB, and BFILE.

local partitioned index — An index on a partitioned table, in which the index is partitioned in the same way and on the same columns as the table.

local user — A user who must enter a password whenever logging into a database. User name and password are managed inside the database. *See also* external user, and global user.

locally managed tablespace — A tablespace that contains an internal bitmap that stores all the details about free space, used space, and the location of extents. Locally managed tablespace is the default setting for tablespaces. *See also* dictionary-managed tablespace.

log switch — An event in which the LGWR process stops writing to one log group and begins writing to another log group. *See also* redo log group.

logging — A parameter specified during the creation of a tablespace that sets the default action for objects in the tablespace so that all transactions that change the objects are recorded automatically. Opposite of nologging.

logical structure — A structure that is composed of orderly groupings of information that allow for the manipulation and access of related data. In Oracle9*i*, a table is a logical structure. *See also* physical structure.

LogMiner — LogMiner is a utility supplied with the Oracle Server that enables the viewing and parsing of both online and archived redo log files.

low cardinality — Low cardinality means that the number of distinct values in an indexed column is low compared to the number of rows in the table. For example, the CUSTOMER table has 50000 rows and the STATE column has 50 distinct values.

management packs — Sets of related tools, wizards, and assistants that are added as a group to the console. For example, the Change Management Pack contains tools for managing, scheduling, and monitoring table changes.

manual undo management mode — Older method of managing redo segments in which the DBA establishes files for storing undo records. Opposite of automatic undo management mode.

media recovery — A recovery that requires restoring the database from a backup, rolling forward through archived redo logs, and finally rolling forward through online redo logs. It is called media recovery because the most common reason for this type of recovery is the failure of a disk drive or some other storage media.

member — Each file in a redo log group is considered a member of that group.

metadata — Data about a specific set of data in the database including how, when, and by whom it was collected, and how the data is formatted. Data Dictionary views display metadata about tables, for example.

methods — Sets of predefined code segments that perform tasks on the data in an object. For example, a GIVE_RAISE method might update a record with a certain percentage increase in pay based on length of service.

migrated — If a row is updated and requires more free space than the block contains, then the entire row is migrated, that is, moved to another block. A pointer remains in the original row location to redirect a process to the correct location.

multibyte character set — A character set that is complex and requires multiple bytes to represent a single character. Chinese is a multibyte character set. *See also* character set.

multiple instance server — With a multiple instance server installation, one computer has multiple instances, each instance with its own database files.

multiplexing — The process of making redundant copies of files to guarantee that if the original file is damaged, another can take its place immediately. Control files and redo log groups are usually multiplexed.

national character set — Each database has a base language (database character set) and a second language that is called a national character set. The national character set stores data in columns that are defined as one of the three NLS data types (NCHAR, NVARCHAR2, and NCLOB).

National Language Support (NLS) — NLS is a database feature that makes it possible for tables to store data in different languages (for example Chinese and French).

nested table — An object type that allows one column to store a table of data.

NOARCHIVELOG mode — A database in NOARCHIVELOG mode does not archive redo log files. The opposite is ARCHIVELOG mode.

node — Any point in a binary tree where two or more lines meet. *See also* branch, leaf, and binary tree.

nologging — A parameter specified during the creation of a tablespace that sets the default action for objects in the tablespace so that transactions that perform mass inserts are not recorded. Opposite of logging.

normalize — Normalizing tables is part of the design process, in which this and other normalizing rules are followed: every table should have a key that contains a unique value for every row.

object privilege — Gives a user or role the ability to perform certain tasks on specific tables or other objects that are owned by a schema. For example the SELECT object privilege allows a user to query a specific table. *See also* system privilege.

object table — A table that holds objects and attributes. An object table is similar to a relational table, except each row is a single unit of data defined by an object type.

object type — An object type has the attributes (which are equivalent to columns) and traits found in object-oriented databases, such as methods and inheritance.

offline tablespace — A tablespace that is not available for use.

online tablespace — A tablespace that is available for use.

open backup — A backup of the tablespace that is made while the tablespace is available for use.

operating system (OS) authentication — When using operating system (OS) authentication, the user logs in without specifying an Oracle user name and password. Oracle derives the Oracle user name from the operating system user name.

optimizer — A process used within Oracle9*i* that selects the fastest access path for the execution of a SQL command. *See also* rule-based optimizer and cost-based optimizer.

Oracle Advanced Security — A feature that encrypts outgoing data before it goes onto the network or across the Internet. It also supports special methods of user authentication such as the programs used with automated tellers, which require that a user possess a valid bank card and know the personal identification number (PIN) to access account data.

Oracle Change Management Pack — A feature that adds several utilities to the Enterprise Manager console for identifying and distributing database changes among multiple databases.

Oracle Data Mining — A feature that supports setting up algorithms and functions that search and retrieve data warehouse information.

Oracle Diagnostics Pack — A feature that brings expert advice with graphs, database monitoring, and analysis tools into the Enterprise Manager console to diagnose bottlenecks, performance problems, and storage usage.

Oracle Enterprise Manager (OEM) — A Database Administrator tool with a Windows-like interface, which was first introduced with Oracle7. The tool integrates many utilities and monitoring tools into a single interface.

Oracle Financials — A set of software functions designed for use in the fields of bookkeeping, accounting, inventory, and sales.

Oracle JDeveloper — An application builder that writes Java code using a Windows-like interface.

Oracle Label Security — A tool for restricting access to rows within a table or within a view. Useful when high security standards must be met. It labels individual rows with a security profile and matches that row with a user's security profile that is stored in the database.

Oracle Management Pack for Oracle Applications — A set of tools that can be added to the Enterprise Manager console. Used to automate the monitoring of form sessions and concurrent managers, which are used when running Oracle Applications.

Oracle Management Pack for SAP R/3 — Enhances the Enterprise Manager console for the detection and monitoring of SAP R/3 application servers and clients that use the Oracle9*i* database.

Oracle On-line Analytical Processing (OLAP) Services — Services that make it possible for standard file directory support to be combined with database delivery of tables or views.

Oracle partitioning — A feature that enables the process of dividing tables across multiple tablespaces. High volume tables, such as historical records in data warehouses, benefit from partitioning because it speeds up data retrieval for queries.

Oracle Real Application Clusters (RACs) — Provide management tools to support database clusters.

Oracle Spatial — A feature that adds programmed packages to the database to handle spatial objects. Geographic mapping is a common use for Oracle Spatial.

Oracle Tuning Pack — A set of tools that can be added to the Enterprise Management console. The tools show which SQL commands use the most resources and provide automated tools to analyze and tune the commands.

ORACLE_BASE — A root directory that stores all Oracle-related files, including the database's data files and configuration files. Database data files reside on subdirectories under ORACLE_BASE.

ORACLE_HOME — The directory tree in which Oracle executable files are stored. Usually a subdirectory under ORACLE_BASE.

Oracle9*i* Application Server — The Web server with special plug-ins for the Oracle database and Oracle applications.

out-of-line constraint — A restriction (or rule) that appears after the full list of columns in the CREATE TABLE command and usually applies to multiple columns. Can also appear in the ALTER TABLE command. *See also* inline constraint.

partitioning key — A range of values in a set of columns that determine how a table or index is stored in partitions.

permanent tablespace — Stores permanent objects such as tables and indexes. Permanent tablespace is the default setting for tablespaces.

physical structure — A structure that is composed of operating system components and has a physical name and location. Physical structures can be seen and manipulated in the computer's operating system. In Oracle9*i*, a datafile is a physical structure. *See also* logical structure.

PL/SQL — A procedural language that Oracle provides to write simple programs in the database. Uses SQL commands along with variable definitions, loops, if-then-else logic, and so on.

plan — A list of steps developed by the optimizer that will be taken to retrieve data for a query.

precompiler — Precompilers (such as Pro*COBOL and Pro*C) support embedded SQL commands within programs in C, C++, or COBOL. The precompiler translates the SQL command into the appropriate set of commands for the program, which is then compiled and ready for executing.

primary key — The column or set of columns that define a unique identifying value for every row in a table. For example, the CLIENT table has a unique identifying column called CLIENT_ID. Usually enforced by defining a primary key constraint on the table.

privilege — A capability, such as the ability to create new users, or an authorization, such as the authority to SELECT on a table. A user or role can be given a privilege. *See also* object privilege and system privilege.

profile — A collection of settings that limit the use of system resources and the database. All users have one profile, usually the DEFAULT profile.

pseudocolumn — Acts like a column in a query, but actually is calculated by the database for the query. For example, ROWNUM is a pseudocolumn containing the sequence number of each row returned in a query.

public synonym — A unique name for an object that allows any user to use the object without prefixing it with the owner name. A public synonym does not give users the privilege to actually select from the table; that privilege must be granted.

query rewrite — The ability of Oracle9*i* to modify a query that is executed and change it into an equivalent but more efficient statement before actually running the query. This ability is enabled or disabled with the QUERY_REWRITE_ENABLED initialization parameter.

quota — The maximum amount of storage space a user can be allocated in a tablespace. Set with the CREATE USER or ALTER USER command.

range partitioning — A method of table partitioning in which the table is stored in partitions according to a specified set of ranges of values in the partitioning key. For example, the first range is all rows with partitioning key values less than 100, the second range is all values less than 1000, and so on.

read consistency — A type of status that allows only the user who issued an update command to see the changes to the data in queries until that user commits the change. Other users see the data as if it had not been changed until the commit occurs.

read-only tablespace — A tablespace mode in which objects in the tablespace can be queried but not changed. No user can execute an INSERT, UPDATE, or DELETE command on the objects in a read-only tablespace.

record sections — Lists of information by category within a control file. For example, one record section stores file names and locations whereas another record section stores recovery information such as the date of the last archive.

Recovery Manager — A utility that automates database recovery after a failure.

redo entry — Synonymous with redo record. The redo log file contains sets of redo entries that are made up of a related group of change vectors that record a description of the changes to a single block in the database.

redo log file — A file containing redo log entries. Redo log files store information critical to the recovery of changed data when the database has lost the data due to a failure of some kind.

redo log group — A file or set of files (members) that store redo entry data. A database must have at least two redo log groups containing at least one file each. *See also* member, log switch, and multiplexing.

reference semantics — External large objects (LOBs) use reference semantics, which means that when copying an external LOB, only the pointer to the location of the external LOB is actually copied. *See also* copy semantics.

relational table — The type of table traditionally used in a relational database as well as in Oracle9*i*. A relational table is usually referred to simply as a "table." A table stores data of all types and is the most common form of storage in a database.

role — A collection of privileges that are given a name and can be assigned to users or even to another role.

rollback segment — Created automatically by Oracle in an undo tablespace when using the Automatic Undo Management feature. Stores information that allows data changes to be undone (rolled back) if the change is not committed.

root — The starting point for searching a binary tree or a b-tree. The root has branch or leaf nodes below it, but no nodes above it.

row data — A table component that consists of bytes of storage used for rows inserted or updated in the data block.

row directory — A group of table components that consist of a list of row identifiers for rows stored in the block.

row header — Stores the number of columns contained in the column data area, as well as some overhead, and the rowid pointing to a chained or migrated row (if any).

rowid — Contains the physical or logical address of the row.

rule-based optimizer — A component of the database still available for backward compatibility that uses static rules to rank possible access paths and select the best path.

SAP R/3 — A popular financial planning and accounting software package, similar in scope to Oracle Financials.

schema — The collection of database objects created by one user, such as a table, index, user-defined attribute, an integrity constraint, or a procedure. A schema has the same name as the user who created the objects.

schema object — An object created by a user (schema) in the Oracle9*i* database.

segment — The set of extents that make up one schema object within a tablespace. For example, a table has one segment containing all its extents. A partitioned table has one segment per partition.

server process — On the database side of a transaction, the process that interacts with a user process is called a server process.

service name — A set of information that Oracle Net uses to locate and communicate with an Oracle database.

service unit — The name for the calculated value of the weighted sum of the resources consumed by the user session. Used when setting the COMPOSITE_LIMIT parameter of a profile.

session — A period of computer use that lasts from the time a user makes a connection to the database until the user ends the connection.

shared server — A mode of database server that connects multiple user processes with one server process. *See also* dedicated server.

silent mode — The Universal Installer provides a silent mode of installation that enables the database administrator to run an installation without any human intervention. Useful when unattended installation processes must be run on many machines.

single instance server — The typical type of installation for Oracle9*i*. A single instance server installation involves one computer with one set of database files accessed by one instance.

single-byte character set — US7ASCII is an example of a single-byte character set, because each character (such as the letter "A") in the language is represented with a single byte code in the base operating system's machine language.

spatial objects — Store data related to time and space in a way that allows database administrators to perform functions including: calculating distance or time between objects; and drawing lines, polygons and points on a map.

standby database — A clone of the current working database that is kept current with the existing database by applying changes stored in the archived redo logs. If the primary database fails, the standby database replaces it automatically.

status (of a tablespace) — Defines the availability of a tablespace to end users and also defines how the tablespace is handled during backup and recovery.

store table — The data in a nested table is actually stored in its own table (called the store table), whereas the main table contains an identifier that locates the associated nested table for that row.

sub-object — An object that is based on another object. For example, an object type can be based on another object type.

subpartition template — Describes all the subpartitions in a table once and then all the partitions that use that template. Used as a shorter method of defining subpartitions in the CREATE TABLE command.

subpartitioning — The partitioning of data within each partition of a table.

subquery — A query that is embedded in another SQL command.

system change number (SCN) — A sequential number that is incremented for each change that modifies the physical database files. Controlled by the CKPT background process.

System Global Area (SGA) — The portion of computer memory (RAM) allocated for database memory by a database instance.

system privilege — Represents the ability to manage some part of the database system. Can be assigned to users or roles. For example, the ALTER TABLESPACE privilege allows a user to modify an existing tablespace using the ALTER TABLESPACE command. *See also* object privilege.

table alias — A shortcut name for a table that is used to prefix a column name in an SQL command in place of using the entire table name.

table constraints — Restrictions (rules) that apply to multiple columns, such as a constraint for a compound foreign key. Table constraints are placed immediately after the list of columns in the CREATE TABLE command. Table constraints can also appear in the ALTER TABLE command. *See also* out-of-line constraints.

table directory — A data block component that consists of information about which table has data in the block.

tablespace — A logical data storage space that maps directly to one or more datafiles. The storage capacity of a tablespace is the sum of the size of all the datafiles assigned to that tablespace.

temporary segment — Created during execution of a SQL command that creates a need for space to perform sorting or other operations. Usually stored in a temporary tablespace.

temporary table — As with standard tables, a temporary table contains data; however the data is private (not seen by other users) and disappears at the end of the user's session or when the user commits a transaction. Multiple users can store data in one temporary tablespace, but each user sees only his own data.

temporary tablespace — Stores objects, such as temporary segments only for the duration of a session.

thread — A link between a user process and a server process that requires processing by the database. The multithreaded server allows one server process to handle many user process threads.

trigger — A program that runs whenever a certain event occurs in a table on which the trigger is defined, such as the insertion of a row into a table.

undo data — Made up of undo blocks. Each undo block contains the before image of the data in the block. Usually stored in an undo tablespace and used to restore changed data to its original state if the change is not committed.

undo data retention — The retention of data for a short time after the data has actually been rolled back or committed. Used for read consistency and flashback queries.

undo extents — The data in the undo tablespace that is added in the form of extents.

undo tablespace — An entire tablespace that is reserved for undo data.

unique index — A unique index requires that every row inserted into a table has a unique value in the indexed column or columns.

universal language character set — A character set that contains codes for every language that is supported by Oracle9*i*. The national character set, UTF8, is an example.

upgrading (a table structure) — A function that causes any object type used in a table to be updated with the most recent version of the object type definition.

UROWID — A data type that stands for universal rowid data type, and can hold any type of rowid, including a record identifier from a non-Oracle database.

user process — Whenever a user runs an application that uses the database, the application creates a user process that controls the connection to the database process.

user-defined data type — A datatype defined by a user. Must be an object type, array type and table type. User-defined types help define object columns or object tables. For example, the user-defined data type ADDRESS_TYPE is made up of three attributes: STREET, CITY, and COUNTRY, which are all VARCHAR2(30) data types.

user-managed redo segments — Groups of files that store all the undo data. Older method of managing redo segments. *See also* automatic undo management mode.

utilities — Programs that handle backup, migration, recovery, and transporting data from one database to another.

variable-length — A multibyte character set can be either fixed-length or variable-length. A variable-length, multibyte character set uses a variable number of bytes for each character in the character set. This saves space but requires more complex retrieval routines that take more time. *See also* character set.

XML table — A table that is created with one column of the XML type data type is an XML table.

Index

D

V

W

X

Managing Undo Data

Exam Objective	Chapter
Describe the logical structure of segments within the database	6
Describe the purpose of undo data	6
Implement Automatic Undo Management	6

Managing Tables

Exam Objective	Chapter
Identify the various methods of storing data	7
Describe Oracle data types	7
Distinguish between an extended versus a restricted ROWID	7
Describe the structure of a row	7
Create regular and temporary tables	7
Manage storage structures within a table	7
Reorganize, truncate, drop a table	8
Drop a column within a table	8

Managing Indexes

Exam Objective	Chapter
Describe the different types of indexes and their uses	9
Create various types of indexes	9
Reorganize indexes	9
Drop indexes	9
Get index information from the data dictionary	9
Monitor the usage of an index	9

Maintaining Data Integrity

Exam Objective	Chapter
Implement data integrity constraints	10
Maintain integrity constraints	10
Obtain constraint information from the data dictionary	10

Managing Password Security and Resources

Exam Objective	Chapter
Manage passwords using profiles	11
Administer profiles	11
Control use of resources using profiles	11
Obtain information about profiles, password management and resources	11

Managing Users

Exam Objective	Chapter
Create new database users	11
Alter and drop existing database users	11
Monitor information about existing users	11

Managing Privileges

Exam Objective	Chapter
Identify system and object privileges	12
Grant and revoke privileges	12
Identify auditing capabilities	12

Managing Roles

Exam Objective	Chapter
Create and modify roles	13
Control availability of roles	13
Remove roles	13
Use predefined roles	13
Display role information from the data dictionary	13

Using Globalization Support

Exam Objective	Chapter
Choose a database character set and national character set for a database	14
Specify the language-dependent behavior using initialization parameters, environment variables, and the ALTER SESSION command	14
Use the different types of National Language Support (NLS) parameters	14
Explain the influence on language dependent application behavior	14
Obtain information about Globalization Support usage	14